Galactic X-Ray Sources

Based on the Proceedings of the
NATO Advanced Study Institute on
Galactic X-Ray Sources
held in Sounion, Greece

Galactic X-Ray Sources

Edited by

Peter W. Sanford

Department of Physics and Astronomy
University College, London
Mullard Space Science Laboratory

Paul Laskarides

Department of Astronomy
University of Athens

Jane Salton

Department of Physics and Astronomy
University College, London
Mullard Space Science Laboratory

A Wiley–Interscience Publication

JOHN WILEY & SONS

Chichester · New York · Brisbane · Toronto · Singapore

British Library Cataloguing in Publication Data:

Galactic X-ray sources,
   1. Extraterrestrial radiation—Congresses
I. Sanford, Peter W.
II. Laskarides, Paul
III Salton, Jane
523.01'92   QB461

ISBN 0 471 27963 3

Printed in Great Britain

Contents

<u>Part 4 - Globular clusters and burst sources</u>

<u>Part 5 - Recent results</u>

Foreword

It is a pleasure to be invited to write a foreword to "Galactic X-ray Sources". It was widely reported that the conference was very successful and we should be especially grateful to Peter Sanford and his associates not only for bringing together such a diverse and representative group of X-ray astronomy enthusiasts but also for carrying through the much less exciting task of collecting and editing a really adequate "proceedings" of the Institute.

It was, I am confident, a special happiness that Bruno Rossi was able to participate, for it was in a discussion with him in September 1959 that Ricardo Giacconi was interested in the idea of looking for cosmic X-ray sources. That year just two decades before the Institute may have some claim to the birth of practical action to explore the possibility of X-ray astronomy. I well recall a meeting of the British National Committee on Space Research in May of that year in which the discussion is minuted "Current theories suggested that there may be objects in the sky with strong X-ray emission although inconspicuous visually. A search for these is a matter of great interest and importance".

Yet that itself is a reminder of how much the subject owes to Herbert Friedman, for I have no doubt that interest in the idea, at any rate on the European side of the Atlantic, was sparked off by his report to the IAU meeting in Moscow. Whatever the events at that time none of us could have foreseen the huge impact on astronomy as a whole made by the first discovery and identification of X-ray sources in the early sixties. I imagine that the thinking of Giacconi and Friedman, like my own, before Scorpius X-1 changed its scale, was concentrated mainly on the possible occurrence of coronal like phenomena perhaps particularly associated with the release of magnetic energy. It is therefore pleasing that after two decades which have seen, largely as a result of the power of X-ray cosmic diagnostics, the recognition of the tremendous role of gravitational energy in the context of neutron stars and black holes now at last the study of X-rays from stars fired by nuclear fusion is becoming important. Some energetic plasma region, contained and powered by magnetic fields, seems to be a normal stellar phenomenon and even planets are found to be X-ray sources - a reminder perhaps that Giacconi's historic discovery was made by a flight ostensibly aimed to study the moon.

The Institute provided a comprehensive overview of the work in progress on understanding the several hundred X-ray sources seen so far in our galaxy. Time has been made available for the contributors to update their work both in response to the lively disscusions at the meeting and to take account of the natural evolution in the subject up to mid-april 1980.

The proceedings begin with the observational and theoretical work on binary star systems. Hutchings gives an overview of the optical studies and is careful to point out the difficulties involved in obtaining accurate data for the masses and orbits from these.

Ilovaisky and Cowley complement this with papers on their optical studies which have concentrated on the photometry of the visible counterparts of X-ray binary stars. Ilovaisky's work on the Scorpius X-1 system is of special interest as measurements were obtained closely comparable with those made by the SAS-C satellite. The remarkable correlation between X-ray and optical fluctuations indicates that this particular system generates much of its fluctuating visible light very close to the high temperature region which emits the X-rays. The review by Bradt also provides an account of variability measurements and their interpretation. Predictions made several years ago by McCray, that the plasma around intense X-ray sources can be studied most effectively with high quality X-ray spectrometers are now regarded as basic. McCray's lecture describes the models developed in the last few years to explain the mechanisms for the production of atomic emission lines. Resonance lines for highly stripped heavy elements fall in the X-ray region and are shown to be of importance both in the design of future spectrometers and in the interpretation of the existing data from the Einstein Observatory.

There are two papers on the dynamics of binary X-ray stars. Sutantyo describes the effects on the evolution of two gravitationally bound stars, of instantaneous ejection of matter in the supernova phases. The changes to their orbits are estimated and arguments are given to show that because of mass transfer in the binary system, the mass of a star, preceeding an explosion, is less than 7 solar masses. Savonije discusses the effects of mass transfer on the evolution of a source. Parmar discusses a recent observation of Hercules X-1 by the Ariel 5 satellite which monitored the behaviour of this source for many days. This has led to better discrimination between the intrinsic changes in output and extrinsic changes in the absorbing material in the line of sight. Cordova and Chester describe some observations by HEAO-1 of X-ray emission from dwarf novae.

The second part of the proceedings is concerned with conditions very close to compact objects those characteristics of the X-ray pulsars which are an indicator of the rotation of the neutron star, and observational and theoretical work on accretion. Rappaport's comprehensive review describes how the changing rotational rate can give information on the accretion phenomena and related luminosities. He also describes the determination of the orbital parameters of the binary system 4U0900-40. His analysis, which is in the tradition of the classical work by optical astronomers who early this century derived the masses of stars, uses Doppler changes in the pulsation frequency, together with optically determined radial velocity curves for the primary star.

Professor Trumper's invited lecture introduced his pioneering work on quantised cyclotron radiation in neutron star aurorae. His co-workers, Kendziorra, Ventura and Kirk provided substantial contributions on the physics of plasmas in these intense magnetic fields.

The twisted accretion disc hypothesis described by Petterson is an attempt to explain the characteristics of the Hercules X-1 system. The early life of pulsars in binary systems was described by Maraschi and Treves and their colleague Chiappetti reported some preliminary studies of data from the Ariel 5 satellite on the transition between active and quiescent states of the Cygnus X-1 black hole source. X-ray emission theories from white dwarfs are outlined in a short paper by Kylafis and Lamb.

The Einstein Observatory has given us an insight into the nature of low-luminosity X-ray sources in the galaxy, X-ray emission having been measured from stars of all spectral types on the main sequence. Giants, supergiants and white dwarfs have also been detected.

Part 3 records the invited lecture by Cassinelli which describes our current understanding of stellar winds and the evidence for coronae around the early type stars and Dupree reviews the ultraviolet spectroscopy of galactic X-ray sources. Both papers provide important material for workers involved in what is likely to become a major area observationally and theoretically.

The probable supernova origin of the soft X-ray background, is reviewed by Bleeker. Raymond and Wolff describe ultraviolet measurements made with the IUE of one such supernova, the Cygnus Loop. Charles describes discoveries of RS CVn systems, flare stars and A stars as soft X-ray sources.

Two major reviews of globular clusters by Grindlay, and Lewin and Clark are contained in part 4. Preliminary work by the IUE satellite to investigate X-ray emitting globular clusters is reported by Hartmann.

The proceedings conclude with a section (Part 5) on recent observations both from ground and space. Einstein Observatory results are reported by Schreier, and Heise and Brinkman. Margon has provided an extensive paper on the peculiar object SS433. Raymond has a paper on ultraviolet studies of AM Her with the IUE satellite and Blissett has re-analysed data from the Ariel 5 satellite and reports variability in the spectrum of the Cygnus X-3 source.

No one can read these papers without recognising that the study of galactic sources in X-rays is now an integral part of modern astronomy. The scope of this Institute was limited to our Galaxy. Developments in extragalactic X-ray astronomy have been no less dramatic. The image of M31, our nearest neighbouring galaxy, from Schreier's paper, must stand as a symbol of extragalactic achievements undescribed here and of the promise of high resolution observations yet to be made both on our galaxy and to the furthest reaches of the Universe.

R. L. F. Boyd

Preface

The era of modern observational astronomy which began about twenty
five years ago has a number of highly specialised disciplines each
with their own techniques and terminology. However the opportunity to
exchange ideas and understand the problems of other workers is often
limited by a shortage of time and travel funding for younger
scientists.

The Advanced Study Institute on Galactic X-ray Sources was an attempt
to form bridges between specialists and young astronomers in a relaxed
setting at Cape Sounion. Mainland Greece was selected for the
Institute at the suggestion of the NATO sponsors; the enthusiastic
participation of the local astronomers was a reflection of their
advice.

Astronomers are always aware of the need to relate photometric and
spectral measurements from the entire electromagnetic spectrum. The
ingenuity of a theory can be well tested and even lead to occasional
adjustment, albeit subtle. Participants at the Institute were
fortunate to be presented with measurements from the radio to the
gamma-ray regions of the spectrum. A common theme of these studies
involved correlation of photometric measurements on timescales ranging
from less than a millisecond to several years.

The remarkable images from the Einstein observatory, of X-ray emitting
stars were displayed by Ethan Schreier in his presentation of recent
results from the satellite, for example, the early picture of the
luminous X-ray stars in the nearby galaxy M31.

The Institute was also sponsored by the Greek National Committee for
Astronomy and partly financed by the Greek Ministry of Culture and
Civilisation (Minister Professor D. Nianias). The Local Organiser was
Dr. Paul. G. Laskarides and the Local Organising Committee consisted
of Professor John Xanthakis, Professor G. Contopoulos, Dr. M. Zikides,
Professor S. N. Svolopoulos, Dr. P. Niarchos and Mr. E. Theodossiou.
The Minister appointed Mr. N. Zorogiannides and Mr. A. Vassilopoulos
as observers. Dr. N. Shakura (Moscow) attended at the invitation of
the Greek National Committee.

The seminars and working sessions took place in the "cool" dancing
room of the "Cape Sounion" Hotel where most of the participants were
accommodated (a few of them were staying in nearby hotels). The Greek
participants were transferred daily to and from Athens by a special
bus.

Professor John Xanthakis, member of the Academy of Sciences of Athens,
and President of the Greek National Committee for Astronomy, welcomed
the participants and noted the scientific interest aroused with the
discovery of X-ray sources. Dr. G. Contopoulos, National
Representative of Greece in the NATO Scientific Division, addressed
the participants on behalf of NATO. The ceremonies were concluded
with a reception of local dishes.

The contributions received in time for publication, are arranged under broad classes of objects - some synthesis, such as the galactic soft X-ray background, and supernova remnants, was necessary. A section of recent results and discoveries has been included for convenience.

Professor Bruno Rossi's recollections of the dramatic discovery of Scorpius X-1, (the birth of galactic X-ray astronomy) are regretfully not in the proceedings. However the participants will recall his presence at the Institute with affection.

The manuscript has been typed by Pamela Garland and Jocelyn Bell Burnell has helped with the arduous task of proof reading. Derek Hoyle was responsible for the artwork. We thank them for their help and the encouragement of all our colleagues in these works.

We acknowledge the generous support and advice of Howard Jones, Ian Johnston and their colleagues at John Wiley on the production of the camera ready manuscript. The Advanced Study Institute owes much to the support and hard work of our Organising Committee, Martin Rees, John Hutchings, Andrea Dupree and Herbert Gursky. The help of Jane Salton, secretary for the Institute, and associate editor of these proceedings is greatly appreciated.

Paul Laskarides
Peter Sanford

List of Contributors

J.A.M. BLEEKER, Cosmic Ray Working Group, Huygens Laboratorium, Leiden, The Netherlands.

R.J. BLISSETT, Mullard Space Science Laboratory, University College London, Holmbury St. Mary, Dorking, Surrey, UK (now at ESTEC, Noordwijk, The Netherlands)

H.V. BRADT, Department of Physics and Center for Space Research, Massachusetts Institute of Technology, Cambridge, MA 02139, USA

A.C. BRINKMAN, Laboratorium voor Ruimte-Onderzoek, Beneluxlaan 21, Utrecht, The Netherlands.

J.B. CASSINELLI, Washburn Observatory, University of Wisconsin-Madison, 475 North Charter Street, Madison, Wisconsin 53706, USA.

P.A. CHARLES, Space Sciences Laboratory, University of California, Berkeley, California 94720, USA. (now at University of Oxford, Dept of Astrophysics, South Parks Road, Oxford OX1 3RQ, UK)

T.J. CHESTER, California Institute of Technology, Pasadena, California 91125, USA

L. CHIAPPETTI, Mullard Space Science Laboratory, University College London, Holmbury St. Mary, Dorking, Surrey, UK. also at Istituto di Fisica dell´Universita and Laboratorio di Fisica Cosmica e Tecnologie, Relative del CNR, Milano, Italy.

G.W. CLARK, Massachusetts Institute of Technology, Department of Physics and Center for Space Research, Cambridge, MA 02139, USA.

F.A. CORDOVA, California Institute of Technology, Pasadena, California 91125, USA

A.P. COWLEY, University of Michigan, Astronomy Dept, Ann Arbor, MI 4828109, USA & Dominion Astrophysical Observatory, 5071 North Saanich Road, Victoria, BC, V8X3X3, Canada.

A.K. DUPREE, Harvard/Smithsonian Center for Astrophysics, 60 Garden Street, Cambridge, MA 02139, USA.

J.E. GRINDLAY, Harvard/Smithsonian Center for Astrophysics, 60 Garden Street, Cambridge, MA 02139, USA.

xvi

L. HARTMANN, Harvard/Smithsonian Center for Astrophysics, 60 Garden
          Street, Cambridge, MA 02139, USA.

J. HEISE, Laboratorium voor Ruimte-Onderzoek, Beneluxlaan 21,
          Utrecht, The Netherlands.

J.B. HUTCHINGS, Dominion Astrophysical Observatory, Herzberg
          Institute of Astrophysics, 5071 Saanich Road, Victoria,
          BC, Canada.

S.A. ILOVAISKY, LA173, DAPHE, Observatoire de Paris, 92191, Meudon,
          France.

R.L. KELLEY, Department of Physics and Center for Space Research,
          Massachusetts Institute of Technology, Cambridge, MA
          02139, USA

E. KENDZIORRA, Astronomisches Institut der Universitat Tubingen,
          Waldhauserstrasse 64, 7400 Tubingen, West Germany.

J.G. KIRK, Max-Planck-Institut fur Physik und Astrophysik, 8000
          Munich 40, West Germany.

N.D. KYLAFIS, Department of Physics, University of Illinois at
          Urbana-Champaign, Urbana, Illinois 61801, USA

D.Q. LAMB, Department of Physics, University of Illinois at
          Urbana-Champaign, Urbana, Illinois 61801, USA.

W.H.G. LEWIN, Massachusetts Institute of Technology, Department of
          Physics and Centre for Space Research, Cambridge, MA
          02139, USA.

L. MARASCHI, Istituto di Fisica dell'Universita and Laboratorio di
          Fisica Cosmica e Tecnologie, Relative del CNR, Milano,
          Italy

B. MARGON, Department of Astronomy, University of California, Los
          Angeles, CA 90024, USA (now at Astronomy Dept FM-20,
          University of Washington, Seattle WA 98195, USA)

R. McCRAY, JILA, University of Colorado and National Bureau of
          Standards, Boulder, Colorado, 80309, USA.

A.N. PARMAR, Mullard Space Science Laboratory, University College
          London, Holmbury St. Mary, Dorking, Surrey, UK.

L.D. PETRO, Department of Physics and Center for Space Research,
          Massachusetts Institute of Technology, Cambridge, MA
          02139, USA

K.A. PETTERSON, Department of Physics, University of Illinois at
          Urbana-Champaign, Urbana, Illinois 61801, USA.

W. PIETSCH, Max-Planck-Institut fur Physik und Astrophysik, Institut
     fur Extraterrestrische Physik, 8046 Garching, West
     Germany.

S. RAPPAPORT, Department of Physics and Center for Space Research,
     Massachusetts Institute of Technology, Cambridge, MA
     02139, USA.

J.C. RAYMOND, Harvard/Smithsonian Center for Astrophysics, 60 Garden
     Street, Cambridge, MA 02139, USA.

C. REPPIN, Max-Planck-Institut fur Physik und Astrophysik, Institut
     fur Extraterrestrische Physik, 8046 Garching, West
     Germany.

G.J. SAVONIJE, Institute of Astronomy, Madingley Road, Cambridge,
     UK.

E.J. SCHREIER, Harvard/Smithsonian Center for Astrophysics, 60
     Garden Street, Cambridge, MA 02139, USA.

R. STAUBERT, Astronomisches Institut der Universitat Tubingen,
     Waldhauserstrasse 64, 7400 Tubingen, West Germany.

W. SUTANTYO, Bosscha Observatory, Bandung Institute of Technology,
     Indonesia.

A. TREVES, Istituto di Fisica dell´Universita and Laboratorio di
     Fisica cosmica e Tecnologie, Relative del CNR, Milano,
     Italy.

J. TRUMPER, Max-Planck-Institut fur Physik und Astrophysik, Institut
     fur Extraterrestrische Physik, 8046 Garching, West
     Germany.

J. VENTURA, Max-Planck-Institut fur Physik und Astrophysik, Institut
     fur Extraterrestrische Physik, 8046 Garching, West
     Germany.

W. VOGES, Max-Planck-Institut fur Physik und Astrophysik, Institut
     fur Extraterrestrische Physik, 8046 Garching, West
     Germany.

M.G. WATSON, X-ray Astronomy Group, University of Leicester,
     Leicester, UK

R.S. WOLFF, Bell Laboratories, Box 400, Holmdel, New Jersey 07733,
     USA

PART 1
X-ray binary stars

Optical review:
X-ray binaries

J. B. Hutchings

Dominion Astrophysical Observatory,
Herzberg Institute of Astrophysics,
5071 Saanich Road,
Victoria,
B.C.

1.   Overview

2.   The Supergiant Primaries

3.   Spectroscopic Complications

4.   Other System Quantities

5.   Line Emission

6.   The Be Primary Systems

7.   Light Curve Analysis

8.   Summary

1.   OVERVIEW

The X-ray binary stars with massive hot primaries form a very distinct group; they can be treated very well on their own for optical
studies. The reason for the grouping is that these primaries lose
their mass at a rate sufficient to power, but not blanket, X-rays
from a companion, by mechanisms other than strict Roche lobe overflow. Less massive stars (<2 $M_\odot$) can only power a source at a
suitable rate by Roche lobe overflow. We will speak of Roche lobe
in the massive systems, but the phenomenon is more one of overflow
by the extended envelope (the stellar wind), than by overflow of the
photosphere, which may itself lie a few percent within this radius.
The paper by Savonije in this volume deals with this point in depth.
A second group of massive primaries lose their mass by stellar
rotation, and these generally power X-ray sources at a much reduced
level, presumably because of a low accretion rate.

Table 1 summarises the systems for which we have reasonably extensive optical data in these two groups. The mass limitations are
described above but, for completeness, we should note the narrow
range of spectral types are a result of evolutionary factors:
Evolution beyond the hydrogen burning phase is very rapid in such

A) Short period, high energy ($E_x \sim 10^{37}$ erg/sec, $E_{opt} \sim 10^{39}$ erg/sec)

NAME	PERIOD (DAYS)	PULSE PERIOD	Opt*	$\dfrac{M_{opt}}{M_\odot}$	$\dfrac{M_x}{M_\odot}$	ECLIPSE?
Cen X-3 (Krz' star)	2.1	4 sec	O7	17	1.0	E
1538-52 (Star 12)	3.7	9 min	O9	20	2.0	E
1700-37 (HD 153919)	3.4	95 min?	O6	27	1.3	E
Cyg X-1 (HDE 226868)	5.6	-	BO	15	5	
1653-40 (HD 152667)	7.8		BO	25	7	E
0900-40 (HD 77581	9.0	5 min	BO	22	1.6	E
Cir X-1	16.6	-	B	?	?	
1223-62 (Wra 977	>22	10 min	B1	25:	2.5:	
See also: SMC X-1 (Sk 160)	4.0	0.7 sec	BO	16	1.4	E
LMC X-4	1.4			O9	24	2.4	E

B) Long period, low energy ($E_x \sim 10^{34}$ erg/sec, $E_{opt} \sim 10^{38}$ erg/sec)

NAME	PERIOD (DAYS)	PULSE PERIOD	Opt*	$\dfrac{M_{opt}}{M_\odot}$	$\dfrac{M_x}{M_\odot}$	ECLIPSE?
0115+63	24	3 sec	BOe	15?	?	
0535+26 (HDE 245770)	28	100 sec	BOe	15?	?	
0352+30 (X Per)	590?	13 min	BOe	15?	?	
0045+60 (γ Cas)	1000?		BOe	15?	?	
1145-60 (HD 102567)		10 min	BOe	15?	?	
1258-61 (GX 304-1)	>20	4.5 min	$\sim$BOe	?	?	
0114+650 (LSI+65° 010)	?	?	$\sim$BOe	?	?	
1118-615	?	7 min	$\sim$BOe	?	?	
0236+610 (LSI+61° 303)	?		BIe	?	?	
See also: SMC X-2, 3	?		O9e			

TABLE I: HARD GALACTIC X-RAY SOURCES WITH HIGH MASS COMPANIONS. APPROXIMATE OR PROBABLE SYSTEM PARAMETERS

stars, and we would not therefore expect to see many systems in the subsequent stages of evolution. (Wra 977 (1223-62) is the most extreme case.) We note that the binary period, which reflects the component separation, increases with the stage of stellar evolution for the primary, throughout the systems of group A. (See Figure 1). This again indicates that the powerful X-ray emission occurs only when the primary fills (or nearly fills) its Roche lobe.

Figure 2 shows a typical supergiant binary system. Only two systems differ from this: Cyg X-1 and HD 152667 (1653-40) (if it is an X-ray system). In these, the X-ray companion is thought to have a higher mass, so the centre of mass of the system lies further from the primary, and the tidal distortion of the primary is larger. In most cases we have some evidence for an accretion disk and a region of emission at $\lambda = 4686A$, near the X-ray source. We will discuss these points later.

2.   THE SUPERGIANT PRIMARIES

TABLE 2

OPTICAL PRIMARIES

System	V-E slope (1)	M.L. B.P. (2)	Criteria	$\dot{m}$ (3)	$\dfrac{V_{rot}}{V_{sync}}$	$\dfrac{M^*}{M_{norm}}$	V_{o} (km s^{-1})
SMC X-1	O	O	1	$\sim\!10^{-7}$	1.0	16/32 = .50	180
LMC X-4	O	O	O	$<\!10^{-7}$	0.5	25/35 = .71	280
Cen X-3	10	O	3	$\sim\!10^{-6}$	1.0	17/45 = .38	40
0900-40	8	50	4	5×10^{-6}	0.8	21/50 = .42	-5
1700-37	6	70	6	$\sim\!10^{-5}$	1.0?	28/55 = .51	(-60)
1538-52	O	O?	>1	2×10^{-7}	1.0	20/30 = .67	-170
1653-40	4(V)	O(V)	4	7×10^{-6}	>.9	30/50 = .60	-30
Cyg X-1	15	5	2.5	$\sim\!10^{-6}$	0.5:	15/34 = .44	-7

(1)   = 2x(eV/km s^{-1})
(2)   Mass Loss Balmer Prog. = H β - H limit, (km s^{-1})
(3)   = $M_\odot$/year

Table 2 summarises the available information on the spectra of the primary stars. As is normal for stars of this type and luminosity, all spectra indicate mass loss. (See e.g. Hutchings 1976). The first 3 columns refer to the indicators of mass loss:

   a)   a radial velocity progression with line excitation, indicating an outward acceleration with falling temperature through the stellar envelope;

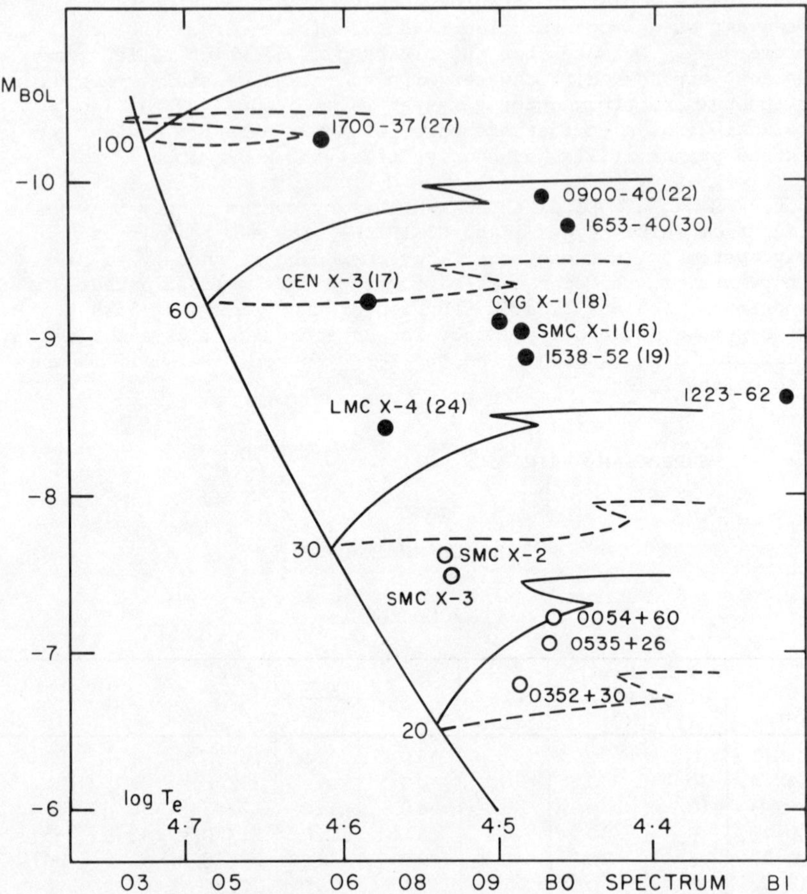

Figure 1:   H-R diagram with positions of supergiant (filled) and Be
            (open) primaries. Masses in $M_\odot$ given in parentheses.
            Lines are evolution tracks with mass-loss (dashed) and
            loci if constant mass with mass-loss evolution (solid).
            ZAMS initial masses given in $M_\odot$.

b)   a similar progression down the Balmer series, arising in
     the same way; and

c)   the number of mass loss indicators (such as P Cyg profiles)
     present in the spectrum. Rates of mass loss (column 4) are
     estimated, as in the reference above, by comparison with
     stars with detailed models for their stellar winds, and
     these should at least form a self-consistent set.

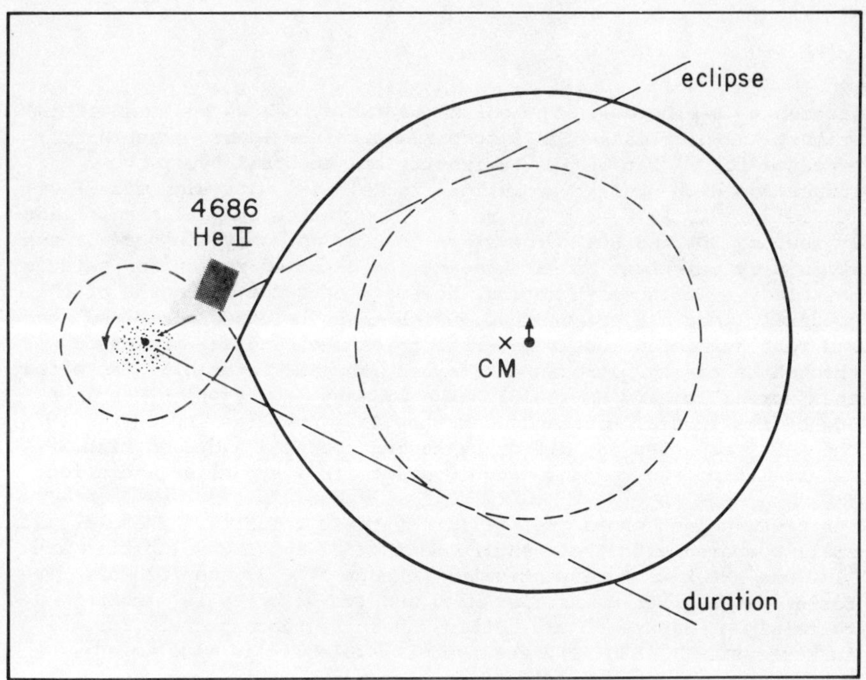

Figure 2: Sketch in orbital plane of SMC X-1 system. This is typical of all supergiant X-ray binaries.

Note that the least luminous star, LMC X-4, has no indicators of a stellar wind in the visible spectrum, and the most luminous (1700-37) has one of the highest rates for mass loss known for stars in the galaxy. Clearly, these circumstances must affect the accretion process onto the compact companion.

A further important parameter is the rotation of the primary. Transfer of mass depends on whether the rotation is slower than, or synchronous with the orbit. (Presumably faster rotation would lead to unstable mass loss which would slow the star down quickly, or blanket the source). The values given above are derived from estimates of $V \sin i$ from the optical spectra. There is some disagreement here since the stellar wind can itself distort line profiles and several of the systems are very faint. However, there is evidence that rotation is slow in some cases, and this affects the light curve analysis (see below) and the formation of wakes.

The most important quantity is the mass of the primary since this relates to the X-ray star mass, and the evolution of the system. The mass is well determined in the systems which have orbits determined from measurement of the X-ray pulsations (see Saul Rappaport's contribution in these proceedings), measured eclipses, and orbits from optical observations: (SMC X-1, Cen X-3, 0900-40, 1538-52). We also have good values for the remainder from other arguments (see below). The column in table 2 shows the mass of the primary in solar masses $(M_\odot)$ compared with the lowest mass (M_{norm}) expected for a star of its luminosity and spectral type, by comparison with stellar evolution tracks, even with high mass-loss (Chiosi et al. 1978: see figure 1). The values are all significantly low, by 30% and 60%. The stars thus do not obey the normal mass-luminosity relation, and at present the detailed reason for this is not clear. It is worth noting, however, that the strengths of the spectral lines are not unusual, which might be expected if the stars had lost sufficient outer material to expose material processed through to carbon, nitrogen and oxygen (the CNO material). Note that the "normal" binary HD 163181 has a primary (BO I) which has lost > 60% of its initial mass and does show CNO anomalies (Hutchings 1975). The X-ray binaries may differ by having accreted hydrogen rich material from the companion, now compact, in its earlier evolution.

The final column shows the stellar systemic velocity. These are all small compared with their environments (SMC and LMC velocities are $\sim$170 and 290 km s^{-1} respectively), except 1700-37 and 1538-52. The former has an extreme stellar wind and its velocity is almost certainly a measure of the outflow near the photosphere. For the latter system, (1538-52), the system's velocity is $\sim$100 km s^{-1}, high compared with its galactic environment. This then may be the only case in which the presumed supernova (in which the compact star was formed) may have given the whole system an appreciable rebound. This in turn may be relevant data for testing the theory of supernova outbursts.

3.   SPECTROSCOPIC COMPLICATIONS

Tables 3 and 4 summarise the deductions made largely on the basis of the optical measurements on these systems, but in some cases in conjunction with X-ray information. In order to assess their reliability, we look in this section at some of the problems encountered and assumptions made in interpreting the optical spectra.

The star is non spherical, with a temperature and intensity gradient over its surface, and this leads to a distortion of the line profiles, which can give errors in the radial velocities. These effects have been calculated by Wilson and Sofia (1976), van Paradijs et al. (1977) and Hutchings (1977), and can be large (see Figure 3). However, in all cases the predictions are for a spurious value of ω of 90° or 270°. In practice ω is almost zero for all systems. Thus these effects, while presumably present, are weak and are not the dominant one. We do, however, see systematic changes in He I, He II line strengths with phase in SMC X-1, LMC X-4 and Cen X-3, due presumably to this distortion, and X-ray heating.

A different effect is one of a variable stellar wind. Radial velocities will appear to have negative values at phases when the wind is enhanced along the line of sight. In a circular orbit, this will occur where the surface gravity is lowest, and the resulting

TABLE 3:   MASSIVE BINARIES - SPECTROSCOPIC QUANTITIES

SYSTEM	SPEC-TRUM	ORBIT PERIOD	q	M_x	L_x (Log_{10})	M_{BOL}	$\dfrac{L_{opt}}{L_x}$	e
Cir X-1	B	16.6 d				- 5.4	~ 1	
SMC X-1	BO	4.0 d	16	1.0	38.6	- 9.0	4	0
LMC X-4	O9	1.4 d	9	2.6:	38.3	- 8.3	30	
Cen X-3	O7	2.1 d	17	1.0	37.1	- 9.3	100	0
Cyg X-1	BO	5.6 d	3	6	37.1	- 9.1	100	0.04
1223-62	B1	>22 d	10:	2.5:	37:	- 9.0:	100	
1538-52	BO	3.7 d	10	2.0	36.2	- 9.1	500	
0900-40	BO	9.0 d	12	1.7	36.3	-10.0	1300	0.10
1700-37	O6	3.4 d	20	1.4	35.8	-10.2	5000	0.05
1653-40	BO	7.8 d	3.5	7	~ 34	-10.1	$\sim 10^5$	<0.05
0352+30	O9	581 d?			~ 33	- 5.7	$\sim 10^5$	

TABLE 4:   MASSIVE BINARIES - SYSTEM QUANTITIES

SYSTEM	i	$\dfrac{Sep}{R}$	E.W. emis (Å)		WAKE?	PULSE PERIOD	SPUR-IOUS e?
			4686	Hα			
SMC X-1	$65°$	1.6	1.2	~ 2	-	0.7 s	0.36
LMC X-4	$75°$	1.7	0.7		Yes		
Cen X-3	$80°$	1.6	1.2	1.7	?	4 s	
0900-40	$>75°$	1.6	0.2:	2:	Yes	283 s	0.2,*
1700-37	$87°$	1.5	~ 4	~ 15	Yes	94 m	.25?*
1538-52	$70°$	1.7	0.2		?	9 m	.14?
1653-40	$\sim 90°$	2.0	0	6	Yes		.08
Cyg X-1	$\sim 60°$	2.1	1.2	1.3	-		real
1223-62	$\sim 90°$	$\leqslant 4$				11 m	

*part real

 Blank denotes quantity unknown.

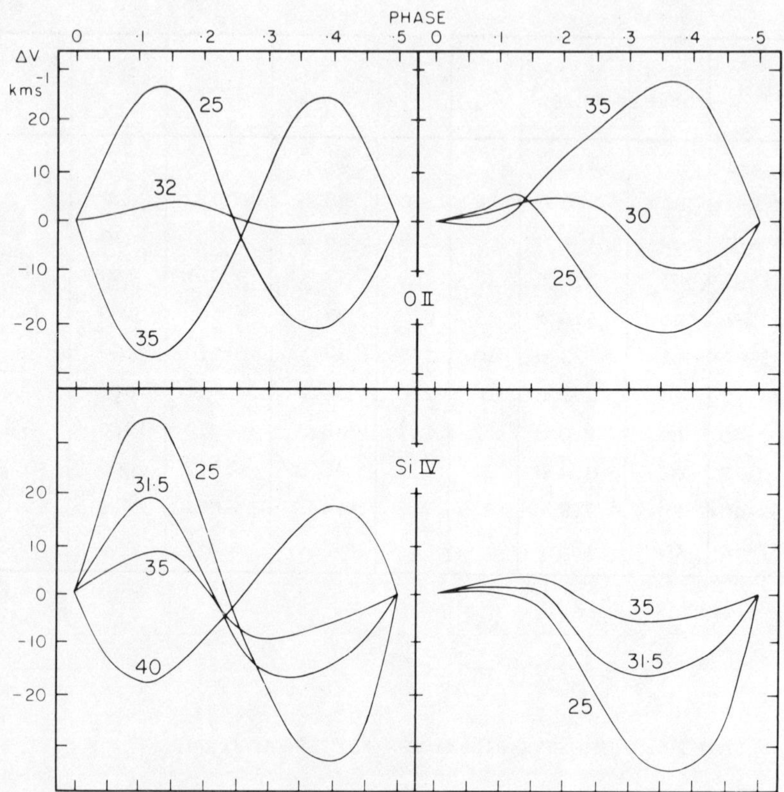

Figure 3:   Calculated distortion of radial velocities for lines
            indicated due to Roche lobe distortion of primary is
            SMC X-1.   Numbers are polar temp in 1000 K units.
            Left: without X-ray heating; right: with heating.
            Similar results apply to all distorted primaries.

velocity curve will have $\omega \sim 0^{\circ}$. Milgrom (1978) has shown that
this is a very likely explanation for most systems, but it is
difficult to account for quantitatively. An empirical approach is
to correct the observed radial velocities to a sine wave in the
systems where X-ray orbits show that $\underline{e} = 0$ (SMC X-1, Cen X-3).

In cases where $\underline{e} = 0$ there is a further complication. Mass flow
(and hence velocity) is enhanced at periastron, when tidal forces
are greatest, and this can give various distortions to the radial
velocities, depending on the value of ω. We can see this effect in
the non X-ray binaries HD 47129 and 57060, as well as in 0900-40
and 1700-37 in our group. In the latter case, Hensberge (1978) has
reduced the value of e from ~ 0.2 to ~ 0.5 by removing the phase

dependent stellar wind effects. We show below that the lower
value of $\underline{e}$ may be real in this system.

A further complication is in the "wake" phenomenon. In several
systems, there are additional absorptions or line broadening at
$\emptyset_x \sim 0.75$ (LMC X-4, 0900-40, 1700-37, 1653-40). These appear to
arise in a gas stream or shock trailing the compact body in its
passage around the massive primary. Similar effects are present in
some "normal" binaries, such as HD 47129. These may further distort the velocity curves.

Finally, there appear to be occasional irregular deviations from
orbital velocity which must be caused by instabilities in the whole
outer atmosphere of these stars. Such an event was seen e.g. in
0900-40 by Wallerstein (1974).

Much thought and care has been put into allowing for these problems,
but they inevitably make the orbital and mass determinations uncertain in systems where the amplitude is low, as in most of these
X-ray binaries. As we see later, the problems are even worse for
the Be star primaries.

4.   OTHER SYSTEM QUANTITIES

The relative X-ray and optical luminosities are of interest, since
they appear to vary very widely. To determine this, the distance
must be estimated, and the optical extinction. These have been
discussed carefully for most systems, by looking at interstellar
reddening and absorption lines in the stellar spectra of the
primaries and in stars in the field, or by looking for membership
in clusters and associations. There is general agreement on most
of these, with the probable exception of Cir X-1, and the
appropriate bolometric corrections yield the X-ray and optical
luminosities shown in table 2. These quantities and their ratios
have a bearing on the understanding of the accretion processes and
the heating of the disk and even of the optical primary by the
X-rays, as well as processes of quenching and X-ray variability.
The figures derived for L_x show no significant correlation with
L_{opt}, the separation of the components, or the mass ratio.

Another system parameter of importance is the inclination, $\underline{i}$, of
the orbit to the line of sight. Figure 4 shows how this can be
estimated from spectroscopic constraints and the eclipse duration,
assuming a point source for the X-rays. From these, we find the
values in table 4.   The case of Cyg X-1 is less certain because
it does not eclipse, and the value depends on consideration of
several less well defined quantities (light curve analysis, soft
X-ray "eclipse" and hard X-ray modulation, polarization). However,
whatever its value, we appear to have an excess of high inclination
systems for the distribution to be isotropic. The distribution of
$\underline{i}$, as well as M_x and M_{opt}, is shown in Figure 5. Clearly eclipsing
systems are much easier to detect and identify, but there is still

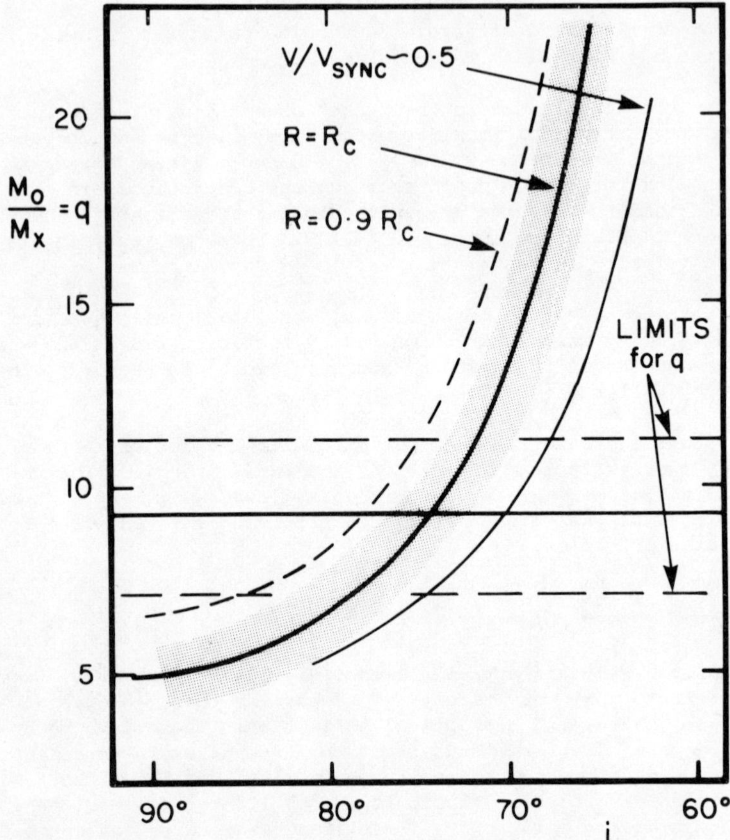

Figure 4: Estimation of q and i for LMC X-4, limits to q from
optical mass function. Curves for known eclipse
duration for various primary star shapes. Similar
arguments apply to all eclipsing systems.

an excess of eclipsing systems with bright optical counterparts,
both in the galaxy and the Magellanic Clouds. This suggests a lack
of isotropy in the X-ray beaming, and clearly needs to be explained.

The masses of the X-ray stars all fall in or near the classical
range of neutron stars ($< \sim 2\ M_\odot$) so that we are obliged to
regard the more massive objects Cyg X-1, 1653-40 as serious
candidates for black holes.

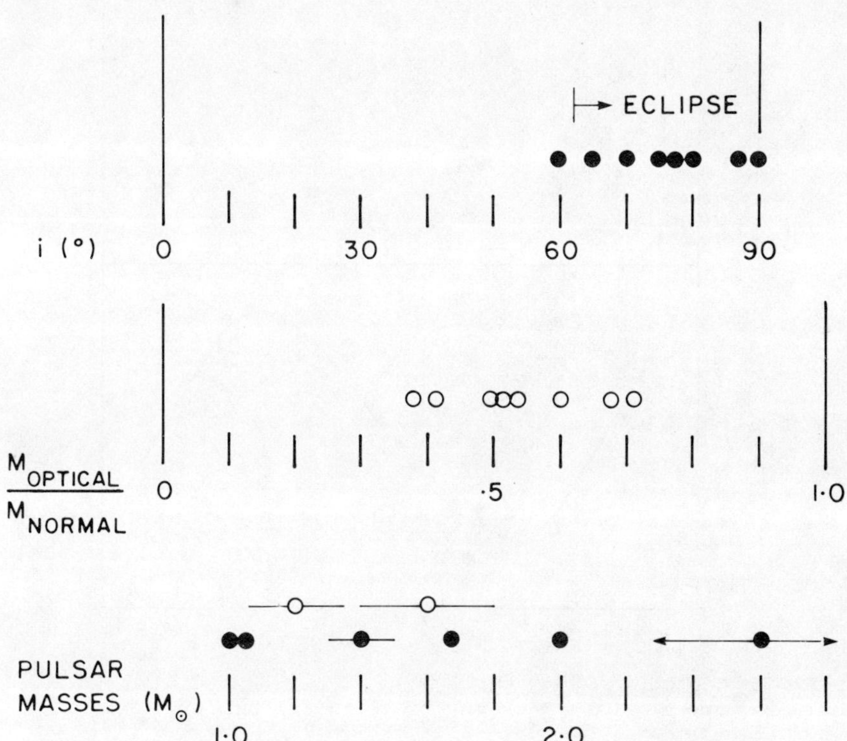

Figure 5: Distributions of $\underline{i}$, M_{opt} as fraction of 'normal' value, and M_x. $\underline{i}$ values are significantly grouped above 60°. M/M values are all similarly less than unity. M_x values are all consistent with upper limit of $\sim$ 2 $M_\odot$ for neutron stars. Open symbols are for Her X-1 and Cyg X-2.

5.   LINE EMISSION

The characteristic He II λ 4686 emission line (and N III + C III λ 4640 blend) is present in all these systems, with the possible exceptions of 1223-62 and 1653-40.   The mean strength of this line is given in Table 4, together with that of Hα. Both of these emissions are at least partly associated with the mass transfer streams and are thus useful in studying the phenomenon. In the stars Cen X-3 and 1700-37 the He II emission is largely formed in the Of envelope of the primary, and the smaller contribution from the stream is hidden. In the other systems the He II emission shows large orbital motion, approximately antiphased with the primary (see e.g. Figure 6.) However, from the X-ray pulsar orbits

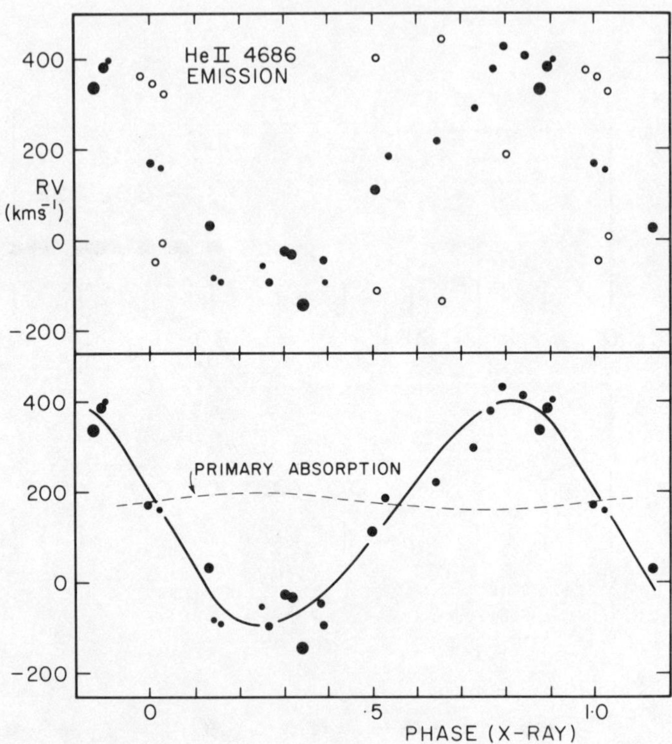

Figure 6: Typical velocity curve of He II λ 4686 (SMC X-1). Note approximate antiphasing with primary and large amplitude.

SYSTEM	q	K_x	K_{4686}	$\Delta\phi_{4686}$	$\dfrac{K_{4686}}{K_x}$
SMC X-1	16	300	245	0.04	.82
77581	12	270	140:	?	.52
LMC X-4	9	∿500:	500	?	–
Cyg X-1	3	200:	120:	0.15:	.60:
153919	20	385	–*	–	
152667	3.5	300:	–	–	
Cen X-3	17	410	–*	–	
1538-52	10	325	302	?	0.93

* Of primary

TABLE 5 - HE II EMISSION

or other mass-ratio indicators, it is clear that the emission
arises between the stars, and probably lagging somewhat behind the
X-ray star. This corresponds to where a gas stream might hit an
accretion disk, and is analogous to cataclysmic variables. It has
been proposed that the emission may arise on the heated face of the
primary, but the observations in these systems do not support this
idea. There is a weak correlation between λ 4686 intensity and L_x
and L_x/L_{opt}.

The behaviour of Hα is generally more complex, since this emission
arises to some extent in the wind of the primary star and is often
centrally reversed by the wind or stream. However, the width of the
feature and the velocity behaviour also suggest that some of it
originates in an accretion disk with large Keplerian velocities.
Figure 7 shows the behaviour of Hα in Cyg X-1 in 1977. Very similar
behaviour is evident in 0900-40 and 1653-40, and in all cases the
profile can show profound changes between seasons, presumably as the
wind and stream structure alter. The shift of the absorption in
phases near 0.75 is reminiscent of the wake effects seen in other
lines.

The detailed understanding of these lines are particularly closely
associated with the accretion processes, and the data should provide
useful constraints on models for these.

6.   THE Be PRIMARY SYSTEMS

These are the most difficult to study, optically, in spite of their
relative brightness, for several reasons. They are not powered by
Roche lobe overflow - thus periods are long and the radial velocity
amplitudes are small. None are eclipsing, presumably because of
their wide separation, so that the X-ray data provide no periodic
clock to look for. X-rays are weak and very variable, presumably
because of the weak mass flow, and the pulse periods are long (see
van den Heuvel in this publication) making X-ray orbital determinations very difficult, in the presence of spin up rates comparable
with doppler orbital changes. Finally, the optical behaviour in
these objects is characteristically very variable and irregular, so
that persistent underlying periodic changes are very masked. To
date no optical orbit exists for any one of these objects, and an
X-ray orbit exists for only one (0115 + 63). Thus we do not yet
have confirmation of the proposed model, or an estimate for the
masses. X Per has an apparent 580 day radial velocity period which
may or may not be real and whose amplitude is probably not orbital,
since it implies a very high mass for the (pulsing) companion. The
recurrent transient 0535 + 26 shows some pattern in its outburst
times (occasional multiples of 28-day - or longer - intervals) and
both optical and X-ray velocities suggest periods in the range 28-
100 days, with large uncertainties (Figure 8). The X-ray outbursts
of 1145-61 appear to have a 188 day periodicity.

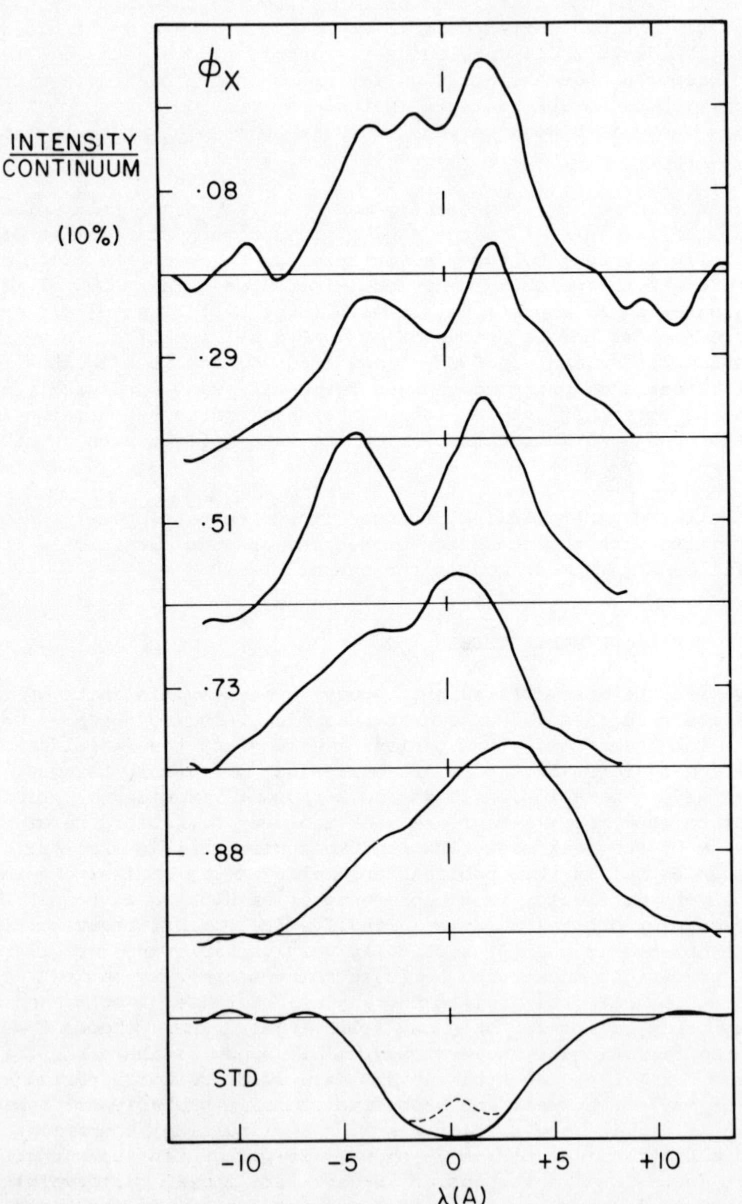

Figure 7: Cyg X-1 Hα emission corrected for photospheric ('STD') absorption in 1977. Note shift of absorption to blue in phases 0.5 - 1.0.

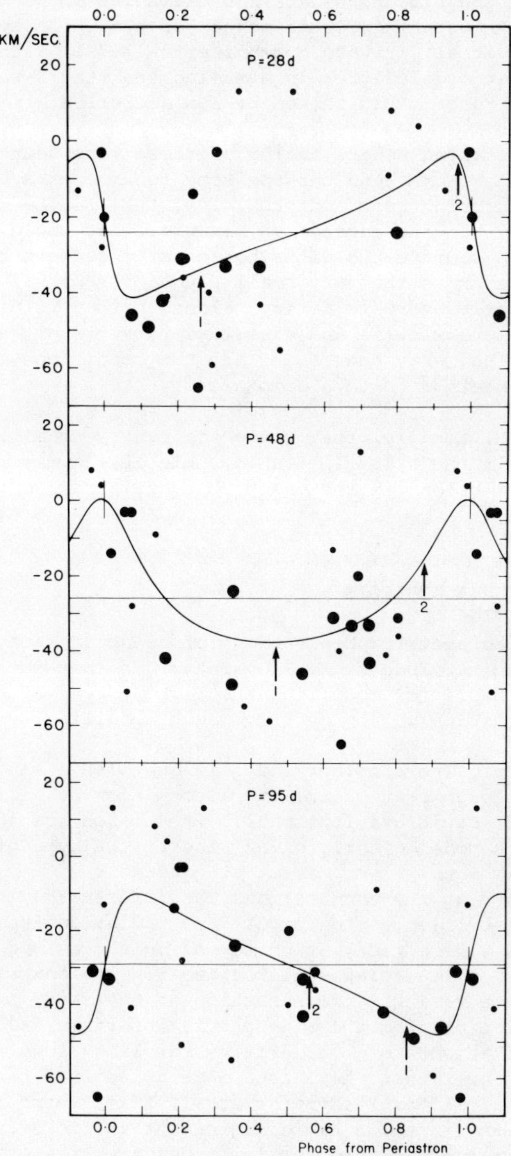

Figure 8: Radial velocities and orbital solutions for Be star primary HDE 245770 for three "best" periods under 100 days. Note large scatter in data.

It is likely that systems like these have quite eccentric orbits, since tidal interactions are small. This may cause significant modulation of the mass transfer rate (and hence X-ray luminosity), especially if the circumstellar ring fills its Roche lobe at periastron. An eccentric orbit model has been proposed for Cir X-1 (Murdin, Holt et al. private communication 1979), although the details of that model differ in assuming that the X-rays are cut off by absorption rather than choked by low accretion.

Clearly the detailed understanding of these weak sources needs a lot of observational work, and perhaps some lucky breaks. Considering their relative brightness and proximity, they may be more numerous in the galaxy than the supergiant binaries. It would also be interesting to compare the distribution of i between the samples. Finally, we may note the very small range of spectral type in these stars. Evidently there is a fine balance between stellar mass (O stars rotate more slowly) and mass loss rate (late B stars have weaker winds) and only near to BO are the conditions met to produce sufficient accretion on to a companion.

We should note, finally, that since the binary model is only assumed to apply to all the stars listed in table 1B, it may be that some objects do not belong in the group, but are X-ray active for some other reason.

7.   LIGHT CURVE ANALYSIS

The optical photometric observations of galactic X-ray sources are reviewed in this volume by S. Ilovaisky. In this section we review the physical basis for modelling the light curves of X-ray binaries, and the information we can get from such modelling.

The basic model. The simplest model is one with a single luminous body, the non-degenerate star, and a companion of negligible size, whose only effect is gravitational. The bodies are in a circular orbit. Such a model clearly gives rise to the type of variations observed: two equal light maxima at quadratures, and minima of different depth at the conjunctions, as illustrated in Figure 9. The numerical problem is to compute the radiation from the stars viewed from any direction, which may be characterised by phase angle and the orbital plane inclination i. The computation must account for the variable cross-section of the star, the emergent angle of radiation from the non-spherical surface, and the surface variations of temperature and gravity resulting from the tidal distortion. Such models have been made by several people (e.g. Avni and Bahcall 1975, Zuiderwijk et al. 1977, Petro 1977, Hutchings 1978) and there is negligible disagreement in the results. Differences are mostly due to the stellar model atmospheres and limb darkening used, and to a lesser extent to the integration code or algorithm. The results quoted here are derived from the code described by Hill (1979) and are discussed in several cases in

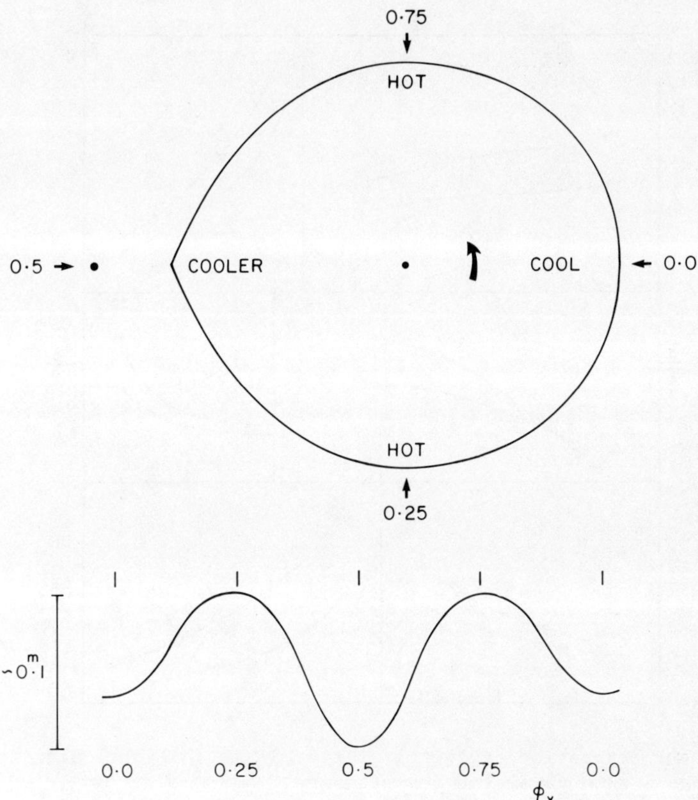

Figure 9:   Schematic of light curve in circular orbit and
            definition of phases.

more detail by Hutchings (1978). The parameters defining the simple
model are then $\underline{i}$, T_{eff} (usually polar), $R_\ast$/separation, q, g (polar)
and the stellar rotation as a fraction of synchronous. Roche
geometry is generally assumed to apply to synchronous (or greater)
rotation, and a tidal lobe for slower values.

In fitting a model to observations, any of these parameters may be
fixed or constrained by information from other sources (spectral
type, mass function, eclipse duration, V sin $\underline{i}$ et.). Limitations
are further imposed by the errors in the observations and the
relative sensitivity of the light curve to the different parameters.
A good idea of these sensitivities in the region of parameter space
occupied by the massive X-ray binaries is given by the grid of
models shown in Figure 10.

20

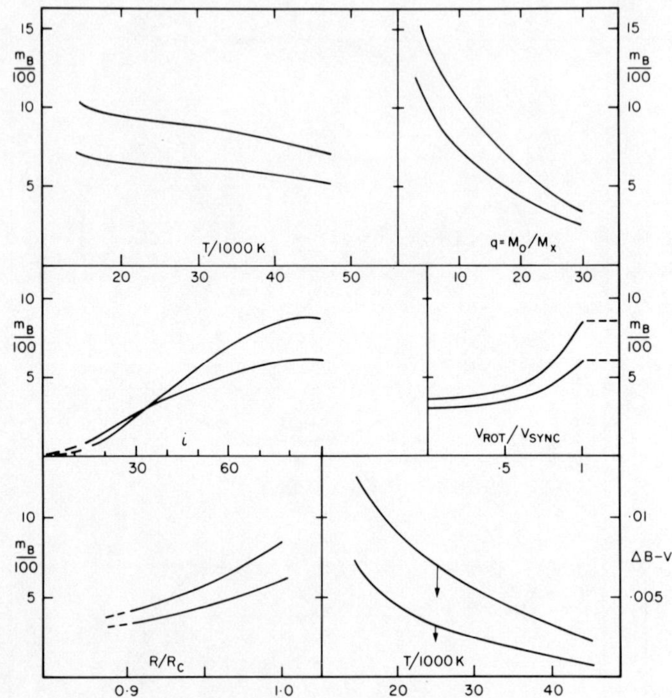

Figure 10:   Variation of depths of two minima in light curve for
             each parameter for standard T = 30000 K, q = 13,
             e = o, $\underline{i}$ = 90°, R = R_{crit} model.

If we now attempt to model the light curves with this simple model
we quickly find that it is impossible. In no case do we see an
observed light curve with equally spaced maxima of equal intensity
and equally spaced minima of the expected depth ratio. Several
workers have pointed this out (e.g. Petro 1977, Zuiderwijk et al.
1977). We must therefore include further complications. In the
systems Cyg X-1 and 0900-40 we have clear evidence for non-circular
orbits, so that this is a good next step. The introduction of e >
o is described by Hutchings (1974, 1978) and involves three
assumptions:

1)   The primary may not exceed its Roche limiting surface at
     periastron. Thus tidal distortion at all other phases
     is smaller, and the effect is to <u>reduce</u> the light curve
     amplitude - except at periastron - below that for an e =
     o orbit with Roche lobe filling. Inspection of Figure 11
     shows that this can be noticeable for very small values of
     e - as small perhaps as 0.01. Note, incidentally that the
     spacing of the turning points and levels of the maxima

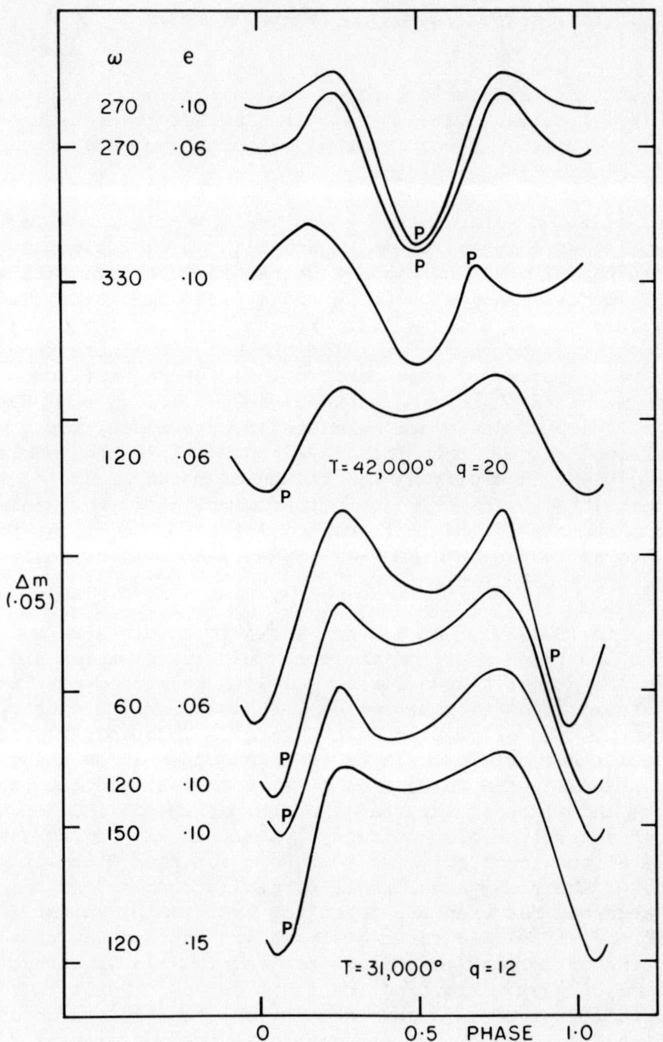

Figure 11:   Examples of e > o orbit light curves.   Note decreased range away from periastron.   P marks phase of periastron passage.

will in general not be even for e > o since the orbital velocity becomes a function of phase.

2) The primary star is always at its equilibrium shape.

In practice the star may take some hours to react to a
changing tidal force and if this is an appreciable fraction
of the orbit a phase lag can be introduced.

3) Since the rotation of the primary is considered to be
uniform, a phase dependent orbital velocity will cause the
tidal bulge to lead or lag the line of centres, in addition
to effect 2 above. This may also be inserted as a phase
dependent time change.

Thus, the additional parameters $\underline{e}$ and ω allow much more flexibility
in the model, as illustrated in Figure 11. If we apply the model to
the curves for 1700-37 and Cyg X-1 we can fit the data well with
reasonable models, and values of $\underline{e}$ and ω quite compatible with
spectroscopic values (Figure 12). We can also constrain the model
for 0900-40 by using the new well determined parameters from the
X-ray pulses (Rappaport, this publication). Here excellent
agreement with the shape is achieved (Figure 13), at some times,
but the full amplitude is not permitted by the model, for q values
in the allowed range. Note that in Hutchings 1978 the requirement
of fitting the full amplitude led to the adoption of e = o, which is
now not permitted. The case is further complicated by changes in
the mean photometric behaviour, especially at $\emptyset \sim 0.75$, as pointed
out in several instances (LMC X-4, 1538-52, SMC X-1 as well) by
Ilovaisky (this publication).

The orbits for SMC X-1, Cen X-3 are known to be circular and the
amplitudes observed in all of the other massive binaries are
larger than can be fit even for e = o, with permitted mass ratios.
Thus, while we have had some success, it is clearly necessary to
add further to the original model. There is independent evidence
for the existence of a disk in SMC X-1 and Cen X-3 (Hutchings et al.
1977, van Paradijs and Zuiderwijk 1977, Bonnet-Bidaud and van der
Klis 1979) and hints in the shapes of the minima of SMC X-1 and
Cen X-3 of an eclipse discontinuity. A simple way of increasing the
amplitude of the light curve is to have a luminous disk which is
eclipsed, or which may even itself partially obscure the primary.
Such an argument has been presented for V861 Sco by Walker (1972)
and Hutchings (1979). Unfortunately we are now faced with the
introduction of so many parameters to describe the disk that we have
a multitude of ways of fitting the light curves. Until the disk can
be standardised either by other theoretical constraints or other
observational inferences, we are unable to make meaningful detailed
light curve models.

A further effect is that of X-ray heating on the primary, which is
clearly present in Her X-1 and probably important in SMC X-1.
These effects, too, can be modelled, but depend to some extent on
the disk model, as well as complex soft X-ray reprocessing in the
outer layers of the primary. Again, we do not have a meaningful
standard model.

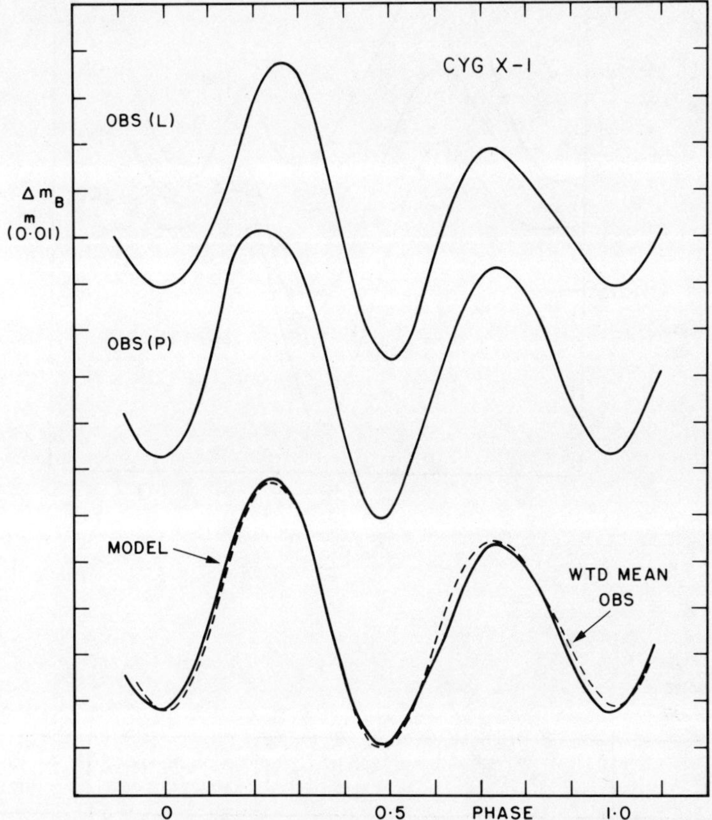

Figure 12:   Observed and fitted blue light curves of Cyg X-1,
requiring e > o model.

I have therefore simply tried to examine all the light curves for
the presence of unmodelled effects, and to estimate them where
possible, by comparing the simple calculated model with the data.
This is summarised in Table 6.

There is one more qualitative correlation to consider. Several
light curves have marked variability near ϕ = 0.75 and some at
ϕ = 0.25 as well. The former phase is where we have observed the
wake effects in spectroscopic data and we may well be seeing
distortion of the light from the primary when a shock or wake is
superposed on it. In addition, if rotation is slower than synchronous the gas stream will lag the X-ray companion and may give
rise to variable opacity in phases following phase 0.5.

24

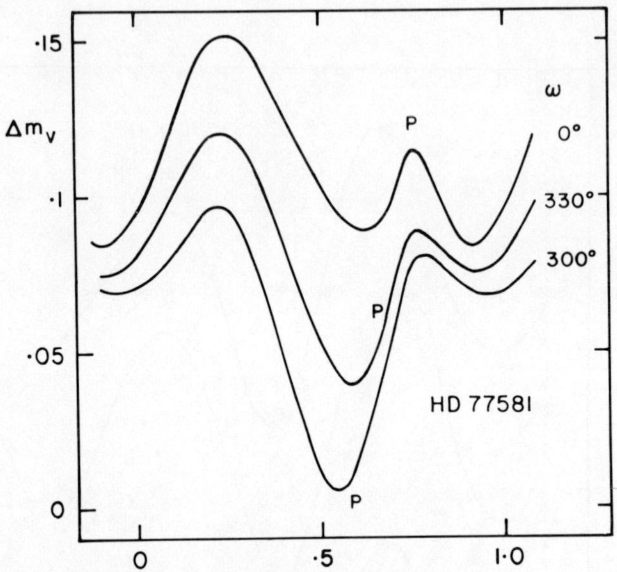

Figure 13: Grid of light curves for HD 77581 with e = 0.095,
T = 30000 K, i = 90°. Best fit is ∿ 330°.

SYSTEM	M_v	EXTRA LIGHT (m)	@ phase 0.75 VARIATION	WAKE PRESENT?
LMC X-4	-5.3	0.10	0.08	√
1538-52	-6.3	0.05	0.05	?
SMC X-1	-6.5	0.05	0.05	×
1700-37	-6.8	0	0	√
Cen X-3	-7.1	0.02	0	?
0900-40	-7.4	0.02	0.04	√
152667	-7.6	0.03 :	0	√

TABLE 6

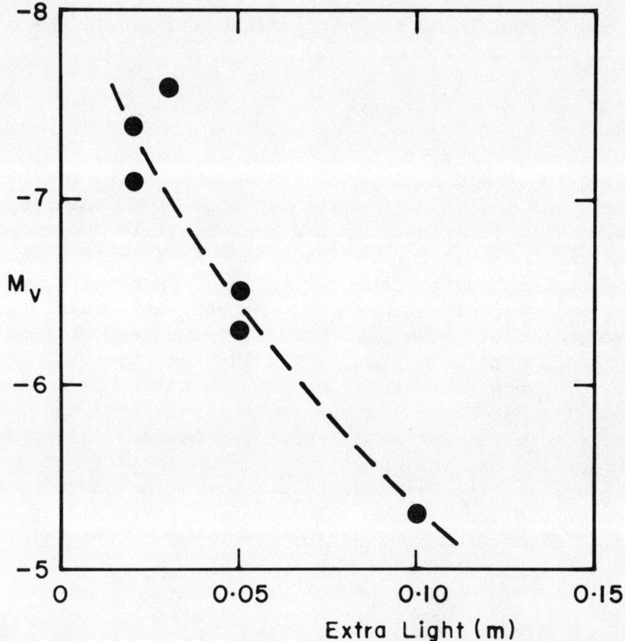

Figure 14: Extra light required to match model to observed light
curve correlated with luminosity of primary. This may
imply disk a standard luminosity.

Variability at phase 0.25 could arise from streaming effects (e.g.
Prendergast and Taam 1974) but it is not clear as a general
argument. Table 6 presents the relevant observational data for
consideration of these points.

One interesting correlation is suggested. The amount of extra
light, presumably from the disk or stream, is larger for the
intrinsically less luminous stars, and Figure 14 indicates a
relationship that measures this. This diagram suggests that the
disk light is approximately independent of the primary and if it is
thermal at 12000 - 20000 K its energy is in the region 10^{36} - 10^{37}
erg/sec. No correlation with L_x is seen. The disk light for Her
X-1 and Sco X-1 can also be estimated from IUE fluxes and the m_V
eclipse depth in Her X-1 and the assumption that all of m_V is disk
in Sco X-1. These numbers are also in the same energy range. We
thus have the suggestion that the disk luminosity is comparable
with X-ray luminosity and it is a relevant question to examine
whether this can be accounted for in accretion luminosity and/or

reprocessing of X-rays. While the modelling problems are large, the work is still potentially possible and valuable and needs to be pursued. It may also be of considerable value to derive light curves in the UV and IR to further isolate and define the contribution from the disk and stream.

8.   SUMMARY

The foregoing is a necessarily biassed and truncated review of these objects and individual references should be consulted for wider views and information. In particular, while I have pointed to problems and uncertainties, I have made no attempt to quantify these. The numbers in the tables therefore represent my own preferences and prejudices to a large extent, and contain no assessment of uncertainties. The main thrust of the presentation has been to regard the objects as a group and study their general properties and behaviour, since these have now been established reasonably firmly. Specific areas of concern are the primary masses, the distribution of $\underline{i}$, and the mass transfer processes, as evidenced by wakes, extra light and variable line emission. It is of particular interest to examine the extragalactic systems for comparison (e.g. the MC sources are more luminous in X-rays), particularly the new objects being discovered by the Einstein Observatory.

REFERENCES

Avni, Y., and Bahcall, J., 1978. Ap. J., 197, 675.
Bonnet-Bidaud, J.M., and van der Klis, M., 1979. A. and Ap., 73, 90.
Chiosi, C., Nasi, E., and Sreenivasan, S.R., 1978. A. and Ap., 63, 103.
Hill, G., 1979. Publ. Dom. Astr. Obs., (in press).
Hutchings, J.B., 1974. Ap. J., 188, 341.
Hutchings, J.B., 1975. PASP, 87, 245.
Hutchings, J.B., 1976. Ap. J., 203, 438.
Hutchings, J.B., 1977. Ap. J., 217, 537.
Hutchings, J.B., 1978. Ap. J., 226, 264.
Hutchings, J.B., 1979. MNRAS, 187, 53P.
Hutchings, J.B., Cowley, A.P., Crampton, D., Osmer, P., 1977. Ap. J., 217, 186.
Milgrom, M., 1978. A. and Ap., 70, 763.
Petro, L., 1977. Thesis, Univ. of Michigan.
Prendergast, K., and Taam, R., 1974. Ap. J., 189, 125.
van Paradijs, J., and Zuiderqijk, E., 1977. A and Ap., 61, L19.
van Paradijs, J., Takens, R., Zuiderwijk, E., 1977. A. and Ap., 57, 221.
Walker, N., 1972. MNRAS, 159, 253.
Zuiderwijk, E., Hammerschlag-Hensberge, G., van Paradijs, J., Sterken, C., and Hensberge, H., 1977. A. and Ap., 54, 167.

On the mass of the exploding stars in massive close binary systems

W. Sutantyo

Bosscha Observatory,
Bandung Institute of Technology,
Indonesia.

1.   ABSTRACT

We study the effects of asymmetric instantaneous mass ejection on
the orbital parameters of massive close binary systems in the first
and second supernova explosions. We present three independent
arguments which show that the mass of the exploding stars in the
explosions preceding the X-ray and binary pulsar stages is not more
than about 7 $M_\odot$.

2.   INTRODUCTION

Two supernovae are expected to occur during the course of evolution
of a massive close binary system (cf. van den Heuvel, 1976). The
first occurs after the first stage of mass transfer which leads to
a system consisting of a massive non-degenerate star and a compact
star. The compact star becomes a strong X-ray source when the
normal companion has evolved to a stage where it possesses a strong
stellar wind. The second supernova happens after the second stage
of mass transfer following the X-ray stage. If the system is not
disrupted by the explosion, it becomes a binary with two compact
star components similar to the binary pulsar PSR 1913+16.

In the earlier studies of the effects of a SN (supernova) explosion
in a close binary system, it was generally assumed that the
explosion occurs in a spherically symmetric way (Boersma, 1961; van
den Heuvel, 1968; McCluskey and Kondo, 1971; Sutantyo, 1974). In

28

fact, there is no particular reason that the explosion should happen
symmetrically. An asymmetry of a few percent in the mass ejection
will be enough to give a kick velocity of the order of $\sim$ 100 km s^{-1},
which is of the order of magnitude of the orbital velocity of both
components. Furthermore, the orbital parameters of the binary
pulsar PSR 1913+16 indicates that the explosion which has happened
in the system should have been asymmetric if the mass of the
exploded star was larger than $\sim$ 3 - 4 $M_\odot$, as one expects from the
evolutionary point of view (Flannery and van den Heuvel, 1975; De
Loore et al. 1975; Sutantyo, 1978).

In this paper we will study the effects of asymmetric supernova
explosions on the orbital parameters of close binary systems. A
constraint on the mass of the exploding stars will be derived.

3.   ASYMMETRIC SN EXPLOSION

We will represent the degree of asymmetry of the explosion with a
kick velocity V_k gained by the exploding star due to the asymmetric
mass ejection. We consider the relative orbital motion of the
exploding star (star 1) with respect to the unexploding star (star 2).
The geometry of the explosion is illustrated in Figure 1. Let M_1
and M_1^f be the mass of star 1 before and after the explosion,
respectively, and M_2 be the mass of star 2. The impact of the
supernova shell will impact a momentum to star 2 which will result
to an impact velocity V_i in a radial direction. As is seen from
star 2, this velocity is reflected as the velocity of star 1 to the
other direction. The impact velocity can be calculated from the
formula given by Colgate (1970). However, Fryxell and Arnett (1979)
have argued that Colgate's formula largely overestimates the effects
of impact. We will, therefore, assume two extreme cases, i.e. the
impact velocity calculated by Colgate's formula and zero impact
velocity.

Since the velocity of the mass ejection is several orders of magnitude higher than the orbital velocity, we assume that the
explosion happens instantaneously. We also assume that the orbit
is circular with $r = a_0$ before the explosion. After the explosion,
the radial and the tangential velocity of star 1 are,

$$V_r = V_k \sin \theta \cos \phi + V_i , \tag{1}$$

$$V_t = (V_o^2 - 2V_o V_k \cos \theta + V_k^2 (\cos^2 \theta + \sin^2 \theta \sin^2 \phi))^{1/2} \tag{2}$$

respectively, where V_o is the initial orbital velocity of star 1.
Using the momentum and energy equations we can derive the eccentricity and the semi-major axis of the orbit after the explosion,

$$a = a_o m^f / (2 m^f - (V_r^2 + V_t^2) a_o) , \tag{3}$$

$$e^2 = 1 - a_o^2 V_t^2 / a m^f , \tag{4}$$

where $m^f = G(M_2 + M_1^f)$.

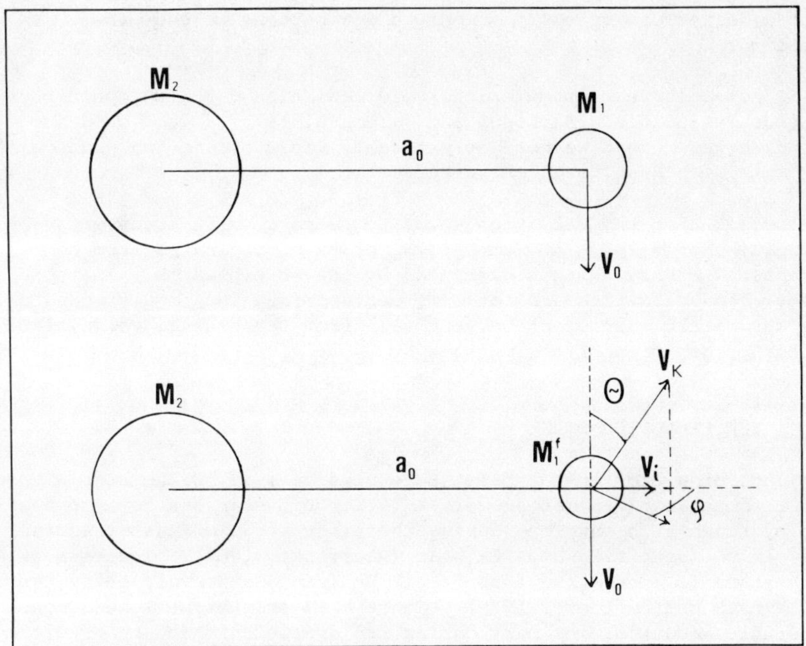

Figure 1:   Geometry of the explosion.

4.   THE KICK VELOCITY PRODUCED BY THE ASYMMETRIC EXPLOSION

Some pulsars are observed to have large transverse velocities. The
observed mean transverse velocity of such pulsars is 180 km s^{-1}
(Taylor and Manchester, 1977). If the velocity vector is randomly
oriented in space, this value corresponds to a mean space velocity
of 180 $\sqrt{3/2}$ km s^{-1} = 220 km s^{-1}. We cannot, however, adopt this
value as a mean velocity of all pulsars since a large selection
effect might exist in those data. Hanson (1979) has modelled the
large selection effects in the present proper motion data and found
values of 70 - 140 km s^{-1} for the mean velocity of pulsars. This
range of velocity is consistent with the observed < z > of pulsars,
i.e., $\sim$ 230 pc (Taylor and Manchester, 1977).

Most pulsars may have originated in single stars (Smith, 1977;
Sutantyo, 1978; Clark et al., 1979), therefore those velocities might
reflect the kick velocities gained by the pulsar due to the asymmetric SN explosion. For the above reasons, we adopt that the kick
velocity is $\sim$ 70 - 150 km s^{-1}.

The observed velocities of matter ejected by a supernova are $\sim 10^4$
km s^{-1}. This means that only a small asymmetry in the mass ejection
is required to produce the above range of the kick velocity. We
conclude therefore that most supernovae should be highly symmetric

with a degree of asymmetry of only a few percent in the mass ejection.

It is possible that asymmetric dipole radiation may also contribute to those velocities (Harrison and Tademaru, 1975). It should be noted, however that we need an extremely rapidly rotating neutron star for this mechanism to be efficient.

We suggest that the very high velocity pulsars (with transverse velocities of $\sim$ 200 - 500 km s^{-1}) might have originated from components of binary systems disrupted by the SN explosion. In those cases, the velocities are not only derived from the impulse due to the asymmetric mass ejection, but also from the orbital velocities of the progenitors of the pulsars.

5.   THE EXPLODING STAR

The exploding star in the first SN explosion is a helium star with an OB companion (van den Heuvel, 1976). This star has lost most of its hydrogen rich envelope during the first stage of mass transfer and can be identified as a WR star (Paczynski, 1967). The mass of such stars is in the range of 5 - 12 $M_\odot$ (van den Heuvel, 1973; Vanbeveren and De Loore, 1979). Since a WR star might lose a considerable amount of its mass during its evolution through a stellar wind mechanism (with a rate of 10^{-6} - 10^{-4} $M_\odot$ yr^{-1}, Conti, 1978, and the references therein), the mass of such stars might be somewhat lower than the above quoted values at the time of the explosion.

Van den Heuvel (1976) suggests that the exploding star in the second SN explosion is also a helium star which can be identified as a WR star. The companion of this star is a compact star which is the remnant of the first SN explosion. During the second stage of mass transfer, the compact star cannot accept most matter transferred by the mass-losing star due to the strong radiation pressure in the surrounding of the compact star. A large amount of mass and angular momentum is therefore expected to have been lost from the system during the second stage of mass transfer. As a result the system is expected to become a short (a few hour) period binary system (cf. Taam et al. 1978). It is difficult to observe the binary nature of such a system, so that most of such helium stars can only be identified as single WR stars. No direct information on the mass of such stars can be obtained. We expect, however, that the mass of such stars is very likely similar to the helium stars produced by the first stage of mass transfer. Therefore, as in the first SN explosion, we expect that the mass of the exploding star in the second SN explosion is somewhat lower than 5 - 12 $M_\odot$.

6.   THE BINARY PULSAR PSR 1913+16

It seems very likely that the binary pulsar PSR 1913+16 is a binary system in the post second SN stage (Smarr and Blandford, 1976;

Flannery and van den Heuvel, 1975; De Loore et al. 1975). The
orbital parameters of the system are accurately known (Taylor et al.
1979). We can, therefore, deduce fairly accurate information on the
parameters of the second SN explosion which has produced the system.
This is particularly true because the cross section of the unexploded
star, which is a compact star, is very small so that any effect of
impact of the SN shell can be neglected. We can also be sure that
the tidal interaction between the two stars has not affected the
orbital parameters of the system since the explosion. The changes
of the orbital parameters due to the gravitational radiation is also
negligibly small (Taylor et al. 1979).

Sutantyo (1978) has studied combinations of parameters of the SN
which can produce the observed orbital parameters of the binary
pulsar PSR 1913+16. He finds that there are minimum values of $\cos \theta$
and the kick velocity V_k as a function of M_1, above which the
present parameters of the PSR 1913+16 system can be obtained. The
result is shown in Figure 2. As we can see from the figure, for a
larger M_1 we need a larger V_k and $\cos \theta$ to produce the observed
parameters. This means that if we allow a large M_1 we need a large
kick velocity directed almost opposite to the orbital motion, otherwise the system will have too large eccentricity or is even disrupted.

We also see from Figure 2 that if $V_k \sim 70 - 150$ km s^{-1} as is discussed above (cf. Section 4),   the mass of the exploding star
should be $\sim 4 - 5$ $M_\odot$ which falls very nicely in the expected range
of M_1 (cf. section 5).

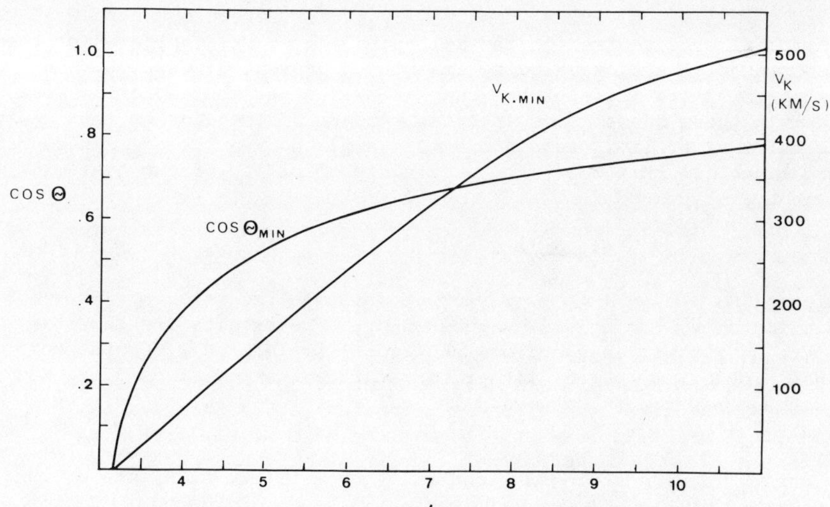

Figure 2:   Lower limits of $\cos \theta$ and V_k required to produce the
observed orbital parameters of PSR 1913+16 as a function
of M_1 (Sutantyo, 1978). Note that if $V_k \sim 75 - 150$ km s^{-1}
then $M_1 \lesssim 4 - 5$ $M_\odot$. *(Copyright (c) 1978 by D. Reidel Publishing Company, Dordrecht, Holland)*

32

7. THE SURVIVAL PROBABILITY OF SYSTEM IN THE SECOND SN EXPLOSION

There is only one binary pulsar among 304 known pulsars (Manchester et al. 1979). The rare occurrence of binary pulsars does not necessarily imply that the survival probability in the second SN explosion is low (Smith, 1977; Clark et al. 1979). This is based on the fact that not all pulsars have originated in close binaries. The lower mass limit for having a SN explosion in close binary systems is $\sim 15\ M_\odot$ for the original primary (van den Heuvel, 1976). This limit is considerably larger than that in single stars or wide systems ($\sim 2.5 - 4\ M_\odot$, Taylor and Manchester, 1977; Clark et al. 1979). It also implies that binary pulsars must have originated from much rarer massive type stars. In other words, most pulsars must have originated in single stars rather than in close binary systems.

Using the Salpeter's mass function, Clark et al. (1979) has estimated that the survival probability of a binary system in the second SN explosion is $f \approx 0.10 - 0.23$. This means that one out of 4 to 10 binary systems will survive the explosion. We will derive the parameters of the explosion which can give the above value of f.

Sutantyo (1978) shows that a binary system will remain bound after the explosion only if $\cos \theta > \cos \theta_{cr}$, where the critical angle θ_{cr} is given by

$$\cos \theta_{cr} = \frac{(1 + \alpha)\ (v_k^2 + 1) - 2(\alpha + \beta)}{2(1 + \alpha)v_k} \tag{5}$$

where $\alpha = M_2/M_1$, $\beta = M_1^f/M_1$, and $v_k = V_k/V_o$. Therefore, there will be a critical cone centred on the axis parallel to V_o, inside which the direction of the kick velocity is such to prevent disruption. We denote the solid angle of the cone as $\Delta\omega$. Since the explosion most likely happens due to some internal process independent of the orbital motion, it is reasonable to assume that V_k takes a random direction. In this case, the survival probability of the system is given by,

$$f = \Delta\omega/4\ \pi = (1 - \cos \theta_{cr})/2. \tag{6}$$

Adopting $M_2 = M_1^f = 1.4\ M_\odot$ (Taylor et al. 1979) we calculate f as a function of v_k for several values of M_1. The results are shown in Figure 3. As we can see from the figure, for $M_1 > 4.2\ M_\odot$, the survival probability is low either for small or large values of v_k and reaches a maximum in between. The value of $f \gtrsim 0.10$ can only be obtained if $M_1 \lesssim 15\ M_\odot$.

We can put a more stringent constraint on M_1 if we take into account the limits on the allowed values of V_k. As is discussed in Section 5, we have adopted that the most likely value of V_k is $\sim 70 - 150$ km s^{-1}. To find the corresponding v_k we must know the value of V_o. This, in turn, needs a knowledge of a_o. The only information on a_o we can have is from the binary pulsar PSR 1913+16. It has been

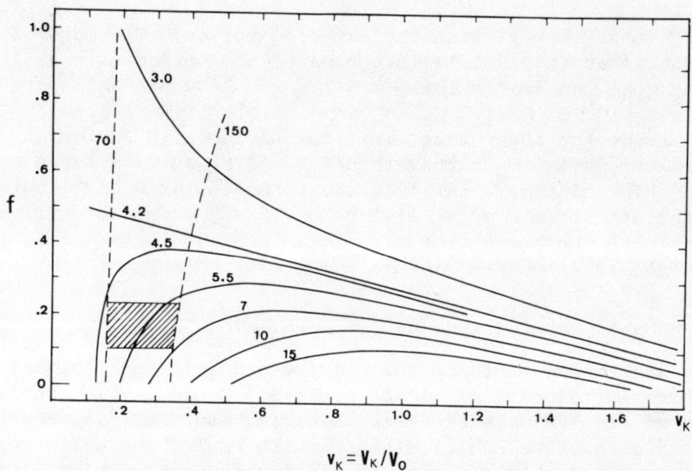

$$v_k = V_K/V_0$$

Figure 3: Survival probability of a binary system in the second SN explosion as a function of v_k for various values of M_1 (solid lines, label on each line indicates the value of M_1 in solar masses). The dashed line labelled by 70 and 150 give the boundary of v_k within which $70 \lesssim V_k \lesssim 150$ km s^{-1} can be obtained.

derived that the ratio a_0/a should be in the range,

$$(1 - e) \leq a_0/a \leq (1 + e), \tag{7}$$

(Flannery and van den Heuvel, 1975, Mitalas, 1976, Sutantyo, 1978). We know that for PSR 1913+16, $e = 0.617$ and $a = 1.942 \; 10^{11}$ cm. Therefore, we have, $7.439 \; 10^{10} \leq a_0 \leq 3.1408 \; 10^{11}$ cm. We assume that this range of a_0 represents the pre-explosion orbital separation for all second SN explosions.

From the upper limit of a_0, we derive, for a given M_1, a value of v_k above which V_k is always larger than a certain value. On the other hand, a boundary of v_k which gives an upper limit to V_k can also be derived from the lower limit of a_0. Using this procedure, we can determine, for a given M_1, a region of v_k, where a kick velocity between 70 - 150 km s^{-1} can be obtained. Outside this range, the velocity will always be less than 70 km s^{-1} or always be larger than 150 km s^{-1}. This region of v_k is bounded by the dashed lines by 70 and 150 in Figure 3.

As is discussed above, the most likely value of the survival probability is $0.10 \lesssim f \lesssim 0.23$. The domain of f and v_k which satisfies both the constraints of f and v_k is indicated by the shaded area. As one can see from the figure, those constraints can only be satisfied if $4.5 \lesssim M_1 \lesssim 7 \; M_\odot$. This gives the range of the mean mass of the exploding stars in the second SN explosion. Again,

34

those values fit well with the expected values. The mass of the
exploding star in PSR 1913+16 estimated in Section 6 is a little
smaller than the mean value.

If we assume the lower mass limit for having a SN explosion in a
single star is ~ 6 $M_\odot$ rather than $2.5 - 4$ $M_\odot$ as is assumed above
*(Romanishin and Angel, 1977), f can be as low as O.05. In this
case the upper limit of M_1 will be ~ 7.5 $M_\odot$, which is still consistent with the above view.
(*Romanishin, W., Angel, J.R.P., 1977. Bull Am. Astron. Soc., 9, 567)

8.   RUNAWAY VELOCITY OF MASSIVE X-RAY BINARIES

There is a strong tendency for massive galactic X-ray binaries to be
concentrated towards the galactic plane. Table 1 gives the list of
the known massive galactic X-ray binaries and their distances to the
galactic plane. As can be seen from the table, the galactic z
distribution of those systems is similar to that of population I
stars ($< z > \stackrel{\sim}{\scriptstyle<} 70$ pc). If the lifetime of such systems is $\sim 5 \cdot 10^6$
yr, this implies excess z velocities less than ~ 30 km s^{-1}, or excess
space velocities less than ~ 50 km s^{-1}.

X-RAY	OPTICAL	SPECTRUM	d(kpc)	b^{II}	z(pc)
Cyg X-1	HD 226868	O 9.7 Iab	2.5	3°.1	135
Vela X-1	HD 77581	B O.5 Iab	1.4	3.9	95.2
Cen X-3	Krzeminski	O 6.5 III	8	0.3	42
1700-37	HD 153919	O 6 f	1.7	2.2	65
1538-52	-	B O I	6	2.1	220
1223-62	Wra 977	B 1.5 Ia	2	O.O	O
O352+30	X-Per	O 9.5 pe	0.350	-17.0	-103
1653-41	V 861 Sco	B O I ae	2	1.5	52
1516-57 (Cir X-1)	-	OB	8-10	O.O	O
O115+63	-	B O e	≤ 7	1.0	≤ 122
O535+26	HDE 245770	B O e	1-3	-2.6	-45-136
OO54+60	γ Cas	B O e	0.3	-2.1	-11
1145-60	-	B O e	1.5	-0.2	-5

TABLE 1: Massive galactic X-ray binaries and their distances to the
          galactic plane (Bradt et al., 1979, and the references
          therein).

X-ray binaries are expected to be binaries in the post SN stages.
One expects that a SN explosion will give an additional space
velocity to the system, called the runaway velocity, which makes the
system   leave   the galactic plane. A question might be raised,
therefore, whether such low galactic z's are consistent with the view
that SN explosions have happened in the systems. Sutantyo (1975) has
studied the runaway velocity of massive X-ray binaries resulting from
symmetric explosions. He concludes that the runaway velocity of
those systems is not necessarily high provided that the mass of the
exploding stars is not large ($M_1 \lesssim 10\ M_\odot$). We will extend that work
for the cases of asymmetric SN explosions.

A scheme of the changes of the orbital parameters of a close binary
system before and after the first SN explosion is given in Table 2.
Note that we assume a circular and synchronous orbit (the rotation
of star 2 is synchronous with the orbital motion) at the pre-explosion stage. Circular and synchronous orbits are common among close
binary systems (Gibbons and Hawkins, 1971; Levato, 1975) as one
expects if tidal evolution has been operative in the systems (Kopal,
1972; Zahn, 1977). We also assume that the tidal forces have
circularized and synchronized the orbit at the present stage.

The total angular momentum of the system immediately after the
explosion is given by,

$$H = A\ P_f^{1/3}\ (1 - e^2)^{1/2} + 2\pi\ C_2/P_o\ , \tag{8}$$

where,

$$A = \left(\frac{G^2}{2\pi\ (M_2 + M_1^f)} \right)^{1/3} M_2 M_1^f \tag{9}$$

and C_2 is the moment of inertia of star 2. We neglect the rotation
angular momentum of the compact star because the radius of such a
star is negligibly small. Using equations (3) and (4) we can trans-
form equation (8) into,

$$H = A\ P_o^{1/3}\ \left(\frac{1 + \alpha}{\alpha + \beta}\right)^{2/3} (1 - 2\ v_k \cos\theta + v_k^2\ (\cos^2\theta + \sin^2\theta\ \sin^2\phi))^{1/2}$$

$$+ 2\pi\ C_2/P_o. \tag{10}$$

Note that this formula is independent of P_f, e_f and the effects of
impact. This equation can be used to solve P_o if the other parameters
are given. From the calculated P_o we can determine P_f, e_f and the
runaway velocity of the system V_g. We can, therefore, derive V_g as
a function of M_1 if M_2, M_1^f, H and the vector $\vec{V}_k$ are given.

We assume that $V_k \sim 100$ km s^{-1} (cf. Section 4). If the kick velocity
is exactly in the same direction as the orbital motion ($\theta = 180°$),
the effects of the explosion will be maximum. While if V_k is in the
opposite direction ($\theta = 0°$), the effects are minimum. For this
reason we will only consider those two extreme cases. All other
solutions will have values between those two limits.

PARAMETERS	EXPLOSION		PRESENT STAGE
	BEFORE	AFTER	
Mass of star 1 (the exploding star)	M_1	M_1^f	M_1^f
Mass of star 2 (the unexploding star)	M_2	M_2	M_2^n
Orbital period	P_o	P_f	P_n
Eccentricity	O	e_f	O
Rotation period of star 2	P_o	P_o	P_n
Moment of inertia of star 2	C_2	C_2	C_2^n
Total angular momentum	H_o	H	H_n

TABLE 2: Scheme of the changes of the orbital parameters of a massive close binary system before and after the first SN explosion.

The remaining problem is to set the values of M_2, M_1^f and H for a particular system. The simplest case is if there has been no mass and angular momentum loss from the system since the explosion (conservative evolution). In this case we simply set M_2, M_1^f and H to have the present observed values (i.e., $M_2 = M_2^n$, $H = H_n$, while M_1^f always remains the same, regardless of conservative or non conservative evolution).

There is, however, some evidence that some OB stars are losing mass with fairly large mass loss rate ($10^{-6} - 10^{-5}$ $M_\odot$ yr^{-1}, Conti, 1978, and the references therein). Therefore, a considerable amount of mass and angular momentum is expected to have been lost from the system during a few million years after the explosion. For this reason, we will also consider the nonconservative evolution. We will assume several values for the amount of mass loss,

$$\Delta M = M_2 - M_2^n \, . \tag{11}$$

The angular momentum loss is assumed to be linear with the mass loss, i.e.

$$\Delta H = \frac{\Delta M}{(M_2 + M_1^f)} \, H \tag{12}$$

We will apply the above method to the X-ray binary Cen X-3. The observed masses of the components are $M_2^n = 18\ M_\odot$ and $M_1^f = 1.4\ M_\odot$ (Hutchings et al. 1979). The present total angular momentum of the system is calculated by assuming that the radius of star 2 is nearly equal to the Roche lobe radius and that star 2 is rotating synchronously with the orbital motion (Hutchings et al. 1979). Figure 4 gives the runaway velocity of the system as a function of M_1 for $\theta = 0^\circ$ (solid lines) and for $\theta = 180^\circ$ (dashed lines). The impact velocity V_i is calculated by the formula given by Colgate (1970). We do the calculations for conservative case ($\Delta M = 0$) and for nonconservative cases ($\Delta M = 6$ and $18\ M_\odot$). No solution can be obtained for $\theta = 180^\circ$ in case $\Delta M = 18\ M_\odot$.

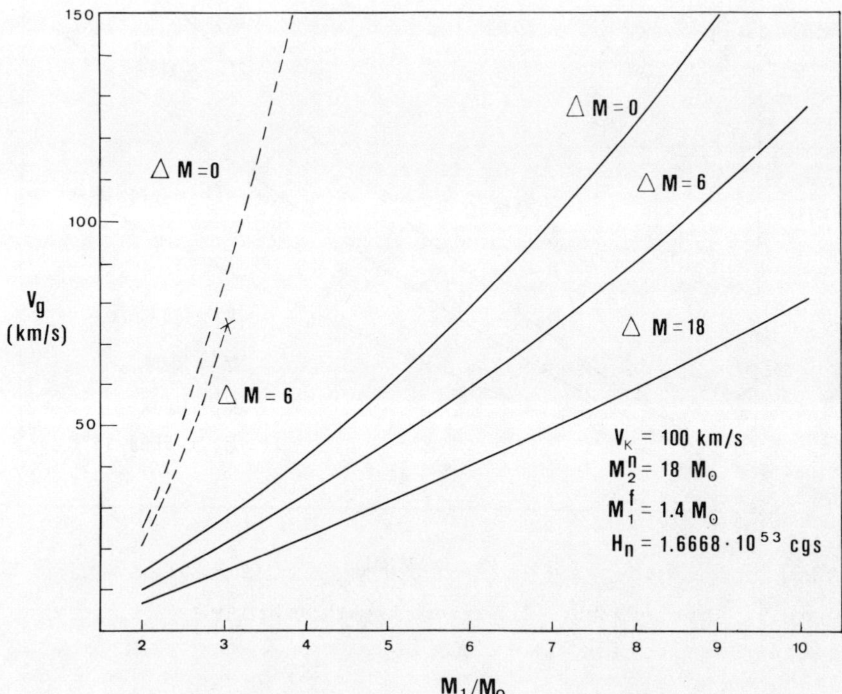

Figure 4: Runaway velocity calculated from the parameters of Cen X-3 as a function of M_1. ΔM indicates the amount of mass loss from star 2 during the period after the explosion up to the present stage. The kick velocity is assumed, i.e. $\theta = 0^\circ$ (solid lines) and $\theta = 180^\circ$ (dashed lines). The velocity due to the impact of the SN shell is calculated by the formula of Colgate. Above the point marked x, no solution exists for $R_2 \le a_0$.

Fryxell and Arnett (1979) have argued that Colgate's formula might overestimate the effects of impact. For this reason, we also consider the case where the impact velocity is zero. The results are given in Figure 5. Note that the results for those two assumptions of V_i do not differ significantly, except for $\theta = 180^\circ$.

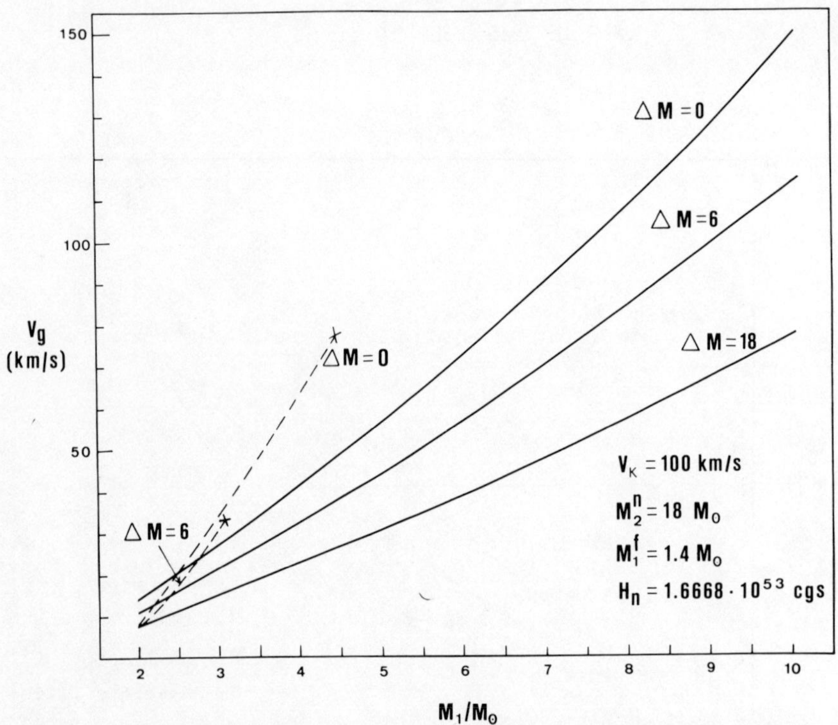

Figure 5: Same as Figure 4 for zero impact velocity.

In order to obtain $V_g \lesssim 50$ km s^{-1} (as one expects from the low galactic latitudes of all known massive galactic X-ray binaries), the mass of the exploding star should be less than ~ 4.5 $M_\odot$ for a conservative case, and less than ~ 7 $M_\odot$ if we allow mass loss of ~ 18 $M_\odot$. This limit is consistent with the upper limit of M_1 in the second SN explosion (cf. Sections 6 and 7). This supports the view that the exploding stars in the first and second SN explosions are of a similar type, i.e., helium stars resulted by heavy mass transfers.

ACKNOWLEDGEMENTS

Part of this work is carried out during my stay at the Astronomical
Institute, University of Amsterdam. I thank Professor E.P.J. van
den Heuvel for the hospitality and for the helpful advice. My
thanks is also due to S. Siregar, P. Wiyanto and J.P. De Cuyper for
discussions. I express my gratitude to the Leids Kerkhoven-Bosscha
Foundation for a financial support which has enabled me to visit
the University of Amsterdam and the Nato ASI on Galactic X-ray
Sources.

REFERENCES

Boersma, J., 1961. Bull. Astron. Inst. Neth., 15, 291.
Bradt, H.V., Doxsey, R.E., Jernigan, J.G., 1979. Adv. Space Expl.
    In press.
Clark, J.P.A., van den Heuvel, E.P.J., Sutantyo, W., 1979. Astron.
    Astrophys., 72, 120.
Colgate, S.A., 1970. Nature, 225, 247.
Conti, P.S., 1978. Ann. Rev. Astron. Astrophys., 16, 371.
De Loore, C., De Greve, J.P., De Cuyper, J.P., 1975. Astrophys.
    Space Sci., 36, 219.
Flannery, B.P., van den Heuvel, E.P.J., 1975. Astron. Astrophys.,
    39, 61.
Fryxell, B., Arnett, W.D., 1979. Enrico Fermi Institute Preprint,
    No. 78-64.
Gibbons, G.W., Hawking, S.W., 1971. Nature Phys. Sci., 238, 7.
Hanson, R.B., 1979. Yal University Observatory preprint.
Harrison, E.R., Tademaru, E., 1975. Astrophys. J., 201, 447.
Hutchings, J.B., Cowley, A.P., Crampton, D., van Paradijs, J., White,
    N.E., 1979. Astrophys. J., 229, 1079.
Kopal, Z., 1972. Astrophys. Space Sci., 17, 161.
Levato, H., 1975. Astrophys. J., 203, 680.
Manchester, R.N., Lyne, A.G., Taylor, J.H., Durdin, J.M., Large, M.I.,
    Little, A.G., 1978. MNRAS, 185, 409.
McCluskey, G.E., Kondo, Y., 1971. Astrophys. Space Sci., 10, 464.
Mitalas, R., 1976. Astron. Astrophys., 46, 223.
Paczynski, B., 1967. Acta Astron., 17, 355.
Shklovsky, I.S., 1968. Supernovae, Interscience, New York.
Smarr, L.L., Blandford, R., 1976. Astrophys. J., 207, 574.
Smith, F.G., 1977. Pulsars, Cambridge University Press, Cambridge.
Sutantyo, W., 1974. Astron. Astrophys., 31, 33.
Sutantyo, W., 1975, Astron. Astrophys., 41, 47.
Sutantyo, W., 1978. Astrophys. Space Sci., 54, 479.
Taam, R., Bodenheimer, P., Ostriker, J.P., 1978. Astrophys. J., 222,
    269.
Taylor, J.H., Manchester, R.N., 1977. Astrophys. J., 215, 885.
Taylor, J.H., Fowler, L.A., McCulloch, P.M., 1979. Nature, 277, 437.
Vanbeveren, D., De Loore, C., 1979. Preprint (sub. to Astron. Astro).
Van den Heuvel, E.P.J., 1968. Bull. Astron. Inst. Neth, 19, 432.
Van den Heuvel, E.P.J., 1973. Nature Phys. Sci., 242, 71.
Van den Heuvel, E.P.J., 1976. Proc. IAU Symp No. 73, p. 35.
Zahn, J.P., 1977. Astron. Astrophys., 57, 383.

Some aspects of mass transfer in massive X-ray binaries

G.J. Savonije

Institute of Astronomy, Cambridge, U.K.

1.   Introduction to Mass Transfer Processes in Binary Stars.

2.   The Initial Phase of Roche Lobe Overflow.

3.   The "Roche Lobe Process" of Overflow.

4.   Estimates for the Rate of Transfer of Mass by the Roche Lobe Process.

5.   The Expected Life time of Massive X-ray Binaries when powered by Roche Lobe Overflow.

6.   The Evolutionary Status of the Early Type components of Massive X-ray Binary Systems.

7.   Stabilization of Roche Lobe Overflow by Stellar Winds.

8.   Mass Ejection from Non-synchronous Contact Stars.

9.   Periodic Phenomena in Non-synchronous Binary Stars.

1.   INTRODUCTION

Binary X-ray stars are thought to be close binary systems in which mass transfer from the primary onto the collapsed secondary component (neutron star or black hole) causes the high energy emission. For an accreting neutron star with mass $M \simeq 1 M_\odot$ and radius $R \simeq 10^6$ cm the liberated gravitational energy is of order $L \simeq GM\dot{M}/R \simeq 10^{37}$ ($\dot{M} \neq 10^{-9}$ $M_\odot$ yr^{-1}) erg s^{-1}, which is emitted mainly as X-rays from a small region near the neutron star's magnetic poles. The accretion onto the compact secondary can result from two main types of mass transfer, i.e. 'stellar wind' or 'Roche lobe overflow'. We will use the term 'Roche lobe overflow' rather loosely to describe the process in which the gravitational attraction by the secondary, combined with the centrifugal force due to the rotation of the primary, is sufficiently strong to cause mass ejection from the face of the primary closest to the compact star. Stellar wind ejection from massive stars, on the other hand, is not caused by the companion but thought to be the result of radiation pressure (cf. Castor et al., 1975).

For a general introduction to the characteristics and evolutionary aspects of massive X-ray binaries the reader is referred to the contribution by John Hutchings. (Page 3 ff)

The distinction between 'Roche lobe overflow' and 'stellar wind ejection' seems rather artificial for massive X-ray binaries, because both types of mass ejection will occur simultaneously and will interact with each other. However, for simplicity we will regard them as independent mechanisms. It has been argued (e.g. van den Heuvel, 1975) that Roche lobe overflow cannot be responsible for the X-ray emission from massive X-ray binaries because of the high mass transfer rates calculated for massive stars which are overflowing their Roche lobes. When this rate of flow is higher than the critical accretion rate of the neutron star, assuming that all the transferred matter is actually accreted, matter will pile up around the X-ray source and then absorb all the X-rays causing the emission from the source to become much softer and to peak outside the X-ray band. But we will see below that, provided the massive primary star is still in the core hydrogen burning phase, the initial mass transfer rate from Roche lobe overflow is by no means as high as thought previously. At present one does not know whether the massive X-ray binaries are powered primarily by stellar wind accretion or by Roche lobe overflow, in fact this may vary from system to system. It has been argued (e.g. Lamers et al., 1976; Petterson, 1978; Savonije, 1978) that the intrinsically bright X-ray binaries Cen X-3 and SMC X-1 are more likely to be powered by Roche lobe overflow than by accretion from a stellar wind. The relatively weak X-ray sources, Vela X-1 and 3U1700-37, for which the stellar wind from the primary is particularly strong, seem more easily explained in terms of accretion from a stellar wind. However the evidence for this distinction between sources powered by the stellar wind and Roche lobe overflow is inconclusive, because the accretion rate is very sensitive to the speed of the stellar wind near the X-ray source and this is very difficult to estimate: most, if not all, of the primaries in the best studied massive X-ray binaries seem to be close to their 'critical lobes' (Avni, 1976) and we can be sure that these systems, if not doing it at present, will eventually initiate Roche lobe overflow.

In the following we will consider what happens when the massive primaries reach their 'Roche lobe' and start to transfer matter onto their compact companions.

2.   THE INITIAL PHASE OF ROCHE LOBE OVERFLOW

Before discussing the initial phase of Roche lobe overflow let us first consider briefly how a massive early type star responds to mass loss. When we instantly remove a non-negligible amount of mass from the surface of a star its outer layers will expand almost adiabatically to restore hydrostatic equilibrium. Massive early type stars have envelopes in radiative equilibrium: the temperature gradient in their envelope is sub-adiabatic. Therefore the star will be somewhat smaller than before the mass was removed. But, though the star is now in hydrostatic equilibrium it will, except for the extreme outer layers in which the thermal timescale is comparable to the dynamic timescale, not yet be in thermal equilibrium. The gas which has expanded adiabatically will thus absorb energy from the only available energy source: the outwardly diffusing radiation flux.   After thermal relaxation the structure of the expanded layers will again be compatible with the rest of the star which is now almost as large as before the mass loss. The important point to remember is that a massive star will be essentially smaller than it was before the mass loss only during the limited time before thermal equilibrium is restored.

The nuclear burning in the star tends to upset the thermal equilibrium of the star by continuously forcing it to change its structure. But these structural changes occur on a timescale of nuclear reactions which is much longer than the thermal timescale, so that they cannot drive the star out of thermal equilibrium. In general the 'nuclear' evolution phase forces the star to expand its outer layers. It is this evolutionary expansion which causes the stars to overflow their Roche lobes in close binary systems.

3.   THE ROCHE LOBE PROCESS OF OVERFLOW

Mass transfer by Roche lobe overflow in close binaries is a very complicated phenomenon (e.g. Paczynski, 1971) which can only be studied by making rather bold simplifying assumptions. Let us assume for the moment (we will comment on the more general case in section 7) that the binary system under consideration is tidally relaxed, i.e. the rotation axes of the binary components are perpendicular to the orbital plane, the orbit is circular and the contact* component rotates synchronously with the orbital revolution of its companion. The total gravitational potential of the binary system which is considered to consist of two mass points - is in that case independent of time (neglecting the small mass involved in the mass transfer) and such a stationary case can be treated with the well-known Roche lobe concept.

The 'Roche lobe' is the equipotential surface $\psi = \psi_1$ (where ψ is the total gravitational potential of the binary system corrected for centrifugal forces) which pass through the inner Lagrangian point L_1. At L_1, which lies on the line joining the centres of the two binary components, the effective gravity (corrected for centrifugal forces) vanishes on this surface. When one of the binary components expands beyond L_1 the pressure gradient in its outer layers - which is no longer balanced by gravity in the vicinity of L_1 - will push the stellar gas through the gravitational saddlepoint. Close to the stellar surface the resulting outward flux of matter will change into a horizontal flow directed towards the region close to L_1 where the actual mass transfer occurs.

These (slow) surface currents have never been studied in detail, but one can argue (e.g. Lubow and Shu, 1975) that the streamlines will, at a first approximation, lie on equipotential surfaces and, close to L_1, pass through a sonic point. The average radius of the Roche lobe which is a convenient parameter to use in evolutionary calculations (e.g. Plavec, 1970), is a function of the mass ratio q, of the binary, and has been tabulated by Kopal (1959). Paczynski (1971) has derived the following expression.

*The author uses the term "contact component" to describe the normal
 star filling its Roche Lobe. (Eds)

$$R_C = a(0.38 + 0.2 \,^{10}\!\log \frac{M_1}{M_2})$$

where $0.3 < M_1/M_2 < 20$; a is the orbital separation of the binary, M_1 the mass of the contact component and M_2 the mass of the companion. This estimate, for R_C is accurate to a few percent.

It is generally assumed in calculations of binary evolution via mass transfer between the components that the average radius R_* of the contact star remains precisely equal to the average radius of the Roche lobe R_C. At first sight this seems a reasonable approximation enabling one to determine the rate of mass transfer during the over-flow phase. It was on the basis of such calculations, that one con-cluded that Roche lobe overflow could not be the cause of the X-ray emission in massive X-ray binaries: it would result in super-critical accretion rates and suffocate the X-ray source almost immediately.

But how realistic are the results obtained with the boundary condit-ion $R_* = R_C$? Applying the condition $R_* = R_C$ to a star when it reaches its Roche lobe, means that we force it to stop its evolution-ary expansion and force it even to shrink immediately. This is because conservation of orbital angular momentum of the binary system causes the binary separation and thus the Roche lobe to shrink when some mass is transferred to the less massive companion (e.g. Paczynski, 1971). The expanding contact star can only accommodate this sudden restriction on its size by shedding mass from its surface, i.e. by Roche lobe overflow. But we have seen above that mass loss from a massive star can only cause shrinkage during the time the star has not yet recovered thermal equilibrium. The condition $R_* = R_C$ thus forces the contact star to increase its mass loss in order to remain out of thermal equilibrium. The continued mass transfer makes the Roche lobe even smaller and results inevitably in an even stronger restric-tion on the star's size. In other words, the contact star immediately starts to transfer matter on a timescale comparable to its thermal timescale, in order to deviate progressively from its thermal equi-librium state. This means for instance that a 20 $M_\odot$ main sequence star would immediately start to transfer mass at a rate

$$\dot{M} \simeq M_*/\tau_{KH} \simeq (20 M_\odot / 10^4 \text{ yr}) \simeq 2 \times 10^{-3} \; M_\odot/\text{yr}$$

This is much larger than the critical accretion rate for a typical neutron star, which is thought to be about $10^{-7.5}$ $M_\odot/\text{yr}$ (e.g. Rees, 1974; Thorne, 1974). But will a star in real life remain precisely inside its Roche lobe? Obviously not, because there can only be a finite mass transfer rate when the contact star is bigger than its Roche lobe and expands beyond it. The layers above the Roche lobe

will indeed flow away through the gravitational saddlepoint at L_1.
However to obtain realistic results one should not assume a priori
that the contact star will adhere precisely to the Roche geometry.
One can better determine the mass transfer rate onto the companion
by integrating the mass flux density of the outflowing excess
layers over the area of the gravitational throat at L_1. Only that
will show to what extent the contact star is willing to conform
itself to the Roche geometry. The first attempt to proceed in a way
like this was made by Jedrzejec (1969). He derived an expression
for the mass transfer rate as a function of the star's excess size
R_*-R_C in order to calculate Roche lobe overflow from red giant stars
which have convective envelopes. Such stars have a (super) adiabatic temperature gradient in their envelopes and do not shrink at
all when some mass is removed from their surface. It was therefore
impossible to determine the mass transfer rate with a condition
$R_*=R_C$, because the red giants cannot even fulfil such a constraint.

More realistic calculations of Roche lobe overflow from massive
stars, which we will discuss in section 3 and 4, show that the
initial mass transfer rate is relatively small so that the contact
star remains initially very close to thermal equilibrium and
continues to expand at about the same rate as before the contact
phase.

Initially the expansion rate of the star is in fact much larger than the
shrinking rate of the Roche lobe, so that the rate of mass transfer
is proportional to the evolutionary expansion rate. This results in
a 'slow phase' of Roche lobe overflow intermediate between the
nuclear and the thermal timescale of the star. During most of this
time the mass transfer rate is below the critical accretion rate for
X-ray emission from a neutron star. The slow phase is followed by a
shorter phase of rapid mass loss during which the bulk of the mass
transfer takes place. This 'rapid phase' of Roche lobe overflow
occurs after the continued expansion of the contact star has brought
denser layers above the Roche lobe, so that the mass transfer rate
becomes continuously larger and starts to result in a non-negligible
shrinking rate of the Roche lobe. This process will obviously
accelerate itself and give rise to a thermal runaway during which
the mass transfer rate approaches the thermal rate $\dot{M} \simeq M/\tau_{KH}$. The
contact star will then lose mass on a thermal timescale and shrink
together with the Roche lobe.

But the important point is that this thermal runaway does not occur
immediately, as thought previously, but only after a relatively long
period since the beginning of the overflow phase. In the next
section we will examine the initial, slow phase of Roche lobe overflow in more detail and estimate what sort of mass transfer rates
one may expect from massive contact stars.

4.   ESTIMATE FOR THE RATE OF TRANSFER OF MASS BY ROCHE LOBE OVERFLOW

The velocity of the overflowing matter near L_1 is approximately
equal to the local speed of sound (c_s). As the flow is subsonic,

the mass density does not change appreciably (according to Bernoulli's
equation only by a factor of $\sim \frac{1}{2}$) along the streamlines towards L_1.
The mass transfer rate onto the seconday is thus approximately given
by

$$\dot{M} \simeq \int \rho c_s \, \frac{dA}{d\psi} \, d\psi \, ,$$

where $\rho = \rho(\psi)$ is the mass density in the outer layers of the star,
A the area of the gravitational throat at L_1 and ψ the gravitational
potential. Obviously, A depends on the difference $\Delta\psi = \psi_s - \psi_1$
between the gravitational potential of the stellar surface ψ_s and
that of the Roche lobe ψ_1. For a synchronously rotating contact star
one can show (e.g. Savonije 1978, 1979a) that

$$\frac{dA}{d\psi} = 2\pi \, (1-\phi)^{-\frac{1}{2}} \, \phi\Omega^{-2} \, ,$$

where Ω is the orbital angular velocity and $\phi = \phi(q)$ is a dimensionless function of the binary mass ratio q. The potential difference
between the stellar surface and the Roche lobe is approximately
$\Delta\psi = \psi_s - \psi_1 \simeq -(GM_1/R_c^2) \, H$, where H is the atmospheric scale height
and R_c the average radius of the Roche lobe.

Let us derive a typical value for the area of the throat A, by substituting typical values for a massive X-ray binary system into the
expression for $A \simeq \frac{dA}{d\psi} \, \Delta\psi$. With $q = M_1/M_x \simeq 20$ ϕ comes out to be

$$\sim 0.2 \text{ and } \Omega^2 \simeq (GM_1/a^3) \simeq 5(GM_1/R_c^3) \, ,$$

so that a typical value $\Delta\psi \simeq -(GM_1/R_c^2)H$ yields $A \simeq 2\pi R_c H \simeq 10^{22} \text{cm}^2$ for
a somewhat evolved massive main sequence star (for which the atmos-
pheric scale height H is typically a few times $10^{-3}R$).

With a typical value for the sound speed $c_s \simeq 2 \times 10^6 \text{ cm s}^{-1}$, one
therefore requires mass densities of only $\sim 10^{-11} \text{ g cm}^{-3}$ in order to
obtain a mass transfer rate $\dot{M} \sim \rho c_s A$ of $\sim 2 \times 10^{17} \text{ g s}^{-1}$. Hence mere
atmospheric Roche lobe overflow may be sufficient to power a bright
X-ray source. This would mean that Roche lobe overflow may well be
important even when the photospheres of the massive optical components are still inside the Roche lobe.

5.   THE EXPECTED LIFE TIME OF MASSIVE X-RAY BINARIES WHEN
        POWERED BY ROCHE LOBE OVERFLOW

When the massive optical component expands beyond its Roche lobe the
overflowing layers become increasingly denser, whereas the flow
speed ($\sim c_s$) and throat area ($A \sim 2\pi RH$) remain of the same order.
The rate of mass transfer, $\dot{M} \simeq \rho c_s A$ will therefore increase at a
rate proportional to the rate at which the density of the layers
just above the Roche lobe increase. The latter rate is given by the
evolutionary expansion rate of the massive star. The growth in the
rate of the mass transfer ($\dot{M}$) is therefore of the order $\dot{R}/H$, where
$\dot{R}$ is the evolutionary expansion rate of the optical component and H
is again the atmospheric scale height. If we define the X-ray lifetime to be the period during which the mass transfer rate increases
from $10^{-10} \text{ M}_\odot \text{yr}^{-1}$ to $10^{-7.5} \text{ M}_\odot \text{yr}^{-1}$, it follows that $\tau_x \sim 6H/\dot{R}$. This

is because the optical component must expand over some 6 scale
heights in order to increase its mass transfer rate by a factor
$10^{-7.5}/10^{-10} \sim e^6$. By substituting a typical value H/R (a few times
10^{-3}), for a somewhat evolved massive MS-star, one finds that the
X-ray lifetime is of the order of a few percent of the evolutionary
expansion time $(R/\dot{R})$. For a massive main sequence star the evolutionary expansion time is typically $\sim 5 \times 10^6$ yr, so for a massive
X-ray binary, we obtain a crude estimate for the life time τ_x
$\sim 5 \times 10^4$ years, of an X-ray source powered by Roche lobe overflow.

More detailed numerical calculations of the Roche lobe overflow
process have been performed by Savonije (1979a). These give X-ray
lifetimes in the range $10^4 - 10^5$ yr depending on the system parameters adopted (mass of contact star and binary period). For wider
binaries, in which the massive component reaches its Roche lobe after
it has depleted the hydrogen in its core and expands on a thermal
rather than a nuclear timescale, the expected X-ray lifetime is a
factor $\sim 10^3$ smaller and thus negligible. The existence of a
relatively long lasting 'slow phase' of Roche lobe overflow, during
which the mass transfer rate is just right to produce a bright X-ray
source, is thus crucially dependent on whether the massive stars
remain on the main sequence. It seems that the same condition
applies to X-ray sources which accrete from a stellar wind. For if
the optical stars have left the main sequence and expand on a thermal
timescale ($\sim$ few 10^3 yr) they will very soon reach their Roche lobe
and initiate Roche lobe overflow. That would presumably almost
immediately finish the stellar wind powered X-ray phase by embedding
the source in a dense layer of transferred matter.

From the statistics of galactic X-ray binaries, and their probable
progenitor systems (Wolf-Rayet binaries) van den Heuvel (1976) has
derived an average X-ray lifetime which can be expressed as

$$\tau_x \sim 5 \times 10^{+4} f^{-1} \text{ yr,}$$

where f is the fraction of the WR-stars which are found in binary
systems and which evolve to collapsed components in massive X-ray
binaries. It is difficult to estimate f, but one can argue that it
is probably not much smaller than unity because at least half of the
WR-stars appear to have OB-type companions (Conti 1979). The two
derived values for the X-ray lifetime (remembering that both theoretical and observational values are nothing more than crude order of
magnitude estimates) seem consistent. That is provided that the
massive primary components are still in the hydrogen core burning
phase and expand on a nuclear timescale.

In sections 6 and 7 we will discuss effects which could increase
the average X-ray life time of an X-ray binary powered by Roche lobe
overflow. But let us now consider the crucial question of whether
there is any evidence that the early type primary components of
massive X-ray binaries are still on the main sequence.

6.   THE EVOLUTIONARY STATUS OF THE EARLY TYPE COMPONENTS

The observations of the early type components in the massive X-ray
binaries show that these stars are substantially overluminous for
their mass (e.g. Hutchings' contribution to these proceedings).
Ziolkowski (1977) first pointed out that this overluminosity can (at
least qualitatively) be explained by stellar wind losses during their
previous evolution on the main sequence. Mass losing MS-stars also
become appreciably oversized for their mass, i.e. they cover a much
wider strip in the HR-diagram (e.g. de Loore et al. 1977, 1978,
Chiosi et al. 1978). The latter has a very important effect on the
expected average X-ray lifetime for massive X-ray binaries.

When massive MS-stars become oversized for their mass they can
reach their Roche lobe even in relatively wide binary systems (with
$P_b \lesssim 5$ days) without having to wait for the rapid expansion phase
which begins after the exhaustion of hydrogen in their cores. Unfortunately, the stellar wind loss rates from massive MS-stars are
very uncertain, so that the calculations for the evolution of mass
losing stars can only be performed on an ad hoc basis with mass loss
rates lying in the observationally estimated range of roughly 10^{-7}
- 10^{-5} $M_\odot$/yr. It is generally believed and consistent with observations that these stellar winds are produced by the absorption of
line radiation (cf. Lucy and Salomon 1970, Castor et al. 1975). But,
though the actual mass loss rates through stellar winds (M_W) are
uncertain, we can still derive a fair upper limit by considering the
momentum balance.

We find that $\dot{M}_W \lesssim (\frac{c}{v_\infty}) \, {}^L/_c 2$, where L is the total luminosity of the
star, v_∞ the terminal wind velocity and c the speed of light. Sub-
stituting a typical (observed) value of $v_\infty \sim 2 \times 10^8$ cm s^{-1} yields:

$$\text{Effective number of lines (N)} = \frac{c}{v_\infty} \sim 10^2.$$

This upper limit compares well with observational estimates of mass
loss rates obtained from observations of a number of stars (e.g. de
Loore et al. 1977). Thus, allowing for some multiple scattering of
photons in extended atmospheres, one can say that N $\sim$ 300 represents
a fair upper limit to radiation pressure driven   mass loss. Calculations with N = 300 show (de Loore et al. 1977) that the stars
losing mass evolve along almost horizontal lines (of constant
luminosity) in the HR-diagram and they lose nearly 50% of their
original mass during the main sequence phase.

However, if one compares evolutionary tracks of evolving mass MS-stars
- losing mass at a maximum rate (N$\sim$300) - with the observationally
estimated position in the HR diagram, the agreement appears to be very
poor (eg. De Loore et al., 1978; Savonije, 1979b). The overluminosities
appear far too high to be explainable in terms of the above calculations. It seems unlikely that the optical components have in fact undergone a substantially larger mass loss than assumed here (N$\sim$300), because
this heavy loss of mass would show up as abundance anomalies in the

surface layers and such anomalies are not observed in the early
type components of X-ray binaries (see Hutchings' contribution).

The conclusion, therefore, is that the reported large overluminosities are probably a spurious result caused by errors in the
determination of mass and or luminosity of the early type components. The accuracy of the mass determination, for instance, is
probably much poorer than is sometimes suggested (e.g. Bahcall 1978
for a critical review of the mass determination). This may be
because the actual optical components do not obey the 'Roche' or
'tidal' geometry usually assumed (see section 8). In view of the
above uncertainty in the true position of the optical components in
the HR- diagram, it is impossible to draw a definite conclusion
regarding their evolutionary status. But the estimated position in
the HR- diagram of many of these stars is at least consistent with
the assumption that they are still burning hydrogen in their cores,
allowing for the effects of strong stellar wind losses ($N \sim 300$).
HD77581 (the companion of Vela X-1) appears to be the only obvious
post main sequence star in the sample considered by Savonije (1979b).
HD77581 is thought to be close to filling its critical lobe (Avni
1976). Therefore this star cannot have been considerably bigger in
the past without having started a process of runaway mass loss which
would have terminated the X-ray phase. The star must thus be in the
phase of shell hydrogen burning just after the depletion of
hydrogen in the core and is expanding on a thermal timescale. One
would thus expect that Vela X-1 would be powered by the strong
stellar wind ($\sim 5 \ 10^{-6} M_{\odot}/yr$) from its optical companion until that
star starts to overflow its Roche lobe and the X-ray source would
then fade away from the sky, probably within a few thousand years.

7.   STABILISATION OF ROCHE LOBE OVERFLOW BY STELLAR WINDS

Accretion of stellar wind material, ejected by the massive component,
may under certain circumstances (i.e. dense enough stellar wind and
not too high a wind velocity near the X-ray source) result in a
bright X-ray source (Davidson and Ostriker 1973). But we know that
at some stage the massive component of a close binary must eventually
expand beyond its Roche lobe and start to overflow. Provided the
expanding star is at that time still burning hydrogen in its core
(i.e. expanding on a nuclear timescale) this will result in a
relatively low rate of mass transfer for duration of 10^4 to 10^5 yr,
depending on the precise binary parameters. For this period we would
expect the compact companion receiving mass to be a bright source of
X-rays. Before the stage of Roche lobe overflow it may have been
persuaded to emit X-rays by the accretion of stellar wind material.
But the ejection of the stellar wind material may still go on during
the overflow phase and this will tend to stabilise the Roche lobe
overflow by diminishing the rate at which the Roche lobe shrinks
(Basko et al. 1977). The reason for this is as follows. The centre
of gravity of the binary system nearly coincides with that of the
massive star ($M_1 \gg M_x$), so this star will have a very low orbital
angular momentum. The stellar wind which is ejected by this star

therefore provides a sink of matter, but hardly any sink of orbital angular momentum for the binary system. This tends to increase the orbital separation and thus to widen the Roche lobe. The matter escaping in the stellar wind will therefore counteract the shrinking of the Roche lobe which is itself a result of the mass transfer to the low mass compact star. We obtain an upper limit for this stabilising effect by assuming that the matter in the wind carries no orbital angular momentum at all. If we also neglect the mass accreted by the secondary from the stellar wind, we find, by differentiating the approximate expression for the average radius of the Roche lobe (Paczynski 1971),:

$$R_c = a(0.38 + 0.2 \log (M_1/M_x)), \text{ that } (\dot{R}_c/R_c) \sim (\dot{M}_w/M_1).$$

But, as discussed in section 3, the growth in the rate of mass transfer, during the initial phase of Roche lobe overflow, is primarily determined by the evolutionary expansion rate of the massive star and not by the rate at which the Roche lobe shrinks (the latter rate is initially negligible compared with the evolution- ary expansion rate). The stabilising effect of the stellar wind is therefore not important unless it is strong enough to result in an expansion of the Roche lobe which can compete with the rate at which the massive star expands. But that requires a very strong stellar wind loss rate of the order:

$$\dot{M}_w \sim M_1 \; (\dot{R}/R)$$

where $\dot{R}/R$ is the expansion rate of the massive star.

Savonije (1979b) has estimated the stellar wind loss rates required to stabilise possible Roche lobe overflow from the companions of the five best studied massive X-ray binaries. For the companions of SMCX-1 and Cen X-3 the required wind loss rates are estimated to be more than a few times 10^{-6} $M_\odot$ yr^{-1}, which seems rather high for these two stars which show no sign of heavy mass loss. However the strong stellar wind ($\sim 10^{-5}$ $M_\odot$ yr^{-1}) thought to be ejected by the Of star companion of 3U 1700-37 seems certainly capable of preventing the star from rapidly overflowing its Roche lobe, provided that the star is still on the main sequence. With such a strong stellar wind one can easily explain the observed (relatively low) X-ray luminosity of this source in terms of accretion from a stellar wind. But the stellar wind stabilisation becomes completely inefficient when the massive stars have evolved away from the main sequence and have envelopes which expand on a thermal timescale. (e.g. Cox and Giuli 1968). In the case of the Vela X-1 system, if we assume the normal component has left the main sequence as seems very likely, it would require a stellar wind loss rate of more than $\sim 4 \; 10^{-3} M_\odot$/yr to stabilise the beginning of Roche lobe overflow. That is roughly a factor 10^3 higher than the estimated mass loss rate in the wind from this star.

The stellar wind mass loss rates required to stabilise or prevent
Roche lobe overflow therefore seem, in at least several systems,
higher than the observations indicate. However, it must be
remembered that the observational estimates are quite uncertain and
do not permit any definite conclusion. Besides that the expected X-ray
lifetime of a massive X-ray binary, powered by Roche lobe overflow,
may be longer than estimated in section 4, even without any
stabilisation by a stellar wind. The reason for this is that the
material ejected by the massive component is not necessarily
completely captured by the secondary. The compact star may capture
only a small fraction of the ejected material and this depends on
whether the rotation of the optical components is synchronised with
the orbital motion of the compact companion, as was assumed
hitherto.

8.   MASS EJECTION FROM NON-SYNCHRONOUS CONTACT STARS

There are many subtleties not taken into account in the usual
idealised treatment of Roche lobe overflow. For example, the
assumption that the stars ejecting mass rotate synchronously is
likely to be wrong. Rough estimates of the rotation rate of the
outer layers of the massive components of X-ray binaries, based upon
the observed widths of absorption lines (e.g. Hutchings in these
proceedings) indicate that many of these stars may rotate appreciably
slower than synchronously. There are many possible causes for the
non-synchronism. In the first place, these stars may simply not have
enough time to synchronise themselves after the supernova explosion
which produced the collapsed component. The fact that several of
these binaries seem to have almost circular orbits does not contradict this because the time for tidal circularisation may in certain
circumstances be shorter than the synchronisation time. There are
several effects which counteract the tidal forces. The expansion
of the envelope of a massive component will tend to disturb the
tidal synchronisation by a spin-down of the outer layers. Mass
transfer to the low mass collapsed star will also tend to increase
the non-synchronism because it makes the binary separation smaller
and results in a higher orbital velocity of the companion. The
rotation of the outer layers of the mass ejecting star will
undoubtedly be influenced to some extent by the flux of matter from
the envelope (e.g. Pratt and Strittmatter (1976)). The details of
the mass ejection become much more complicated than for synchronous
stars. This may be seen from calculations of particle trajectories
(Kopal 1959, and Kruszewski 1966). These neglect the hydrodynamic
character of the mass flow and are therefore only of limited
importance to our problem. Because the orbiting companion is not at
rest in the frame corotating with a non-synchronous component, the
latter star will experience a time varying gravitational potential
with a frequency of $\Omega-\omega$. (We use Ω for the orbital angular velocity
and ω for the rotational angular velocity of the non-synchronous
component). This lack of synchronism induces a tidal wave which
propagates along the surface of the primary. It is in contrast to
the synchronous case ($\omega = \Omega$), where the tidal disturbance is

52

stationary with respect to the primary. The Roche lobe concept,
based on the stationary situation, loses its meaning where $\Omega \neq \omega$ and
is often replaced by the term 'critical lobe'. It should be
stressed, however, that this 'critical lobe' or limiting surface for
mass transfer is no longer a simple equipotential surface (e.g.
Limber, 1963) and we cannot determine it in a relatively simple way,
as was possible for the synchronous case.

The difficulty with mass ejection from a non-synchronous binary
component (or from a component of an eccentric binary) is that the
companion moves (with a relatively high speed) with respect to the
surface material of the primary. It is therefore impossible to
determine, without solving the complete hydrodynamics whether the
tidally ejected material will actually be captured by the secondary.
Low mass secondaries in massive X-ray binaries ($M_X \ll M_1$) may find
it difficult to capture much of the tidally ejected material (e.g.
Petterson, 1978). Critical lobes for mass ejection, have therefore
nothing to do with the actual mass transfer and accretion rates
(e.g. Kruszewski, 1966). It may for instance very well be that a
slowly rotating star which has reached its 'tidal lobe' (that is the
analog of the Roche lobe for $\omega = 0$), though ejecting matter pro-
lifically, does not really transfer any matter to its low mass
companion. Most or all of the ejected matter may fall back onto the
non-synchronous star. The process of mass transfer and accretion in
close binaries therefore remains very much open to speculation until
detailed hydrodynamical investigations have been made.

9.   PERIODIC PHENOMENA IN NON-SYNCHRONOUS BINARY STARS

The surface of a star in a close binary system is tidally deformed
and this alters the distribution in the flux of the radiation over
its surface (see Hutchings' contribution). For non-synchronous
binary stars the flux distribution will depart from the usual von
Zeipel gravity darkening law (e.g. Lubow, 1979). In contrast to
the 'equilibrium tide' which occurs in synchronous binary stars,
non-synchronous stars will be subject to the so-called 'dynamical
tide', which appears as a result of the interaction between the free
oscillation modes of the star and the tidal wave.

The existence of such 'dynamical tides' was first pointed out by
Cowling (1941), who noted that the frequency of the free gravity
modes of a binary component can be comparable to the tidal frequency.
Binary stars are therefore subject to forced oscillations which may
result in resonances.

The amplitude of this dynamical tide is in general smaller than that
of the equilibrium tide (except close to a resonance) but, in contrast
to the equilibrium tide can have any (fixed) orientation with respect to the orbiting companion, depending on the tidal frequency
$\Omega - \omega$ (e.g. Zahn, 1977).

One may also expect low amplitude variations on timescales longer

than the binary period. Papaloizou (1979) has shown that the tidal
flow in the surface of a non-synchronous binary star is likely to be
unstable and give rise to modes which propagate in the binary frame
with a frequency comparable to the tidal frequency $\Omega-\omega$. We may
therefore expect a weak modulation of the light curve of a non-
synchronous star on a typical timescale of $\Pi \simeq P_b / (1-\omega/\Omega)$. As
mentioned by Papaloizou, such propagating disturbances have
interesting implications for the mass ejection from non-synchronous
binary stars. If the non-synchronous star is close to or filling
its 'critical lobe' one may expect the mass ejection rate to increase
when the disturbance passes along the critical point region. For
low values of the azimathal index (m) of the unstable modes, this
enhanced mass ejection will last for a non-negligible fraction
(say 1/3) of the total cycle. It is tempting to link this modulated
ejection of mass with the extended high and low X-ray intensity
states observed in X-ray binaries like Cen X-3 and Her X-1.

It has been suggested (Schreier et al. 1976) that the mass transfer
in Cen X-3 is modulated by variations in the density of the stellar
wind ejected by the optical component. Although such an effect
could also be linked with the propagating instabilities described
above, modulated Roche lobe overflow seems an interesting alternative
possibility. It is likely that the mass transfer to Cen X-3 occurs
through the mediation of an accretion disc, which seems indicated
for instance by the rapid spin-up of this X-ray pulsar (cf. Rappaport
and Joss 1977, Mason 1977).. An additional timescale exists,
connected with the changes in the X-ray intensity of Cen X-3, namely
the time it takes for accreting matter to spiral down through the
disc. The viscous timescale (τ_ν) of an accretion disc is of order:

$$\tau_\nu \sim R^2/_{2\nu} \; ,$$

where R is the radius and ν the typical kinematic viscosity of the
disc. This latter quantity is unknown, but we may try to estimate
τ_ν for Cen X-3 by simply scaling up the viscous decay time of
accretion discs in dwarf novae (e.g. Bath 1976), (assuming that the
kinematic viscosities are comparable in the two types of disc). For
a dwarf nova we have crudely :

$$R \sim 5 \times 10^{10} cm$$

and, as can be derived from the quasi-periodic optical outbursts,
typically $\tau_\nu \sim 2^d$. This would then yield:

$$\tau_\nu \sim (\frac{2 \times 10^{11}}{5 \times 10^{10}})^2 \times 2^d \sim 30^d \; ,$$

for a crude value of the viscous timescale for an accretion disc of
the size expected around Cen X-3.

It is interesting that Bonnet-Bidaud and van der Klis (1979) have
suggested a viscous decay time of $\sim 20^d$ on the basis of their
extended X-ray observations of Cen X-3 with the Cos-B satellite. But

54

there is no hope of getting a better understanding of the quasi-
periodic variations in X-ray sources like Cen X-3 without much more
extended X-ray observations. In particular one would like to have
more observations of the spectral change which (sometimes?) occur
in Cen X-3 during the transitions from one X-ray intensity state
to another (e.g. Schreier et al., 1976; Tuohy and Cruise, 1975). The
spectral changes, which indicate changes in the amount of absorbing
matter, could be an important factor in determining the actual
mechanism behind the intensity changes. When the X-ray intensity
changes of Cen X-3 are a result of a modulated Roche lobe overflow,
one would expect the absorption effects to occur during the phase
of enhanced mass transfer. Some time (τ_ν) after the end of such a
(quasi) periodic phase of enhanced mass transfer, when all the
additional matter has spiralled down and is accreted, the X-ray
source is expected to fade. Such a transition to a lower X-ray
intensity state would presumably occur without absorption effects.
In the stellar wind model by Schreier et al., (1976), on the other
hand, the low X-ray state is caused by a sudden increase in the density
of the wind. A transition to the low state would, in that case, be
linked with strong absorption and a softening of the X-ray spectrum.

It is a pleasure to thank Drs. P. Eggleton, A. Fabian, E. van den
Heuvel, J. Papaloizou and J. Pringle for fruitful discussions. I
am grateful to the European Space Agency for financial support.

REFERENCES

Avni, Y., 1976. Invited review presented at the joint discussion on
     X-ray binaries and compact objects at the XVI General Assembly of
     IAU, Grenoble, France.
Bahcall, J., 1978. Ann. Rev. Astron. Astrophys., 16.
Basko. M.M., Hatchett, S., McCray, R., Sunyaev, R.A., 1977. Astrophys.
     J., 215, 276.
Bath, G., 1976. IAU Symposium, 73, eds. P. Eggleton et al.
Bonnet-Bidaud, J.M., and van der Klis, M., 1979. Astron. and Astro-
     phys., 73, 90.
Castor, J.I., Abbott, D.C., and Klein, R.I., 1975. Astrophys. J., 195,
     157.
Chiosi, C., Nasi, E., and Shreenivasan, S.R., 1978. Astron. Astrophys.,
     63, 103.
Conti, P.S., 1979. private communication.
Cowling, T.G., 1941, MNRAS, 101, 367.
Cox, J.P., and Giuli, R.T., 1968. Principles of Stellar Structure.
Davidson, K., Ostriker, J.P., 1973. Astrophys. J., 179, 583.
De Loore, C., De Greve, J.P., and Lamers, H.J., 1977. Astron. Astro-
     phys., 61, 251.
De Loore, C., De Greve, J.P., Vanbeveren, D., 1978. Astron. Astrophys.,
     67, 373.

van den Heuvel, E.P.J., 1975. Astrophys. J., 198, L109.
van den Heuvel, E.P.J., 1976. IAU Symp., 73, eds. P. Eggleton et al.
Jedrzejec, E., 1969. MS Thesis, Warsaw University.
Kopal, Z., 1959. Close Binary Systems, Chapman and Hall, London.
Kruszewski, A., 1966. Adv. Astron. Astrophys., 4, 223.
Lamers, H.J., van den Heuvel, E.P.J., Petterson, J.A., 1976. Astron. Astrophys., 49, 32.
Limber, D.N., 1963, Astrophys. J., 138, 1112.
Lubow, S.H., 1979. Astrophys J., 229, 1008.
Lubow, S.H., and Shu, F.H., 1975. Astrophys. J., 198, 283.
Lucy and Solomon, P.M., 1970. Astrophys. J., 158, 879.
Mason, K.O., 1977. MNRAS, 178, 81p.
Paczynski, B., 1971. Ann. Rev. Astron. Astrophys., 9, 183.
Papaloizou, J., 1979. MNRAS, 186, 791.
Petterson, J.A., 1978. Astrophys. J., 224, 625.
Plavec, M., 1970. IAU Colloq., 4, ed. A. Slettebak.
Pratt, J.P., and Strittmatter, P.A., 1976. Astrophys. J. (Letters), 204, L29.
Rappaport, S., and Joss, P.C., 1977. Nature, 266, 683.
Rees, M., 1974. Proc. 16th Solvay Conference, Brussels.
Savonije, G.J., 1978. Astron. Astrophys., 62, 317.
Savonije, G.J., 1979a, Astron. Astrophys., 71, 352.
Savonije, G.J., 1979b. Astron. Astrophys, in press.
Schreier, E.J., Swartz, K., and Giacconi, R., 1976. Astrophys. J., 204, 539.
Thorne, K.S., 1974. Astrophys. J., 191, 507.
Tuohy, I.R., and Cruise, A.M., 1975. MNRAS, 171, 33p.
Zahn, J.P., 1977. Astron. Astrophys., 57, 383.
Ziolkowski, J., 1977. Proc. of the 8th Texas symposium on Relativistic Astrophysics, Boston (1976), Ann. N.Y. Acad. Sci., 302, 47.

Pulsed dwarf novae
X-ray emission

F.A. Cordóva and T.J. Chester

California Institute of Technology,
Pasadena, California,
USA.

ABSTRACT

Pulsed soft X-rays (0.1-0.5 keV) with a period of 9 s and a pulsed
fraction that varies between 0 and 100% were detected from the
dwarf nova SS Cygni at the peak of an optical outburst (Cordóva et
al, 1980). This detection confirms for the first time the supposed
high-energy origin of optical pulsations seen in erupting dwarf
novae. The pulse shape is remarkably sinusoidal for such a large
amplitude oscillation. The X-ray pulsation observed in this outburst
is not coherent, in contrast to previous claims for the related
optical oscillations. Instead, the phase of the pulsation apparently
executes a random walk with a Q of $\sim$ 25, whereas the period is quite
stable: $\dot{P} \sim -1 \times 10^{-5}$s s^{-1}. There is no evidence for any other
periodic behaviour on timescales between 160 ms and 3 h. A blackbody
fit to the observed spectrum yields $T \sim 3.5 \times 10^5$K and a flux at the
Earth of $\sim 4.5 \times 10^{-11}$ erg cm^{-2} s^{-1} in the 1/4 keV band. This in
turn implies a total luminosity (integrated over all frequencies) at
the source of $\sim 1.8 \times 10^{33} \times$ (d/200 pc)^2 erg s^{-1}.

A soft X-ray pulsation has also been detected from the dwarf nova U
Geminorum on the declining portion of an optical outburst (Cordóva
et al, 1979). The pulsation is quasi-periodic with periods ranging
from 20 to 30s and present throughout all binary phases, including
eclipse, with an average amplitude of $\sim$ 15%. Although superficially
the quasi-coherent pulsation from U Gem seems qualitatively different from the more coherent soft X-ray pulsation of SS Cygni, we find
that a change in only the strength of the random walk in phase can
account for the difference between the two stars. We have developed
a simple, powerful technique for analysing the unstable pulsations;
our analysis shows that the phase noise in U Gem is from 10 to 30
times greater than in SS Cyg. A blackbody spectral fit to the U
Gem data gives the best-fit parameters kT $\sim$ 25 eV and $N_H \lesssim 5 \times 10^{18}$
cm^{-2}, which imply a flux at the Earth of 3×10^{-10} erg cm^{-2} s^{-1} in
the energy interval 0.13 to 0.5 keV.

Previous models for the optical oscillations fail to account for the
properties of the X-ray pulsations. It is possible that an as yet
unspecified instability in the boundary layer between the white
dwarf and the accretion disk is the origin of the pulsation.

"

REFERENCES

Cordóva, F.A., Chester, T., and Garmire, G.P. 1979, in preparation.
Cordóva, F.A., Chester, T.J., Tuohy, I.R., and Garmire, G.P., 1980.
    Astrophys. J., 235, (Jan 1, 1980), in press.

Low mass, optically identified galactic X-ray sources

Anne P. Cowley

University of Michigan,
Astronomy Department,
U.S.A.

1. Abstract
2. Introduction
3. Sco X-1 Type Systems
4. Systems in which the non degenerate star dominates the optical spectrum
5. AM Her-Type Optical Systems
6. X-Ray Novae and Transients

1. ABSTRACT

In this paper I will consider only those sources with $\Sigma M \lesssim 5\ M_\odot$, while the high mass systems will be covered by J.B. Hutchings. Low mass sources which are principally soft sources such as hot white dwarfs, RS CVn stars, and SS Cyg - type systems will also be covered by other speakers. The remaining low mass hard X-ray sources include

    1) those in which the light of the disk dominates the optical spectrum, such as V818 Sco/Sco X-1,

    2) systems in which the non-degenerate star dominates the optical spectrum such as HZ Her/Her X-1 and V 1341 Cyg/Cyg X-2.

    3) the strongly magnetic cataclysmic variables, such as AM Her/1814 + 498, and

    4) some of the X-ray novae or transients, such as Nova Mon 1975/0620-00.

There is increasing evidence that the burst sources which occur both in and out of globular clusters are probably closely related to the Sco X-1 type objects with most of the optical light coming from a bright disk.

In the following paper characteristics of each group are discussed with particular emphasis on the individual objects for which there is good spectroscopic and photometric data.

An accompanying table lists all of the low mass systems which have
been firmly identified at this time together with a few references
to the literature. A more detailed bibliography may be found in the
excellent catalogue by Bradt et al. (1978).

2.   INTRODUCTION

What constitutes a low mass X-ray source? In Table 1, we list
those hard X-ray sources for which reasonable optical identifications
have been found. In only a few cases are we able to directly
determine the masses involved, but it is reasonable to expect that
other optically similar systems may also be low mass ($\Sigma\ M \lesssim 5\ M_\odot$).
The globular clusters are not included since it is likely that the
X-ray sources associated with them are specific stars within the
clusters which have not yet been isolated.

We can subdivide the low mass sources into four broad classes,
depending on their optical properties:

  1) a bright disk dominates the optical spectrum (like
      Sco X-1 and probably the bursters),

  2) the non-degenerate companion dominates the optical
      spectrum (like Her X-1 and Cyg X-2),

  3) the strongly magnetic short period variables (like
      AM Her), and

  4) the X-ray/optical novae or transients (like 0620-00).

3.   SCO X-1 TYPE SYSTEMS

In the optical region these systems are characterized by a very
blue continuum with weak He II (λ4686) emission, occasionally the
C III - N III blend (λ4640), and sometimes weak H emission. In a
U-B, B-V diagram these systems lie on or near the black body line,
or along an appropriate reddening line away from it. That is to say
they lie above the relation for normal stars. In Table 1, we have
included the burst sources in this category. The optical properties
of the optically identified burst sources are very similar to the
other Sco X-1 type systems (McClintock et al. 1978a). The important recent observations of simultaneous optical and X-ray bursts
from two of these systems (MXB 1735-44: McClintock et al. 1978b and
MXB 1837 + 05: Hackwell et al. 1979) not only strengthen the
identification but allow one to deduce further physical parameters.
It appears highly likely that most of the optical light we observe,
both in the bursting and non-bursting Sco X-1 type systems, comes
predominately from reprocessed X-rays in a luminous disk around the
neutron star. The unseen companion is presumed to be a low mass non-

degenerate star which transfers material to the neutron star by
Roche lobe overflow. It is possible that the companion may no
longer be on the main sequence, as in the case of Sco X-1 (see below).

In Figure 1, we show the galactic distribution of the various types
of low mass X-ray sources. The majority of the Sco X-1 type systems
lie near the galactic centre and most are thought to be at a distance
of ~ 10 kpc (Sco X-1 itself lies only ~ 1 kpc from us and therefore
appears to lie a bit above the galactic plane).

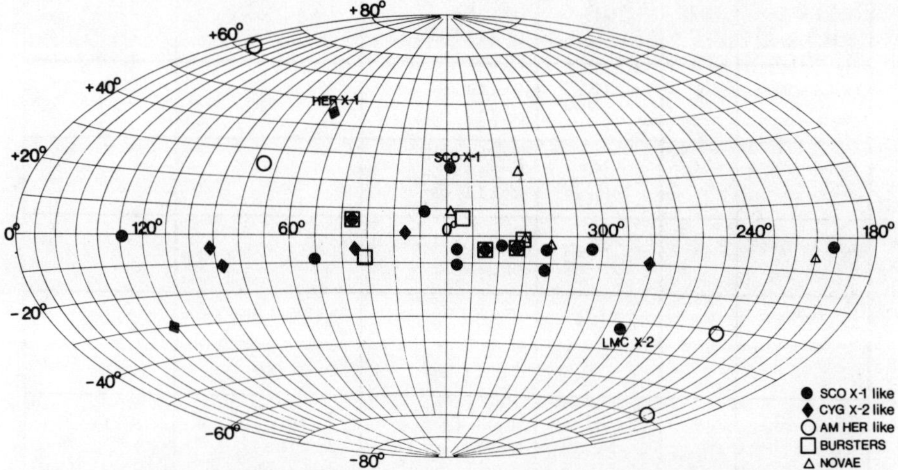

Figure 1: Galactic distribution of the optically identified, low
mass X-ray sources. Various symbols, as shown in the key,
denote different types of systems. Note the high galactic
latitude of the very nearby AM Her-like sources compared
to the Sco X-1-like systems which are concentrated to the
galactic centre. A few of the sources are identified, if
they are close, and therefore at an unexpectedly high
galactic latitude or belong to the LMC.

In Table 2, we summarize some of the characteristics of the low mass
systems. The galactic distribution of the Sco X-1 type sources
together with their low masses is consistent with their being Pop II
objects. Note the high mean L_x/L_{op} which probably accounts for the
brightness of the disk and thus our inability to observe the non-
degenerate companion.

Sco X-1 itself is the only system in the group for which an orbital
period, velocities and masses have been determined directly, although
van Paradijs (1978) has convincingly argued that the bursting sources
are low mass systems containing a neutron star. The orbital period
of 0.787 days in Sco X-1 was found independently from photometric
variability (Gottlieb et al. 1975) and from radial velocity
variations (Cowley and Crampton 1975). The light curve shows one
maximum per cycle which is consistent in phase with X-ray heating on

Table I - OPTICALLY IDENTIFIED LOW MASS X-RAY SOURCES

I. Disc Dominates Optical Spectrum (Sco X-1 like)

SOURCE	STAR	MAG.	SPECTRUM	REMARKS	REFERENCES
0142+614	3	~20	weak C III - N III em?	ID not certain	1
0521-720 (LMC X-2)	22	~18.5	Hα, He II, N III em	ID not certain	2,3,4
0614+091	D	18.5	C III - N III em		5,6
1254-690	30	~19	C III - N III em		7
1543-624	6	>20	strong UV excess		8
1617-155 (Sco X-1)	V818 Sco	12-14	H, He II, C III - N III em	light curve spec. orbit $P=0.76^d$	9,10
1627-673	4	18.7	weak He II, N III em	7 sec pulsar optical pulses	11,12
1636-536	3	17.5	He II em, UV excess	also burst source	13,14
1659-487 (GX339-4)	V	16.6V	Hα, He II, C III - N III em	highly var. X-rays	15
1728-169 (GX9+9)	DMB	16.4	He II, C III - N III em		16,17
1735-444	5	17.6V	Hβ, He II, N III em	simult. opt. & X-ray bsts.	14,11,18
1755-338	6	~19	UV excess		8,19
1822-371	5	16.3	He II, N III em		7
1837+049 (Ser X-1)	comp to D	19.2	weak He II em	simult. opt. & X-ray bsts.	20,21
1957+115	M	18.7	featureless		22

II. Non-Degenerate Star Dominates Optical Spectrum

SOURCE	STAR	MAG.	SPECTRUM	REMARKS	REFERENCES
0042+327	3	19.3	G0 + He II em?	ID not certain	23
0921-630	5	17	Hβ, He II em + K abs.		24,25
1656+345 (Her X-1)	HZ Her	15V	B-A var. + He II em	$P=1.7^d$ orb, pulse 1.2^s, eclipsing, $M_X=1.3$, $M_{op}=2.2\ M_\odot$	26,27 28,29
1813-140 (GX17+2)	28	17.5	dK	radio position	30,31
1908+005 (Aql X-1)	5	17-20	dK0 + weak He II, N III em (at min)	correlated optical & X-ray outbursts	32,33,34
2129+470		17V	He II, C III - N III em + G -K abs.	$P=5.2$ hours $\Delta m\sim 1.5^m$	35
2142+380 (Cyg X-2)	V1341 Cyg	15V	A - F var + He II, N III em	$P=9.8^d$ orbit	36,37,38

1. Bradt et al. 1978
2. Pakull 1979
3. Pakull & Swings 1979
4. Cowley et al. 1978
5. Davidsen et al. 1974
6. Murdin et al 1974
7. Griffiths et al. 1978
8. McClintock et al. 1978b
9. Gottleib et al. 1975
10. Cowley & Crampton 1975
11. McClintock et al. 1977
12. Grindlay 1978
13. Jernigan et al. 1977
14. Grindlay et al. 1978
15. Doxsey et al. 1979
16. Davidsen et al. 1976
17. Charles et al. 1977
18. Bond 1977
19. Canizares et al. 1979b
20. Charles et al. 1979b
21. Hackwell et al. 1979
22. Margon et al. 1977
23. Charles et al. 1978
24. Li et al. 1978
25. Charles 1979
26. Crampton & Hutchings 1974
27. Nelson et al. 1977
28. Gerend & Boynton 1976
29. Middleditch & Nelson 1976
30. Hjellming 1978
31. Margon 1978
32. Thorstensen et al. 1978
33. Puetter & Smith 1978
34. Watson 1976
35. Thorstensen et al. 1979
36. Cowley et al. 1979
37. Ilovaisky et al. 1979
38. Marshall & Watson 1979

TABLE II - CHARACTERISTICS OF STELLAR GALACTIC X-RAY SYSTEMS

TYPE	GALACTIC DISTRIBUTION	DISTANCE	M_v	POPULATION	L_x/L_{op}	PERIOD	ΣM ($M_\odot$)
Disk dominates (Sco X-1)	center	1 - 10 kpc	$\sim$0 to +3?	II?	$\sim 10^2 - 10^4$	$\sim$1 day	$\sim$1 - 2?
Star dominates (Cyg X-2)	halo?	2 - 10 kpc	$\sim$0 or fainter	II	$\sim 10 - 10^3$	days	$\sim$1 - 4?
Cataclysmic variables (AM Her)	nearby (high lat.)	few 100 pc	$\sim$+5 to +7	old disk?	$\sim$1	hours	$\sim$ 1.5?
High Mass (Vela X-1)	Gal. plane (spiral arms)	2 - 10 kpc	-3 to -7	I	$\sim$0.01	days	$\sim$20

the non-degenerate star. The emission lines must be formed in a
region around or near the X-ray source. The suggestion of Milgrom
(1978) that the emission lines might be formed on the inner face of
the non-degenerate star requires the centre of mass to lie inside the
non-degenerate star to reproduce the observed phasing of the velocities. However, in such a case the heated star should then have a
high enough luminosity to be observed over the bright disk (as in
Her X-1), because of its necessarily large radius. Thus this model
seems untenable.

From the observed velocities, assuming the emitting material
reflects the motion of the degenerate star, we infer a secondary of
$\sim$ 1 $M_\odot$. The primary is probably a neutron star, although a white
dwarf is not excluded by these data. The orbital inclination must
be $\sim$ 25 - 30°. This leads to the interesting implication that the
non-degenerate star must be somewhat evolved if it fills its Roche
lobe (see Cowley and Crampton 1975). Perhaps this is why the
present rate of mass transfer is high as compared to the cataclysmic
variables where the mass-losing star is a lower main sequence object.

An important problem in understanding Sco X-1 results from trying to
model the light curve. Perrenod (1976) has shown the inclination of
the system must be very low (i $\lesssim$ 10°) to explain the low amplitude
of the light curve. However this inclination also implies masses so
large that the secondary star should be easily observed. Thus we
must conclude either there is a non-isotropic X-ray flux or that
there is substantial X-ray absorption in the orbital plane between
the component stars. Whatever the cause, the contradiction between
the light curve and radial velocities shows the companion star does
not "see" as high an X-ray flux as the observer on earth detects.

Finally it should be noted that although 7^S X-ray pulsations are
observed from 1627-673, no orbital motion has been detected (Pravdo
et al. 1979, Rappaport et al. 1977). The period must be either

$<1^{hour}$ or $>300^{days}$, if the system is a binary. It seems unlikely
that we are viewing the system exactly pole on (Rappaport 1979).
Thus the nature of this system may be quite different from Sco X-1.

4.   SYSTEMS IN WHICH THE NON-DEGENERATE STAR DOMINATES THE
      OPTICAL SPECTRUM

Probably the best known examples of this type are Her X-1 and
Cyg X-2. In Table 1 we list seven such systems, although it is
likely that new ones will be found as higher resolution spectra of
the faint sources become available. We note, for example, that
0921-630 had been thought to be similar to Sco X-1 (Li et al. 1978),
but a recent observation by Charles (1979) shows the presence of
absorption lines from a late type star.

The mean-galactic latitude of these sources is slightly higher than
for the Sco X-1 like sources. Strong arguments can be made that at
least Cyg X-2 belongs to the halo population (Cowley et al. 1979),
although the more massive Her X-1 must be a younger system. The
average L_x/L_{op} is approximately a factor of 10 smaller than for the
Sco X-1 systems, probably thus accounting for the fainter disk and
resulting visibility of the non-degenerate star.

Thorstensen et al. (1979) have recently discovered that 2129 + 470
shows ~ 1.5 mag variation in light through a period of 5.2 hours,
due to heating by the X-ray source, similar to the variation in
Her X-1. The absorption line object appears to be later than HZ Her,
judging from the strong Ca II lines, but as yet no detailed spectroscopy is available from which to determine masses and luminosities
of the components.

A detailed investigation of Cyg X-2 (V1341 Cyg) has recently been
completed (Cowley et al. 1979) yielding new information about the
distance, masses, luminosity and the parameters of the system. The
spectrum is that of an early F giant with variable, generally weak
emission of He II and sometimes H and the N III - C III (λ4640)
blend. There is a very strong overlying blue continuum, which gives
the absorption lines in the blue and UV a filled-in appearance. The
orbital period of 9.87^{days} is found not only in the radial velocity
variations of both the stellar absorptions and the emission lines,
but also in a small spectral type variation from X-ray heating of
the inward stellar hemisphere (~A5 - F5) emission line strengths
and profiles, and in the optical light curve. As in HZ Her some of
the emissions are formed on the heated face of the optical star.
From the radial velocities and modeling of the ellipsoidal
variations a rather small range of possible masses is derived:

$1.3 < M_x < 1.8\ M_\odot$ and $0.5 < M_{op} < 1.1\ M_\odot$. Assuming the optical star
nearly fills its Roche lobe one derives M_{bol} ~ -0.5 which is in good
agreement with a purely spectroscopic absolute magnitude. Previously
the star had been thought to have a significantly lower absolute
magnitude, based on the assumption it was a dwarf or subdwarf, and

thus the distance had been thought to be only a few hundred parsecs. The newly derived luminosity implies a distance of $\sim$ 8 kpc and a height above the galactic plane of $\sim$ 1.5 kpc. Further the systemic radial velocity of -222 km s^{-1} is nearly the reflex of solar motion in that direction. Thus it seems reasonable to conclude both from its location and velocity that the star belongs to the halo population (II) and does not share in galactic rotation. The low mass of the optical star is also consistent with this hypothesis. One of the greatest difficulties in studying Cyg X-2 is that the system shows large erratic variations in both optical and X-ray luminosity on time scales from minutes to weeks and months. Many of these outbursts are as large as the underlying periodic variation so that many cycles or observations must be combined to average out these fluctuations. Apparently the inclination is too low for the system to eclipse, and Holt et al. (1979) find no evidence for any modulation of the X-rays with the orbital period, while Marshall and Watson (1979) report a significant variation in X-ray intensity. Further observations are needed to clarify this discrepancy. In the case of Cyg X-2 correlated optical spectroscopy, photometry and X-ray observations might lead to a better understanding of the non-periodic behaviour in the system which may arise from fluctuations in the mass transfer rate.

5.   AM HER-TYPE SYSTEMS

At the present time four AM Her-type X-ray sources are known (see Table 1), although there are several optically similar systems, such as VV Pup, from which X-ray emission has not yet been detected. All are characterized by strong emission lines of H, HeI, He II and C III-N III with no certain stellar absorptions. They also show orbital periods in the range of a few hours and high polarization from which the presence of large magnetic fields are inferred. These systems are thought to contain a white dwarf and a low mass dwarf which fills its Roche lobe. Two systems (0311-227, Williams et al. 1979, and 0526-326, Charles et al. 1979c) have only recently been recognized, and as yet no period or polarization has been found for the latter.

From Figure 1 we see that these systems are at high galactic latitudes, but this undoubtedly reflects the fact that they are relatively nearby objects at only a few hundred parsecs. Both their mean luminosities and L_x/L_{op} are much lower than for the other systems discussed above.

Because the emission lines apparently arise primarily in material streaming between the stars rather than from a disk around the white dwarf, it has not yet been possible to determine the masses directly. Arguments about the mass ratio, q, based on the P/q relation for cataclysmic variables have been used, but these of course may not be correct for the AM Her systems. Certainly many observable characteristics differ between the two types of systems. For example, when AM Her drops to its low luminosity state its emission lines are greatly weakened, whereas in dwarf novae the emission are strongest at minimum light. One infers that the emitting stream may be directly related to the enhanced luminosity of AM Her, while the weakly emitting disk in the CV's is relatively unchanged by the stellar outburst.

A number of models have been put forward to explain AM Her. (See for example, Chanmugan and Wagner 1977, 1978, Stockman et al. 1977, Michalsky et al. 1977, and Kruzewski, 1978).

6.   X-RAY NOVAE AND TRANSIENTS

The X-ray novae seem to divide into two types:

1)   those associated with luminous Be stars, such as O535+26, with the optical star showing no pronounced changes during the X-ray outburst, and

2)   those associated with a star which shows a simultaneous optical outburst, such as O620-00. It is now thought that the latter are low mass systems, and thus here we only discuss those.

At maximum light these X-ray novae show a nearly continuous optical spectrum with weak He II (λ4686) and H emission. (see references in Table 1.) Although the rise in optical luminosity (Δ m $\sim$ 7 magnitudes for O620-00) is comparable to classical recurrent novae, other observed properties are not similar. The classical novae show material is ejected with velocities of $\sim$ 2 x 10^3 km s^{-1} following outburst, while O620-00 showed no evidence of substantial mass outflow. Observational details and possible models have been given by Whelan et al. (1977) and Kaluzienski et al. (1977). Furthermore, the classical nova V1500 Cyg (1975) brightened about the same time as Nova Mon 1975 (O620-00), but the former was not detected as an X-ray source.

When O620-00 had faded to $\sim$ 18^m Oke (1977) was able to detect the presence of a late type ($\sim$ K5) dwarf, which is likely to be the non-degenerate companion in the presumed close binary system.

Several other X-ray/optical novae are listed in Table 1. Because of their relatively faint magnitudes even at maximum light, considerably less is known about them, but all seem to show similar, Sco X-1-like, spectral characteristics at maximum light.

Most recently Cen X-4 has been detected as an X-ray/optical/radio nova (Kaluzienski and Holt 1979, Canizares et al. 1979a, Hjellming 1979). At maximum light the spectrum shows weak He II, λ4686, emission as did O620-00. It will be especially interesting to follow its further development.

REFERENCES

Bradt, H.V., Doxsey, R.E., and Jernigan, J.G., 1978. IAU/COSPAR Symp.
Bond, H.E., 1977. IAU Circ. 3085.
Bond, H.E., and Tifft, W.G., 1974, PASP, 86, 981.
Canizares, C.R., McClintock, J., and Grindlay, J., 1979a. IAU Circ.
     3362.
Canizares, C.R., McClintock, J., and Grindlay, J., 1979b. in press.
Chanmugan, G., and Wagner, R.L., 1977. Ap.J. (Letters), 213, L13.
Chanmugan, G., and Wagner, R.L., 1978. Ap.J., 222, 641.
Charles, P., 1979. Private Communication.
Charles, P.A., Thorstensen, J.R., and Bowyer, S., 1977. IAU Circ. 3096.
Charles, P.A., Thorstensen, J.R., and Bowyer, S., 1978. MNRAS, 183,
     29P.
Charles, P.A., Thorstensen, J.R., and Bowyer, S., 1979a. IAU Circ.
     3298.
Charles, P.A., Thorstensen, J.R., Bowyer, S., Clark, G.W., Li, F.K.,
     van Paradijs, J., Remillard, R., Holt, S.S., Kaluzienski, L.J.,
     Junkkarinen, V.T., Puetter, R.C., Smith, H.E., Pollard, G.S.,
     Sanford, P.W., Tapia, S., Vrba, F.J., 1979b. Ap.J. in press.
Charles, P., Thorstensen, J., Bowyer, S., and Middleditch, J., 1979c.
     Ap.J., (Letters), in press.
Cordova, F., Garmire, G., and Tuohy, I., 1978. IAU Circ. 3235.
Cordova, F.A., and Riegler, G.R., 1979. MNRAS, 188, 103.
Cowley, A.P., and Crampton, D., 1975. Ap. J. (Letters), 201, L65.
Cowley, A.P., and Crampton, D., 1977. Ap. J. (Letters), 212, L12.
Cowley, A.P., Crampton, D., and Hutchings, J.B., 1978. Astron. J.,
     83, 1619.
Cowley, A.P., Crampton, D., and Hutchings, J.B., 1979. Ap. J., in
     press.
Crampton, D., and Hutchings, J.B., 1974. Ap. J., 191, 483.
Davidsen, A., Malina, R., and Bowyer, S., 1976. Ap. J., 203, 448.
Davidsen, A., Malina, R., Smith, H., Spinrad, H., Margon, B., Mason,
     K., Hawkins, F., and Sanford, P., 1974. Ap.J. (Letters), 193,
     L25.
Doxsey, R., Grindlay, J., Griffiths, R., Bradt, H., Johnston, M., Leach,
     R., Schwartz, D., and Schwarz, J., 1979. Ap. J. (Letters), 228,
     L67.
Fabbiano, G., Gursky, H., Schwartz, D.A., Schwarz, J., Bradt, H.V.,
     Doxsey, R.E., 1978. Nature, 275, 721.
Fergelson, E., Dexter, L., and Liller, W., 1978. Ap. J., 222, 263.
Gerend, D., and Boynton, P.E., 1976. Ap. J., 209, 562.
Gottlieb, E.W., Wright, E.L., and Liller, W., 1975. Ap. J. (Letters),
     195, L33.
Griffiths, R.E., Bradt, H., Doxsey, R., Friedman, H., Gursky, H.,
     Johnston, M., Longmore, A., Malin, D.F., Murdin, P., Schwartz,
     D.A., and Schwarz, J., 1977. Ap. J. (Letters), 221, L63.
Griffiths, R.E., Gursky, H., Schwartz, D.A., and Schwarz, J., 1978.
     Nature 276, 247.
Griffiths, R.E., Ward, M.J., Blades, J.C., Wilson, A.S., 1979. IAU
     Circ. 3326.
Grindlay, J.E., 1978. Ap.J., 225, 1001.
Grindlay, J.E., and Liller, W., 1978. Ap.J. (Letters), 220, L127.
Grindlay, J.E., McClintock, J.E., Canizares, C.R., van Paradijs, J.,
     Cominsky, L., Li, F.K., and Lewin, W.H.G., 1978, Nature, 274, 567.

Hackwell, J.E., Gehrz, R.D., Grasdelen, G.L., Cominsky, L., van Paradijs, J., and Lewin, W.H.G., 1979. IAU Circ. 2221.

Hearn, D.R., and Marshall, F.J., 1979. Ap. J. (Letters), in press.

Hjellming, R.M., 1978. Ap.J., 221, 225.

Hjellming, R.M., 1979. IAU Circ. 3369.

Holt, S.S., Kaluzienski, L.J., Boldt, E.A., and Serlemitsos, P.J., 1979. preprint.

Holt, S., Kaluzienski, L., Mushotzky, R., Boldt, E., and Serlemitsos, P., 1978. IAU Circ. 3292.

Ilovaisky, S.A., Chevalier, C., Motch, C., and Janot-Pacheco, E., 1979. IAU Circ. 3325.

Jernigan, J.G., Apparao, K.M.V., Bradt, H.V., Doxsey, R.E., McClintock, J.E., 1977. Nature, 270, 321.

Kaluzienski, L., and Holt, S., 1979. IAU Circ. 3360 and 3362.

Kaluzienski, L.J., Holt, S.S., Boldt, E.A., and Serlemitsos, P.J., 1977, Ap. J., 212, 203.

Kruzewski, A., 1978. Non Stationary Evolution of Close Binaries, ed. A.N. Zytkow (Warsaw: Polish Sci. Pub.), p. 55.

Krzeminski, W., and Serkowski, K., 1977. Ap. J. (Letters), 216, L45.

Li, F.K., van Paradijs, J.A., Clark, G.W., Jernigan, J.G., 1978. Nature, 276, 799.

Liller, W., 1977. IAU Circ. 3110.

Margon, B., 1978. Ap.J., 219, 613.

Margon, B., Szkody, P., Bowyer, S., Lampton, M., and Paresce, F., 1978. Ap. J., 224, 167.

Margon, B., Thorstensen, J.R., and Bowyer, S., 1977. Ap. J., 221, 907.

Marshall, N., and Watson, M.G., 1979. IAU Circ. 3318.

Mason, K.O., Lampton, M., Charles, P., and Bowyer, S., 1978. Ap. J. (Letters), 226, L129.

McClintock, J.E., Canizares, C.R., and Backman, D.E., 1978a. Ap. J. (Letters), 223, L75.

McClintock, J.E., Canizares, C.R., Bradt, H.V., Doxsey, R.E., Jernigan, J.G., and Hiltner, W.A., 1977. Nature, 270, 320.

McClintock, J., Canizares, C., Hiltner, W.A., and Petro, L., 1978b. IAU Circ. 3251.

Michalsky, J.J., Stokes, G.M., and Stokes, R.A., 1977. Ap. J., (Letters), 216, L35.

Middleditch, J., and Nelson, J.E., 1976. Ap. J., 208, 567.

Milgrom, M., 1978, Munich "Texas Symposium".

Murdin, P., Griffiths, R.E., Pounds, K.A., Watson, M.G., and Longmore, A.J., 1977. MNRAS, 178, 27P.

Murdin, P., Penston, M.J., Penston, M.V., Glass, I.S., Sanford, P.W., Hawkins, F.J., Mason, K.O., and Willmore, A.P., 1974. MNRAS, 169, 25.

Nelson, J.E., Chanan, G.A., and Middleditch, J., 1977. Ap. J., 212, 215.

Oke, J.B., 1977. Ap. J., 217, 181.

Pakull, M., 1979. IAU Circ. 3313.

Pakull, M., and Swings, J.-P, 1979. IAU Circ. 3318.

Perrenod, S.C., 1976. Ap. J., 206, 876.

Pravdo, S.H., White, N.E., Boldt, E.A., Holt, S.S., Serlemitsos, P.J., Swank, J.H., Szymkowlak, A.E., Tuohy, I., and Garmire, G., 1979. preprint.

Puetter, R.C., and Smith, H.E., 1978. IAU Circ. 3243.

Rappaport, S.A., 1979. This conference.
Rappaport, S.A., Markert, T., Li, F.K., Clark, G.W., Jernigan, J.G.,
    and McClintock, J.E., 1977. Ap. J. (Letters), 217, L29.
Robinson, E.L., 1976. Ann. Rev. Astr. Astrophys., 14, 119.
Stockman, H.S., Schmidt, G.D., Angel, J.R.P., Liebert, J., Tapia, S.,
    and Beaver, E.A., 1977. Ap. J., 217, 815.
Swank, J.H., Boldt, E.A., Holt, S.S., Rothschild, R.E., and Serlemitsos,
    P.J., 1978. Ap. J. (Letters), 226, L133.
Szkody, P., and Brownlee, D.E., 1977. Ap. J., (Letters), 212, L113.
Tapia, S., 1979. IAU Circ. 3327.
Tapia, S., 1977. Ap. J. (Letters), 212, L125.
Thomas, R., Greenhill, J., and Watts, D., 1979. IAU Circ. 3334.
Thorstensen, J.R., Charles, P.A., and Bowyer, S., 1978. Ap. J. (Letters),
    220, L131.
Thorstensen, J., Charles, P., Bowyer, S., Briel, U.G., Doxsey, R.E.,
    Griffiths, R.E., Schwartz, D.A., 1979. Ap. J. (Letters), in
    press.
van Paradijs, J., 1978. Nature, 274, 650.
Ward, M.J., Allen, D.A., Smith, M.G., and Wright, A.E., 1979. IAU
    Circ. 3335.
Watson, M.G., 1976. MNRAS, 176, 19P.
Watson, M.G., Ricketts, M.J., and Griffiths, R.E., 1977. Ap. J.
    (Letters), 221, L69.
Whelan, J.A.J., Ward, M.J., Allen, D.A., Danziger, I.J., Fosbury,
    R.A.E., Murdin, P.G., Penston, M.V., Peterson, B.A., Wampler,
    E.J., and Webster, B.L., 1977. MNRAS, 180, 657.
Williams, G., Johns, M., Price, C., Hiltner, W.A., Boley, F., Maker,
    S., and Mook, D., 1979. Nature, in press.

Spectral formation
in galactic X-ray sources

Richard McCray

Joint Institute for Laboratory Astrophysics,
University of Colorado and National Bureau of Standards,
Boulder, Colorado, U.S.A.

1.   Abstract

2.   Introduction

3.   The coronal model

4.   The nebular model

5.   Diffusion models

6.   Conclusions

1.   ABSTRACT

Theories for the formation of the spectra from galactic X-ray sources
are reviewed, with emphasis on the mechanisms for the production of
atomic emission lines and the interpretation of the observations.
Three classes of idealised models for sources are discussed. The
first is a "coronal model", in which electron collisions control the
ionisation and emissivity of the gas. When departures from ionisation
equilibrium in a shock are included, such models may describe the
X-ray emission from a supernova remnant. The second "nebular model"
is for the distribution of gas around a compact source of continuum
X-rays, in which photoionisation and recombination of atoms in the
surrounding cloud control the form of the spectrum. The third is a
"diffusion model", a variant of the nebular model, in which the opacity is so great that electron scattering plays a significant role.
The two latter models describe aspects of spectral formation in binary X-ray sources.

2.   INTRODUCTION

Thanks to X-ray telescopes with improved spectral resolution, such
as the proportional counters on Ariel V, OSO-8, and HEAO-1 and more
recently the solid state and crystal spectrometers on the Einstein
Observatory, we now know that galactic X-ray sources of all kinds
show features in their X-ray spectra due to atomic transitions to
abundant elements such as Si, S, and Fe. For details, see the article
by Steve Holt in this volume. Analysis of these spectra permits the
determination of physical characteristics of the emitting region -
such as density, temperature, element abundances, and velocity -
impossible to infer from X-ray photometric data alone. As our

technical capability advances towards higher sensitivity and spectral
resolution, we can expect that the X-ray spectra will reveal the kind
of information about compact objects and supernova remnants (SNR's)
that optical spectroscopy has extracted from the spectra of stars and
nebulae.

What is the best strategy for understanding X-ray source spectra? One
approach would be to include all we know about the structure of an
X-ray source, and do our best to estimate the emergent X-ray spectrum.
In my opinion, such an approach is premature. I think that a better
strategy is to consider highly idealised models for the X-ray sources,
ignoring much of the complexity that we know is characteristic of the
galactic X-ray sources. Once we have gained a physical understanding
of a variety of qualitatively different idealised source models, we
can sensibly hope to synthesize these models into a reasonable approximation of a real source.

In this spirit, I shall discuss three very different, idealised models,
each of which illustrates different processes involved in the formation of X-ray spectra. In the first, a "coronal model", the gas is
assumed to be transparent to all radiation, and the ionisation and
spectral emission are dominated by electron collisions. This model,
suitably modified, may be a reasonable approximation for the X-ray
emission from stellar coronae and SNR's; but it is probably inappro-
priate for the bright compact galactic sources. The second, a
"nebular model", is the X-ray analogue of a planetary nebula or H II
region: continuum X-rays from a central point source propagate through
the surrounding gas cloud which imparts spectral features to the
emergent spectrum. In this model, X-ray photoionisation controls the
ionisation balance, and fluorescence and recombination lines dominate
the emission spectrum. The nebular model may represent the propagation of X-rays through a stellar wind or through interstellar gas
surrounding a compact X-ray source. The third model, a "diffusion
model", is like the nebular model except that it applies to environments sufficiently opaque that electron scattering - "Comptonisation"
 - strongly affects the emergent spectrum. Such a model may represent
a semi-opaque distribution of gas surrounding an accreting magnetised
neutron star or white dwarf star.

3.   THE CORONAL MODEL

The coronal model has existed for a long time; it is by far the most
highly developed and has been used successfully to interpret the X-ray
spectra of the solar corona and solar flares. In this model the gas
is assumed to be transparent to all radiation. Its temperature T,
is a parameter controlled by external processes, not necessarily
specified. The densities, n_i^j, of each ion stage j of element i are
controlled by rate equations having the form:

$$\frac{dn_i^j}{dt} = n_e n_i^{j-1} I_i^{j-1} + n_e n_i^{j+1}\left[R_i^{j+1}+D_i^{j+1}\right] - n_e n_i^j\left[I_i^j+R_i^j+D_i^j\right] \qquad (1)$$

where I_i^j, R_i^j, and D_i^j

are respectively, rate coefficients for electron collisional ionisation

$$(n_i^j + e \rightarrow n_i^{j+1} + 2e),$$

radiative recombination

$$(n_i^j + e \rightarrow n_i^{j-1} + h\nu),$$

and dielectronic recombination

$$(n_i^j + e \rightarrow n_i^{j-1*}).$$

Dielectronic recombination leaves the ion in an autoionising state

$$n_i^{j-1*},$$

which can decay by the emission of a "satellite line". In the coronal model a stationary local balance is assumed:

$$dn_i^j/dt = 0.$$

Provided that the electron density

$$n_e < 10^8 \text{ cm}^{-3},$$

all these rate coefficients are functions only of temperature. Therefore, the fraction

$$f_i^j = n_i^j / \sum_j n_i^j$$

of element i in ionisation stage j is a function of temperature alone:

$$f_i^j(T).$$

However, when $n_e \gtrsim 10^8$ cm^{-3}, the dielectronic recombination rate coefficients, and hence ionisation fractions, become density dependent. The stationary ionisation fractions

$$f_i^j(T, n_e)$$

have been calculated by many authors; for recent results see Jordan (1969, 1970), Landini and Fossi (1972), Summers (1974a,b), Shapiro and Moore (1976), and Jacobs et al. (1977a, b).

74

In the coronal model, the emergent spectrum is dominated by emission
lines excited by electron collisions. Typically, the strongest X-ray
emission lines from each abundant element (except Fe) are the

$$(2\,^1P_1,\ 2\,^3P_1,\ 2\,^3S_1 \rightarrow 1\,^1S_0)$$

triple systems of the helium-like ions, with the Lyα lines of the
hydrogenic ions somewhat weaker. With sufficient spectral resolution
($\lambda/\Delta\lambda \sim 100$) to resolve the triple systems, it should be easy to ver-
ify whether collisional excitation is dominant, because the relative
emission line strengths should be proportional to the collisional
excitation rates.

As Gabriel and Jordan (1969a, b) and Blumenthal, Drake and Tucker
(1972) have described, the forbidden $2s\,^3S$, member of the helium-like
triple system may be used as a density diagnostic, since it is suppressed in favour of the 2^3P intercombination line by electron collisions at electron densities $n_e \gtrsim 0.6\ Z^{12}$ cm^{-3}, where Z is the nuclear
charge of the relevant ion.

The emergent X-ray line spectrum is calculated from an expression of
the form

$$L_\nu \;=\; EM \;\sum_i A_i G_i(\nu,T), \tag{2}$$

where

$$G_i(\nu,T) \;=\; \sum_{j,k} f_i^j(T) C_i^{jk}(T) h\nu_{jk}, \tag{3}$$

where the

$$C_i^{jk}(T)\text{'s}$$

are rate coefficients for excitation of level k of each ion by
electron collisions and where

$$EM = \int n_e^2 dV$$

is the volume emission measure. Note that the spectral shape is
determined entirely by the assumed element abundances, A_i, and the
temperature. Figure 1, taken from Shapiro and Moore (1976), shows
a typical coronal spectrum for standard cosmic abundances. Note
that it is dominated by emission lines. It is common practice in
X-ray astronomy to determine A_i, T, and EM by varying these parameters
until the resulting coronal model best fits the observed spectrum.
If no combination of A_i and T produces an acceptable fit, one might

try a composite of two or more regions with different emission measures
and temperatures. By permitting such flexibility, it is always possible
to find some combination of coronal model parameters that fits an
observed spectrum arbitrarily well.

An important caveat is the accuracy of the atomic physics behind the
theory. We must rely largely on theoretical calculations, because
very few of the relevant collisional excitation and ionisation rate
coefficients are known experimentally. Thus, abundance determinations
based on X-ray emission line ratios are no more accurate than the
approximate theoretical rate coefficients for the emitting ions, which
may differ from the true rates by a factor of up to 3. It is possible
to obtain more accurate theoretical results (say, within $\pm$ 20%), but
this requires laborious calculations which have only been performed
for a few selected ions.

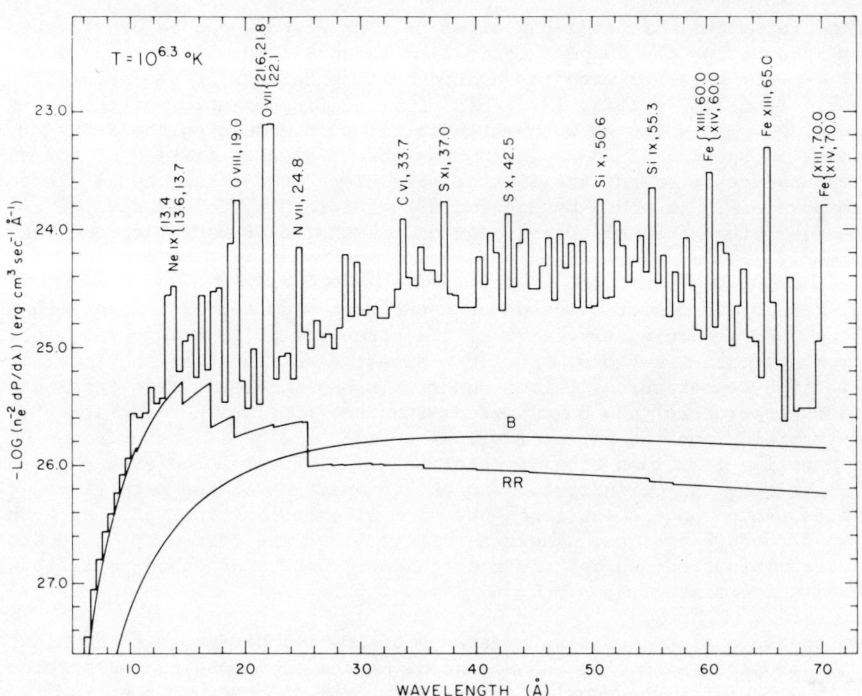

Figure 1:   Theoretical X-ray emission spectrum from an optically thin
coronal model with gas temperature $T = 10^{6.3}$ K. Emission
lines are plotted with a resolution of 0.5 Å (from
Shapiro and Moore, 1976).

Another important caveat is the use of the coronal model in situations where it is not really applicable - in particular, in the case of SNR's. An assumption, built into the coronal model, is that the ionisation of the elements is determined by a stationary, local balance between ionisation and recombination. While this is probably a good approximation for the solar corona and for gas in clusters of galaxies (flow times should be long compared to recombination times), it most likely breaks down in supernova shells, in which a blast wave propagates into interstellar gas. Even in the idealised case of a SNR propagating into a uniform medium, the shocked gas has a rapidly changing temperature profile, and the elements do not reach the stationary ionisation balance determined by the local post-shock temperature. Changes in the ionisation level tend to lag behind the changes in temperature. If radiative cooling is dynamically negligible, as is likely in a young hot SNR, the elements in gas at a given temperature are likely to be less ionised than the coronal model would imply, and the line emission is likely to be stronger. By fitting coronal models to the X-ray spectra observed from such shocks, one could infer lower temperatures and higher element abundances than the true values.

A better model for the X-ray emission from a shock can be constructed by integrating the coupled ionisation and collisional rate equations (1) - (3), together with the hydrodynamic equations of the shocked gas. Itoh (1977, 1978, 1979) has provided fine examples of this sort of model. It would be worthwhile to fit such models to the X-ray emission spectra of SNR's, using the shock velocity instead of the gas temperature as one of the fitting parameters (the others being element abundances). It would be particularly interesting to see how the derived element abundances depend on the choice of model (coronal or shock).

We can obtain direct evidence of departures from ionisation equilibrium. For example, the ratio of the strengths of Ly-β and Ly-α lines from hydrogen-like ions, provide a direct measurement of the local electron temperature (if line opacity is negligible). This "excitation temperature" can be compared with the "ionisation temperature", determined from the ratio of emission line strengths from hydrogenic and helium-like ions of a given element. Similarly, the electron temperature can be inferred from the intensities of the satellite lines, found on the low frequency side of emission lines of ions with two or more electrons. These satellite lines are formed by dielectronic recombination, which requires a higher temperature than collisional excitation of the same ion.

Of course optical and X-ray pictures of SNR's indicate that the blast wave propagates into an inhomogeneous medium, compounding the problem to how to infer the physical conditions from an observed spectrum. The power available through X-ray spectroscopy to infer physical conditions and element abundances in SNR's will improve steadily with improving spatial and spectral resolution, and with better theoretical models.

4.   THE NEBULAR MODEL

Most of the bright, celestial X-ray sources are compact objects: binary systems containing a white dwarf star, a neutron star, or a black hole; or, in the case of extragalactic sources, the nuclei of active galaxies. The primary X-ray emission from these objects probably comes from a region of such small size and great electron density that the emitted spectrum is a continuum with few spectral features. However, the compact X-ray source is likely to be embedded in a distribution of gas, which may impart spectral features to the X-rays which propagate through it. By understanding the mechanisms by which such spectral features are formed, we may hope to use the observed X-ray spectra to infer the conditions in the surrounding gas. This is important, because such gas is likely to supply the flow of accreting material responsible for the X-ray emission.

The simplest model one can construct for a compact X-ray source embedded in gas is the X-ray analogue of a "planetary nebula", in which a point source of continuum X-rays, with a luminosity spectrum $L_\nu = L\,g(\nu)$, is surrounded by a spherically symmetric distribution of gas with given element abundances and radial density distribution $n(r)$ (Hatchett, Buff and McCray, 1976). This model differs dramatically from the coronal model. The temperature of the gas is not a free parameter, but is determined by the absorption and emission of radiation. The ionisation of the elements is dominated by photoionisation, not electron collisions, and the formation of the emergent X-ray spectrum is dominated by photoabsorption, radiative recombination, and fluorescence, rather than by collisional excitation.

Assuming a stationary local balance between ionisation and recombination, we determine the ion densities

$$n_i^j(r)$$

by equations of the form:

$$\sum_{j'<j} n_i^{j'}(r) \int d\nu \; \frac{4\pi J_\nu(r)}{h\nu}\, \sigma_i^{j'}(\nu)\, a_i^{j'j} + n_i^{j+1}(r)\, n_e(r)\, R_i^{j+1}(T)$$

$$- n_i^j(r) \left[\int d\nu \; \frac{4\pi J_\nu(r)}{h\nu}\, \sigma_i^j(\nu) + n_e(r)\, R_i^j(T) \right] = 0, \qquad (4)$$

where $J_\nu(r)$ is the mean intensity of the radiation field, the

$$\sigma_i^j(\nu)\text{'s}$$

are photoionisation cross sections, and the

$$a_i^{j'j}\text{'s}$$

78

are Auger yields, giving the fractional probability that an ion in stage j' will relax to stage j following the creation of an inner shell vacancy by X-ray photoionisation. Typically, creation of a K-shell vacancy will be followed by the emission of a few Auger electrons.

The coupled set of equations (4) are solved together with an energy equation, expressing the condition that the luminosity of radiation emitted at a given place equals that absorbed. Thus, in this model the temperature $T(r)$ is part of the solution.

The mean intensity, $J_\nu(r)$ is determined by an equation of transfer, with emissivity and opacity being functions of

$$n_e(r), \quad n_i^j(r), \quad \text{and } T(r).$$

However, in some environments these coefficients are small enough to be neglected. This permits a great simplification - the optically thin approximation:

$$J_\nu(r) \;=\; Lg(\nu)/(4\pi r)^2.$$

Then, the solutions of the coupled set of equations (4) can be expressed in the form:

$$f_i^j \;=\; f_i^j(\xi) \text{ and } T(r) = T(\xi) \text{ where } \xi \equiv L/nr^2.$$

The functional dependences of ξ depend sensitively on the shape of the input spectrum, $g(\nu)$, but once calculated they may be used to describe a great variety of environments using this simple scaling law.

When the opacity of the surrounding gas becomes important, the transfer equation must be solved to find $J_\nu(r)$, making the theory much more complicated. However, the equations can be solved to a reasonable level of accuracy by making suitable approximations to the transfer equation. Hatchett et al., (1976) use an "outward-only integration" approximation, which is known to give a reasonably accurate description of a planetary nebula. For a given source spectrum, $g(\nu)$, the resulting ionisation and temperature no longer depend only on the single parameter, ξ, but also on a second "opacity parameter", q. In the case of a constant density gas cloud, $q = Ln$.

The temperature and ionisation structure and the emergent X-ray spectrum of a nebular model are very different from those of a coronal model. Because photoionisation dominates collisional ionisation, highly ionised atoms tend to be found at a much lower temperature than in the coronal model. For example, according to the coronal model oxygen is mostly hydrogenic at a temperature $T \sim 10^{6.5}$ K,

whereas in a typical nebular model oxygen is mostly hydrogenic at
$T \sim 10^5$ K. Because the ions are at low temperature, collisional
excitation is a much less important radiation mechanism; instead,
recombination and fluorescence lines dominate the emission spectrum.

The latter process, X-ray fluorescence, is particularly important in
the case of iron K lines. When an X-ray with $h\nu \gtrsim 8$ keV creates a
K-shell vacancy in an iron ion with three or more electrons, the
excited ion has a substantial probability of emitting a K_α fluorescence photon. The energy of the K_α photon depends on the particular
transition and ionisation stage, and ranges from 6.40 keV for Fe^{+1}
to 6.65 keV for Fe^{+23}. Iron is unique among the abundant elements
in having a large fluorescence yield, $y \sim 30\%$; elements such as O,
Si, S, have $y < 3\%$. (The more likely consequence of inner shell
photoionisation is Auger emission.) This may explain why the iron
K_α lines are by far the strongest X-ray emission lines observed in
the spectra of compact X-ray sources.

Figure 2, taken from Hatchett, Buff and McCray (1976), shows a typical
emergent X-ray spectrum from a nebular model. In this example, the
central source has an exponential "optically thin bremsstrahlung"

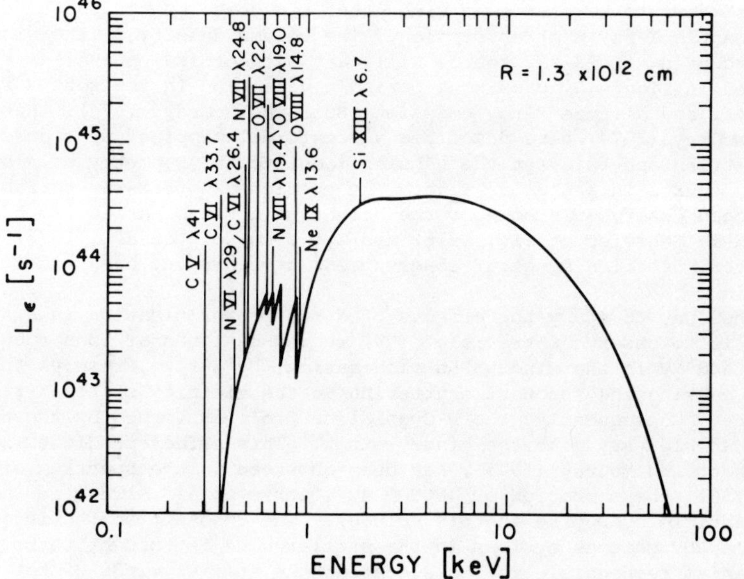

Figure 2: Theoretical X-ray spectrum for a nebular model of a compact
source of continuum X-rays with luminosity $L = 10^{37}$ ergs/s,
surrounded by a spherical gas cloud of atomic density
$n = 10^{11}$ cm^{-3} and radius $R = 1.3 \times 10^{12}$ cm. Emission lines
and recombination continua are plotted with a resolving
power $\lambda/\Delta\lambda = 100$ (from Hatchett et al., 1976).

spectrum with kT = 10 keV and luminosity $L = 10^{37}$ ergs/s. The
surrounding gas has atomic density $n = 10^{11}$ cm^{-3} and radius
$R = 1.3 \times 10^{12}$ cm, typical of a strong stellar wind surrounding a
compact binary source. This spectrum differs dramatically from the
coronal spectrum of Figure 1. The prominent spectral features with
$h\nu \lesssim 1$ keV typically appear in the spectral range where the soft X-rays
from the primary source are photoabsorbed. Recombination lines, such
as Ly-α of C, N, O, Ne, and very narrow recombination continua, appear
in emission. The width of the emission continua are indicative of
electron thermal velocities: typically $\Delta\nu/\nu \sim 10^{-2}$ $(T/10^6$ K$)^{1/2}$.
The expected spectral structure is sensitive to the column density of
the surrounding gas. More gas would absorb (photoelectrically) the
spectral features in Figure 2, but would create new ones of Si, S,
etc., at higher energies. Such photoabsorption by ionised gas near the
the X-ray source, can be distinguished from interstellar photoabsorption because the energies of the K-edges depend on ionisation stage.
Of course, we cannot observe such spectral features if the column
density of interstellar gas exceeds that of the gas around the source.

The nebular models have many applications to cosmic X-ray sources;
I will only discuss a few here. The first is the propagation of
X-rays from a compact source through a dense stellar wind of a binary
companion star. Several of the bright galactic X-ray sources (e.g.
Cyg X-1, Cen X-3, 4U1700-37, 4U0900-40), have early-type companion
stars known to have stellar winds that are dense enough to cause
noticeable X-ray photoabsorption. The column density of the wind
intercepting the X-ray source will vary with orbital phase, as illustrated in Figure 3, causing a periodic variation in the soft X-ray
cut-off and diffuse X-ray emission (Buff and McCray, 1974). Hatchett
and McCray (1977) have described a "conformal mapping" procedure
permitting one to adapt the nebular model to the geometry of a binary
X-ray source in a stellar wind. The predicted soft X-ray variability
has been observed in some of the bright galactic binaries, such as
Cen X-3 (Schreier et al., 1976) and Cyg X-1 (Holt et al., 1976).
However, detailed spectral observations have not yet been made.

Another way to study the effects of X-ray photoionisation in a stellar
wind is to observe ultraviolet (UV) resonance lines of ions such as
C IV and NV in the wind. Photoionisation, by X-rays, removes the
ions causing the resonant scattering in the vicinity of the X-ray
source. Consequently, the P-Cygni line profiles formed by these
ions should vary with the binary orbit. This effect predicted by
Hatchett and McCray (1977), has been observed in the spectrum of
HD 77581, the companion of 4U0900-40 (Dupree et al., 1980; see also
the article by Dupree in this volume). The resonant scattering of
stellar UV photons by ions in the stellar wind is thought to be the
mechanism responsible for accelerating the stellar winds of hot stars
(Cassinelli, 1979). Therefore, the analysis of UV and X-ray spectral
observations of binary X-ray sources will be a powerful tool for
studying the structure and dynamics of strong stellar winds.

A second application of the nebular model is the prediction of zones
of highly ionised atoms in the interstellar gas in the vicinity of a
galactic X-ray source. According to calculations by McCray, Wright

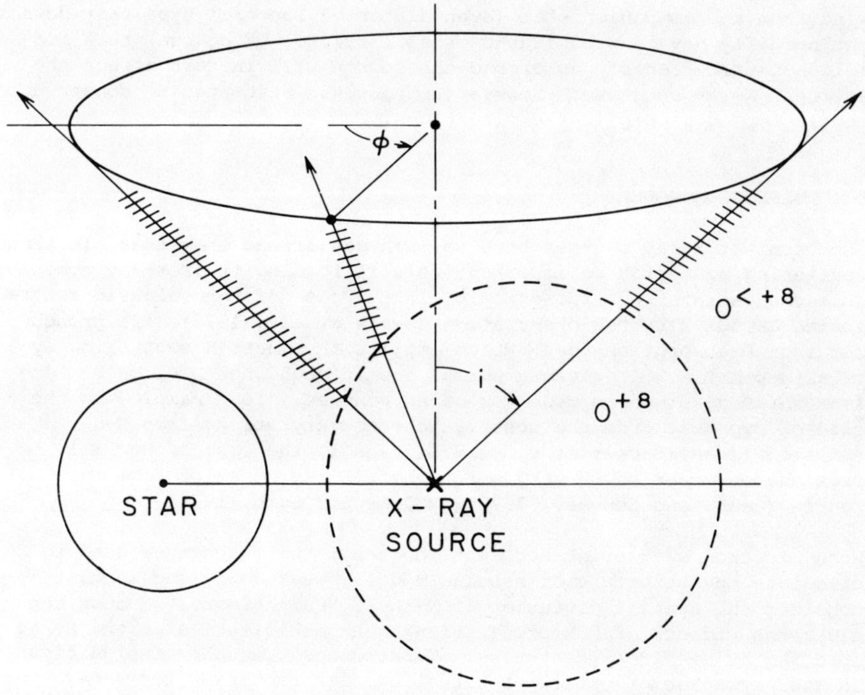

Figure 3:   Model for variable soft X-ray absorption due to a stellar
            wind in a binary X-ray source. The dashed circle indicates
            the spherical zone where oxygen is almost fully ionised
            and does not photoabsorb X-rays. The angle i is the
            inclination of the binary system, and the angle φ is the
            orbital phase. The ray from the X-ray source to earth
            traces out the cone during the binary period. X-ray
            photoabsorption by oxygen occurs beyond the dashed circle,
            and is greatest near the star, where the wind density is
            greatest. The zones where other elements photoabsorb
            X-rays have similar geometry but different sizes.

and Hatchett (1977), the bright galactic X-ray sources should be
surrounded by shells of interstellar C IV and N V with column densities
are in the range 10^{12} - 10^{14} cm^{-2}, easily observable with the IUE
spacecraft if the companion star is bright enough. Observations by
Dupree et al. (1978) do show indications of enhanced C IV and Si IV
in the UV spectrum of HD 153919 (4U1700-37), but the interpretation
of these observations will remain ambiguous until similar observations
of more stars are analysed.

A third, and perhaps the most important application of the nebular
model, is its implication for gas flows in the vicinity of X-ray
sources. Although discussion of this subject is beyond the scope of
this article, I should at least mention that heating of gas by X-rays
can drastically modify the accretion flows onto the X-ray source
(Ostriker et al., 1976), and can also induce a substantial stellar
wind from the companion star (even a star of spectral type that does
not normally have a wind (Basko et al., 1977)). Of course, this modified distribution of gas around the source will in turn affect the
emergent X-ray spectrum. However, no one has attempted to construct
models for such complicated situations.

5.   DIFFUSION MODELS

Some compact X-ray sources have so much gas around them that electron
scattering must play an important role in forming their X-ray spectra.
This can be inferred, for example, in the case of the galactic centre
source GX 301-2 by the observation (Swank et al., 1976) of a pronounced iron K-absorption edge, which implies an electron scattering optical depth $\tau_{es} \gtrsim 1$ (given a normal cosmic abundance of iron). Similar inferences can be made for other sources: for example, in the
case of Cyg X-3, from the observations of very strong iron K-α
emission lines (Becker et al., 1978); and in the case of Her X-1,
from the observation of very strong soft X-ray emission from the
source (Basko and Sunyaev, 1976; McCray and Lamb, 1976).

When electron scattering becomes important, the techniques used to
calculate the structure of nebular models break down, and we must
consider the spatial diffusion of X-rays. Furthermore, we must consider the effects of "Comptonization", the modification of the X-ray
spectrum by electron scattering. This process, usually insignificant
in the formation of optical spectra, becomes important for X-ray spectra
for two reasons: firstly, because photoelectric cross sections decrease rapidly above threshold ($\sim\nu^{-3}$), electron scattering can be the
dominant opacity for hard X-rays; secondly, the fractional energy
shift of a photon upon electron scattering is of the order 2% for a
10 keV X-ray, but $\sim10^{-3}$ for an optical photon (typically formed where
$T \lesssim 10^4$ K).

Because of the importance of Comptonization in the formation of X-ray
spectra, considerable effort has been devoted during the past few
years to the development of theories capable of describing its
effects accurately. There are at least four distinct kinds of models
in which Comptonization has qualitatively different consequences.
The first is that of a cool source of relatively soft photons surrounded by a distribution of very hot gas. At temperatures greater than
10^8 K, radiative processes become so inefficient that the gas becomes
"photon-starved". In this case, the emergent X-ray spectrum may be
dominated by the hardening of photons from the cool source as they
propagate through the hotter gas. The result is a hard continuum
spectrum that may resemble a power law (Sunyaev and Titarchuk, 1979).
Such models have been proposed to explain the X-ray spectra of quasars
(Katz, 1976) and of accretion disks around black holes (Shapiro et al.,
1976).

The second model is a variant of the first in that soft photons are hardened by Comptonization in a gas of very high temperature. However, in this case the source of soft photons is presumed to be external, and the hard spectrum is created by diffuse reflection of the soft photons by the hot gas (Lightman and Rybicki, 1979a, b). In the third model, an external source of hard photons is incident on a cool atmosphere, which reflects a Comptonized spectrum that may contain fluorescent lines. This model describes the albedo of the companion star to illumination by a compact X-ray source in a binary system (Basko, 1978).

The fourth kind of model is that of a compact source of hard X-rays surrounded by a distribution of relatively cool gas. Such "diffusion models" are similar to the nebular models, except that the column density of gas is great enough (> 1 g cm^{-2}) that Comptonization is important. They are likely to be important in describing compact sources such as Cyg X-3, which shows strong iron K emission, or Her X-1, which shows a strong soft X-ray excess. The periodic X-ray variability of Cyg X-3 may result from orbital motion of a compact X-ray source embedded in a very dense stellar wind of its companion (Hertz, Joss and Rappaport, 1978). The soft X-ray excess of Her X-1 may result from a dense shell of gas around the accreting neutron star, perhaps supported by its rotating magnetic field (Basko and Sunyaev, 1976; McCray and Lamb, 1976). In each case, spectral features in the emergent X-rays are likely to show the effects of Comptonization. Because this fourth kind of model is likely to be most relevant for the interpretation of spectral line features in compact galactic sources, I shall discuss it in more detail.

The kinetic theory of Comptonization has been discussed thoroughly by Pozdnyakov et al., (1979), Illarionov, et al., (1979), and Sunyaev and Titarchuk, (1979); I shall summarize some of the main results here. X-ray photons with $h\nu > kT$ which propagate through a region of thickness R and electron scattering optical depth $\tau_e = n_e\sigma_T R$ will have an average fractional energy loss

$$<\Delta\nu/\nu> \sim -h\nu\tau_e^2/m_e c^2$$

and a dispersion

$$<(\Delta\nu/\nu)^2>^{1/2} \sim (7h\nu/5m_e c^2 + 2kT/m_e c^2)^{1/2} \tau_e.$$

As a result, a spectral line propagating through such a medium will be spread into a skewed profile whose centroid is shifted to lower frequency by $\sim<\Delta\nu/\nu>$. An input spectrum of hard X-rays will obtain a high energy cutoff at

$$h\nu_{max} \sim m_e c^2/\tau_e^2 .$$

To go beyond these rough estimates and construct an actual model for the spectrum requires a kinetic theory for Comptonization. Three approaches have proved useful.

The first, and most straightforward, technique is to do a Monte Carlo calculation. In this method, the computer actually simulates the

photon transport process and the effects of Compton scattering may be
described as precisely as desired. However, the Monte Carlo technique
has its limitations. The first is that the calculations can consume
great quantities of computer time, particularly if accurate results
are desired. The second limitation is that the results so obtained
are model-specific; it is not readily apparent how they depend on
the parameters defining the model. The third limitation is that the
method is poorly suited for problems in which the medium depends on
the spectrum of radiation propagating through it. The gas near a
real X-ray source is likely to have this non-linear behaviour because
its temperature and ionisation are controlled by the X-rays propagating through it.

Despite these shortcomings, the Monte Carlo method remains a powerful
and important tool for understanding the formation of X-ray spectra,
particularly in models with complex geometries that are hard to
describe analytically. A fine example of a Monte Carlo calculation
is the model for Cyg X-3 by Hertz, Joss and Rappaport (1978). These
authors have simulated the transport of X-rays from a compact source
through a presumed dense stellar wind of the companion star. They
have included the effects of Comptonisation and the production of
iron K fluorescence lines, and have shown how the observed X-ray
spectrum should vary with the 4.8 hour orbital period of the system.
The approximation of a passive medium is probably a good one for this
model, because the presumed stellar wind is sufficiently dense and
extended that heating and ionisation by the X-ray source should not
substantially alter its properties.

The second technique for describing Comptonisation is the use of a
frequency diffusion (Fokker-Planck) approximation; this results in
the famous equation of Kompaneets (1957). The Kompaneets equation is
non-linear, having a term proportional to the square of the photon
intensity to describe induced Comptonisation; but this non-linear
term is usually unimportant for the case of X-rays, which are likely
to be very dilute. Ross, Weaver and McCray (1978) have shown that
the Kompaneets equation should be modified to describe accurately the
Comptonisation of X-rays by low temperature electrons. The resulting
equation can be incorporated into a spatial diffusion equation,
including line emission and photoelectric absorption. This equation
can then be solved together with ionisation rate equations and an
energy balance equation, to give a comprehensive description of the
radiative transfer of X-rays through an optically thick distribution
of gas. Such solutions have been generated by Ross et al. (1978) and
by Ross (1979).

Figure 4, taken from Ross et al. (1978), illustrates the formation
and Comptonisation of iron lines in a model consisting of a dense
($n_e \sim 10^{16}$ cm^{-3}) spherical shell of gas with $\tau_C = 6$ surrounding a
compact source of continuum X-rays at a radius $R = 7 \times 10^8$ cm. The
dashed curve is the source spectrum, presumed to be an exponential
with $kT_x = 40$ keV and luminosity $L = 10^{37}$ erg/s. The shell is assumed
to have a temperature $T = 10^7$ K, but its ionisation is calculated
self-consistently. The dotted curve shows the spectrum that emerges
when Comptonisation is neglected in the theory. Strong photoelectric

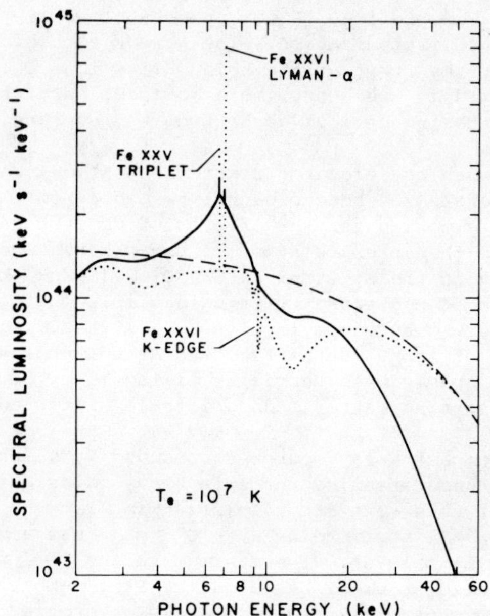

Figure 4:   Theoretical X-ray spectrum from a diffusion model of a
compact X-ray source of continuum X-rays surrounded by a
spherical shell of atomic density $n = 10^{16}$ cm^{-3},
temperature $T = 10^7$ K, electron scattering optical depth
$\tau_c = 6$, and radius $R = 7 \times 10^8$ cm. Dashed curve:
assumed spectrum of central source. Dotted curve:
emergent spectrum neglecting the effects of Comptonization.
Solid curve: emergent spectrum including the effects of
Comptonization. Narrow cores of spectral lines are
plotted with resolving power $\lambda/\Delta\lambda = 35$ (from Ross et al.
1978).

absorption by iron is evident in the range 8-30 keV. The absorption
profile is complex because it is caused by many ions of iron, each
of which has a different photoelectric threshold, and because it is
contaminated by recombination continuum emission. Weaker photoabsorption due to other elements such as Si, S, etc., is evident
below 5 keV. Very strong emission lines due to recombination of
hydrogenic and helium-like ions of iron are also evident.

The solid curve in Figure 4 is the spectrum which emerges when
Comptonization is included. Three pronounced effects are evident:
first, the high energy photons are degraded by Comptonization,
greatly suppressing the high energy tail of the emergent spectrum;
second, the photoelectric absorption is largely filled in; and

third, the strong emission lines are merged and spread into a broadened and skewed profile. A few photons do escape in narrow line cores without Comptonization. The narrow core of emergent Lα photons is weaker than that of the helium-like ions because the Lα photons are emitted in the inner, more ionized, part of the shell, at a greater scattering optical depth from the surface.

It follows that when the electron scattering optical depth is substantial, accurate inferences cannot be made from the observed spectrum without allowing for the effects of Comptonization. For example, consider the interpretation of the observations of strong iron K absorption in the spectrum of GX 301-2 (Swank et al. 1976). If Comptonization is neglected, fitting a simple photoelectric model to the observed spectrum implies $\tau_e \gtrsim 1$, assuming a normal cosmic abundance of iron. But if so, the actual photoelectric absorption has probably been partially filled in by Comptonization, implying more absorbing matter than originally inferred.

The model of a compact X-ray source surrounded by a spherical shell may be of use in understanding the soft X-ray emission from Her X-1. The brightness of this soft X-ray emission implies the presence of dense gas with $\tau_e > 1$ at a radius $R \gtrsim 10^8$ cm (Basko and Sunyaev 1976; McCray and Lamb 1976). The minimum radius is inferred from a simple thermodynamic argument based on the Stefan-Boltzmann law, that can be circumvented only by invoking some exotic radiation mechanism like an X-ray laser on the neutron star (unlikely, I think!). Because the soft X-rays pulsate (cf. the article by S. Holt in this volume), the opaque shell cannot be spherically symmetric, and probably corotates with the neutron star.

Ross (1979) has constructed idealized models for X-ray transfer through such an opaque shell, including Comtonization as well as emission and absorption of X-rays by abundant elements. The shell absorbs energy from the hard X-rays that diffuse through it and re-radiates this energy as soft X-rays at lower temperature. One important result of such models is that the soft X-rays are unlikely to have a Planck spectrum, but may have an abrupt photoelectric cutoff at photon energies in the range 0.3-0.4 keV due to ions of carbon. An analogous effect occurs in model atmospheres of hot stars (Hummer and Mihalas 1970), which emit a smooth spectrum below the He II ionization threshold at 54 eV, but almost no radiation above this energy.

A limitation of the Fokker-Planck approximation used to make such calculations, is that it cannot accurately describe the effects of small numbers of scatterings. Therefore, the method breaks down near the line core and when the scattering optical depth is low (say, $\tau_e < 2$)). Ross et al. (1978) and Ross (1979) have shown how the method can be modified to describe the escape of photons in the narrow line cores.

A third technique to describe Comptonization is an analytical stat-istical theory developed by Illarionov et al. (1979). This theory

is capable of describing the effects of small as well as large
numbers of scatterings, and shows what can be learned from high
resolution observations of Comptonized line profiles. One important
effect is that the once-Comptonized line profile, with $\Delta\lambda \sim 0.05$ Å,
shows noticeable thermal distortion at very low temperatures (10^5K),
because it reflects electron rather than ion thermal velocities.
X-ray resonance lines are likely to emerge with once-Comptonized
profiles because the unscattered line cores are resonantly trapped
(Pozdnyakov et al. 1979). However, Langer (1979) has shown that
some details of the emergent line profiles still require Monte
Carlo calculations.

6.   CONCLUSIONS

The collection of X-ray sources in the galaxy is diverse: some
look like regions of hot transparent gas; some, like compact objects
surrounded by diffuse gas; some, like compact objects surrounded by
nearly opaque gas. In reality, the sources are very complex.
Theoretical studies indicate that, depending on the source model,
very different physical processes may dominate the spectral formation: electron collisions, photoionization and recombination, or
fluorescence and Comptonization.

There are two morals to be learned from these studies. The first is
that to deduce the physical properties of X-ray sources correctly
from their X-ray spectra, one must understand the assumptions and
limitations of the theoretical models for spectral formation. Some
inferences are fairly model-independent; others are very sensitive
to the assumptions of the model. The second moral is that the great
scientific potential of X-ray spectroscopy depends as much on the
development of good theoretical tools to interpret the spectra as it
does on the development of better telescopes and detectors.

This work was supported by a grant from the National Aeronautics and
Space Administration.

REFERENCES

Basko, M.M., 1978. Ap. J., 223, 268.
Basko, M.M., Hatchett, S., McCray, R., and Sunyaev, R.A., 1977. Ap.
    J., 217, 276.
Basko, M.M., and Sunyaev, R.A., 1976. MNRAS, 175, 345.
Becker, R.H., et al., 1978. Ap. J. (Letters), 224, L113.
Blumenthal, G.R., Drake, G.W.F., and Tucker, W.H., 1972. Ap. J.,
    172, 205.
Buff,, J., and McCray, R., 1974. Ap. J. (Letters), 188, L37.
Cassinelli, J., 1979. Ann. Rev. Astr. Ap., 19, 275.
Dupree, A.K., et al., 1978, Nature, 275, 400.
Dupree, A.K., 1980. Ap. J., 238, 969.
Gabriel, A.H., and Jordan, C., 1969a. Nature, 221, 947.
Gabriel, A.H., 1969b. MNRAS, 145, 241.

Hatchett, S., Buff, J., and McCray, R., 1976. Ap. J., 206, 847.

Hatchett, S., and McCray, R., 1977. Ap. J., 211, 552.

Hertz, P., Joss, P.C., and Rappaport, S., 1978. Ap. J., 224, 614.

Holt, S.S., et al., 1976. Nature, 261, 213.

Hummer, D.G., and Mihalas, D., 1970. MNRAS, 147, 339.

Illarionov, A., Kallman, T., McCray, R., and Ross, R., 1979. Ap. J.,
     228, 279.

Itoh, H., 1977. Publ. Astr. Soc. Japan, 29, 813.

Itoh, H., 1978. ibid., 30, 489 (erratum: ibid., 31, 429).

Itoh, 1979. ibid., 31, 541.

Jacobs, V.L., Davis, J., Kepple, P.C., and Blaha, M., 1977a. Ap. J.,
     211, 605.

Jacobs, V.L., 1977b. Ap. J., 215, 690.

Jordan, C., 1969. MNRAS, 142, 501.

Jordan, C., 1970, MNRAS, 148, 17.

Katz, J.I., 1976. Ap. J., 206, 910.

Kompaneets, A.S., 1957. Soviet Phys. -JETP., 4, 730.

Landini, M., and Fossi, B.C.M., 1972. Astr. Ap. Suppl., 7, 291.

Langer, S., 1979. Ap. J., 232, 891.

Lightman, A.P., and Rybicki, G.B., 1979a. Ap. J. (Letters), 229, L15.

Lightman, A.P., 1979b. Ap. J., 232, 882.

McCray, R., and Lamb, F.K., 1976. Ap. J. (Letters), 204, L115.

McCray, R., Wright, C., and Hatchett, S., 1977. Ap. J. (Letters),
     211, L29.

Ostriker, J.P., McCray, R., Weaver, T., and Yahil, A., 1976. Ap. J.
     (Letters), 208, L61.

Pozdnyakov, L.A., Sobol, I.M., and Sunyaev, R.A., 1979. Astr. Ap.,
     75, 214.

Ross, R.R., 1979. Ap. J., 233, 334.

Ross, R.R., Weaver, R., and McCray, R., 1978. Ap. J., 219, 292.

Schreier, E.J., et al., 1976. Ap. J., 204, 539.

Shapiro, P.R., and Moore, R.T., 1976. Ap. J., 207, 460.

Shapiro, S.L., Lightman, A.P., and Eardley, D.M., 1976. Ap. J., 204,
     187.

Summers, H.P., 1974a. MNRAS, 169, 663.

Summers, H.P., 1974b. Appleton Lab., Report CR 74.102, Culham, UK.

Sunyaev, R.A., and Titarchuk, L.G., 1979. Astr. Ap., 86, 121.

Swank, J.H., et al., 1976. Ap. J. (Letters), 209, L65.

Rapid aperiodic X-ray variability

Hale V. Bradt
Richard L. Kelley and
Larry D. Petro

Department of Physics and Center for Space Research
Massachusetts Institute of Technology,
Cambridge.

ABSTRACT - Data from at least ten "persistent" galactic X-ray
sources, (Cyg X-1, Cir X-1, GX339-4, 4U0900-40, 4U1700-37, Sco X-1,
LMC X-4, GX301-2, GX304-1, and 4U1626-67) have exhibited rapid,
aperiodic (non-burst) X-ray variability with time constants ranging
from 20s down to at least tens of milliseconds in some cases. Random
variability on time scales of 30-50 ms has also been detected in
both Type I and Type II X-ray bursts and <0.4 s rise times have been
observed in "fast transient" precursors. Recent X-ray and optical
results from these sources may make possible further theoretical
progress on the understanding of the emission processes.

1.   Introduction

2.   Recent results
    a)   Cyg X-1
    b)   Cir X-1
    c)   MXB 1728-34
    d)   GX 339-4
    e)   MXB 1730-335 (Rapid Burster)
    f)   4U0900-40
    g)   Fast Transient = ? MX1716-31 (2S1715-321)
    h)   4U1700-37
    i)   Sco X-1
    j)   LMC X-4
    k)   GX301-2 (2S1223-624)
    l)   GX304-1
    m)   4U1626-67

3.   Discussion

1.   INTRODUCTION

Rapid aperiodic variability as a class of X-ray phenomena has
received relatively little attention in recent years. The 50 ms
(e.g. Oda et al., 1971; Rappaport et al., 1971) and 1 ms (Rothschild
et al., 1974, 1977; but also see Weisskopf and Sutherland, 1978)

variability of the prime black-hole candidate Cyg X-1 attracted
considerable interest early in this decade. Subsequent attention
was directed largely to the X-ray pulsars which provide ready insight into the nature of the emitting systems (Rappaport 1981, this
volume) and to the X-ray bursters which appear to have yielded
significant insight into the emission mechanisms (e.g. thermonuclear
flashes, Joss 1978; Lewin and Clark 1980, this volume).

In contrast, aperiodic variability, a feature (on one or more time
scales) of almost all discrete galactic X-ray sources, does not
appear to submit easily to theoretical insight. This is due in part
to the difficulty of describing the variability with physically
meaningful parameters, although a shot-noise model appears to
explain a fair portion of the Cyg X-1 data (Rothschild et al. 1977).
Hope for important insight depends, at least in part, upon the
accumulation of data from a variety of sources during their various
intensity and spectral states. Furthermore, the study of concurrent
variability in the optical, infrared, and radio bands can provide
important diagnostic tools.

The time scales of the aperiodic variability of galactic X-ray
sources range from years (the infrequent transients), to days and
months (the rise and fall times of the slow transients), to hours
(irregular variations of the "steady" emission and some "fast"
transients), to minutes, seconds, and probably milliseconds. A
substantial amount of data on the slower events has been provided by
the scanning instruments on the Uhuru, OSO-7, and Ariel V satellites
and to the continuously viewing All Sky Monitor on Ariel V.

Systematic searches for rapid temporal variability in the $\sim$20
brightest galactic sources have been carried out with the Uhuru (for
0.1-1 s variability; Forman, Jones, and Tananbaum 1976), ANS (1 s
and 16 s; Parsignault and Grindlay 1978), and SAS-3 (8-500 ms; Li
1979) satellites. The analyses were based upon statistical (χ^2)
tests, the interpretation of which has been discussed and questioned
by the latter two groups. The latter author (Li 1979) believes that
there is no compelling evidence for rapid (<16 s) aperiodic
variability from any of these sets of data, with the exception of
the 0.1 s aperiodic variability observed from the Crab (pulsar?) by
Forman, Jones, and Tananbaum (1976). (Due to the unique nature of
the Crab pulsar, we will not discuss it further).

We now know that the nonstationary characteristics of the
variability of galactic X-ray sources require that observations be
carried out for sustained periods with detectors of substantial
areas if the variability is to be properly studied. It turns out,
therefore, that SAS-3 has yielded a substantial amount of data of
this type, because it could point toward a celestial source with
about 250 cm^2 of proportional counter area for days or weeks at a
time. HEAO-1 also has yielded important new results on aperiodic
variability from pointings with durations of a few hours.

2.   RECENT RESULTS

Sources for which results on rapid ($\lesssim 20$ s) aperiodic variability are available include the three black hole candidates Cyg X-1, Cir X-1, and GX339-4, the pulsing X-ray sources (and hence presumably magnetic neutron stars) GX301-2, GX304-1, 4U1626-67, and 4U0900-40, the eclipsing X-ray binaries LMC X-4 and 4U1700-37, and finally the enigmatic Sco X-1. These are sources wherein the variability appears to be closely associated with the "persistent" or "steady" (albeit highly variable on longer time scales) emission because of its large duty cycle (e.g. continual flaring) or modest amplitude (factors of a few) relative to the steady emission.

Rapid temporal variability is also a feature of X-ray bursts and some transient events wherein the X-ray flux density increases by one or more orders of magnitude above the persistent emission. We particularly note here (1) the $\lesssim 1$ s rise times of X-ray bursts (e.g. Grindlay et al. 1976. Hoffman et al. 1979), (2) the $<0.4-10$ s rise times in two fast transient events (Hoffman et al. 1978), (3) the ~ 30 ms variability during a Type 1 burst from MXB1728-34 (Hoffman et al. 1979), and (4) the 50 ms structure in Type II "rapid" bursts (Ulmer et al. 1977).

Some characteristics of these sources are summarized in Table 1, where we list the sources in order of the shortest reported time constants. In the following paragraphs, we present some of the highlights of recent results. The broader implications of the transient/burst phenomena and extensive references are presented by Lewin and Clark (1980). Recent developments on the pulsing sources are presented by Rappaport 1981. Data from some of the "persistent" sources are currently under analysis by our group; in these cases we will only mention a few features of the results to illustrate their developing richness. References to earlier work on all of these sources and to their optical characteristics may also be found in Bradt, Doxsey, and Jernigan (1979).

a) Cyg X-1

This remarkable source, the leading black-hole candidate, exhibits high and low states of different spectral character and appears to be variable on all measurable time scales from ~ 1 ms to ~ 10 s (see review by Boldt et al. 1974 and Oda 1977). An active phase with continual 1-10 s flaring (Canizares and Oda 1977) increased the previously observed time constants by a factor of 2-4 (Fig. 1). Both hard and soft flares were observed, but the mean flare spectrum showed no measurable change from the non-flare (low-state) flux. Also, evidence for a 20 ms characteristic time scale was obtained from autocorrelation analyses of data with high time resolution (Canizares and Oda 1977; also see Ogawara et al. 1977). The interpretation of the ~ 1 ms flaring (Rothschild et al. 1974, 1977) remains in doubt (Weisskopf and Sutherland 1978). Recent correlation analyses and comparisons of Cyg X-1 and Cir X-1 data from Uhuru are now available (Weisskopf and Sutherland, 1980).

	PERIODIC PULSES DETECTED	BINARY PERIOD	APERIODIC BEHAVIOUR
Cyg X-1	No	5.6d	High & Low States; 0.5 s shots (low state); rapid flaring 1-10 s
Cir X-1	No	16.6d (binary?)	On-Off (Mos.); Bright Flaring State; Steady State
MXB 1728-34	No		∿30 ms structure in Type 1 burst
GX339-4	No		High/Low/Off States (∿100d)
MXB 1730-335	No		∿50 ms structure in Type II (rapid) bursts
4U0900-40	Yes 283s	8.97d	3 s flare during 2 hour flare
Fast transients (MX1716-31? + others)	No		∿3 s precursors; Durations 150-1500 s
4U1700-37	No	3.4d	Erratic 1 s - 10 min
Sco X-1	No	0.8d	Active state when optically bright
LMC X-4	No	1.4d	20 s flares in high state
GX301-2	Yes 700s	35d	Burst-like Flare; Active flaring state
GX304-1	Yes 272s		∿100 s flares, 4 times steady flux
4U1626-67	Yes 7.68s	<0.3d	1000 s quasi-periodic flaring

TABLE 1*

FASTEST VARIABILITY REPORTED	X-RAY SPECTRUM IN BRIGHT STATE RELATIVE TO LOW STATE (1-10 keV)	SPECTRUM OF RAPID FLARING RELATIVE TO STEADY FLUX (1-10 keV)	CORRELATED OPTICAL FLARES REPORTED
$\sim$1 ms (20 ms)	Softer	Same (1-10s)	No
$\sim$1 ms (20 ms)	Softer	Same (1-10s)	Delayed IR, radio
30 ms	Harder		(No opt. ident.)
40 ms (low state)	Softer		No
50 ms			No
<0.4 s			No
<0.4 s	Harder (10 min flare)		(No opt. ident.)
$\sim$1 s			No
$\sim$1 s		Harder	Yes
$\sim$2 s	Softer		No
$\sim$3 s	$\sim$Same	Harder	No
$\sim$10 s		Softer	No
$\sim$20 s	Softer (10 min flares)		Yes

*For references see: text of this paper; Bradt, Doxsey, and Jernigan, 1979; Rappaport, 1981 (this volume); Lewin and Clark, 1981 (this volume).

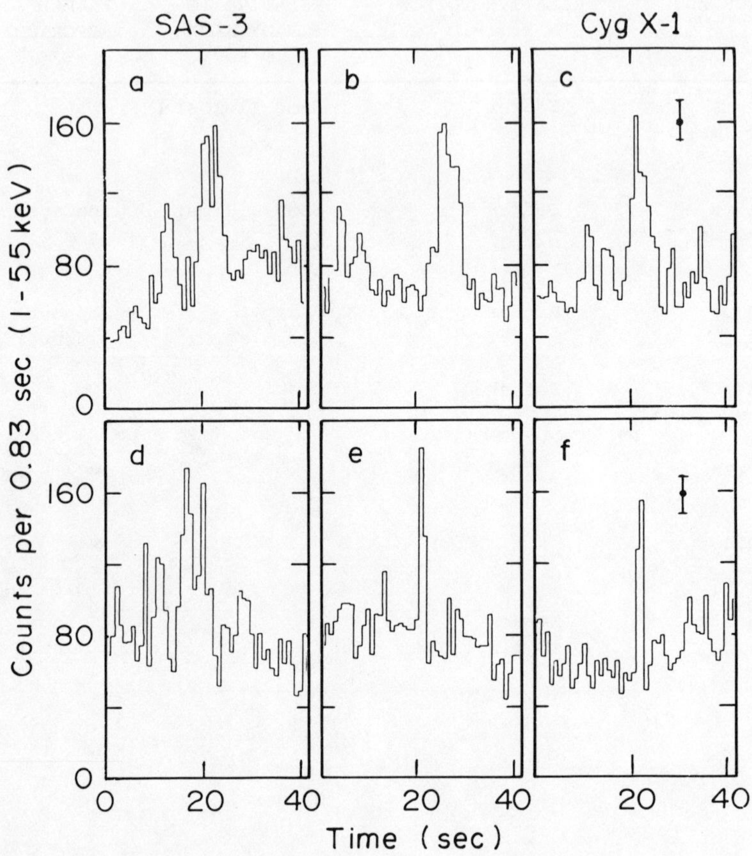

Figure 1: X-ray flaring from Cyg X-1 with SAS-3.   From Canizares
and Oda (1977).

   b) Cir X-1

This source exhibits downward intensity transitions with a 16.6 d
periodicity except when it is in a sustained quiescent ("off")
state (Kaluzienski et al. 1976).   Like Cyg X-1, it also has exhibited erratic variability at least as short as 0.1 s (Jones et al.
1974) and a transient episode with interflare intervals of $\sim$2.5 s
(Sadeh et al. 1979).   It has been identified with a highly
reddened Hα-emission star (Whelan et al. 1977).

During SAS-3 observations (Fig. 2), a particularly rapid transition
was observed: a factor of $\sim$4 decrease in intensity in 60 s

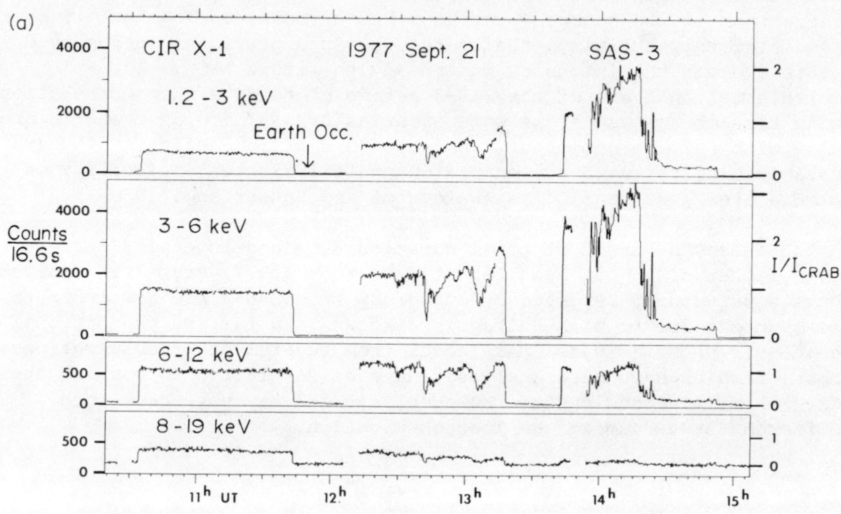

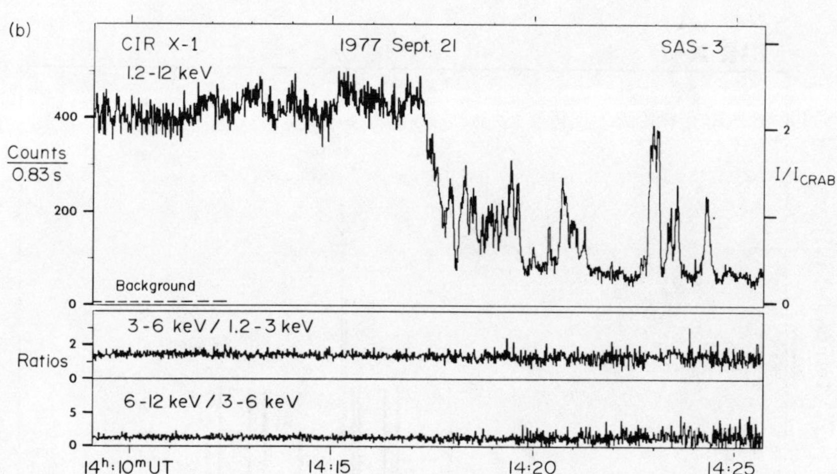

Figure 2:   Unusual and rapid transition of Cir X-1 from SAS-3. (a)
three orbits of data showing steady states, activity,
and brightening, and finally the rapid transition; (b)
expanded view of the transition with hardness ratios that
show no marked spectral variability. From Dower et al.
(1981).

(Dower, Bradt, and Morgan, 1981). This event was preceded during the previous 4 hours by erratic variability and by a few large intensity dips (factor of 2 in about 30 min) and was followed for ∿7 min by large burst-like flares (up to a factor of 4 for 60 s). None of these features were associated with any significant variations in the source spectrum. Furthermore, during the ∿19 hours preceding this activity, the source was in a previously unobserved state wherein the intensity was steady to within ∿10%. Auto-correlation analyses of the SAS-3 active state data show correlations with time constants of the same order as Cyg X-1 (∿0.5 s) and in one case with a 20 ms component. A cross-correlation analysis of Uhuru data from different X-ray energy channels indicates variability on time scales less than 0.3 s (Weisskopf and Sutherland, 1980).

A "millisecond burst" of total duration 8 ms and intensity increase of a factor of 16 has been reported by Toor (1977; Fig. 3). Similar bursts previously reported by our group from SAS-3 are now known to be spurious due to high-voltage breakdown in a failing proportional counter. Uncontaminated SAS-3 data from several other observations, most of which have been analyzed, have so far failed to confirm the Cir X-1 millisecond burst. However, Cir X-1 may have been in a different state during the Toor observation.

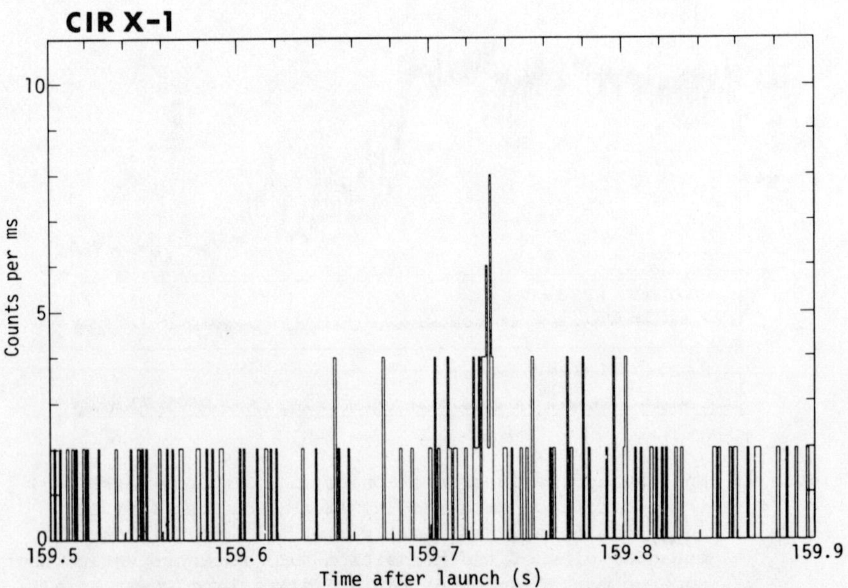

Figure 3: Millisecond burst from Cir X-1. From Toor (1977).

c)   MXB1728-34

A single (Type 1) burst from this source has been observed with the
large aperture instruments on HEAO-1 (Hoffman et al. 1979). The NRL
data (A-1, LASS experiment) with intrinsic 5 ms resolution in the
5-17 keV band showed ~30 ms variability at a level of ~30% of the
peak burst flux (Fig. 4). The measured e-folding rise time of the
burst is 150 ms. The as yet unidentified steady source has a 2-11
keV flux density which ranges from 35 to 210 µJy[*] (see Bradt,
Doxsey, and Jernigan 1979) and a 2-11 keV spectrum which is softer
than that of the bursts (Hoffman et al. 1976).

d)   GX339-4

This source has exhibited long-term high, low, and off states
(Markert et al. 1973). During a low state, erratic variability of
a factor of 3 in 40 ms was observed (Fig. 5) with the NRL HEAO-1
experiment (Samimi et al. 1979). SAS-3 observations in the high
state show no rapid variations (Li, Clark, and Rappaport 1978). An
emission-line optical counterpart (Doxsey et al. 1979) has been
interpreted by Grindlay (1979) as either a main-sequence B star or
as being like the blue-continuum counterparts of X-ray bursters
(e.g., McClintock et al. 1977a).

[*] $1.0 \ \mu Jy = 0.242 \times 10^{-11} \ \text{ergs cm}^{-2} \ \text{s}^{-1} \ \text{keV}^{-1}$. The flux density of
the Crab Nebula, averaged over 2-11 keV, is 1060 µJy.

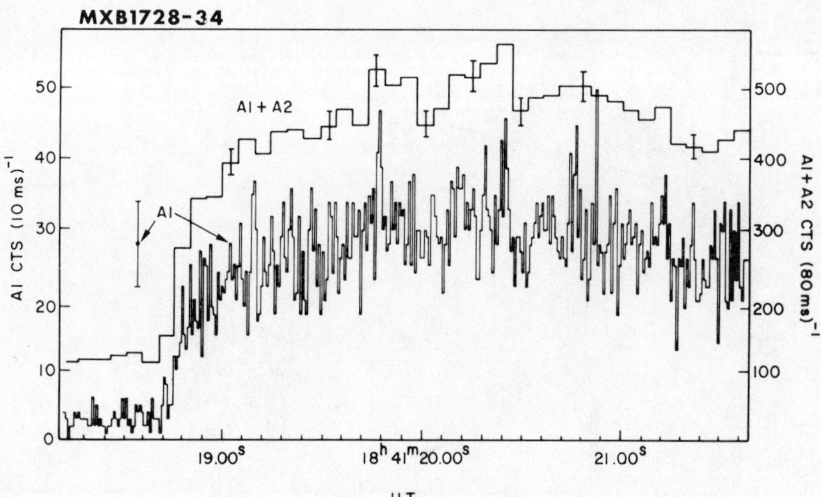

Figure 4:   Early portion of X-ray burst (Type I) from MXB1728-34
            showing ~30 ms variability and 150 ms e-folding rise
            time. From Hoffman et al. (1979).

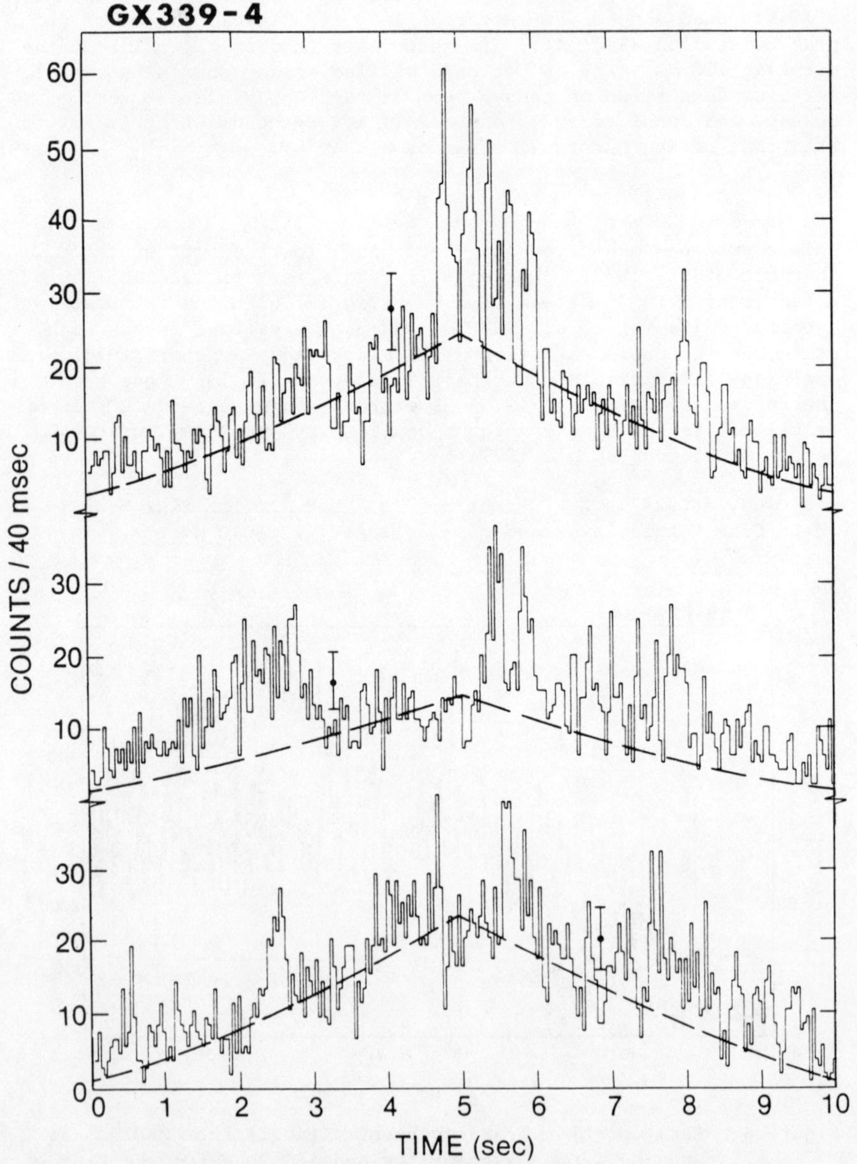

Figure 5:   Rapid variability of GX339-4 during three scans with HEAO-1.   From Samimi et al. (1979).

e)   MXB1730-335 (Rapid burster).

Ulmer et al. (1977) have presented evidence for 50 ms variability in at least 8 and up to 56 rapid type II bursts (Fig. 6).

f)   4U0900-40

A flare of 3 s duration, 24 $\pm$ 4% increase in intensity, and <0.4s rise time was observed from this source in 1971 with the Uhuru satellite (Forman et al. 1973). It occurred at the peak of an intense ~2 hour enhancement from 60 to 1800 µJy (2-6 keV; 0.04 to 1.2 I_{crab}). This source is an X-ray pulsar (P = 283 s) with a binary period of 8.97 d and a B0.5Ib optical counterpart (see references in Bradt, Doxsey, and Jernigan 1979 and Rappaport 1981).

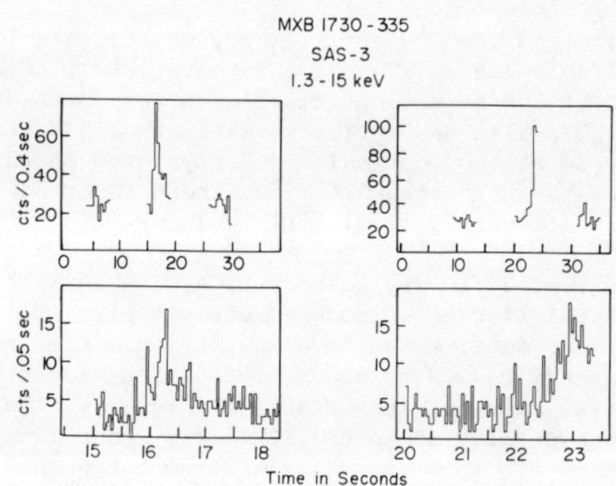

Figure 6:   Two "rapid" (Type II) bursts from MXB1730-335 showing 50 ms structure (lower figures). The compressed scales for the same events (upper figures) show the sampling of the SAS-3 data mode and the earlier and later behaviour of the source. From Ulmer et al. (1977).

g)   Fast transient =? MX1716-31(2S1715-321)

The fast transient event of June 28, 1976 observed with SAS-3 by Hoffman et al. (1978) exhibits a precursor of duration 3 s (rise <0.4 s) and a main pulse with rise ~10 s and duration 150 s (see Figs. 14, 15 in Lewin and Clark 1981). A 10-min transient of about the same intensity (~2000 µJy, 2-11 keV, if MX1716-31 is the source) had been observed in 1972 with OSO-7 from the same region of the sky (Markert, Backman, and McClintock 1976). Both events had positional uncertainties of about 1.4 deg^2 (Markert, Backman and McClintock 1976; Jernigan et al. 1978), consistent with the location of the relatively weak (25 µJy) steady source MX1716-31 (2S1715-321) and, in our view, are probably associated with it. The spectrum of the

OSO-7 flare indicated increased absorption compared to the MX1716-31 spectrum. The steady source has, so far, resisted optical identification. Both the SAS-3 and OSO-7 papers report an additional similar fast transient (e.g. Figs. 12, 13 in Lewin and Clark 1981 p.344/5) from apparently different regions of the sky.

h) 4U1700-37

This eclipsing source (P=3.4d) with an O6.5f optical companion (see refs in Bradt, Doxsey, and Jernigan 1979) exhibited four-fold fluctuations in 10 min, two-fold increases in 1 s, and indications of 0.1 s variability during Uhuru observations in 1972 (Jones et al. 1973).

i) Sco X-1

This, the first discovered celestial X-ray source, has long been known to be variable (Lewin, Clark, and Smith 1968). Correlations between the $\sim$10-min X-ray and optical flares have been observed (e.g. Pelling 1973; also see review by Miyamoto and Matsuoka 1977). Quite recently, correlated optical and X-ray (with SAS-3) flares with time constants as short as 20 s have been reported (Ilovaisky et al. 1980; also Ilovaisky, 1981, this volume).

A more recent extended (10 day) observation with SAS-3 has provided data during several flaring episodes (Petro et al. 1981). Preliminary analysis of the data, which have an intrinsic time resolution limited by telemetry to 0.8 s, shows flares with widths as short as 2 seconds and rise times of less than 1 s (Fig. 7). The flare amplitudes were typically 8% in the 1.2-3 keV channel and 45% in the 8-19 keV channel. This is in qualitative agreement with the previous conclusion (White et al. 1976a) that the spectrum is harder during flares.

World-wide optical observations during this March 1979 SAS-3 observation yielded coincident optical flares. A short stretch of optical data from the Mt. Wilson 100-inch telescope obtained by Horne and Gomer (private communication) is shown in Figure 8 together with the SAS-3 data. The general correlation between the X-ray and optical intensities is excellent. However, the optical flares are distinguished by their lack of rapid (<20 s) structure. The longer optical time scale may be due to delays in the reprocessing of X-rays into optical light (e.g. Chester 1979).

The beginning of the rise of the optical flux is coincident with the onset of the X-ray flare to within $\sim$4 s. The minimum expected photospheric optical-flare delay has been computed for the flare shown in Figure 8 by Petro et al. (1981). They used the model of Crampton et al. (1976) for the binary system (optical emission lines from near the X-ray source, $M_X = 1.3\ M_\odot$, $M_{opt} = 0.8$-$1.2\ M_\odot$) and based their calculations only upon the light travel time from the X-ray source to the inner Lagrange point and the aspect of the

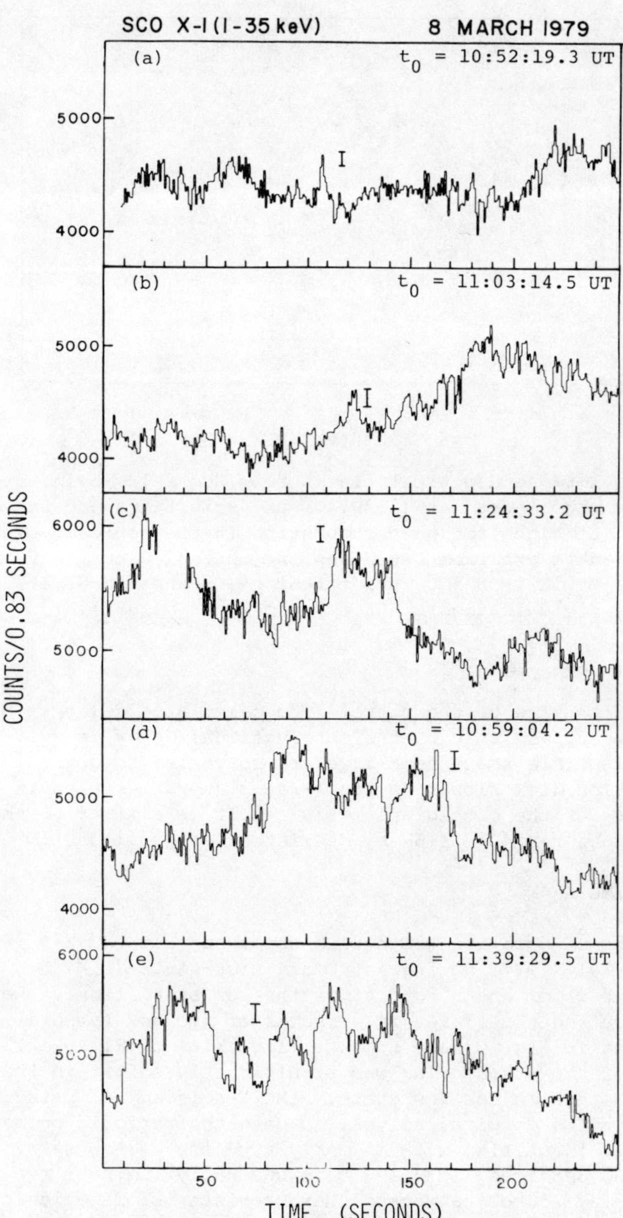

Figure 7:   Flaring from Sco X-1 with SAS-3 showing characteristic times ranging from 1 to 200 sec. From Petro et al. (1981).

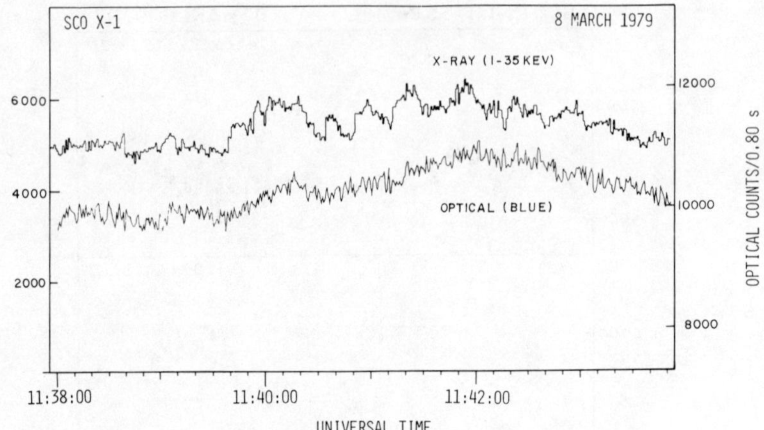

Figure 8: X-ray and optical fluxes from Sco X-1 showing simultaneous onset (at 11:39:40) of 4-min flare and the absence of high frequency modulation in the optical. The X-ray data are from SAS-3 and the optical data are from the Mt. Wilson 100-inch telescope (Horne and Gomer, private communication). From Petro et al. (1981).

system at the time of the flare. The calculated delay is 8-9 s, whereas the upper limit on the observational delay is 4 s. Thus, it is more plausible that the X-rays are converted to optical light in the accretion disk closer to the X-ray star rather than in the photosphere of the companion. This result is similar to the conclusion drawn for 4U1626-67 by McClintock et al. (1980).

j) LMC X-4

Four flares of duration $\sim$20 s were observed from LMC X-4 (Fig. 9; Epstein et al. 1977) during a $\sim$40 min high-state ($L_x = 5 \times 10^{38}$ erg s^{-1}; 2-10 keV) of about four times the low-state flux. The flux increase during the flares is a factor of two and temporal structure is apparent in Fig. 9 down to $\sim$2 s (i.e. rise of Flare 2). The spectrum of the steady flux was significantly softer in the flaring high state than in the low state. LMC X-4 is an eclipsing X-ray binary (P = 1.4 d) with, as yet, no detected periodic pulsations; it has been identified with an early-type star. (See refs. in Bradt, Doxsey, and Jernigan, 1979). It is unusually luminous for a compact X-ray source, a common characteristic of Magellanic Cloud X-ray sources (Clark et al. 1978).

LMC X-4

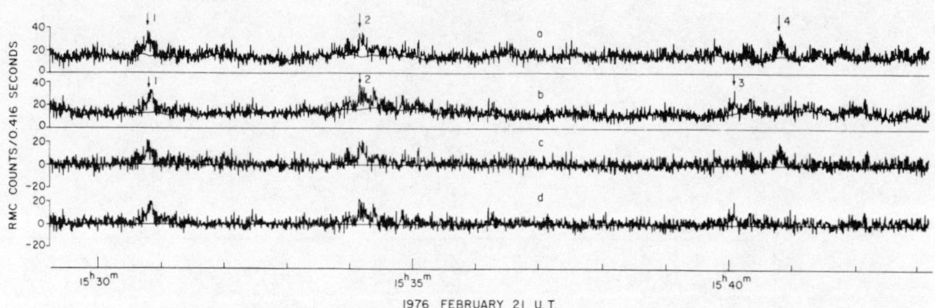

Figure 9:   Four flares from LMC X-4 in two SAS-3 modulation collimator detection systems before (a,b) and after (c,d) the
collimator response is removed.   A given flare (e.g. #4)
may be suppressed by one collimator and not the other.
From Epstein et al. (1977).

k)   GX301-2 (2S1223-624)

A very large burst-like flare (factor of 3 increase; rise time ~3 s,
full width ~6 s) was observed with SAS-3 during a very intense state
of the source in January 1978 (Fig. 10; Kelley and Bradt 1981).   The
large flare appears most pronounced relative to the steady flux (see
Fig. 10 caption) in the central two energy channels.   Pointed
observations ~18^h after the flare showed the source to be in a
continually active flaring state (Fig. 11) although the individual
flares were not quite as pronounced as that of Figure 10.   The
average source intensity (6-12 keV) in the active state was comparable to that of the Crab Nebula.   There was no significant change in
the time-averaged spectral parameters compared to those of the more
frequently encountered low state.   To our knowledge, such a state had
not been observed since early balloon observations (see Ricker et al.
1973, 1976).   The periodic X-ray pulsations (P = 700 s; White et al.
1976b) and its maximum luminosity indicate that the compact object in
the binary system is a neutron star.   This active state is the best
documented example of continual rapid variability from a known
neutron star.   Faster variability ($\lesssim$ 2 s) was not detected during the
SAS-3 observations.

l)   GX304-1

Three flares with rise times of ~10 s, durations of ~100 s, and
intensities ~4 times the 20 µJy average non-flare flux (Fig. 12) have
been observed from GX304-1 (McClintock et al. 1977b).   The two flares
in Figure 12 are, we note, more pronounced relative to the pulsing
steady flux in the low energy channels.   These two flares occurred at
the same phase of the 272 s X-ray pulsation period. (The phase of the
3rd pulse, 2.2 d previously, is not available).   Thus far only limits

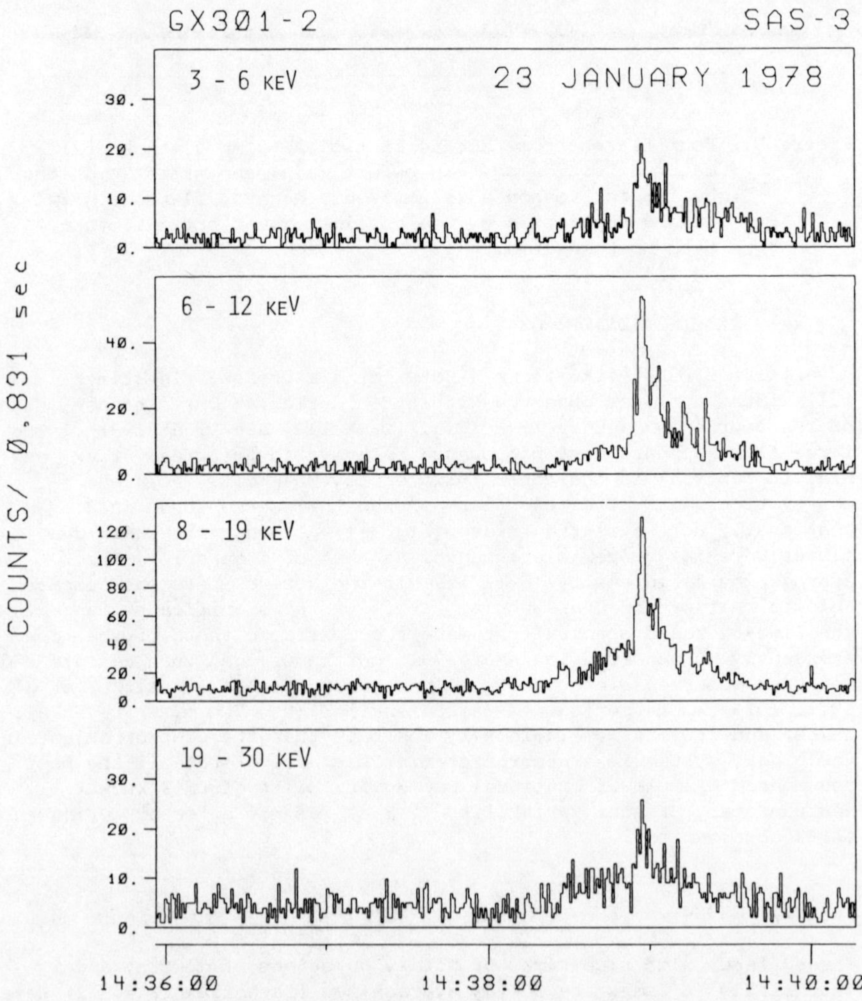

Figure 10: Burst-like flare from GX301-2 in several energy channels,
with SAS-3. Underlying "pedestal" is the steady flux
from GX301-2 modulated by the triangular collimator
response during a SAS-3 scan. From Kelley and Bradt
(1981).

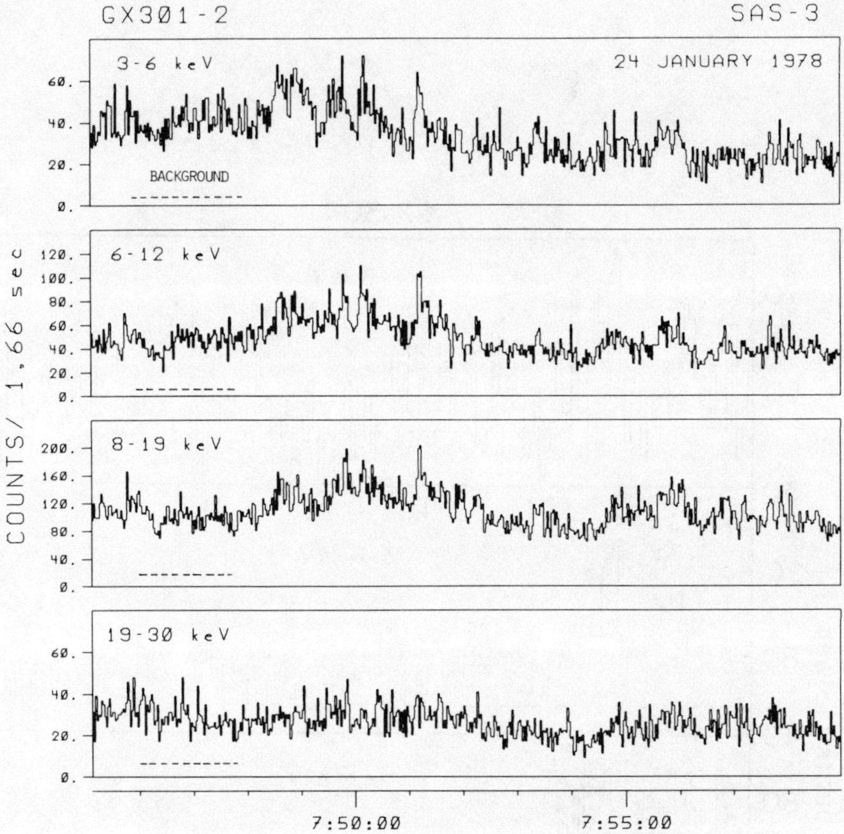

Figure 11: Active state of GX301-2 about 18 h after the flare of
Figure 10, from SAS-3 pointed observations. The underlying 11.7 min pulse (and interpulse) profile is
apparent. From Kelley and Bradt 1981.

on a binary period have been obtained from the X-ray data (McClintock
et al. 1977b), and no evidence of binary motion has yet been detected
in the optical data from the highly reddened optical candidate
(Mason et al. 1978).

m)   4U1626-67.

This is a 7.68 s X-ray pulsar which exhibits no evidence for
Doppler motion indicating a very small and/or highly inclined orbit
(see Rappaport 1981). It has been identified with a faint UV star
(McClintock et al. 1977a) which also exhibits 7.68 s pulsations
(Ilovaisky, Motch, and Chevalier 1978; McClintock et al. 1980).
The X-ray source also undergoes transient oscillations (flares)
with a quasi-period of about 1000 s (Joss, Avni, and Rappaport 1978).
The flux density increases by a factor of two (1.2-12 keV), and the
spectrum softens during the flares.

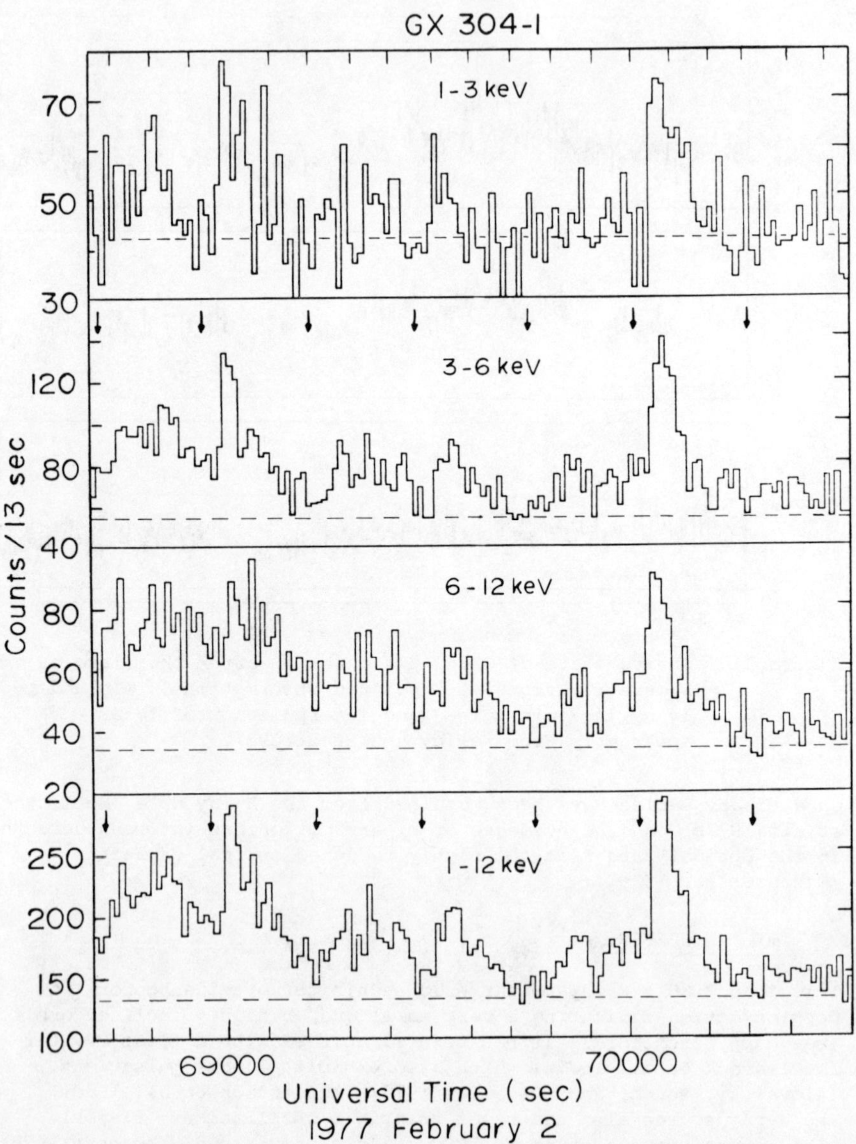

Figure 12: Two ∿100 s flares from GX304-1. The 272 s pulsation is also apparent. From McClintock et al. (1977b).

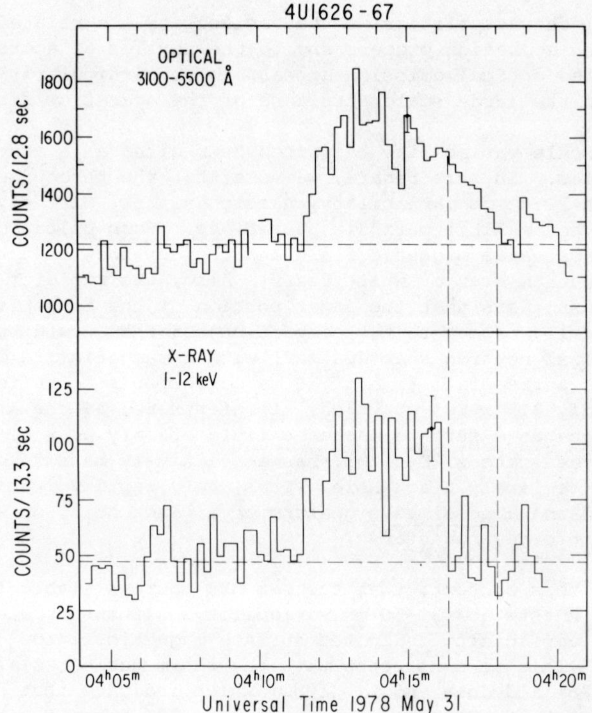

Figure 13:   Optical and X-ray flux (SAS-3) from 4U1626-67 during
             5-minute flare showing 100 s optical tail.   From
             McClintock et al. (1980).

One of the flares obtained from a later SAS-3 observation (McClintock
et al. 1980) is shown in Figure 13.   It has a duration of 5 min and
declines rapidly ($\sim$20 s) at the end of the flare.   Simultaneous
optical data (Fig. 13) shows a coincident optical flare with a
similar rise ($\sim$60 s time scale), but with a noticeable "tail" which
persists for 100 s after the abrupt X-ray decrease.   The authors
conclude, from this event and from the nature of the optical/X-ray
7.7 s pulses, that between 8% and 100% of the optical emission is
due to the conversion of X-rays into light at a location within 0.5
lt.s of the neutron star, probably in an accretion disk.   The tail of
the optical pulse is attributed to the longer processing time for the
higher energy X-rays (e.g. Chester 1979).

3.   DISCUSSION

It appears from the above examples that rapid aperiodic variability
is likely to develop into an important diagnostic tool for the study
of compact X-ray emitting objects and their environs.   The temporal

and spectral characteristics of the variability are related to the nature of the accretion process and to the physics of accretion disks. The correlated optical emission appears to be particularly useful for the study of the large scale structure of the accretion disks.

Short time scale variability has often been cited as a characteristic of black holes. In this regard, we note that the three "persistent" sources with 20-40 ms variability, namely Cyg X-1, Cir X-1, and GX339-4, do not exhibit periodic pulsations. Such pulsations would indicate an off-axis magnetic field and thereby show that the X-ray star is a neutron star or white dwarf. Also, the rapidity of the variability suggests that the inner portion of the accretion disk may be substantially closer to the compact object than would be the case for the nominal neutron star (pulsar) with a magnetic field of $\sim 10^{12-13}$ G. A free-fall time of 20 ms corresponds to an initial radius of only 3.5×10^7 cm which, if interpreted as the Alfven radius, indicates a surface magnetic field of only $\sim 3 \times 10^{10}$ G. One of these three sources, Cyg X-1, has a relatively massive compact member, i.e. a likely black hole. Thus, very rapid variability ($\lesssim 40$ ms) remains a <u>possible</u> signature of a black hole, or at least a weakly magnetic neutron star.

We note, in this context, that the two MXB sources (Table 1) exhibit both Type 1 bursts and 30-50 ms variability. If interpreted as above, this variability indicates surface magnetic fields of less than 10^{11}G. This is consistent with the thermonuclear flash models of Joss (1978) and Joss and Li (1980) which indicate that neutron stars with surface magnetic fields B $\lesssim 3 \times 10^{12}$ G are the source of Type I bursts. [The field for one of these sources, the Rapid Burster, is probably finite because magnetospheric instabilities are the likely origin of the rapid (Type II) bursts (Lewin et al. 1976; Hoffman, Marshall, and Lewin 1978; Lamb et al. 1977).] Of course, the $\sim$30-50 ms variability in these bursts may well arise from different mechanisms or set of physical conditions than similar variability in the "persistent" sources.

The known magnetic neutron stars (e.g. GX301-2, GX304-1, 4U0900-40, and 4U1626-67) also exhibit rapid variability similar in some res- pects to that of the black-hole candidates. However, in these cases, there is as yet no indication of variability with time constants significantly shorter than $\sim$1 s (20 s for 4U1626-67). (Sporadic, low- level variability on $\sim$40 ms time scales cannot be ruled out for GX301-2 because a systematic, sensitive search of the complete data set has not yet been carried out; the low flux densities of GX304-1 and 4U1626-67 ($\lesssim$ 30 µJy; 2-11 keV) may make such searches impractical with the present data base). This, again, is consistent with the above hypothesis, namely that $\sim$40 ms variability indicates non- magnetic or weakly magnetic emitters.

It is not implausible to imagine that the detailed character of the 20-40 ms variability (e.g., spectral evolution or higher Fourier components) might distinguish between the continuously strenthening

gravitational field (down to the event horizon) of a black hole and
the solid surface of a neutron star. The reported ~1 ms bursts
(Rothschild et al. 1974, 1977; Toor 1977) could arise from the final
orbits of the accreting matter (Boldt 1977) and be particularly
informative in this regard. We stress here the need for further
observations of the most rapidly variable (<1 s) sources because the
presence of this variability, in a number of instances, is based
upon rather limited data.

The luminosities of these sources as a function of the spectral
shape (e.g. temperature) should be related to the structure of the
emission region. Interesting features of this region would be the
radius of the compact object (e.g. Kylafis et al. 1980) and the
existence of a magnetosphere. We note (see Table 1) that the three
black-hole candidates have spectra which soften when they undergo
long-term (hours-days) brightening in the 1-10 keV region. The
high luminosity (5 x 10^{38} erg s^{-1}, 2-12 keV), 40-min flaring state
of LMC X-4 also has a softer spectrum than the low state. The
spectrum of one neutron star source (GX301-2) is relatively
independent of intensity and, on a shorter time scale, another
(4U1626-67) softens during its 5-10 min flares.

The spectral character of the more rapid features are given in
Table 1. The ~10 s pronounced flares in Cyg X-1 and Cir X-1 and the
10-30 min dips in Cir X-1 (Figs. 1, 2) do not exhibit marked spectral
changes (averaged over many flares in the case of Cyg X-1) from the
adjacent flux. This lack of pronounced spectral variation on short-
time scales in a given intensity state of Cir X-1 is apparent in
Figure 1 of Weisskopf and Sutherland (1980). In contrast, the flare
spectra of GX301-2, GX304-1, and Sco X-1 appear to be different than
the non-flare fluxes.

Finally, we call attention to the wide range of variability states
that can be observed in a single source. For instance, as noted
above, GX339-4 exhibits long-term high, low and off states, with
rapid variability detected, so far, only in the low state. The high
and low states of Cyg X-1 exhibit different spectral and temporal
characteristics (see Oda 1977). Cir X-1 has shown remarkably
diverse behaviour within a few hours (Fig. 2), namely very steady
emission, followed by bright and variable emission, and finally a
low state with violent flaring. Further study of the different
variability states of a single physical system should help unravel
its intrinsic physical properties.

We have attempted to demonstrate that one can now begin to see
similarities and differences in the characteristics of the rapid
aperiodic X-ray sources, in order to probe the physical conditions
within the systems. However, significant progress and comparison
to theoretical models will be aided by detailed and systematic
descriptions of the observable characteristics of the X-ray and
optical fluxes presently available. It is evident that new data
from Ariel VI, Hakucho, and future missions are very important
for the understanding of these phenomena.

We are grateful to several of our colleagues for helpful conversations and comments, including C. Canizares, R. Dower, P. Joss, D. Lamb, J. McClintock, and S. Rappaport. We also thank T. Welch and A. Ruesga for assistance in the preparation of this manuscript. This work has been supported in part by NASA Grants NAS5-2441 and NAS8-27972.

REFERENCES

Boldt, E., Holt, S., Rothschild, R., and Serlemitsos, P., 1974. Proceedings of the Calgary Conference on X-rays in Space.

Boldt, E., 1977. Ann. N.Y. Acad. Sci., 302, 329.

Bradt, H., Doxsey, R., and Jernigan, J.G., 1979. In (COSPAR) X-ray Astronomy (ed W.A. Baity and L.E. Peterson). Oxford and New York: Pergamon Press, p.3.

Canizares, C.R., and Oda, M., 1977. Astrophys. J. (Letters), 214, L119.

Clark, G., Doxsey, R., Li, F., Jernigan, J.G., and van Paradijs, J., 1978. Astrophys. J. (Letters), 221, L37.

Chester, T.J., 1979. Astrophys. J., 227, 569.

Crampton, D., Cowley, A.P., Hutchings, J.B., and Kaat, C., 1976. Astrophys. J., 207, 907.

Dower, R., Bradt, H., and Morgan, C., 1981. In preparation.

Doxsey, R.E., Grindlay, J.E., Griffiths, R.E., Bradt, H., Johnston, M., Leach, R., Schwartz, D., and Schwarz, J., 1979. Astrophys. J. (Letters), 228, L67.

Epstein, A., Delvaille, J., Helmken, H., Murray, S., Schnopper, H.W., Doxsey, R., and Primini, F., 1977. Astrophys. J., 216, 103.

Forman, W., Jones, C., Tananbaum, H., Gursky, H., Kellogg, E., and Giacconi, R., 1973. Astrophys. J. (Letters), 182, L103.

Forman, W., Jones, C., and Tananbaum, H., 1976. Astrophys. J., 208, 849.

Grindlay, J., Gursky, H., Schnopper, H., Parsignault, D.R., Heise, J., Brinkman, A.C., and Schrijver, J., 1976. Astrophys. J. (Letters), 205, L127.

Grindlay, J.E., 1979. Astrophys. J. (Letters), 232, L33.

Hoffman, J.A., Lewin, W.H.G., Doty, J., Hearn, D.R., Clark, G.W., Jernigan, J.G., and Li, F.K., 1976. Astrophys. J. (Letters), 210, L13.

Hoffman, J.A., Lewin, W.H.G., Doty, J., Jernigan, J.G., Haney, M., and Richardson, J.A., 1978. Astrophys. J. (Letters), 221, L57.

Hoffman, J.A., Marshall, H.L., and Lewin, W.H.G., 1978. Nature, 271, 630.

Hoffman, J.A., Lewin, W.H.G., Primini, F.A., Wheaton, W.A., Swank, J.H., Boldt, E.A., Holt, S.S., Serlemitsos, P.J., Share, G.H., Wood, K., Yentis, D., Evans, W.D., Matteson, J.L., Gruber, D.E., and Peterson, L.E., 1979. Astrophys. J. (Letters), 233, L51.

Ilovaisky, S.A., Motch, C., and Chevalier, C., 1978. Astron. Astrophys. 70, L19.

Ilovaisky, S.A., Chevalier, C., White, N.E., Mason, K.O., Sanford, P.W., Devaille, J.P., and Schnopper, H.W., 1980. MNRAS, 191, 81.

Ilovaisky, S.A., 1981. This volume.

Jernigan, J.G., Apparao, K.M.V., Bradt, H.V., Doxsey, R.E., Dower, R.G., and McClintock, J.E., 1978. Nature, 272, 701.

Jones, C., Forman, W., Tananbaum, H., Schreier, E., Gursky, H., Kellogg, E., and Giacconi, R., 1973. Astrophys. J. (Letters), 181, L43.

Jones, C., Giacconi, R., Forman, W., and Tananbaum, H., 1974. Astrophys. J. (Letters), 191, L71.

Joss, P.C., 1978. Astrophys. J. (Letters), 225, L123.

Joss, P.C., Avni, Y., and Rappaport, S., 1978. Astrophys. J., 221, 645.

Joss, P.C. and Li, F.K., 1980. Astrophys. J., 238, 287.

Kaluzienski, L.J., Holt, S.S., Boldt, E.A., and Serlemitsos, P.J., 1976. Astrophys. J. (Letters), 208, L71.

Kelley, R., and Bradt, H., 1981. in preparation; see also IAU Circ. 3165 (1978).

Kylafis, N.D., Lamb, D.Q., Masters, A.R., and Weast, G.J., 1980. Ann. N.Y. Acad. Sci., 336, 520.

Lamb, F.K., Fabian, A.C., Pringle, J.E., and Lamb, D.Q., 1977. Astrophys. J., 217, 197.

Lewin, W.H.G., Clark, G.W., and Smith, W.B., 1968. Astrophys. J. (Letters), 152, L55.

Lewin, W.H.G., Doty, J., Clark, G.W., Rappaport, S.A., Bradt, H.V.D., Doxsey, R., Hearn, D.R., Hoffman, J.A., Jernigan, J.G., Li, F.K., Mayer, W., McClintock, J., Primini, F., and Richardson, J., 1976. Astrophys. J. (Letters), 207, L95.

Lewin, W.H.G., and Clark, G.W., 1981, This volume.

Li, F.K., Ph.D. Thesis, M.I.T., 1979.

Li, F.K., Clark, G.W., and Rappaport, S.A., 1978. IAU Circ. 3238.

Markert, T.H., Backman, D.E., and McClintock, J.E., 1976. Astrophys. J. (Letters), 208, L115.

Markert, T.H., Canizares, C.R., Clark, G.W., Lewin, W.H.G., Schnopper, H.W., and Sprott, G.F., 1973. Astrophys. J. (Letters), 184, L67.

Mason, K.O., Murdin, P.G., Parkes, G.E., and Visvanathan, N., 1978. MNRAS, 184, 45P.

McClintock, J., Canizares, C., Bradt, H., Doxsey, R., Jernigan, J.G., and Hiltner, W.A., 1977a, Nature, 270, 320.

McClintock, J.E., Rappaport, S.A., Nugent, J.J., and Li, F.K., 1977b. Astrophys. J. (Letters), 216, L15.

McClintock, J.E., Canizares, C.R., Li, F.K., and Grindlay, J.E., 1980. Astrophys. J. (Letters), 235, L81.

Miyamoto, S. and Matsuoka, M., 1977. Space Sci. Rev., 20, 687.

Oda, M., 1977. Space Sci. Rev., 20, 757.

Oda, M., Gorenstein, P., Gursky, H., Kellogg, E., Schreier, E., Tananbaum, H., and Giacconi, R., 1971. Astrophys. J. (Letters), 166, L1.

Ogawara, Y., Doi, K., Matsuoka, M., Miyamoto, S., and Oda, M., 1977. Nature, 270, 154.

Parsignault, D.R., and Grindlay, J.E., 1978. Astrophys. J., 225, 970.

Pelling, R.M., 1973. Astrophys. J., 185, 327.

Petro, L., et al. 1980. In preparation. (See also Petro, L., Bradt, H., Lamb, D., Kelley, R., Jacobs, S., Meiksin, A., Horne, K., and Gomer, R., 1979. Bull. Am. Astron. Soc., 11, 464).

Rappaport, S.A., Doxsey, R., and Zaumen, W., 1971. Astrophys. J. (Letters), 168, L43.

Rappaport, S.A., 1981. This volume.

Ricker, G.R., McClintock, J.E., Gerassimenko, M., and Lewin, W.H.G.,
     1973. Astrophys. J., 184, 237.
Ricker, G.R., Gerassimenko, M., McClintock, J.E., Ryckman, S.G., and
     Lewin, W.H.G., 1976. Astrophys. J., 207, 333.
Rothschild, R.E., Boldt, E.A., Holt, S.S., and Serlemitsos, P.J.,
     1974. Astrophys. J. (Letters), 189, L13.
Rothschild, R.E., Boldt, E.A., Holt, S.S., and Serlemitsos, P.J., 1977.
     Astrophys. J., 213, 818.
Sadeh, D., Meidav, M., Evans, W., Byram, E.T., Chubb, T.A., Friedman,
     H., Wood, K., Yentis, D., Smathers, H., and Meekins, J., 1979.
     Nature, 278, 436.
Samimi, J., Share, G., Wood, K., Yentis, D., Meekins, J., Evans, W.,
     Shulman, S., Byram, E., Chubb, T., and Friedman, H., 1979.
     Nature, 278, 434.
Toor, A., 1977. Astrophys. J. (Letters), 215, L57.
Ulmer, M.P., Lewin, W.H.G., Hoffman, J.A., Doty, J., and Marshall, H.,
     1977. Astrophys. J. (Letters), 214, L11.
Weisskopf, M.C., and Sutherland, P.G., 1978. Astrophys. J., 221, 228.
Weisskopf, M.C., and Sutherland, P.G., 1980. Astrophys. J., 236, 263.
Whelan, J.A.J., Mayo, S.K., Wickramasinghe, D.T., Murdin, P.G.,
     Peterson, B.A., Hawarden, T.G., Longmore, A.J., Haynes, R.F.,
     Goss, W.M., Simons, L.W., Caswell, J.L., Little, A.G., and
     McAdam, W.B., 1977. MNRAS, 181, 259.
White, N.E., Mason, K.O., Sanford, P.W., Ilovaisky, S.A., and Chevalier,
     C., 1976a. MNRAS, 176, 91.
White, N.E., Mason, K.O., Huckle, H.E., Charles, P.A., and Sanford,
     P.W., 1976b. Astrophys. J. (Letters), 209, L119.

Optical photometry of
galactic X-ray sources

Sergio A. Ilovaisky

Observatoire de Paris,
France

1.   Introduction
      a)   Aims and Limitations
      b)   What can be learned from Optical Photometry

2.   Low $L_x/L_{optical}$ Systems

3.   High $L_x/L_{optical}$ Systems

      a) Cygnus X-2
      b) Scorpius X-1
      c) 4U 1626-67

1.   INTRODUCTION

The number of optical identifications of galactic X-ray sources has
now reached about 60 objects, mainly as a result of precise
positions obtained with the Ariel V, SAS-3 and HEAO-1 spacecraft.
Clearly the optical study of these counterparts has already played
and will continue to play an important role in our understanding of
X-ray sources. In this review I will deal with what can be learned
from photometric studies and will discuss in detail some particular
sources as examples. For more general reviews on optical counterparts, see Bahcall (1978) and Margon (1979a, b).

a)   Aims and Limitations

In its broadest sense, photometry is the measurement of the flux
received at Earth from a star in a given wavelength interval. In
the context of X-ray astronomy, the study of the time variability of
this flux is a fundamental aspect. Typically, observations are
limited by three types of handicaps: 1) Time windows: Observations
are made at night and depending on the season and the latitude of
the observatory, a particular object may not be visible all night
through. This is an intrinsic limitation in the search for
periodic behaviour. 2) Atmospheric variability: The content of the
atmosphere at any given location may vary with time (haze, dust,
fog, clouds, etc.) thereby limiting its transmission and increasing
sky brightness. 3) Instrumentation: The size of the telescope, its

114

tracking accuracy, type of detecting system and size of measuring
aperture all may limit the signal-to-noise ratio obtainable for a
star of given magnitude. Most photometry is done with photo-
multipliers as detectors and (usually) glass filters as spectral
window selectors. The <u>ideal</u> photometer should measure the light
from a suspected variable star, substract the sky background,
measure standard stars to correct for changes in the Earth's
atmosphere and (eventually) extrapolate the flux to outside the
Earth's atmosphere. Practically these things are never done
simultaneously nor systematically, resulting in severe limitations
on the study of short-time variations. If a night is photometric
(i.e., no clouds and stable conditions) one can measure sky and
reference stars occasionally and hence spend more time on the
optical candidate. A panoramic-type detector (CCD, CID, etc.) may
be also to measure (almost) everything simultaneously (with certain
restrictions on time resolution).

b) What can be learned from optical photometry?

The X-ray to optical luminosity ratio is a useful measure of the
relative energetic importance of the X-ray phenomenon in a binary
system. For our purposes we can classify optically identified X-ray
sources into two broad groups: low and high L_x/L_{opt} systems, with
the borderline somewhere above unity. Low L_x/L_{opt} systems will be
often characterized as 'normal' in the sense that essentially all
light we perceive comes from the optical companion star with the
effects of the presence of the compact star being relatively minor.
For these systems optical photometry can in principle give us an
idea of the distorted shape of the primary star via the so-called
'ellipticity effect' (see review by J. Hutchings). It can also
indicate the possible presence of other light sources in the system
(disk, streams, etc.) if deviations from the ellipsoidal light
curve are observed. High L_x/L_{opt} systems, on the other hand, are
expected to be dominated by the X-ray luminosity and the optical
light we receive may not come entirely from the companion, often
supposed to be a sub-giant or dwarf. In this instance, optical
photometry may yield evidence for reprocessing of X-rays into
optical light in regions which may be anywhere from the illuminated
face of the companion to matter close to the X-ray source. Eventually it may also show the effects of anisotropic X-ray emission
(due to beaming, shadowing, etc.).

2.   LOW L_x/L_{opt} SYSTEMS

In massive X-ray binaries the 'ellipsoidal' variations show two
equal maxima and two unequal minima per orbital period, resulting
from the changing aspect of the primary, tidally and rotationally
distorted by the compact companion. The exact values for the light
curve amplitudes at phases O.O and O.5, A(O.O) and A(O.5), depend
on the mass ratio of the components, M_{opt}/M_x, while the difference
A(O.5)-A(O.O) depends on the orbital plane inclination i. For the

mass ratios involved (M_{opt}/M_x = 10 to 15), in no case can the amplitude exceed 0.1 mag. In Table 1 are listed, for the seven binaries exhibiting ellipsoidal variations, the X-ray to optical luminosity ratio, the measured mass ratio (when available) and the approximate amplitude in mid-phase.

TABLE 1 -   MEASURED PARAMETERS FOR MASSIVE X-RAY BINARIES

System	L_x/l_{opt} [+]	M_{opt}/M_x [§]	A(0.5)
4U 1700-37	5×10^{-4}		0.06 mag
Vela X-1	3×10^{-3}	13 [1]	0.08
Cyg X-1	2×10^{-2}		0.11
Cen X-3	5×10^{-2}	17 [2]	0.09-0.13
4U 1538-52	1×10^{-2}	10	0.12:
SMC X-1	1	16	0.11
LMC X-4	1		0.20

[+] Maximum values (taken from Bradt et al., 1979).

[§] Ratios derived from optical radial velocity data and Doppler motion of the X-ray pulsar (Bradt et al., 1979).

(1) From van Paradijs et al.,(1978).

(2) From Hutchings et al.,(1979).

In Figures 1 and 2 I have reproduced the light curves of Vela X-1 (HD77581) and 4U 1700-37 (HD153919) from Zuiderwijk et al.(1977) and van Paradijs et al., (1978). Note that the C_1 colour index variations seen in Vela X-1 are compatible with the expected variations in gravity over the surface of the primary. For the sources exhibiting amplitudes in excess of 0.1 mag. it is almost impossible to fit theoretical light curves (see review by J. Hutchings) and quite often fits near phase 0.0 do not correspond to fits near phase 0.5. This is even the case for 4U 1700-37, where the amplitude is the lowest (see Table 1). In Figure 3 we see the first available data for 4U 1538-52 (taken from Ilovaisky et al., 1979) together with a superposed theoretical l light curve corresponding to the observed mass ratio of about 10 (Crampton et al., 1978). The fit is not excellent for the same reasons.

Two binaries form a class apart due to their rather high L_x/L_{opt} ratios. For SMC X-1 (Fig. 4 taken from van Paradijs, 1977), there is evidence for extra light, perhaps from an accretion disk (van Paradijs et al., 1978); moreover, the observed X-ray heating effect (minimum at phase 0.5 brighter than minimum at phase 0.0) is not as strong

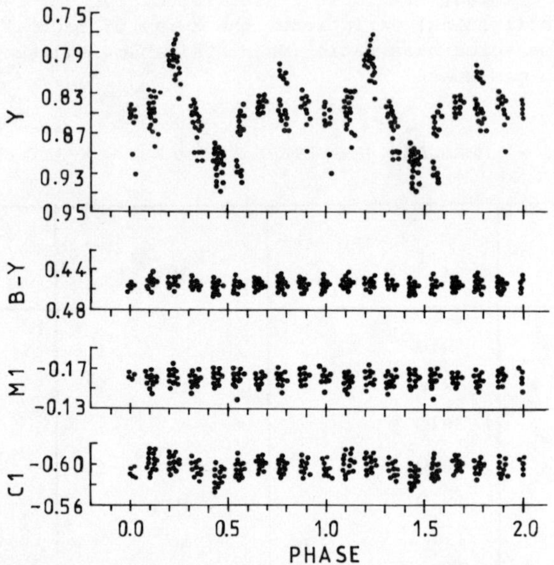

Figure 1: V-light curve and colour indices of HD77581 plotted as a
function of binary phase (P=8.966 d). Phase zero corresponds to X-ray eclipse. From Zuiderwijk et al. (1977).
(Reproduced by kind permission from Astronomy and Astro-
physics, Vol. 54, p167, 1977)

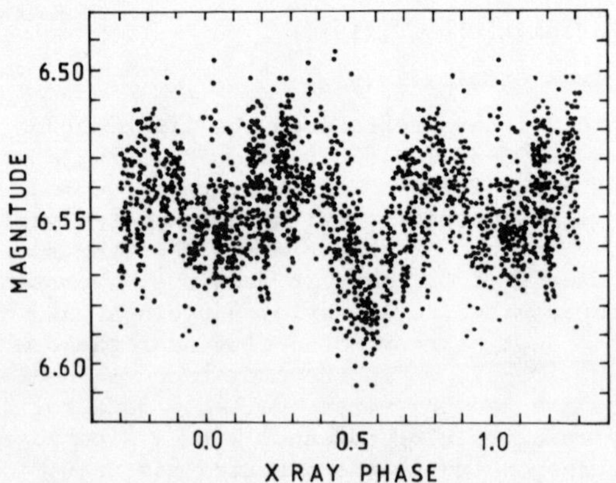

Figure 2: V-light curve of HD153919 based on data obtained by
different observers in South Africa, Brazil and Chile.
Phase zero corresponds to JD 2442231.13 + n 3.41117 d.
Taken from van Paradijs et al. (1978). *(Reproduced by*
kind permission from Astronomy and Astrophysics (Suppl.),
Vol. 31, p189, 1978.)

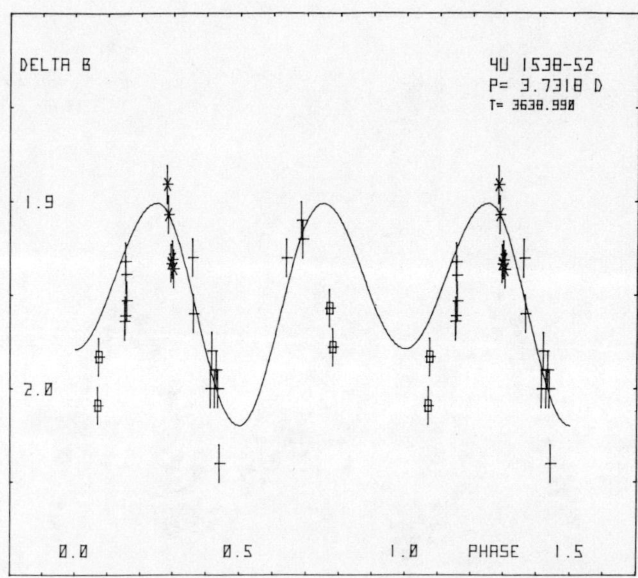

Figure 3: B-light curve of 4U 1538-52 taken from Ilovaisky, Chevalier
and Motch (1979). Magnitudes are differential with respect to a local comparison star. Different symbols identify data from different nights. Continuous curve is a
theoretical fit for a mass ratio of 10 taken from Wickramasinghe and Whelan (1975).

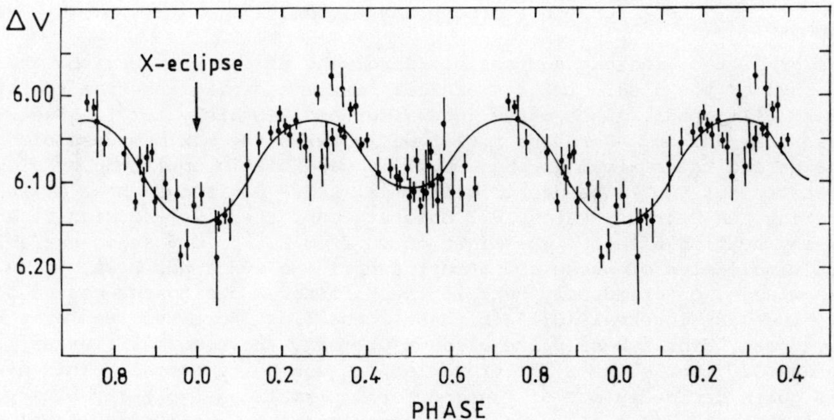

Figure 4: V-light curve of SMC X-1 (Sk 160). The curve drawn is the
average light curve. Observations were made in the Wal-
raven system. Taken from van Paradijs (1978). *(Reproduced
by kind permission from Astronomy and Astrophysics (Suppl.)
Vol. 29, p339, 1978).*

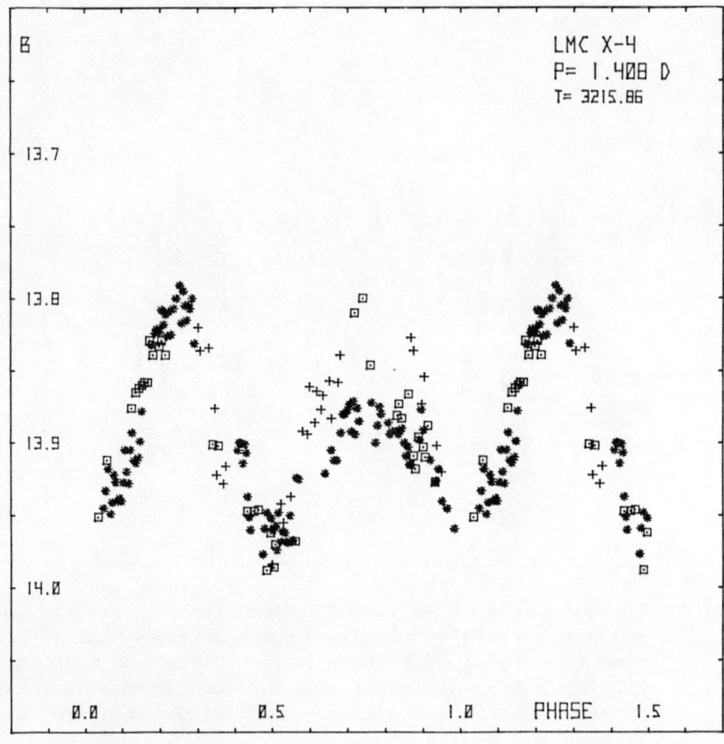

Figure 5a: B-light curve of LMC X-4 taken from Chevalier and Ilovaisky (1977). Different symbols identify three different one-week runs in February and March 1977.

as predicted, indicating possible shadowing of the primary from the X-rays by the disk. LMC X-4 exhibits extreme deviations from the standard ellipsoidal light curve (Chevalier and Ilovaisky 1977). See Figs. 5a and 5b. Here the amplitude is twice the maximum possible value and the maximum at phase 0.74 is variable in amplitude by almost a factor of two. The deeper minimum at phase 0.5 shows, that at least during the February-March 1977 observations, there was no significant X-ray heating effect. The light curve from 0.5 to 0.9 seems variable on time scales of weeks and a most surprising event has been, to my knowledge, observed only once in the U filter: during the course of a five-hour interval (0.15 in phase), the flux increased regularly by 0.25 mag. from a base value already 0.1 mag. above normal flux at phase 0.0, while the B filter variation was quite normal. This event is quite decidedly out of fashion for a normal massive X-ray binary and strongly hints at quite large contributions from 'third light' sources (disk, streams, etc.). The absence of any unusual behaviour in the B and V (not shown) filters favours Balmer continuum emission as the underlying mechanism. I note in passing that medium dispersion spectra taken simultaneously do not reveal any Balmer line emission as far down as H beta.

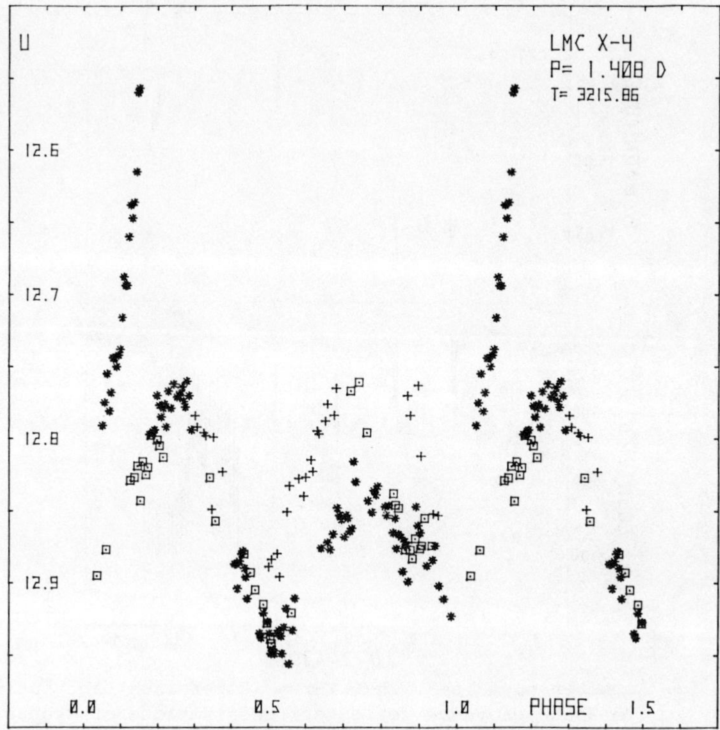

Figure 5b: U-light curve for the same observing runs.

Perhaps the basic and 'intrinsic' light curve for LMC X-4 is the
lower amplitude curve we see from 0.5 to 1.0, the rest (0.1 mag. at
least) being 'third light'.

Obviously much remains to be done to understand the light curves of
massive binaries and, in particular, for the 'extreme' systems
SMC X-1 and LMC X-4, to disentangle the variations due to geometrical
distortion of the primary from the other effects.

3.   HIGH L_x/L_{opt} SYSTEMS

Here we have chosen to discuss three sources, Cygnus X-2, Scorpius
X-1 and 4U 1626-67, the first two identified some twelve years ago
and still rather mysterious, the last being a late newcomer with
bright promise.

    a) Cygnus X-2

This relatively unknown source (L_x/L_{opt} = 450) has been under inten-
sive study recently and while the controversy on its true nature has
not yet been settled I want to present some examples of its optical

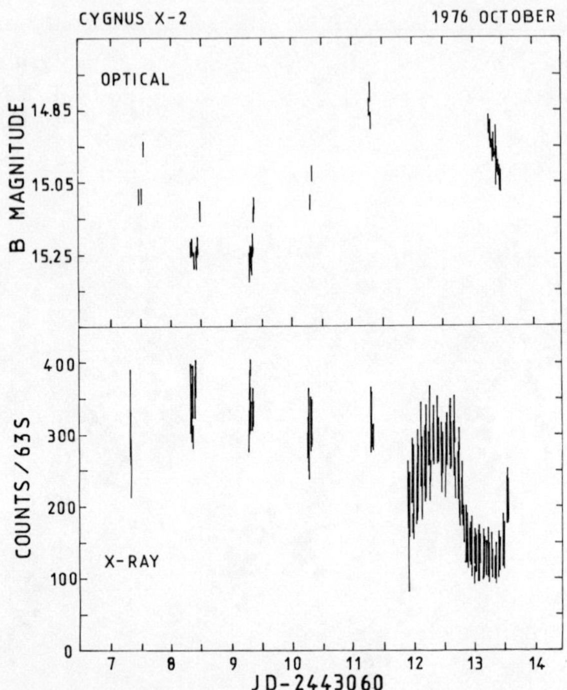

Figure 6: Simultaneous 2.5-7.5 keV X-ray (Copernicus) and optical (2m Haute Provence reflector) observations of Cygnus X-2 made in October 1976. Vertical bars are $\pm$ one sigma errors. Each individual 63 s integration is shown. Optical measurements are complete sky-comparison-star-comparison-sky sequences.

variability. A more comprehensive account will be published elsewhere. As this sytem exhibits a rather large L_x/L_{opt} ratio one would naively expect some evidence for an X-ray heating effect. However, Figure 6 shows the results of one week of simultaneous X-ray (Copernicus) and optical (Haute Provence 2m reflector) observations in October 1976, and it is clear that there is no simple, direct relationship between the fluxes in the two wavebands. In most instances there even seems to be an anticorrelation present (anisotropy, shadowing?). The optical flux from this system varies on time scales as short as a few hours (see Fig. 7) with an amplitude that can amount to as much as half of its full range of variability.

There is one element in favour of X-ray heating of some sort and that is the colour effect observed at all brightness levels (Fig. 8). A clear relation exists between the U-B colour index and the B magnitude, while no such large effect is seen in the B-V index. There has been a recent claim in the literature that the spectroscopic period of 9.843 days can be seen in the published optical photometry. Fig. 9 shows all the optical data obtained at Haute Provence from 1974 to 1978 folded modulo this period and phased

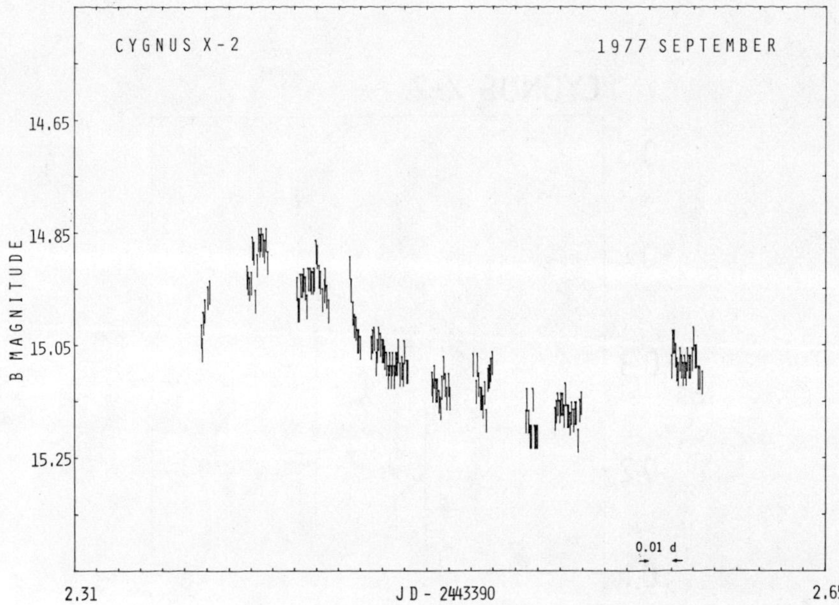

Figure 7:   One complete night (5^{h}30) of photometric observations of
            Cygnus X-2 made with the 1m Chiran telescope (OHP Station)
            on 6 September 1977. Each point shown is an independent
            measurement of the star. Weather conditions were extreme-
            ly stable.

according to the ephemeris of Cowley et al., (1979) with phase 0.0
corresponding to the X-ray source 'behind' the optical star. No clear
modulation is evident, but perhaps further accumulation of data may
resolve this point.

  b)   Scorpius X-1

This sytem (L_x/L_{opt} = 600) was shown in 1975 to be a 0.787 day
spectroscopic binary and the optical photometry exhibits a modulation with this period (one maximum and one minimum per cycle) only
when large amounts of data are averaged together. The nature of the
rapid variability (flaring and flickering) remains a puzzle.
Previous results had shown a marked tendency for the source to be
flare-active only when bright (Pelling, 1973; Canizares et al., 1975).
A recent series of correlated X-ray/optical studies has revealed
some new facts which merit presentation here. A detailed account
will appear in Monthly Notices (Ilovaisky et al., 1979).

Figure 10 shows one-half hour of simultaneous 1-13 keV X-ray data
(SAS-3) and optical (ESO 1m) obtained in March 1977 with a time
resolution (shown) of 5s when the source was near maximum brightness. During the first half of the run one sees highly correlated
rapid variations in both wavebands with a relative amplitude ratio

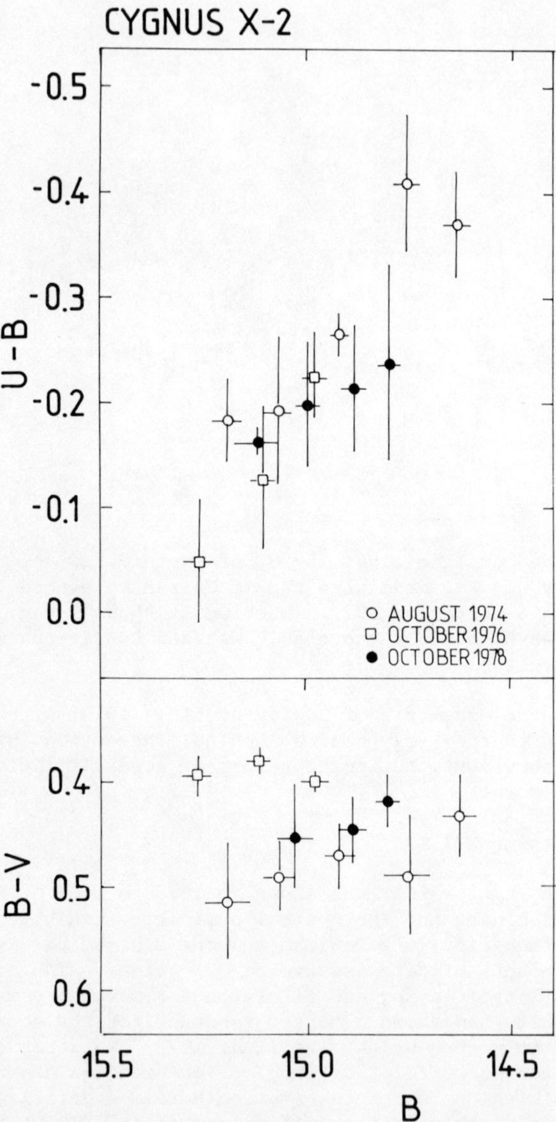

Figure 8: Average U–B and B–V indices for Cygnus X–2 based on three different observing runs in Haute Provence. August 1974: 80 cm reflector; October 1976: 2 m reflector; October 1978: 1 m Chiran reflector. Vertical bars indicate intrinsic scatter of individual measurements. Horizontal bars indicate magnitude intervals over which the means were made.

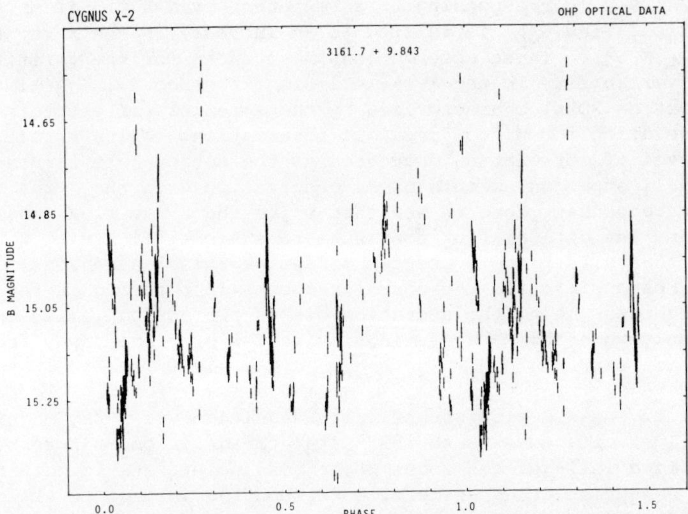

Figure 9: All Haute Provence Cyg X-2 photometry obtained from August 1974 to October 1978 folded with the 9.843 day period of Cowley et al. (1979). Phase zero corresponds to the X-ray source 'behind' the optical star. Error bars have a total length of 1 sigma.

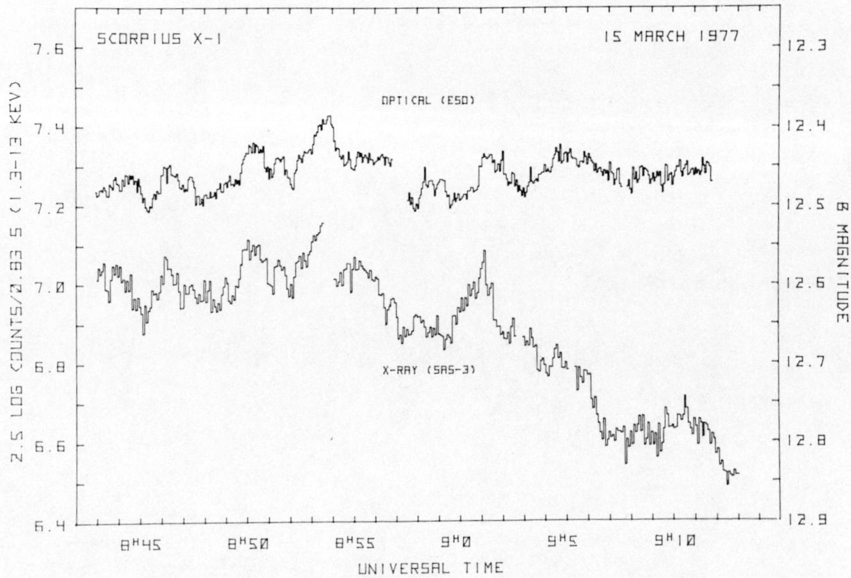

Figure 10: Simultaneous 1-13 keV X-ray (SAS-3) and optical (ESO 1m) observations of Sco X-1 made on 15 March 1977. Each optical point is an average of two 2s integrations. Each X-ray point is the average of 6 consecutive 0.83 s integrations (note the X-ray magnitude scale). Error bars for the X-ray data are 0.02 mag. and for the optical data 0.005 mag. *(Reproduced by permission of the Royal Astronomical Society, Ilovaisky et al., 1980, 191, 81).*

of $A_x/A_o \sim 2$ (corresponding to an exponent of $\alpha \sim 0.5$ if an increase in optical flux S_{opt} is related to an increase in the X-ray flux S_x by $S_{opt} \sim S_x^{\alpha}$). These observations have shown for the first time rapid variability in the X-ray emission from Sco X-1 (on time scales as short as 20 s) characterised by the same intensity-spectral hardness relation found from previous observations (White et al., 1976). The level of correlation decreases as the source gets fainter and Figure 11 shows an example of an observation with the X-ray source near quiescence. Here we see that while the X-ray flux remains constant the optical flux continues to exhibit flickering activity - proof that either there are two different emission regions; one being responsible for the correlated behaviour through X-ray heating, and the other being the accretion disk's own optical emission, detectable only when the X-ray luminosity is low or beamed away from the companion star.

Figure 12 shows a plot of optical B magnitude vs. X-ray magnitude for all simultaneous March 1977 data average in one-minute bins. There is a well-defined locus where most points are found with one clear exception at upper left, corresponding in part to the end of Figure 10, when the X-ray flux is near quiescent levels but the optical flux remains high. This may indicate shadowing effects (i.e. 'something' gets in the way of the observer), although I must say that we have never seen the X-ray source bright and the optical emission low, as in Cygnus X-2.

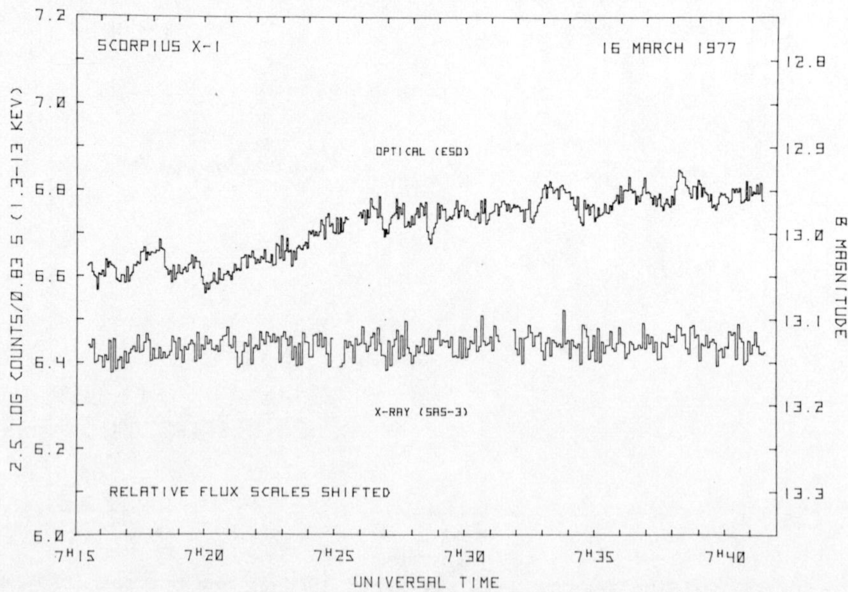

Figure 11: Sco X-1 observations on 16 March 1977. Same remarks as for Figure 10. Relative flux scales have been shifted with respect to the previous figure. *(Reproduced by permission of the Royal Astronomical Society, Ilovaisky et al., 1980, 191, 81.)*

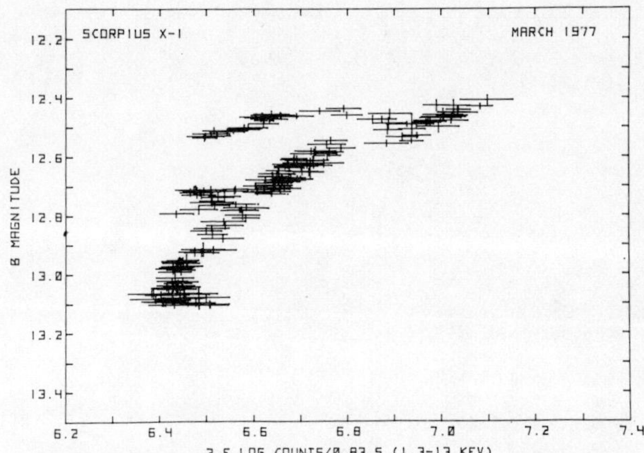

Figure 12: One minute average X-ray and optical measurements of Sco X-1 for all March 1977 runs. *(Reproduced by permission of the Royal Astronomical, Ilovaisky et al., 1980, 191, 81).*

Cross-correlation analysis of the highly correlated segments shows no really significant time delay between the X-ray and optical features although the cross-correlation function peaks systematically for positive lag values (optical following X-rays) around 10s. Note that the light travel time between the primary and the X-ray source in the Sco X-1 system is around 10s for reasonable component masses.

There is little doubt that in this data there are examples of rapid X-ray to optical reprocessing either in the illuminated hemisphere of the dwarf companion (which is, by the way, responsible for the optical modulation at 0.787 days) or near the X-ray source (disk).

c)   4U 1626-67

This object (L_x/L_{opt} = 600) may yet prove to be one of the most interesting of the recent identifications as the existence of regular pulsations in both X-rays and optical (Ilovaisky et al. 1978) can (will?) hopefully be used to probe the nature of this system. The remarkable absence of any Doppler effect in the X-ray pulsations (S. Rappaport, this meeting) has been interpreted in terms of a highly compact X-ray binary model (Joss and Rappaport, 1979) where the X-ray source is the most massive object in the system, the companion being a low-mass dwarf. IF the optical pulsations are produced by rapid X-ray to optical reprocessing in the heated atmosphere of the dwarf, then one would expect to detect its binary motion through an analysis of its arrival times, as is done with other X-ray pulsators.

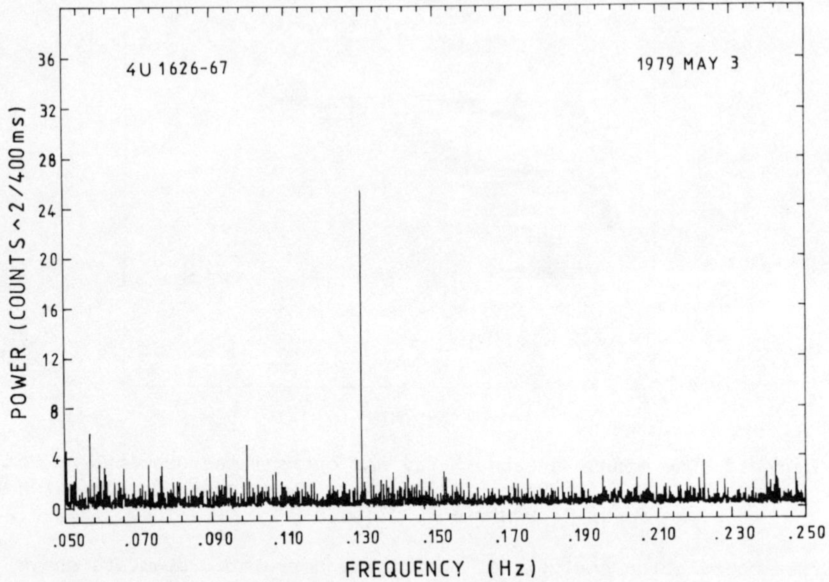

Figure 13: Power spectrum of 4U 1626-67 for the 3 h run of 3 May 1979. Peak power corresponds to a 5% modulation. Observations were made in white light with the ESO 3.6 m telescope using 400 ms integrations.

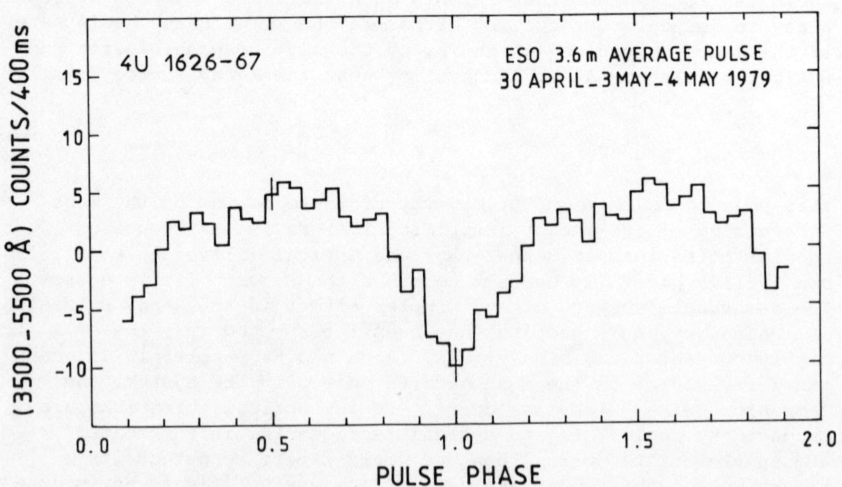

Figure 14: 24 channel grand-average optical pulse profile for 4U 1626-67 based on data obtained on three different nights in 1979. Error bars are ± 1 sigma.

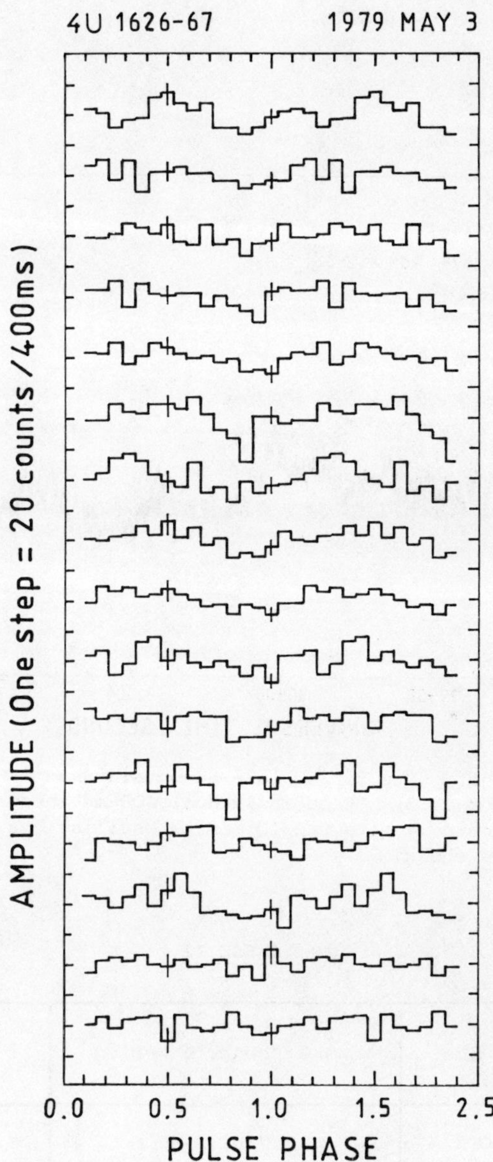

Figure 15:   The 4U 1626-67 data for 3 May 1979 grouped into 10-minute blocks and folded with the best apparent period into 16 phase channels. Error bars are ± 1 sigma. Only one marginally significant Doppler shift larger than 0.32 s is observed (fourth from bottom).

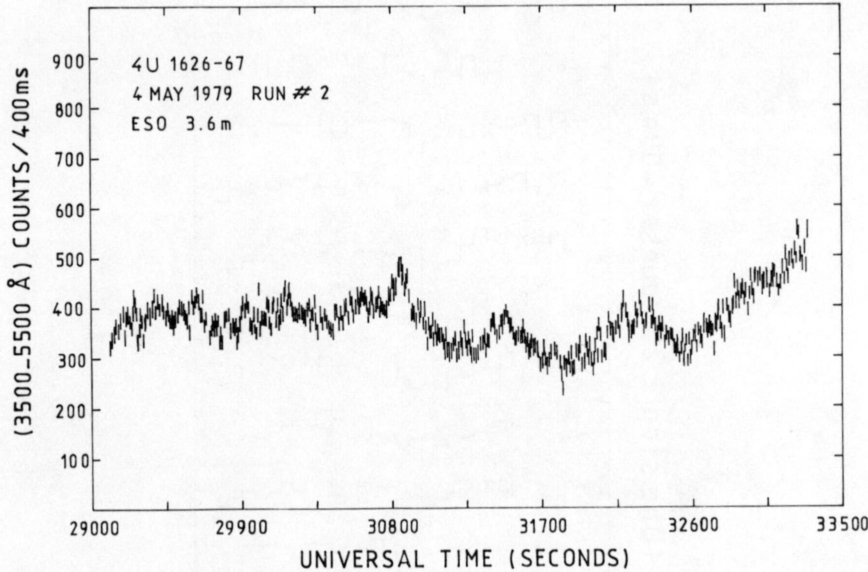

Figure 16: Light curve for 4U 1626-67 obtained on 4 May 1979 (UT
0805-0915). Each individual 400 ms integration is
plotted. Note the 10 minute oscillations. Sky levels
are shown.

TABLE II

UT Date	Heliocentric Period	1 sigma error
1979 April 30.3	7.6773 s	± 0.0007 s
May 1.3	7.6764	0.0070
May 3.3	7.6780	0.0010
May 4.3	7.6773	0.0008
May 5.3	7.6744	0.0020

Optical observations of this object are made difficult due to its faintness (B $\sim$ 19) and only telescopes in the 3-4m class can yield the necessary signal-to-noise ratio. I want to present here some very recent results (May 1979) on this star obtained with the ESO 3.6m telescope in Chile. Although observations were (again) hampered by bad weather, good timing data were obtained on three different nights. Figure 13 shows the power spectrum for the 3^h run of May 3. In Figure 14 we see the average pulse obtained by summing all data obtained on the three nights. Mean pulses from the individual nights do not differ significantly from each other. The profile exhibits a broad maximum and a narrow minimum much like the X-ray profile (Pravdo et al. 1979). The pulse periods obtained on five different nights agree with each other within the errors (see Table II). When combined with the SAS-3 results, we obtain a spin-up rate of -1.9 10^{-4} yr^{-1}, very close to that derived from the 1977 X-ray observations. Using the HEAO-1 period (Pravdo et al. 1979) for 1978 we obtain -7.7 10^{-5} yr^{-1}. The difference may be due to changes in the spin-up rate, as is the case for other pulsars.

Figure 15 shows the May 3 data grouped in 10 min bins and folded with the best period. Only one significant Doppler shift larger than 0.32 s is seen out of sixteen. This restricts radial velocity amplitudes to less than v sin i $\sim$ 50 km s^{-1} for periods between 2 to 3 h. Data obtained on May 4, when weather conditions were excellent, shows definite variations of 30% amplitude on time scales of 10 min. (Fig. 16). Full analysis of all available data is in progress and it is clear that much remains to be done to find out just where the optical pulsations in 4U 1626-67 come from.

REFERENCES

Bahcall, J.N., 1978. Ann. Rev. Astron. Astrophys., 16, 241.

Bradt, H.V., Doxsey, R.E., and Jernigan, J.G., 1979. X-ray Astronomy, Advances in Space Exploration, Vol. 3, page 3. (eds. W.A. Baity and L.E. Peterson), Pergamon Press, Osford.

Canizares, C.R., Clark, G.W., Li, F.K., Murthy, G.T., Bardas, D., Sprott, G.F., Spencer, J.H., Mook, D.E., Hiltner, W.A., Williams, W.L., Moffet, T.J., Grupsmith, G., van den Bout, P.A., Golson, J.C., Irving, C., Frohlich, A., and van Genderen, A.M., 1975. Astrophys. J., 197, 457.

Chevalier, C., and Ilovaisky, S.A., 1977. Astron. Astrophys., 59, L9.

Cowley, A.P., Crampton, D., and Hutchings, J.B., 1979. Astrophys. J., in press.

Crampton, D., Hutchings, J.B., and Cowley, A.P., 1978. Astrophys. J., 225, L63.

Hutchings, J.B., Cowley, A.P., Crampton, D., van Paradijs, J., and White, N.E., 1979. Astrophys. J., in press.

Ilovaisky, S.A., Moth, Ch., and Chevalier, C., 1978. Astron. Astrophys., 70, L19.

Ilovaisky, S.A., Chevalier, C., and Motch, Ch., 1979. Astron. Astrophys., 71, L17.

Ilovaisky, S.A., Chevalier, C., White, N.E., Mason, K.O., Sanford, P.W., Delvaille, J.P., and Schnopper, H.W., 1979, MNRAS, in press.

Joss, P.C., and Rappaport, S., 1979. Astron. Astrophys. 71, 217.
Margon, B., 1979a. in X-Ray Astronomy, Advances in Space Exploration, Vol. 3, page 67 (eds. W.A. Baity and L.E. Peterson), Pergamon Press, Oxford.
Margon, B., 1979b. in Annals of the N.Y. Academy of Sciences, in press.
Paradijs, J. van, 1977. Astr. Astrophys. (Suppl.), 29, 339.
Paradijs, J. van, Hammerschlag-Hensberge, G., and Zuiderwijk, E.J., 1978. Astron. Astrophys. (Suppl.), 31, 189.
Pelling, P.M., 1973. Astrophys. J., 185, 327.
Pravdo, S.H., White, N.E., Boldt, E.A., Holt, S.S., Serlemitsos, P.J., Swank, J.H., Szymkowiak, A.E., Tuohy, I., and Garmire, G., 1979. Astrophys. J., in press.
White, N.E., Mason, K.O., Sanford, P.W., Ilovaisky, S.A., and Chevalier, C., 1976. MNRAS, 176, 91.

Temporal studies of galactic X-ray sources

M.G. Watson

X-Ray Astronomy Group,
University of Leicester.

1.   Introduction

2.   Observational Constraints

3.   Types of Variability:
      a) Periodic Variability
      b) Chaotic Variability
            I)   Transients and Recurring Transients
            II)   Other Flaring Sources
            III)   High-Low States

4.   Mechanisms for Producing long-term variability:
      a) Eccentric Orbit Binaries
      b) Variable Roche-Lobe Overflow
      c) Variable Stellar Wind

1.   INTRODUCTION

Galactic X-ray sources, in contrast with most astronomical objects, display a remarkable range of variability on timescales from milliseconds to years. Studies of such variability have played a crucial role, both in leading to the discovery of entirely new classes of object (e.g. X-ray transients) and in giving important clues to the nature of the systems involved (e.g. eclipsing binaries). In this paper we will concentrate on the medium to long timescale variability (days to years) observed in galactic X-ray sources - a topic which has received relatively little attention. The fascinating shorter timescale phenomena is covered by other contributions in these proceedings (e.g. the papers by Bradt, Rappaport and Lewin).

2.   OBSERVATIONAL CONSTRAINTS

Limits to what can be observed arise from both the modest sensitivity of current instrumentation, and through the constraints on the operation of X-ray satellites. The minimum time resolution normally available, even for the brightest sources, is $\sim$1 ms, a limit set by

photon statistics with detectors of moderate area. A time
resolution in the range 10 ms - 1 s is typical for the current
generation of X-ray astronomy satellites and observations of
variability on timescales from seconds to a few days are usually
possible with these modest instruments.* A large proportion of the
variability studies, which have been published, concern observations
made in this time range. Observations lasting more than a few days
are often difficult to make, both because of a reluctance to devote
extensive periods to monitoring individual objects, and because of
the constraints on keeping the satellite attitude steady. The Ariel
5 satellite was unusual in this respect in that continuous
observations of up to 30 days duration were possible. On the
longest timescales (months-years) the only continuous monitoring
possible is with instruments such as the Ariel 5 All Sky Monitor
(ASM) which had a very large field of view ($\sim$90% of celestial
sphere), but a sensitivity only to the stronger sources with
strengths greater than 100 Uhuru flux units.

Although we have no reason to suppose that the strong galactic
sources are special in any other way, one must be rather cautious
in inferring general properties on the basis of a rather small
sample. We are currently (1980) compiling a catalogue of the light
curves of weaker galactic sources from the Ariel 5 Sky Survey
Instrument (SSI) during the life of the Ariel 5 satellite. This
analysis will help put our knowledge of the temporal variability of
galactic sources on a firmer foundation.

An additional problem exists in the interpretation of the X-ray
observations. Since most X-ray measurements are broad-band, often
with only crude spectra (or none at all), apparent variability can
arise from:

    a)   true variations in the X-ray luminosity;

    b)   variable spectral shape (e.g. due to changing conditions
        in the emitting region);

    c)   variable line of sight absorption (e.g. due to cold matter
        within the system).

Examples of variability due to all these effects have been observed.

3.   TYPES OF VARIABILITY

It is widely accepted that the majority of galactic X-ray sources
(with the exception of supernova remnants and their associated
pulsars) are likely to be binary systems containing a compact object
which is accreting material from its non-degenerate companion. In

*Note that problems often arise with the interruption of observa-
tions once per satellite orbit, which is around 1½ hours for most of
the X-ray astronomy satellites (cf proposed 97 min. period in
4U 1700-37, Matilsky et al., 1978).

this section I discuss both periodic variability, usually associated with modulation of the X-ray flux by the binary motion, and chaotic variability of various types. As emphasised in section 2, our knowledge of long term variability in galactic sources is highly selective, and is also biased towards the study of the more dramatic behaviour e.g. transient outbursts. Nevertheless, the majority of galactic X-ray sources are variable (e.g. Forman et al. 1976, 1978) and one might hope to gain some insight into the mechanisms which produce chaotic variability even by studying the more extreme examples known.

a) Periodic variability

On the timescales discussed here, periodic modulation of the X-ray flux is most commonly associated with the orbital motion of the X-ray source. The binary periods are known for about 15 sources (e.g. Bradt et al. 1979), most being in the range 0.1 - 20 days. Six sources exhibit X-ray eclipses, and several others have features interpreted as partial eclipses (e.g. Cyg X-3, Parsignault et al. 1977). The binary nature of the source is by no means always reflected in its X-ray light curve, and is often revealed only by Doppler studies of the X-ray pulse period, or optical observations (e.g. 4U 0115+63, Rappaport et al. 1978; Sco X-1, Gottlieb et al. 1975). Other sources show X-ray behaviour which is clearly periodic, but requires a more complex interpretation than is demanded for an eclipsing binary. An example is Cir X-1, whose X-ray light curve is shown in Figure 1. These observations display repetitive X-ray flares with a period of 16.6 days which are not yet fully understood but several models have been proposed (e.g. Haynes et al. 1979).

Extensive studies of a number of sources have revealed periodicities which are clearly <u>not</u> orbital in origin. The best known example is the 35-day modulation of Her X-1 (Giacconi et al. 1973) which almost certainly reflects the precession period of the obscuring disc surrounding the neutron star.   Other possible periodicities have been reported on the basis of the Ariel 5 All Sky Monitor (ASM) extensive observations of several bright galactic sources. These include periods of 11.2 d in Cyg X-2, 16.8 d or 33.0 d in Cyg X-3 (Holt et al. 1979b) and 43.0 d in Cen X-3 (Holt et al. 1979a). The nature and indeed reality of these periods is still uncertain (in particular the 9.8 d spectroscopic period for Cyg X-2 proposed by Cowley et al. 1979 is difficult to reconcile with the X-ray modulation).

134

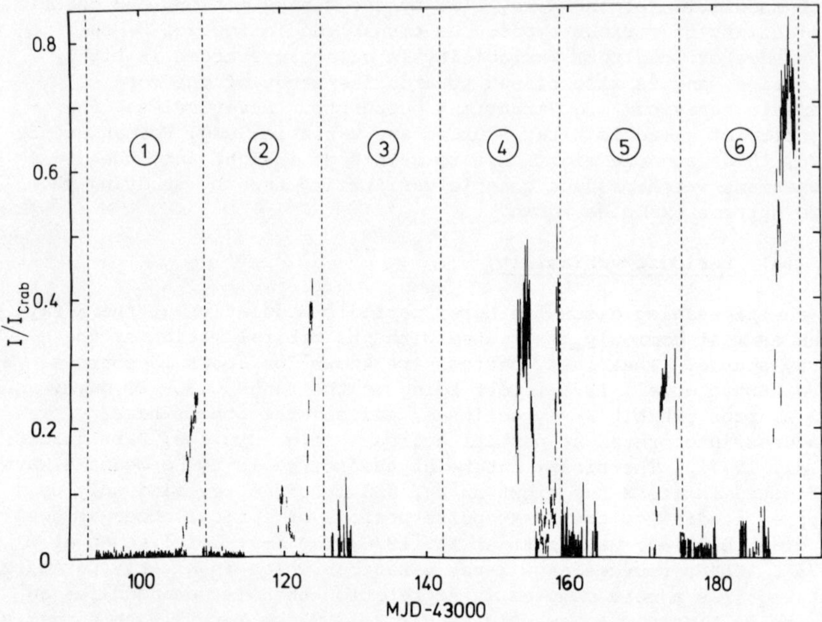

Figure 1:   The light curve of Cir X-1 based on observations with the
            University of Leicester's Sky Survey Instrument (SSI)
            made between 1976 November and 1977 February. Each data
            point is a single-orbit ($\sim$100 min) observation. The
            dashed lines indicate the predicted times of the X-ray
            transitions according to the ephemeris of Kaluzienski and
            Holt (1977). The measurements are normalised with the
            instrument's response to the Crab Nebula.

b)   Chaotic Variability

(i)   Transients and "recurred transients"

The most dramatic variability displayed by galactic X-ray sources is
the impulsive outburst which is characeristic of an X-ray transient.
Fig. 2 shows the complete X-ray light curve for the outburst of
A0620-00 (Mon X-1), to date the brightest X-ray transient. The
X-ray light curve shows the fast rise to peak intensity (timescale
$\sim$days) and a slow, reasonably smooth, decay (timescale $\sim$ months) from
maximum which is typical of "classical" X-ray transients. The list
of such sources includes A0620-00, A1524-61 and H1705-25, all of
which have generally similar light curves, similar maximum
luminosities, and indeed similar optical properties (Watson 1979).
In addition, no subsequent outburst had been recorded from any of
these sources (although it is known that A0620-00 underwent a
similar optical outburst in 1917).

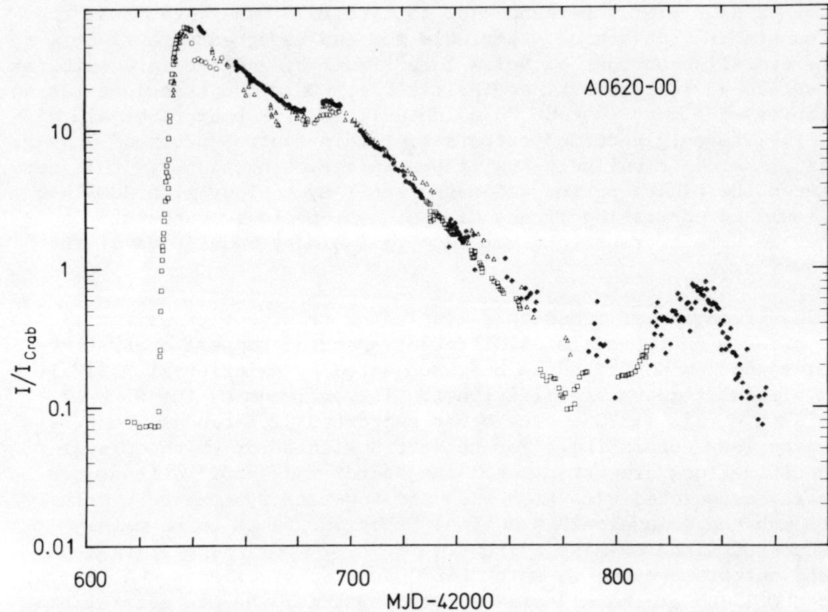

Figure 2: Composite X-ray light curve of the outburst of A0620-00
compiled from observations by the Ariel 5, All Sky
Monitor (ASM), Rotation Modulation Collimator (RMC)
instrument and the Sky Survey Instrument (SSI), together
with SAS-3 coverage. The published light curves have
been digitised at intervals of a few days. No spectral
corrections have been made, and the data has been
arbitrarily normalised to agree between MJD 42700 to
MJD 42740.

According to the working definition of an X-ray transient
(Kaluzienski 1977a, Cominsky et al. 1978) the total number of
recorded transients is ∿25. This definition only requires that the
source shows an outburst by a factor of >10 for an interval short
compared with the time for which it was observed. Thus many show
few similarities with the behaviour of the "classical" X-ray transient. In particular, there are several sources which show
repetitive outbursts (the so-called "recurrent transients") at
intervals ranging from 100 days to several years. Figure 3 shows
the composite X-ray light curve of Aql X-1, one of the better known
recurrent transients. The light curve of an individual outburst
(e.g. Kaluzienski et al. 1977a) does, in fact, show the shape
characteristic of transients, but this is by no means true for
other sources of this type (e.g. 4U1630-47).

Periodicities have been suggested for the intervals between outbursts for Aql X-1 (435 d; Kaluzienski 1977b) and 4U1630-47 (615 d;

Jones et al. 1976). Evidence (up to late 1979) indicates that the
outbursts are not strictly periodic for any recurrent transient
(cf. overall behaviour of Aql X-1 in Figure 3, and the late outburst
of 4U1630-47 (Kaluzienski and Holt 1977). Only one transient has an
established binary period (24 d in 4U0115+63), Rappaport et al.
1978). The only recorded outbursts of this source occurred ∿7 years
apart, making it unlikely that there is any connection, in this case,
between the binary period and outbursts. By analogy with 4U0115+63,
it would be surprising if any of the quasi-periodic outbursts in
other recurrent transients were due to a binary modulation of the
X-ray flux.

A classification of transients (including recurrent transients) into
two classes on the basis of different spectral temperatures, timescales and luminosities has been suggested by Kaluzienski (1977a).
The bimodal temperature distribution is confirmed by the work of
Cominsky et al. (1978). The other suggested differences are
perhaps less convincing. For those transients for which optical
identifications are established the "hard" and "soft" classes are
clearly associated with high-mass and low-mass systems respectively.
Although the "classical" transient behaviour seems to be mainly
associated with low-mass systems (e.g. AO620-00), repetitive transient outbursts occur in both "hard" and "soft" classes. AO535+26,
and 4U0115+63 are both "hard" pulsing systems, whereas Aql X-1 has
a soft spectrum and is identified with a low-mass binary containing
a K dwarf.

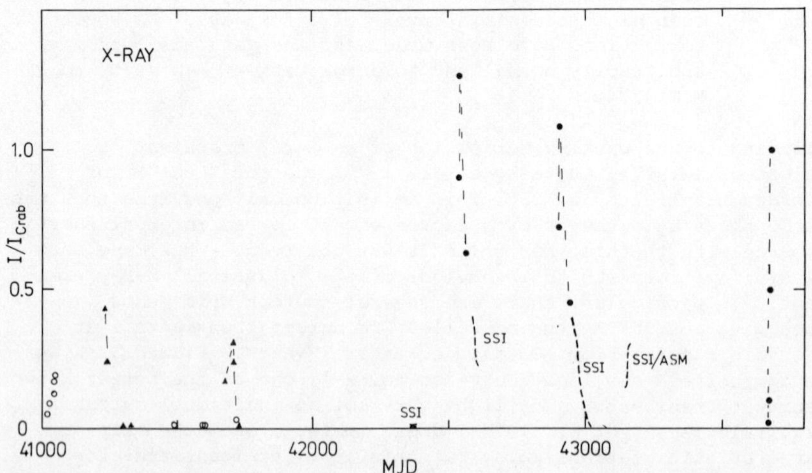

Figure 3: Composite X-ray light curve for the recorded outbursts
of Aql X-1 between 1971 and 1978 based on observations
made by Uhuru, OSO-7 and Ariel 5 ASM and SSI. (See
Watson 1979 for further details).

(ii) <u>Other flaring sources</u>

The behaviour displayed by recurrent transients is also seen (in a less dramatic form) in the variability of several other sources which show repetitive flares lasting a few days. Clearly there can be no rigid distinction between recurrent transients and flaring sources; the division made, simply reflects the way such sources are usually designated. Here we discuss two sources whose behaviour has been clarified by observations made with Ariel 5 SSI (for details of the experiment see Villa et al. 1976.; Cooke et al. 1978). <u>4U1223-62 (GX301-0)</u> is a hard X-ray pulsator with a 699 s period identified with Wray 977, a B1.5 Ia supergiant (Bradt et al. 1979). From both optical and X-ray studies binary periods of 23 d or 41 d have been suggested (e.g. Hammerschlag - Hensberge et al. 1976; White et al. 1978a). Further observations and analysis reported by Rappaport (these proceedings) indicate the true period may be 35 d with an appreciable orbital eccentricity (e = 0.44).

Previous X-ray measurements have demonstrated the large degree of variability of 4U1223-62, including brief flares to an intensity of ∿1000 U.F.U. (Bradt et al. 1979). Figure 4 shows part of the X-ray light curve of 4U1223-62 derived from SSI observations. The main features of the light curve are several flares (lasting 5 - 10 days) reaching the peak intensity of 100 - 300 U.F.U., implying overall variability by a factor of >10. Significant variability is also apparent within each flaring episode. Recurrence of the flares is on a timescale of 20 - 50 days, but the flares do not appear to be strictly periodic at the 35 d, or any other period. Further analysis of the SSI observations is in progress which will clarify the long term behaviour of this source.

<u>4U1145-61</u> is another hard, pulsating source with a 297 s period. Its optical counterpart, Hen 715, is classified as BIVne (Bradt et al. 1979). No binary period has yet been established. Observations with the Ariel V Proportional Counter Spectrometer in 1977 Sept. and Dec. indicated that the source had two distinct pulse periods, p = 292 s as well as 297 s, whose relative amplitude varied between the two observations (White et al. 1978b). This puzzling result is perhaps most easily interpreted if two X-ray pulsars were in the field of view.

Figure 5 shows the X-ray light curve of 4U1145-61 based on observations made over four years. Four widely separated flares are clearly present, lasting ∿10 days in each case, the last of which coincides with the ∿1000 U.F.U. outburst reported from SAS-3 by Jernigan et al. (1978). Analysis of these observations (performed by Martin Ricketts, University of Leicester), has shown that the timing of the flares is entirely consistent with a period of 188±5 days. Because of the length of this period and the small duty cycle of the SSI coverage, it is impossible to rule out the possibility that this periodicity is fortuitous. Nevertheless, the data folded modulo 188 d (shown in Fig. 6) show a rather smooth profile as might be expected if the periodicity were a real feature of the source.

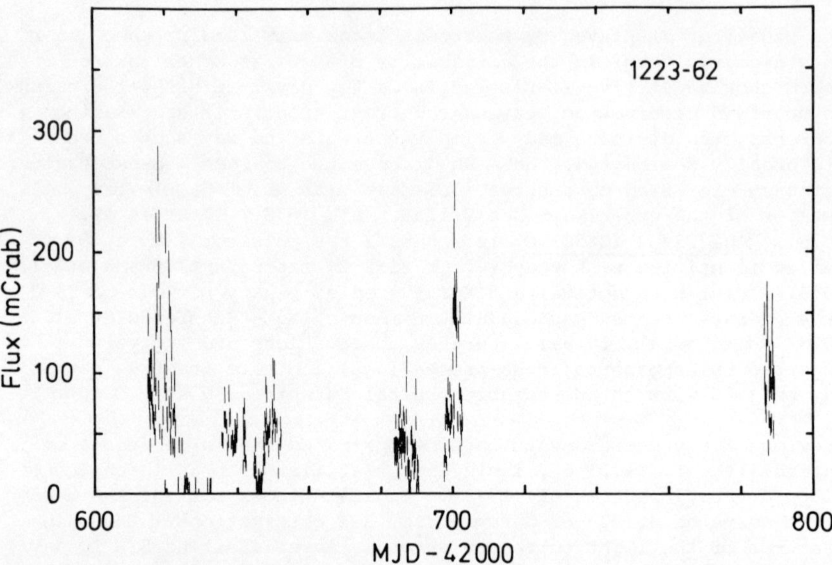

Figure 4: Part of the SSI light curve for 4U 1223-62 (GX301-0).
Each data point is a single-orbit ($\sim$ 100 min.) observation.

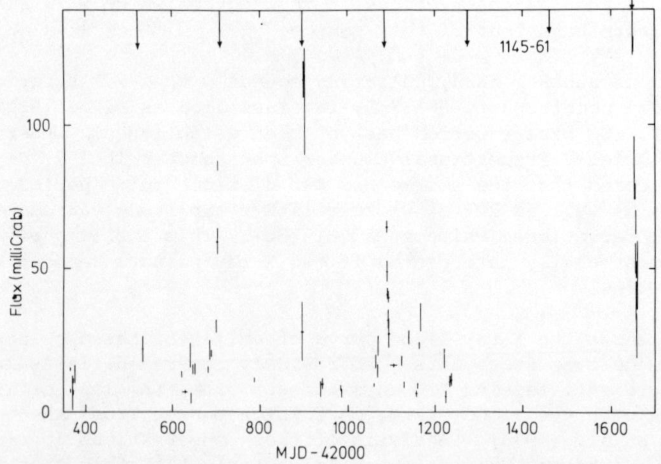

Figure 5: The SSI light curve for 4U1145-61 including all reliable
observations made from launch (1974 October) to 1978
June. Each data point is an average over an observation
lasting typically $\sim$ 3 days (the duration is indicated by
length of horizontal bar). The intensity scale is I/I_{crab}
x 1000.

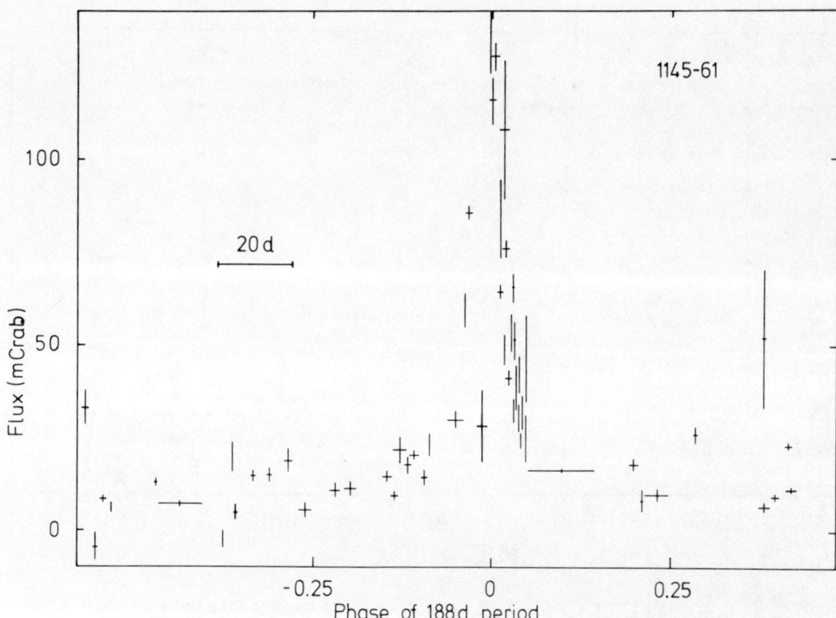

Figure 6:   The SSI observations of 4U1145-61 (shown in Fig. 5)
            folded modulo 188 days. Phase zero has arbitrarily been
            centred near the time of peak intensity. As in Fig. 5,
            the horizontal bars indicate the length of each observa-
            tion.

The interpretation of this source is further complicated by the
positional data from the SSI. Each data point shown in Fig. 5 gives
essentially an independent measurement of the source position (cf.
Cooke et al. 1978). Combining all these measurements gives the SSI
error box for the source. When this is done for 4U1145-61 it is
clear that a number of measurements are inconsistent with the
accurate position of the X-ray source (or Hen 715). The discrepant
measurements all correspond to lower intensity values for the source,
and in fact the situation can be most simply resolved by dividing the
observations into two sets corresponding to I >40 U.F.U. and I >40
U.F.U. respectively.[*] Each set then gives a self-consistent errorbox. The errorbox for observations with I > 40 U.F.U. (i.e. the
flaring component) coincides with the source 4U-145-61/Hen 715; the
errorbox for observations with I <40 U.F.U. (the steady component)
lies ∿30 arcmin NW of Hen 715. (Further details of this analysis
is given in Ricketts et al. (1979).)

This analysis raises the interesting possibility that there are two
distinct X-ray sources in this region. If this is the case then
the new source is reasonably steady, and 4U1145-61 displays the

* Note that such discrepancies have not been found in other SSI
positional work, and there is no known systematic shift in position
correlated with source intensity.

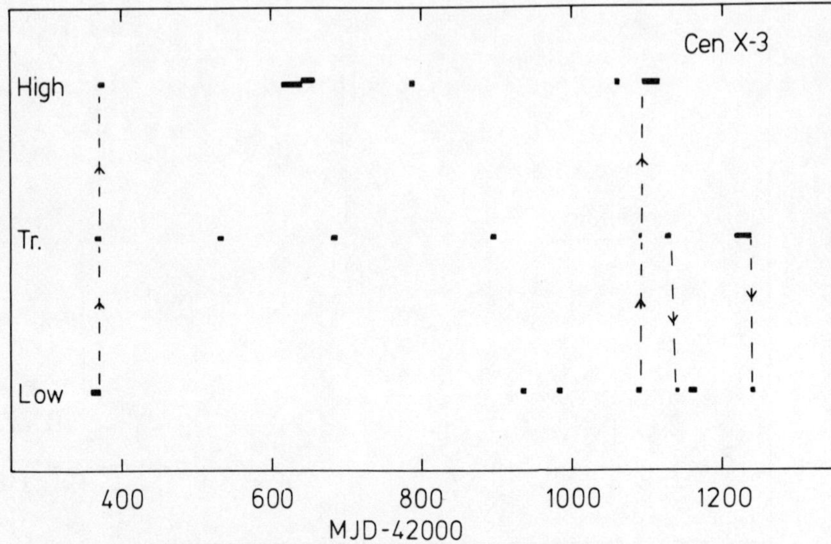

Figure 7: The pattern of high and low intensity states in Cen X-3
derived from SSI observations spanning 1974 November to
1977 March. The length of each block represents the
duration of the observation. The dashed lines indicate
the direction of the transition where the observations
were effectively continuous. Note that "Tr" indicates a
transitional intensity state, intermediate between low and
high states. In some cases there is some ambiguity about
observations labelled "Tr".

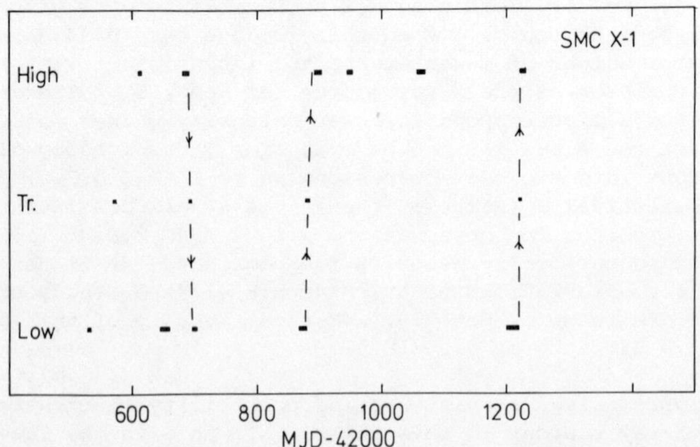

Figure 8: The pattern of high and low intensity states in SMC X-1
derived from SSI observations made between 1975 February
and 1977 February. Details as for Fig. 7.

periodic X-ray flares. We also have a natural, if somewhat unlikely,
explanation for the two pulse periods detected. Although it may seem
improbable that two X-ray pulsars with similar periods should lie so
close together in the sky, it should be noted that this part of the
galactic plane (290° $<l^{II}$ $<305^{\circ}$) contains 5 (possibly 6) X-ray
pulsars (A1118-615, Cen X-3, 4U1145-61, 4U1223-62, 4U1258-61 and
possibly A1239-599).

If the apparent periodicity seen in the SSI observations of 4U1145-61
is confirmed then the most attractive explanation would be that it
represents an extremely long binary period in a system with large
eccentricity, the flares corresponding to greatly enhanced accretion
near periastron passage.

iii)　High-low states

Several well studied galactic X-ray sources exhibit a rather more
subtle variability involving two distinct intensity states. A
typical pattern of behaviour is for the source to persist in one
intensity state for an interval ranging from months to years, with
the transition to the other state occurring on a much shorter timescale ($\sim$days). Such variability is well documented for Cyg X-1 (e.g.
Tananbaum et al. 1972, Holt et al 1976) where the transition to the
high intensity state is accompanied by changes in the X-ray spectrum
and the radio emission.

In Cen X-3 and SMC X-1 the X-ray light curve is also characterised
by high and low intensity states with the source spending approximately equal times in each state. This is illustrated in Figures
7 and 8 which show the SSI observations of Cen X-3 and SMC X-1 for
the period 1974-1978. The pattern of high-low states for Cen X-3 is
similar to that observed by Uhuru (Schreier et al. 1976) with the
source spending $\sim$30 days in each state. For SMC X-1 variability is
clearly very similar despite the small number of SSI observations.
Chester (1978) has suggested that the low states in Cen X-3 exhibit
a 27 d periodicity. This modulation is not apparent in the extensive
ASM observations (Holt et al. 1979a), although a modulation at 43 d
is suggested by their data. Neither periodicity is evident in the
SSI light curve shown in Figure 7, this does not, of course,
preclude these modulations being present at a low amplitude.

Both Cen X-3 and SMC X-1 have early-type, supergiant counterparts.
All the evidence suggests that the counterpart of Cir X-1 is a
similar star (e.g. Whelan et al. 1977). It is interesting to note
that the long term behaviour of Cir X-1 is also dominated by
extended low states lasting, in this case, 200 - 300 days (Kaluzienski
et al. 1976).

4.　MECHANISMS FOR PRODUCING LONG-TERM VARIABILITY

On the shortest timescales, the variability seen in galactic X-ray sources is clearly characteristic of the size of the emitting region near the compact object and the inner parts of the accretion flow surrounding it (e.g. Ostriker 1977). Thus studies of such variability are an important diagnostic of the processes taking place close to the compact object. In contrast, the variability on timescales from days to months is unlikely to be directly associated with the X-ray emitting region and is mainly linked with secular variations in the mass transfer from the mass-losing star, or with other phenomena occurring on the scale of the orbital separation. A variety of mechanisms have been discussed in this context. They include:

a) <u>Eccentric orbit binaries</u> Models involving systems with significant orbital eccentricity, in which outbursts occur as a result of enhanced accretion rates near periastron, have been considered for both transients and recurrent transients (e.g. Avni et al. 1976). Such models are probably only viable for sources which show reliably periodic behaviour over several cycles (e.g. Cir X-1), thus ruling out most recurrent transients. To explain the small duty cycle for X-ray emission in transients, any model of this kind demands extreme orbital eccentricities (e $\sim$ 0.9) which are still regarded as implausible even though several X-ray binaries with non-circular orbits are now known.

b) <u>Variable Roche-lobe overflow</u>. Mass transfer via Roche-lobe overflow is believed to be the dominant mechanism in many low-mass X-ray binaries (e.g. Cowley, this volume). By analogy with dwarf novae, which show quasi-periodic optical outbursts believed to be due to large changes in the mass transfer rate from the Roche-lobe filling companion to the white dwarf (e.g. Bath 1976), several authors have discussed "X-ray dwarf novae" in which the presence of a neutron star, instead of a white dwarf, gives rise to similar variability at X-ray wavelengths. Such models are particularly attractive in explaining the behaviour of both "classical" X-ray transients (like A0620-00) and sources like Aql X-1 which have low-mass optical counterparts, suggesting some similarity to dwarf novae systems (e.g. Wu et al. 1976). It is not clear, however, how far one can take this analogy when discussing the <u>detailed</u> behaviour of low-mass X-ray binary systems.

c) <u>Variable Stellar Wind</u>. High-mass X-ray binaries are believed to be powered by accretion from the strong stellar wind associated with early-type stars (e.g. Davidson and Ostriker 1973, but see Lamer at al. 1976, Petterson 1978 for problems with this interpretation). Thus variability of the stellar wind provides a trivial explanation for variability in sources of this type. The situation may actually be more complex. For example, models for Cen X-3 (Schreier et al. 1976; Hatchett and McCray 1977), involve the obscuration of the X-ray source (when the wind density increases) to explain the low intensity states. In sources like Cen X-3, models of this type might be extended by invoking a feedback mechanism to explain the cycle between low and high states. The X-ray luminosity would, in part, determine the mass transfer rate through heating and ionisation effects in the stellar atmosphere.

The presence of a second source near 4U1145-61, as discussed in section 3(b), has recently been confirmed by observations with the Einstein Observatory (Lamb et al. 1980). The new source is located approximately 20 arcmin north of 4U1145-61 at a position consistent with the SSI errorbox. Additional observations discussed by White et al. (1980) confirm that both sources exhibit X-ray pulsations with periods of 297s and 292s (the new source has the shorter period). White et al. also comment that the period variations observed are consistent with those expected for a system with an orbital period of 100-200 days, as is suggested by the SSI results. (Proof Note: 24 July 1980)

It is a pleasure to thank Martin Ricketts for allowing me to use some of the results on 4U1145-61, and Bob Warwick for reading the manuscript.

REFERENCES

Avni, Y., Fabian, A.C., and Pringle, J.E., 1976. MNRAS, 175, 297.
Bath, G.T., 1976. In "Structure and evolution of close binary systems",
     IAU Symp. No. 73, p. 173.
Bradt, H.V., Doxsey, R.E., and Jernigan, J.G., 1979. Advances in
     Space Exploration, vol. 3.
Chester, T.J., 1978. Ap. J. (Letters), 219, L77.
Cominsky, L., Jones, C., Forman, W., and Tananbaum, H., 1978. Ap.
     J., 224, 46.
Cooke, B.A., et al., 1978. MNRAS, 182, 455.
Cowley, A.P., Crampton, D., and Hutchings, J.B., preprint. 1979.
Davidson, K., and Ostriker, J.P., 1973. Ap. J., 179, 585.
Forman, W., Jones, C., and Tananbaum, H., 1976. Ap. J., 208, 849.
Forman, W., Jones, C., Cominsky, L., et al., 1978. Ap. J. (Suppl.),
     38, 357.
Giacconi, R., Gursky, H., Kellogg, E., Levinson, R., et al., 1973.
     Ap. J., 184, 227.
Gottlieb, E.W., Wright, E.L., and Liller, W., 1975. Ap. J. (Letters),
     195, L33.
Hammerschlag-Hensberge, G., Zuiderwijk, E.J., van den Heuvel, E.P.J.,
     and Hensberge, H., 1976. Astr. Astrophys., 49, 321.
Hatchett, S., and McCray, R., 1977. Ap. J., 211, 552.
Haynes, R.F., Jauncey, D.L., Lerche, I., and Murdin, P.G., 1979.
     Aust. J. Phys., 32, 43.
Holt, S.S., Kaluzienski, L.J., Boldt, E.A., and Serlemitsos, P.J.,
     1976. Nature, 261, 213.
Holt, S.S., Kaluzienski, L.J., Boldt, E.A., and Serlemitsos, P.J.,
     1979a. Ap. J., 227, 563.
Holt, S.S., Kaluzienski, L.J., Boldt, E.A., and Serlemitsos, P.J.,
     1979b. Ap. J., 233, 344.
Jernigan, J.G., Bradt, H., van Paradijs, J., and Rappaport, S., 1978.
     IAU Circ., 3225.
Jones, C., Forman, W., Tananbaum, H., and Turner, M.J.L., 1976. Ap. J.
     (Letters), 210, L9.

Kaluzienski, L.J., 1977a. Ph.D. thesis, University of Maryland.
Kaluzienski, L.J., and Holt, S.S., 1977b. IAU Circ. No. 3104.
Kaluzienski, L.J., Holt, S.S., Boldt, E.A., Serlemitsos, P.J., 1976.
     Ap. J., (Letters), 208, L71.
Kaluzienski, L.J., Holt, S.S., Boldt, E.A., and Serlemitsos, P.J.,
     1977. Nature, 265, 606.
Lamb, R.C., Markert, T.M., Hartman, R.E., Thompson, D.J., and Bignami,
     G.F., 1980. Ap. J., in press.
Lamers, H.J.G.L.M., et al., 1976. Astr. Astrophys., 49, 327.
Matilsky, T., La Sala, J., and Jessen, J., 1978. Ap. J. (Letters),
     224, L119.
Ostriker, J.P., 1977. Ann. N.Y. Acad. Sci., 302, 229.
Parsignault, D.R., Grindlay, J., Gursky, H., and Tucker, W., 1977.
     Ap. J., 218, 232.
Petterson, J.A., 1978. Ap. J., 224, 625.
Rappaport, S.A., Clark, G.W., Cominsky, L., Joss, P.C., and Li, F.,
     1978. Ap. J. (Letters), 224, L1.
Schreier, E.J., Swartz, K., Giacconi, R., Fabbiano, G., and Morin, J.,
     1976. Ap. J., 204, 539.
Tananbaum, H., Gursky, H., Kellogg, E., et al., 1972. Ap. J. (Letters),
     177, L5.
Villa, G., Page, C.G., Turner, M.J.L., Cooke, B.A., Rickets, M.J., and
     Pounds, K.A., 1976. MNRAS, 176, 609.
Watson, M.G., 1979. Ph.D. Thesis, University of Leicester.
Whelan, J.A.J., Mayo, S.K., et al., 1977. MNRAS, 181, 259.
White, N.E., Mason, K.O., and Sanford, P.W., 1978a. MNRAS, 184, 678.
White, N.E., Parkes, G.E., Sanford, P.W., et al., 1978b. Nature,
     274, 664.
White, N.E., Pravdo, S.H., Becker, R.H., Boldt, E.A., Holt, S.S., and
     Serlemitsos, P.J., 1980. Ap. J., in press.
Wu, C-C., Aalders, J.W.G., van Duinen, R.J., Kester, D., and Wesselius,
     P.R., 1976. Astr. Astrophys., 50, 445.

Ariel V observations
of Hercules X-1

Arvind N. Parmar

Mullard Space Science Laboratory,
University College London,
Holmbury St. Mary,
Dorking, Surrey,
U.K.

1.   Abstract

2.   Introduction

3.   Results

1.   ABSTRACT

Detailed spectra obtained with the Ariel 5 satellite for the turn-on
and off of Hercules X-1 from the extended low are presented. We find
that, except for anomalous and pre-eclipse dips, the hydrogen column
density is generally $\sim 10^{22}$ atoms cm^{-2} and the spectral normalisation
increases approximately linearly with time. A light curve of part of
the secondary on-state is also presented.

2.   INTRODUCTION

Hercules X-1 is a pulsing eclipsing binary X-ray source which exhibits a wide range of spectral and temporal phenomena, of which the
least understood is the 35 day cycle (Tananbaum et al., 1972;
Giacconi et al., 1973).

We have used the proportional counter spectrometer (Experiment C) on
the Ariel V satellite with high temporal or spectral resolution to
observe the source for a total of 15 days within one 35 day cycle.
In spectral mode data are accumulated for a satellite orbit (100 minutes) into 31 channels between 2 - 16 keV, the background being estimated from the effects of spin modulation caused by the offset between the experiment and satellite spin axes. In pulsar mode the 5 -
9 keV data are folded into 16 bins modulo a pre-set period and integrated for 32s. The next set of data then fills the next 16 bins,
128 sets being available in all.

3.   RESULTS

The pre-turn on 2 - 16 keV spectrum integrated between days 32.6 -
33.8 (all times are UT 1977) is shown in Figure 1. The best fit was
found to be a power law of index 0.94 $\pm$ 0.11 (all errors are 1σ) and
a hydrogen column density of < 5.10^{22} $\overline{H}$ atoms cm^{-2}. There is

146

evidence for an emission feature between 5 - 8 keV which may be the
iron line emission first seen in the low state by HEAO-1 (Pravdo et
al., 1978). In general, the spectrum is similar to that observed by
Becker et al., (1977).

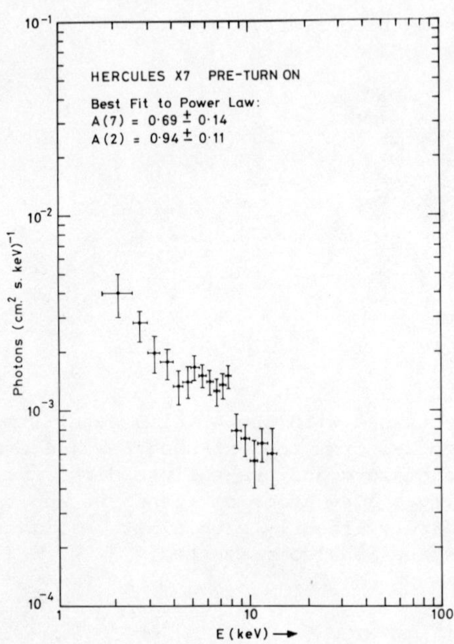

Figure 1:   Pre-turn on spectrum integrated between days 32.6 - 33.9.
*(Reproduced by permission of the Royal Astronomical Soc-
iety, Parmar et al., 192, 311, (1980)).*

The turn-on of the main on-state occurred at $\emptyset_{1.7}$ = 0.24 ± 0.04 which
is consistent with previous observations when the turn-on has always
occurred at $\emptyset_{1.7}$ ∿ 0.2 or 0.7. A power law spectral model was fitted
to the turn-on and off data. The power law index was kept constant
at a value of 0.9 (that of the pre-turn-on and on-states) and the
best fitting hydrogen column (N_H) found by minimising χ^2. The results of this analysis are shown in Figures 2 and 3. It can be seen
that N_H is generally a few 10^{22} atoms cm^{-2} except for an anomalous
dip at $\emptyset_{1.7}$ = 0.5 (similar to those previously observed (Giacconi
et al., 1973)) where it rises to ∿ 2.10^{23} atoms cm^{-2}. The normalisation increases approximately linearly with time suggesting that
electron scattering could be a major source of opacity and that
photo-electric absorption has little effect (except at low energies)
on count rates during the turn-on and off.

A light curve, obtained in pulsar mode, of part of the secondary is
presented in Figure 4. Two total eclipses of duration 0.6 ± 0.2 days

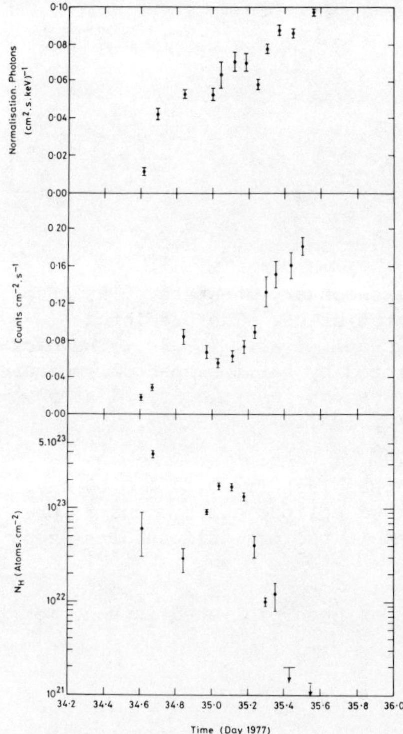

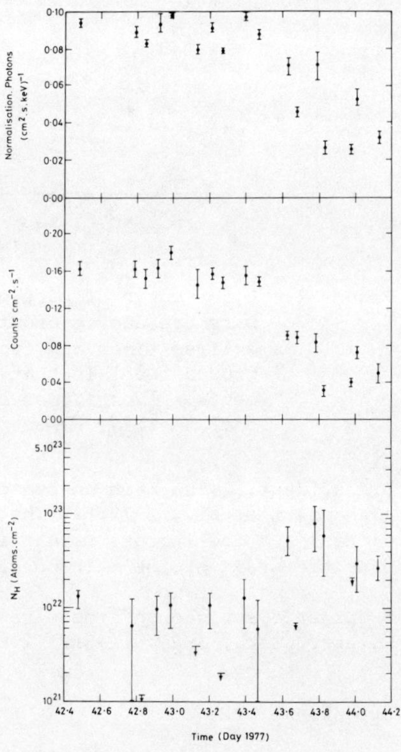

Figure 2: The light curve, spectral normalisation and hydrogen column during the turn-on to the main on-state. The power law index was held constant at a value of 0.9. An anomolous dip (Giacconi et al., 1973) centred on day 35.1 can be seen. *(Reproduced by permission from the Royal Astronomical Society, Parmar et al., 192, 311, (1980)).*

Figure 3: The light curve, spectral normalisation and hydrogen column during the turn-off. The power law index was held constant at a value of 0.9. The increase in hydrogen column on day 43.8 corresponds to the expected time of a pre-eclipse dip. *(Reproduced by permission from the Royal Astronomical Society, Parmar et al., 192, 311, (1980)).*

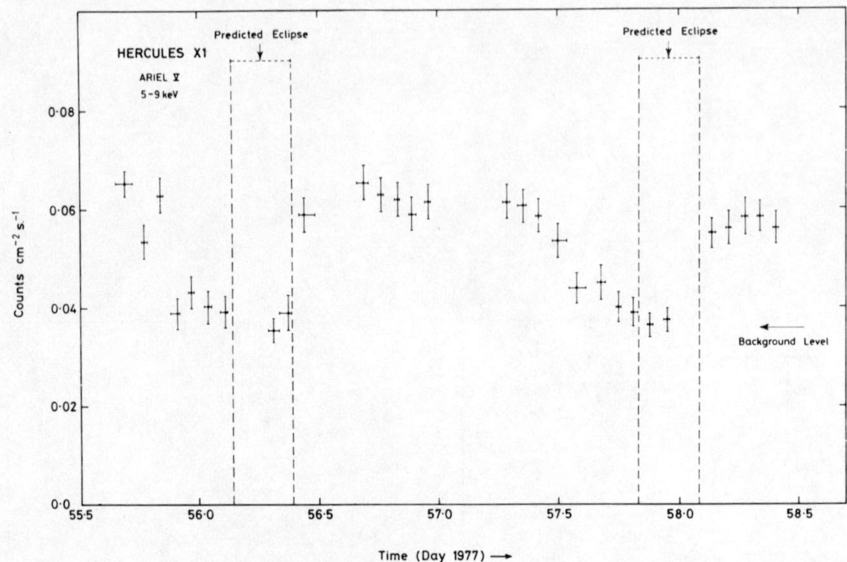

Figure 4: Light curve of part of the secondary on-state. The error
bars are due to counting statistics. The predicted
eclipse times are from Giacconi et al., (1973). The background level is that predicted by earlier spectral measurements. *(Reproduced by permission from the Royal Astronomical Society, Parmar et al., 192, 311, (1980)).*

are evident. (The main on-state eclipses last for 0.24 $\pm$ 0.06 days
(Tananbaum et al., 1972)). The observed eclipse exits occur as predicted, but the ingress is earlier, as in the pre-eclipse dips seen
near the start of the main on-state.

A fuller discussion of these results has been published in Mon. Not.
R. astr. Soc., <u>192</u>, 311-318, 1980.

REFERENCES

Becker, R.J., Boldt, E.A., Holt, S.S., Pravdo, S.H., Rothschild, R.E.,
      Serlemitsos, P.J., Smith, B.N., and Swank, J.H., 1977. Ap. J.,
      214, 879.
Giacconi, R., Gursky, H., Kellogg, E., Levinson, R., Schreier, E.,
      and Tananbaum, H., 1973. Ap. J., 184, 227.
Pravdo, S.H., Boldt, E.A., Holt, S.S., Rothschild, R.E., and
      Serlemitsos, P.J., 1978. Ap. J. (Letters), 225, L53.
Tananbaum, H., Gursky, H., Kellogg, E.M., Levinson, R., Schreier, E.,
      and Giacconi, R., 1972. Ap. J. (Letters), 174, L143.

IUE observations of the
HZ Herculis system

L. Hartmann

Center for Astrophysics
Cambridge,
Mass. 02138.

1. Introduction

2. Observations

3. Interpretation

4. Conclusions

1.   INTRODUCTION

The standard model of the HZ Herculis system is that of a neutron
star accreting material from a late-A star (Oke 1976, Bopp, Grupsmith, and van den Bout 1972). The optical spectral type changes
from early-B to late -A through the 1.7 day orbital period in a
manner that is consistent with X-ray heating of the primary star's
photosphere (Milgrom and Salpeter 1975). However, periodicities
other than 1.7 days are observed in the system. The X-rays are
modulated by the well-known 35 day ON-OFF cycle (Tananbaum et al.
1972), and a 35 day modulation is also present in the optical data
(Deeter et al. 1976). The most favoured explanation of the 35 day
period is that it is the precessional period of the accretion disk
which is twisted out of the orbital plane (Roberts 1974; Petterson
1975; Gerend and Boynton 1976). In this model the disk blocks the
view toward the central, X-ray emitting region over $\sim$ 70% of the
precession cycle. Gerend and Boynton (1976) have shown that the
optical "eclipses" become much narrower at 35-day phases where the
disk is seen edge-on. These authors have managed to reproduce a
great many of the features observed in the optical light curve,
albeit with many model parameters that can be adjusted.

The principal diffulty with optical observations of HZ Her is that
they provide little information on gaseous material at temperatures
> 10^4K. In this paper I shall discuss the preliminary results of a
program of ultraviolet spectroscopy using the International Ultraviolet Explorer (IUE) satellite (Gursky et al. 1980). The observations were made as part of a collaboration between the groups involved
from the three agencies controlling the satellite: The National
Aeronautics and Space Administration (U.S.A.), The Science Research

150

Council (U.K.), and the European Space Agency.

HZ Herculis was observed from IUE on twenty different occasions,
scattered around the orbital phase, and mostly close to the time of
onset of the X-ray emission during the 35-day phase. The phase
coverage is small compared to what has been accumulated at visual
wavelengths. However, we are able to correlate our data with known
features in the light curve. Because of the high temperatures
present in the system, the UV observations are important for testing
models of the hot gaseous emission.

2.   OBSERVATIONS

HZ Her was observed in the low-dispersion mode of the IUE satellite.
Most of the exposures were taken with the short-wavelength (SWP)
camera, covering the spectral region 1150-1950 A with a resolution
$\sim$ 6 A; a few exposures were also taken with the long-wavelength
(LWR) camera, covering the region 2000-3200 A. All spectra were
obtained with the large aperture (10 x 20 arc sec) in order to obtain
absolute spectrophotometry; in some case, two exposures were taken
in the large aperture to save camera preparation time. Further
details of the spacecraft and instrumentation are given by Boggess
et al. (1978a, b).

In Figure 1 we display typical spectra of HZ Her, similar to those
reported by Dupree et al. (1978). The very strong Ly α emission
feature is geocoronal in origin. The dominant features are the
N V $\lambda1240$ and C IV $\lambda1550$ emission lines superposed on a relatively
flat, featureless continuum. He II $\lambda1640$ is also present at a weak
level. There is evidence for Si IV $\lambda1394$, $\lambda1403$ emission as well.
O V $\lambda1371$, N IV $\lambda1719$, and N III $\lambda1750$ may be present but are not
apparent in all spectra. Some later exposures of higher signal-to-
noise suggest that some appreciable fraction of the short-wavelength
"continuum" is made up of weak emission lines. Several spectra show
evidence of interstellar absorption lines at 1300 A (probably
O I $\lambda1302.17$, $\lambda1304.86$, $\lambda1306.03$), and 1335 A (C II $\lambda1334.53$,
$\lambda1335.66$).

The typical long-wavelength spectrum is a featureless continuum,
slightly more intense at 2000 A than at 3000 A. A few long-wavelength spectra show evidence of weak interstellar absorption lines at
2380 A (Fe II $\lambda2382.003$), 2595 A (Fe II $\lambda2599.40$) and 2800 A (Mg II
$\lambda2795.5$ and $\lambda2802.70$). HZ Her is at high galactic latitude and
interstellar reddening is known to be small. We find no sign of the
2200 A interstellar absorption feature (e.g. Nandy et al. 1975),
requiring that $E_{B-V} < 0.^{m}05$.

It should be emphasized that the short-wavelength data described
herein suffer from the ITF calibration problem. However, we do not
expect this error to substantially alter the nature of our conclusions, based on recent exposures taken and reduced with a correct ITF.

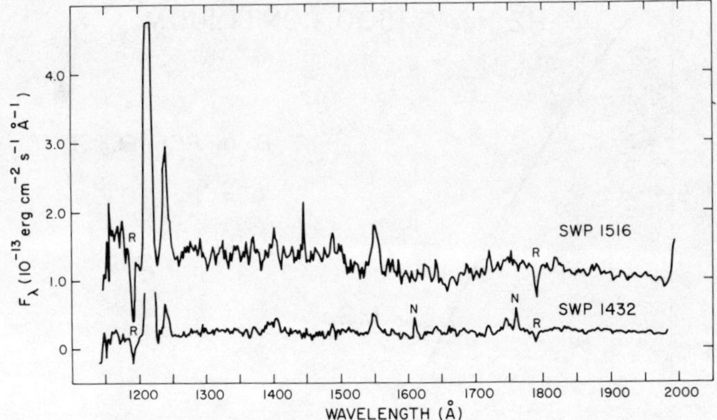

Figure 1: Typical short-wavelength IUE spectra of HZ Herculis: SWP
1432 (ϕ = .143) and SWP 1516 (ϕ = .350). Spectrograph
fiducial marks (reseaux) are indicated by R. Radiation-
induced noise spikes are indicated by N.

3.   INTERPRETATION

In Figure 2 we display the continuum light curve of the HZ Her
system at 1500 A. The data have been reflected about orbital phase
ϕ = 0.5, assuming symmetry; the data shown justify this assumption.
(ϕ is defined by ϕ = 0 at mid-eclipse of the X-rays.) The main
feature of these data is the strong modulation with orbital phase.
The solid line is the prediction of the Milgrom and Salpeter (1975;
MS) heated photosphere model, which matches the magnitude of the B
light variation as well. In Figure 3 a comparison is made between
the observed spectral energy distribution at ϕ = 0.5 and the predicted spectrum of MS. Given the crudity of the wavelength bins in
the calculation, the agreement is fairly good, particularly since no
normal model atmosphere comes close to reproducing the entire spectrum. The strong modulation with orbital phase evident in the
continuum fluxes, and the success of MS in predicting absolute flux
levels and the spectral shape outside of eclipse, make it certain
that the X-ray heated primary atmosphere is a major component of the
ultraviolet emission.

However, it is clear from Figure 2 that another emitting component
must be invoked in addition to the heated photosphere, particularly
to account for the excess flux near orbital phase $\phi \sim$ 0.9 (0.1).
This behaviour is similar to the optical results (Gerend and Boynton
1976; GB) in which emission components with 35 day periodicities are
known to be present, and give rise to excess optical flux over the
MS predictions near eclipse.

In the following we adopt the GB convention of the 35-day cycle,
where phase ψ = 0.85 corresponds to X-ray turn-on and $\psi \sim$ 0.2 is
turn off. An average period of 34.88 days was used with turn-on at

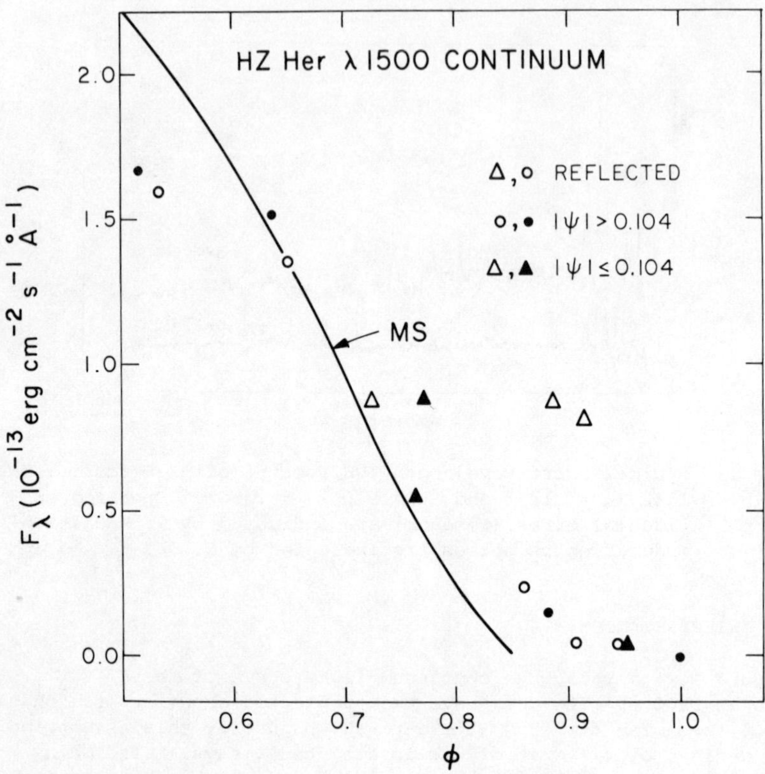

Figure 2: Comparison of the observed and predicted light curves of
HZ Herculis near 1500 A reflected around ϕ = 0.5. Open
circles and triangles are data from ϕ < 0.5 reflected
about ϕ = 0.5.

JD 2,442,652.7 (Giacconi et al. 1973). On the precessing disk
model, $\psi \sim 0$, 0.5 correspond to the disk at its most open-faced
aspects as seen from Earth; at $\psi \sim 0.25$, 0.75, the disk is
approximately edge-on. We have somewhat arbitrarily divided the
plotted data in the figures into two groups, with triangles representing 0.9 < ψ < 0.1, and circles corresponding to data outside
these phases. This division insures that the triangles mark
observations at which the disk is most open towards Earth.

It can be seen from Figure 2 that the departures from the MS model
are in the same sense as observed in the optical fluxes, which show
a dip near ϕ = 0.5 due to occultation of the primary by the disk,
and a narrowing of eclipse, or excess flux near ϕ = 0.9, 0.1 due to
disk emission (Crampton and Hutchings 1972; GB). The triangles at
ϕ = 0.9 show that the flux near eclipse appears to be strongly
modulated with 35 day phase, which favours the disk model rather
than circulation flows of hot gas around the primary (Wilson 1973)
or extended gaseous emitting regions (Joss, Avni and Bahcall 1973) as
an explanation of the excess flux.

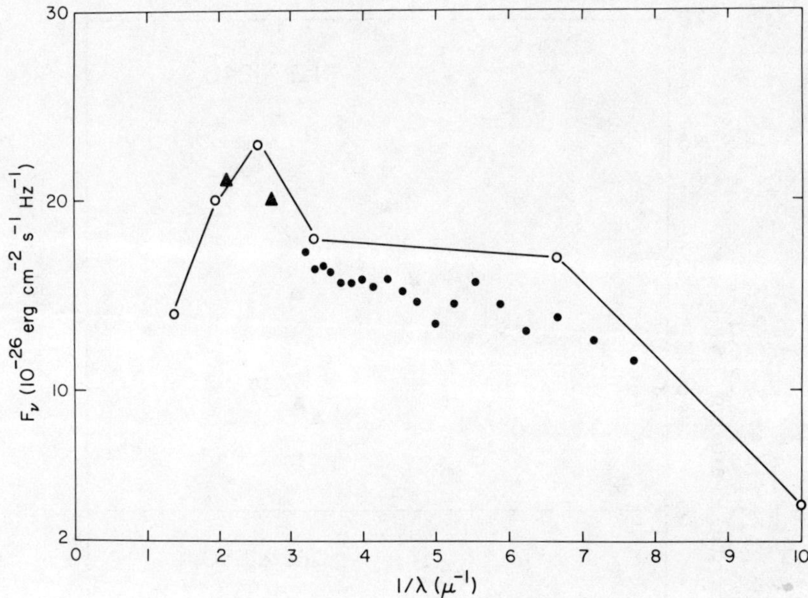

Figure 3: Comparison of the observed and predicted spectra of HZ
Herculis near maximum light. Filled circles: SWP 1962,
LWR 1811 IUE data ($\phi \approx 0.475$). Triangles: U, B data.
Open circles: MS model prediction.

The magnitude of the disk emission at $\phi = 0.9$ suggests that the fluxes
at other phases should be reduced by a considerable factor before
comparison with the MS model. Cherapaschuk et al. (1977) (hereafter,
CGY) and Bisnovatyi-Kogan et al. (1977) have derived average parameters for the disk by following the evolution of the optical colours
through eclipse. With our limited data base, we cannot do this.
However, we can compare the CGY predictions with our observations at
$\phi = 0.1$ (open triangles) when the disk should dominate the light, and
should not be substantially eclipsed by the primary.

The CGY model has an average disk temperature of 22,000 K and a
radius of 10^{11}cm (at the relevant phases). Using the CGY disk
parameters and adopting a distance of 5 kpc to HZ Her, we calculate
a flux at 1500 A of 8 x 10^{-14} erg cm^{-2} s^{-1} A^{-1} for the whole disk, compared to an observed value of 8.6 x 10^{-14} erg cm^{-2} s^{-1} A^{-1} at $\phi=0.113$.
The degree of agreement is probably fortuitous, but still provides an
important confirmation of the essential elements of the CGY model. In
contrast, GB found a disk temperature of $\sim$ 10,000K and radius $\sim$ 1.6 x
10^{11}cm, which shows the importance of the ultraviolet data in
determining disk parameters more precisely.

As shown in Figure 4, the N V and C IV line fluxes vary significantly
as a function of phase. The general behaviour of the lines strongly
suggests that they are formed in or near the heated photosphere as is

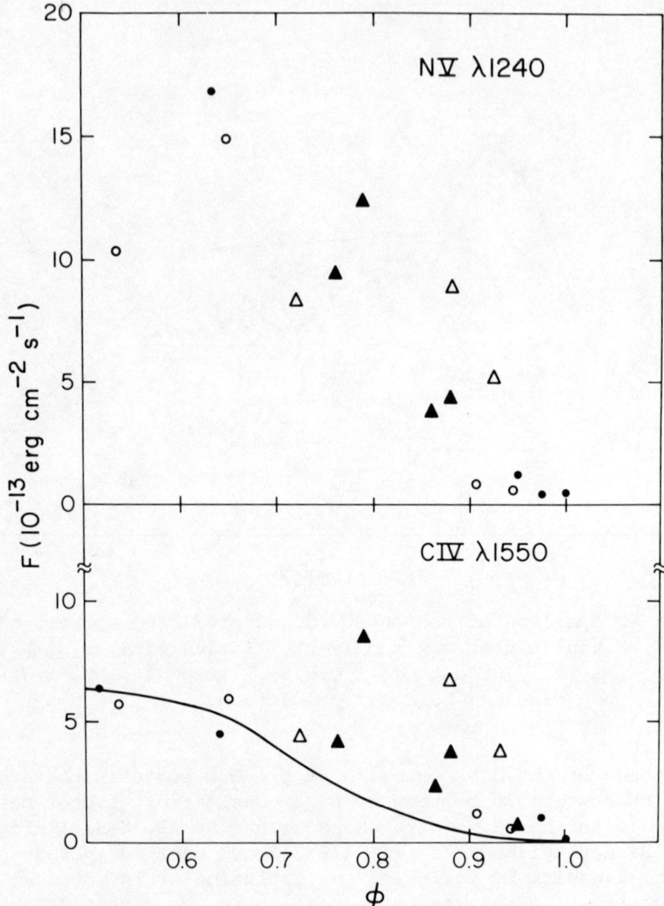

Figure 4: Variations with ϕ of emission-line fluxes for N V λ1240
and C IV λ1550.

the continuum, particularly N V, for which the equivalent width is
roughly constant. The C IV flux does not vary as much, with the
equivalent width increasing as one approaches eclipse. The high
continuum flux exposures at $\phi = 0.8 - 0.9$ also have enhanced line
emission. It is likely that the N V and C IV have a different
origin from the visual emission lines which show no strong correlation with orbital phase (e.g. Crampton and Hutchings 1974; Barbon et
al. 1974).

We begin an approximate analysis by evaluating the emission that may
be produced in the heated photosphere. For a semi-infinite atmosphere, we may use the results of Hatchett, Buff, and McCray (1976) to
compute the C IV and N V luminosities, given $L_x = 1.5 \times 10^{37}$ erg s^{-1}

(Milgrom and Salpeter 1975) and the solid angle of X-rays intercepted by the primary star. Assuming the line flux is emitted into
$\sim 2\pi$ steradians, we compute a flux in N V $\sim$ 9.5 x 10^{-13}erg cm^{-2}s^{-1} at
the Earth, which is comparable to the maximum flux seen; however,
the calculated flux for C IV is $\sim$ 5.8 x 10^{-12}erg cm^{-2}s^{-1}, ten times
the observed maximum; this result suggests that C IV is optically
thick.

The variation of the N V/C IV ratio with phase suggests that near
eclipse the line emission arises from a more extended, optically
thin gas; such a component could be present at ϕ = 0.5 without
affecting the total fluxes from the heated photosphere significantly.

A recent long exposure during eclipse shows weak N V emission which
presumably arises from such an extended region. The fluxes near
eclipse are very uncertain; the best estimate for the flux at 1900A
$\sim$ 2 x 10^{-15}erg cm^{-2}s^{-1}A^{-1}. Comparison to OAO-2 results indicate
that this is consistent with an A7 star (α Aql) normalized to
B $\sim$ 14.7, so that the flux in eclipse appears to be consistent with
the undisturbed optical spectral type.

4.   CONCLUSIONS

The UV observations of the HZ Her system are consistent with the
optical models including a precessing disk, and serve to place
stringent constraints on disk properties. The dominant light of the
system appears to be explained adequately by the X-ray heated photosphere model, but many details remain to be explored. The N V and
C IV emission line strengths are of the same order of magnitude as
those predicted by X-ray photoionization; however, more detailed
calculations including optical depth effects should be pursued.

I would like to acknowledge the assistance of the IUE Observatory
staffs, both at Greenbelt, and Vilspa, in the acquisition and
reduction of these data, and in particular Dr. R. Robin of GSFC and
Drs. A. Holm, S. Schiffer, and C.C. Wu of CSC.

REFERENCES

Barbon, R., Benvenuti, P., Bernacca, P.L., and Ciatti, F., 1974.
    Astr. Ap., 31, 237.
Bisnovatyi-Kogan, G.S., Goncharskii, A.V., Komberg, B.V., Cherepash-
    chuk, A.M., and Yagola, A.G., 1977. Sov. Astron.-AJ, 21, 133,
    (Russian Astr. Jour., 54, 241).
Boggess, A., et al., 1978a, Nature, 275, 372.
Boggess, A., et al., 1978b, Nature, 275, 377.
Bopp, B.W., Grupsmith, G., and Van den Bout, P., 1972. Ap. J. (Letters),
    178, L5.
Cherapaschuk, A.M., Goncharskii, A.V., and Yagola, A.G., 1977. Sov.
    Astron.-AJ, 21, 581.
Crampton, D., and Hutchings, J.G., 1972. Ap. J. (Letters), 178, L65.

156

Crampton, D., and Hutchings, J.G., 1974. Ap. J., 191, 483.
Deeter, J., Crosa, L., Gerend, C., and Boynton, P.E., 1976. Ap. J.,
     206, 861.
Dupree, A.K., et al., 1978. Nature, 275, 400.
Gerend, D., and Boynton, P.E., 1976. Ap. J., 209, 562 (GB).
Gursky, et al., 1980. Ap. J., in press.
Hatchett, S., Buff, J., and McCray, R., 1976. Ap. J., 206, 847.
Joss, P.C., Avni, Y., and Bahcall, J.N., 1973. Ap. J., 186, 767.
Milgrom, M., and Salpeter, E.E., 1975. Ap. J., 196, 589 (MS).
Nandy, K., Thompson, G.I., Jamar, C., Monfils, A., and Wilson, R.,
     1975. Astr. Ap., 44, 195.
Oke, J.B., 1976. Ap. J., 209, 547.
Petterson, J.A., 1975. Ap. J., 201, L61.
Roberts, W.J., 1974. Ap. J., 187, 575.
Tananbaum, H., Gursky, H., Kellogg, E.M., Levinson, R., Schreier, E.,
     and Giacconi, R., 1972. Ap. J., 174, 143.
Wilson, R.E., 1973. Ap. J. (Letters), 181, L75.

PART 2
X-ray pulsars and the physics of accreting material

Binary X-ray pulsars

Saul Rappaport

Department of Physics and Center for Space Research
Massachusetts Institute of Technology
Cambridge, MA   02139

1. Introduction

2. Review of Binary X-Ray Pulsars

3. Three Binary X-Ray Pulsar Systems

1. INTRODUCTION

There are currently over 500 detected galactic X-ray sources (Forman
et al. 1978; Cooke et al. 1978; Markert et al. 1979; Bradt et al.
1979). These include many classes of astronomical objects, but it is
likely that the majority of the more luminous galactic X-ray sources
are binary systems, wherein the X-ray power is derived from accretion
onto a compact object (neutron star, white dwarf or black hole).
Much of our knowledge of these systems has been derived during the
past decade from studies of X-ray pulsars. Optical studies of the
companion stars have also provided valuable information regarding the
nature of X-ray binary systems; however, the study of X-ray pulsars
has yielded data about these systems essential for their definition.
There are presently only ~16 X-ray pulsars known, so that we rely
heavily on this rather small sample to deduce the general properties
of X-ray binaries.

In this paper we first present a brief review of the state of our
empirical knowledge of binary X-ray pulsar systems. Following that,
we report recent results for three X-ray pulsars: 4U1626-67, GX301-2
and 4U0900-40, each of which illustrates a different astrophysical
problem. 4U1626-67 provides us with important clues concerning the
nature of highly compact binary X-ray systems. GX301-2 has been
found to have the widest and most eccentric orbit of any of the X-ray
binaries whose orbit has thus far been determined. Finally, the
apsidal motion test has recently been applied to the 4U0900-40 binary
system in a unique way.

2. REVIEW OF BINARY X-RAY PULSARS

The pulse periods of the known binary X-ray pulsars range over three
decades in period from 0.7 s to 835 s; these are shown in Figure 1
(Rappaport and Joss 1977a, Joss and Rappaport 1980). The apparent
"gap" in the distribution of pulsar periods between 10 s and 100 s

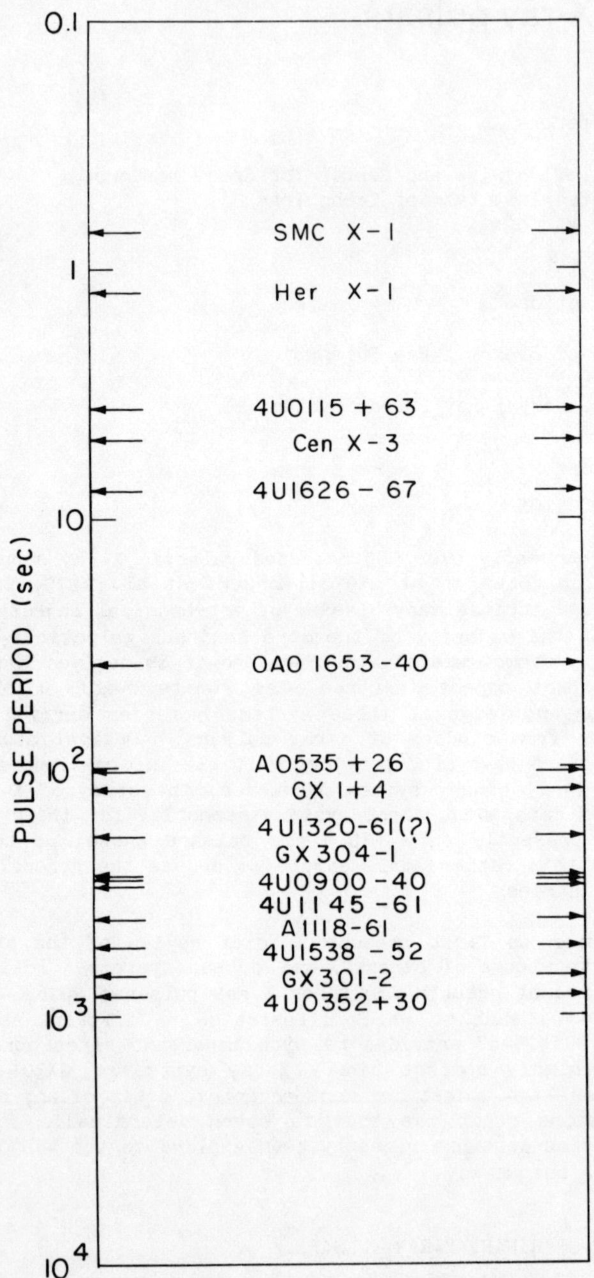

Figure 1: The pulse periods for 16 well established X-ray pulsars (from Joss and Rappaport 1980).

has recently been closed with the discovery of a 38 s pulsar with the
HEAO-1 satellite (White and Pravdo 1979). Sample pulse profiles for
14 of these pulsars are presented in Figure 2. The pulse profiles
are characterized by (a) large "duty cycles" (compared to radio pulsars), (b) modulation factors that range from 25%-90%, (c) shapes
that range from symmetric to highly asymmetric, and (d) no significant trend in profile complexity as a function of pulse period. For
several of the sources (e.g., 4U1626-67, A0535+26 and 4U0900-40) the
pulse profiles vary radically with energy, while for others (e.g.,
Cen X-3, 4U0352+30) the basic pulse profile is retained over a wide
range of energies (see Joss and Rappaport 1980 for references).

To date, no theoretical calculations have accounted for the complex
pulse shapes (Fig. 2) or their dependence on energy. It seems clear,
however, that the pulse profiles result from an X-ray beam pattern
that is misaligned with the rotation axis of an accreting, magnetized
neutron star, and is viewed from different directions as the star
rotates (Pringle and Rees 1972; Davidson and Ostriker 1973; Lamb,
Pethick and Pines 1973; Baan and Treves 1973). The beam pattern, in
turn, is determined by the complex radiative transfer processes for
X-rays propagating from the surface of the neutron star through
regions of accreting, magnetized plasma.

Numerous searches for shorter period (< 0.7 s) X-ray pulsars have not
yielded any positive results (see, e.g., Rappaport and Li 1979).
This is consistent with the notion that for surface magnetic fields
$\gtrsim 10^{12}$ G and luminosities $\lesssim 10^{38}$ ergs s^{-1}, matter could be inhibited
from accreting onto a neutron star with a rotation period $\lesssim 1$ s (see,
e.g., Pringle and Rees 1972). Still other searches have been carried
out to detect pulsars whose emission is only weakly modulated ($\lesssim 5\%$)
(see, e.g., Parsignault and Grindlay 1978). For at least some of the
brighter sources (e.g., the bright galactic center sources) pulsation
limits of $\lesssim 1\%$ have been set (Rappaport and Li 1979).   Thus, the
sample of ~16 pulsars is not likely to grow quickly.

Measurements of the pulse arrival times from X-ray pulsars have been
used very successfully to determine the orbits of several of these
systems. The measured orbits are shown, to scale, in Figure 3, in
order of increasing size of the semi-major axis. In the early 1970s
the relatively smaller orbits of Her X-1 and Cen X-3 were measured
with the Uhuru satellite (Tananbaum et al. 1972; Schreier et al.
1972a), while the larger orbits of 4U0115+63 and GX301-2 required
longer, more extensive pointed observations, and were measured only
recently (Rappaport et al. 1978; White, Mason and Sanford 1978;
Kelley, Rappaport, and Petre 1980). Nominal values for the masses
and radii of the companion stars are also indicated in Figure 3; the
derivation of these parameters will be discussed shortly.   A few
other X-ray binaries that lack detectable X-ray pulsations (e.g., LMC
X-4, 4U1700-37, Cyg X-2) have been sufficiently well studied in the
optical (Chevalier and Ilovaisky 1977; Hutchings et al. 1978;
Hutchings 1976; van Paradijs et al. 1978; Cowley et al. 1979) to
yield important information about the binary system parameters. However, the information is much less complete than for the systems
where the orbit of the X-ray star has been measured.

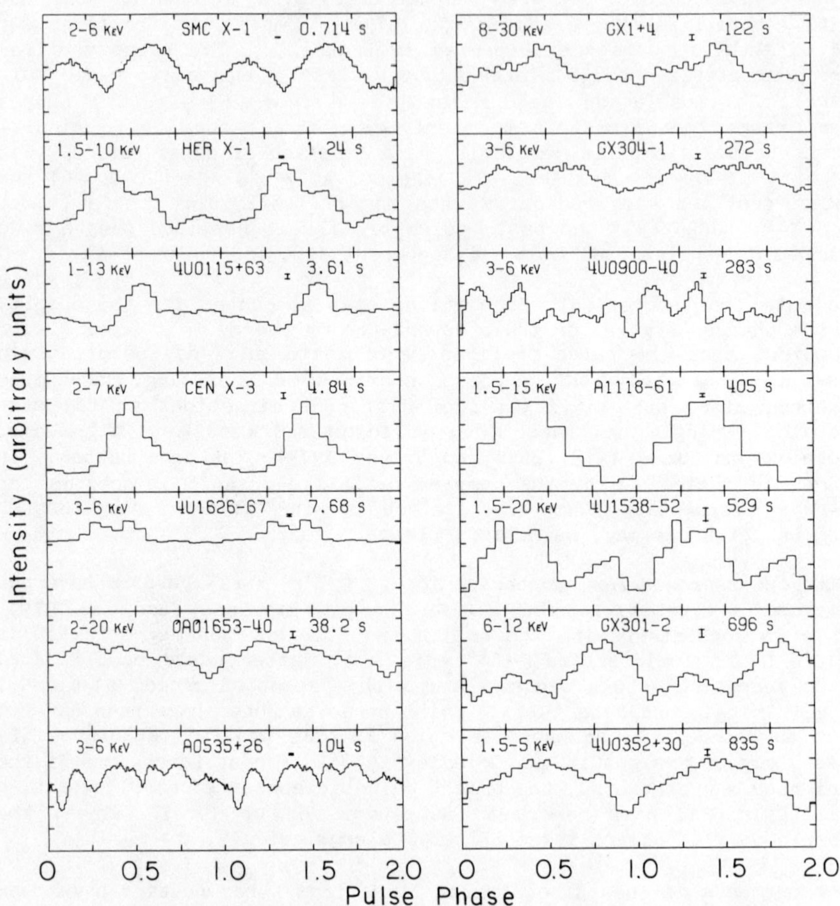

Figure 2: Sample pulse profiles for 14 X-ray pulsars that are thought to be in binary systems (from Joss and Rappaport 1980). The profiles for Her X-1, 4U0115+63, Cen X-3, OAO1653-40, A1118-61 and 4U1538-52 are from Joss et al. (1978), Johnston et al. (1978), Ulmer (1976) White and Pravdo (1979) Ives et al. (1975) and Becker et al. (1977), respectively; the remaining profiles are from SAS-3 data. (For other references see Rappaport and Joss 1977a.) In each case, the data are folded modulo the pulse period and plotted against pulse phase for two complete cycles. The approximate pulse periods and energy intervals are indicated for each pulsar. Non-source background counting rates have been subtracted. A typical $\pm 1\sigma$ error bar, derived from photon counting statistics, is indicated for each pulse profile. *(Copyright (c) 1981 by D. Reidel Publishing Company, Dordrecht, Holland.)*

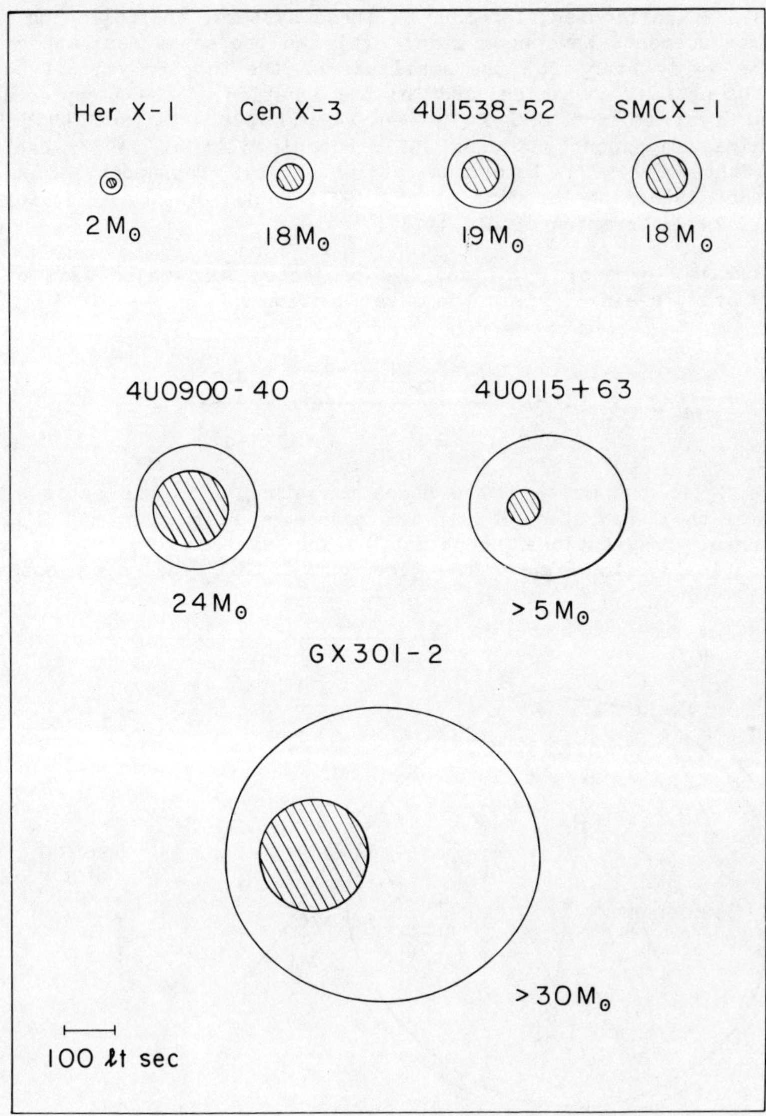

Figure 3: Schematic plot (to scale) of the orbits and companion stars of seven binary X-ray pulsars (from Joss and Rappaport 1980). The approximate mass of the companion star is indicated below each orbit. See text (§2) for a discussion of the derivation of the orbital parameters, and stellar masses and radii. *(Copyright (c) 1981 by D. Reidel Publishing Company, Dordrecht, Holland)*

There are five binary X-ray sources for which sufficient information is available to allow a determination of many of the system param- eters. In particular, for four of these systems, the following three key measurements have been made: (a) the projected semi-major axis of the x-ray star; (b) the amplitude of the Doppler velocity curve for the optical companion; and (c) the duration of the X-ray eclipse. These systems are SMC X-1, Cen X-3, 4U0900-40, and 4U1538-52 (Primini, Rappaport and Joss 1977; Hutchings et al. 1977; Fabbiano and Schreier 1977; Hutchings et al. 1979; Rappaport, Joss and Stothers 1980; van Paradijs et al. 1977; Becker et al. 1977; Davison et al. 1977; Crampton et al. 1978).

The orbital period, P_{orb}, and the projected semi-major axis of the X-ray star, $a_x \sin i$, yield the mass function:

$$f(M) = \frac{4\pi^2 (a_x \sin i)^3}{G\, P_{orb}^2} = \frac{M_c \sin^3 i}{(1+q)^2} \quad , \tag{1}$$

where M_c is the mass of the companion star, q is the ratio of the mass of the X-ray star, M_x, to the companion star mass, and G is the universal gravitational constant. The system geometry is shown schematically in Figure 4. The curve that encircles both the

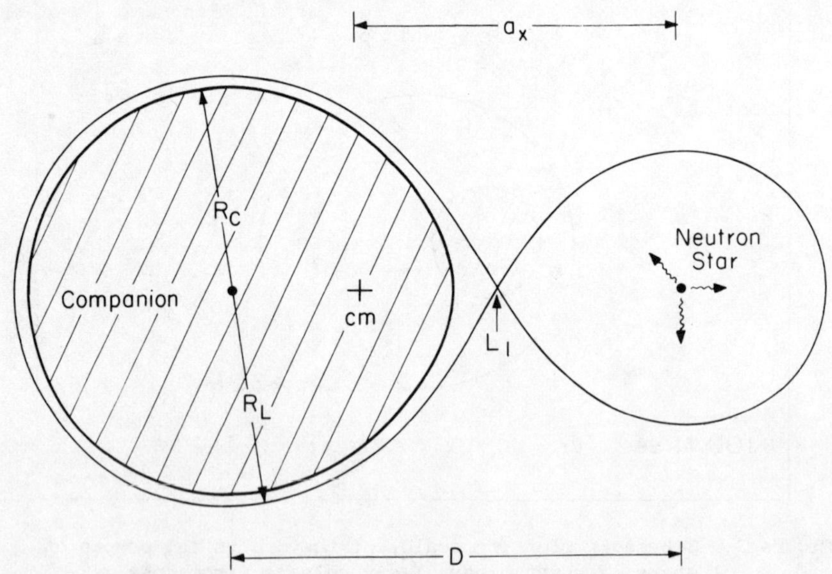

Figure 4: Schematic of a binary X-ray system viewed from above the orbital plane. The companion star (hatched region) and neutron star are shown encircled by their critical potential lobes which join at the inner Lagrange point L_1 (see text).

companion star and the neutron star is a contour of constant potential in the frame of reference that is centered on, and rotates with, the companion star. The particular contour shown is called the critical potential lobe because it represents the largest volume that a dynamically stable companion star can occupy before mass is transferred through the inner Lagrange point (L_1, Fig. 4). The mass ratio q can be determined directly from the ratio of the velocity of the companion star to that of the X-ray star:

$$q \equiv \frac{M_x}{M_c} = \frac{a_c \sin i}{a_x \sin i} = \frac{K_c P_{orb} \sqrt{1-e^2}}{2\pi a_x \sin i} \; , \qquad (2)$$

where K_c is the semi-amplitude of the optical Doppler velocity curve, and e is the eccentricity of the orbit.

The final ingredient for determining the system parameters is the orbital inclination angle. This can be computed approximately, with the aid of a simple model. If we replace the companion star by a sphere of radius R_c whose volume approximately equals the actual volume (shaded region in Figure 4), then the relation between inclination angle and eclipse half angle, θ_{ecl}, is:

$$R_c \simeq D \left[\cos^2 i + \sin^2 i \sin^2 \theta_{ecl}\right]^{1/2} \; , \qquad (3)$$

where D is the stellar separation. The size and shape of the critical potential lobe can be computed from the mass ratio q and the rotation rate of the companion (see, e.g., Avni and Bahcall 1975a). The computed effective radius of the critical potential lobe, R_L, as a function of q has been found (Plavec 1968; Paczynski 1971) to be reasonably well fitted by the expression

$$R_L \simeq D (a - b \log q) \; , \qquad (4)$$

where a and b are constants* that depend on the ratio, Ω, of the rotation frequency of the companion to the orbital frequency. If we define the radius of the companion as some fraction, β, of the radius of the critical potential lobe, then (3) and (4) may be combined to yield an expression for the inclination angle:

$$\sin i \simeq \left[1-\beta^2(a-b \log q)^2\right]^{1/2}/\cos \theta_{ecl} \; . \qquad (5)$$

From the above relations, we see that if K_c, $a_x \sin i$ and θ_{ecl} are measured, and we make certain reasonable assumptions about the

---

* We have computed the constants a and b so as to minimize the difference between the value of i given by eq. (5) and the actual value of i that corresponds to the same values of q and θ_{ecl}. For $0.02 < q < 0.2$, $50^\circ < i < 90^\circ$ and $\Omega = 1$, we find $(a,b) = (0.384, 0.200)$, and for $\Omega = 0$, $(a,b) = (0.405, 0.230)$. These are in excellent agreement with the values given by Avni (1978a) which were computed so that $4\pi R_L^3/3$ best represents the actual volume of the critical potential lobe. The values of a and b given above yield rms errors in i of $1^\circ.3$ and $1^\circ.9$ for the tidal ($\Omega=0$) and Roche ($\Omega=1$) geometries, respectively.

166

rotation rate and fraction of the critical lobe occupied by the companion, all the system parameters discussed above may be determined. The amplitude of the observed ellipsoidal light variations in several of these systems indicates that the companion stars nearly fill their critical potential lobes (i.e., $\beta \gtrsim 0.9$; see, e.g., Avni and Bahcall 1975a, 1975b). It can then be argued on theoretical grounds that these systems would be in approximately synchronous rotation (i.e., $\Omega \simeq 1$; see, e.g., Lecar, Wheeler, and McKee 1976). (For other discussions of these issues, see, e.g., Avni and Bahcall 1976; van den Heuvel 1975; Rappaport and Joss 1977c; Bahcall 1978a and 1978b; Avni 1978; Petterson 1978; Conti 1978).

The nominal values for the system parameters are easily found from equations (1), (2), (3), and (5) by inserting the most probable values for K_c, $a_x \sin i$ and θ_{ecl}, and a best guess for Ω and β. A difficulty arises in propagating both the experimental errors for K_c, $a_x \sin i$ and θ_{ecl}, and the theoretical and experimental uncertainties in Ω and β. We have therefore evaluated the binary system parameters and their corresponding uncertainties by means of a Monte Carlo propagation technique (Rappaport, Joss and Stothers 1980). In 2 x 10^4 trial evaluations, $a_x \sin i$ and K_c were chosen randomly with respect to Gaussian distributions of the appropriate widths, while the values of θ_{ecl} were chosen randomly and uniformly in the appropriate range, to reflect the experimental uncertainties (see Table 1). To simulate the theoretical and observational uncertainties in the values of β and Ω, we also chose values of β randomly and uniformly over the range 0.9-1.0 and Ω randomly and uniformly between 0 and 1.

The ranges of companion star masses and radii, as determined by the Monte Carlo error propagation technique for SMC X-1, Cen X-3, 4U0900-40 and 4U1538-52, are shown in Figure 5 as contours of constant probability. The most probable parameter values are in good agreement with previously reported values (Primini, Rappaport and Joss 1977; Hutchings et al. 1979; van Paradijs et al. 1977; Rappaport, Joss and Stothers 1980; Becker et al. 1977; Davison et al. 1977). The outer contour, which contains 98% of the Monte Carlo events, represents reasonably secure error limits for the masses and radii. These uncertainties are important to bear in mind when trying to fit these companion stars into an evolutionary scenario involving mass-losing supergiant companions (see, e.g., van den Heuvel 1976, 1977; Conti 1978).

In the case of the Her X-1/HZ Her system, a reliable optical Doppler velocity curve is not yet available; however, studies of the optical pulsations from the primary, stimulated by the X-rays, (Middleditch and Nelson 1976) have provided additional information needed to determine the system parameters (Bahcall and Chester 1977).

For the LMC X-4 binary system, the duration of the X-ray eclipse is measured (White 1978; Li, Rappaport and Epstein 1978), as is the velocity amplitude of the optical companion (Hutchings et al. 1978). However, the X-ray source has no observed pulsations, and therefore $a_x \sin i$ cannot be measured directly. Nonetheless, the optical

Table 1

Parameters Used in the Monte Carlo Analysis

Source	$a_x \sin i$* (1t-sec)	K_c* (km s^{-1})	θ_{ecl}[+] (degrees)	References
SMC X-1	53.46 ± 0.03	19 ± 2	26.5-29	1
Cen X-3	39.79 ± 0.01	24 ± 6	37-40.5[‡]	2
4U0900-40	112.3 ± 0.8	21.8 ± 1.2	31-40	3
4U1538-52	55.2 ± 4.3	33 ± 7	25-33	4
LMC X-4	31.6 ± 0.1[†]	50 ± 5	26-33	5

* Quoted uncertainties are $\pm 1\sigma$ error limits.

[+] Extreme range of eclipse half-angles.

[†] This value is inferred from Doppler measurements of the He II $\lambda 4686$ line (Hutchings, Crampton and Cowley 1978), and probably represents only a lower limit to $a_x \sin i$ (see text and Figure 6).

[‡] A more conservative range is probably 35°-$40^{\circ}.5$ (Pounds 1976); however, neither the parameters given for the companion star (Fig. 5) nor the mass of the neutron star (Fig. 9) are significantly altered by this difference.

1. Schreier et al. (1972b); Primini et al. (1976); Primini, Rappaport and Joss (1977); Hutchings et al. (1977).

2. Pounds et al. (1975); Fabbiano and Schreier (1977); Hutchings et al. (1979).

3. Forman et al. (1973); van Paradijs et al. (1977); Watson and Griffiths (1977); Rappaport, Joss and Stothers (1980).

4. Davison, Watson and Pye (1977); Becker et al. (1977); Crampton, Hutchings and Cowley (1978).

5. Li, Rappaport and Epstein (1978); White (1978); Hutchings, Crampton and Cowley (1978).

168

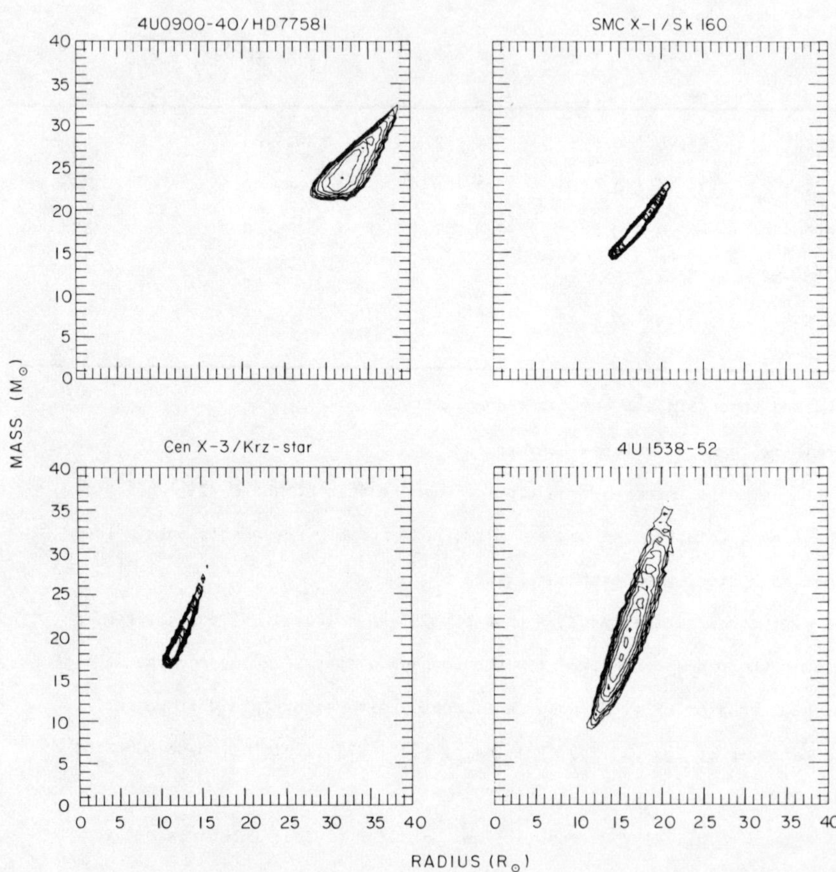

Figure 5: Empirically determined error contours for the mass and radius of four companion stars in binary X-ray pulsar systems (from Joss and Rappaport 1980). The input data are the projected semi-major axis of the neutron star orbit, the X-ray eclipse duration, and the amplitude of the optical Doppler velocity curve (see text). The error contours are derived from a Monte Carlo analysis (Rappaport, Joss and Stothers 1980); the outer contour contains ~98% of the Monte Carlo events. For simplicity, in computing sin i from Eq. (5), we neglected the orbital eccentricity of 4U0900-40. We estimate that this approximation results in an error for M_c, R_c and M_x (Fig. 9) of less than 10%.

observations reveal a He II line (λ4686; Hutchings et al. 1978) that is Doppler shifted with a phase and amplitude suggestive of an origin near the X-ray star. If we assume this Doppler velocity ($\sim$490 km s^{-1}) represents the velocity of the X-ray star, we can apply the same analysis procedure as outlined above. The range of values obtained for the mass and radius of the companion is shown in Figure 6.

Other parameters of these binary systems also follow directly from the above analysis. In Figure 7 we show, for example, the probability distributions of M_x, q, i and R_c/D for the SMC X-1 system.

It is usually assumed that the compact objects in the binary X-ray pulsar systems are neutron stars. Arguments against white dwarfs as the X-ray star include the fact that some of the pulsars are rotating faster than the breakup speed (e.g., SMC X-1; Ostriker and Bodenheimer 1968), or that the luminosity is higher than can plausibly be produced by accretion onto a white dwarf (e.g., Cen X-3; Katz 1977; Kylafis and Lamb 1979). Perhaps the most compelling argument in favor of the neutron-star hypothesis comes from a study of the effects of the torques exerted on the compact stars by the accreting matter. Calculations of these torques (see, e.g., Lamb, Pethick and Pines 1973; Rappaport and Joss 1977b; Ghosh and Lamb 1979) show that over a wide range of conditions, the rate of change of the intrinsic pulse period is related to the luminosity and the parameters of the compact star by the following relation:

$$\frac{\dot{P}}{P} \simeq -3 \times 10^{-5} \left(\frac{\xi v_r}{v_{ff}}\right)^{1/7} M_x^{-10/7} R_6^{6/7} Rg_6^{-2} \mu_{30}^{2/7} L_{37}^{6/7} P^{-1} \; yr \;, \quad (6)$$

where ξ represents the fractional solid angle subtended at the compact star by the infalling matter at the magnetopause; v_r/v_{ff} is the ratio of the average radial infall velocity of a particle to its free-fall velocity just outside the magnetopause; R, Rg, μ, and P are the radius, gyroradius, magnetic dipole moment and rotation period of the X-ray star; and L is the accretion-driven luminosity. [The units for M_x, R, Rg, μ, L and P are $M_\odot$, 10^6 cm, 10^6 cm, 10^{30}G cm^3 and s, respectively, as indicated by the subscripts in the equation.] The overall minus sign in equation (6) is explicitly for the case where the sense of the angular momentum in the accreted matter is the same as for the orbital motion and for the rotation of the X-ray star. The quantity $(\xi v_r/v_{ff})^{1/7}$ is not expected to differ greatly from unity (see, e.g., Lamb, Pethick and Pines 1973).

The rate of change in the intrinsic pulse period has been measured for ten X-ray pulsars (see, e.g., Rappaport and Joss 1977b, Mason 1977; Joss and Rappaport 1980). For most of these sources, a reasonably reliable estimate for the distance is available, and the X-ray luminosity can thus be determined. For some of the pulsars (e.g., GX301-2; Kelley, Rappaport, and Petre 1980) the value of $\dot{P}$ varies with time and even reverses sign. (For a discussion of these effects see, e.g., Ghosh, Lamb and Pethick 1977; Ghosh and Lamb 1978 and 1979; Davies, Fabian and Pringle 1979, and references therein.)

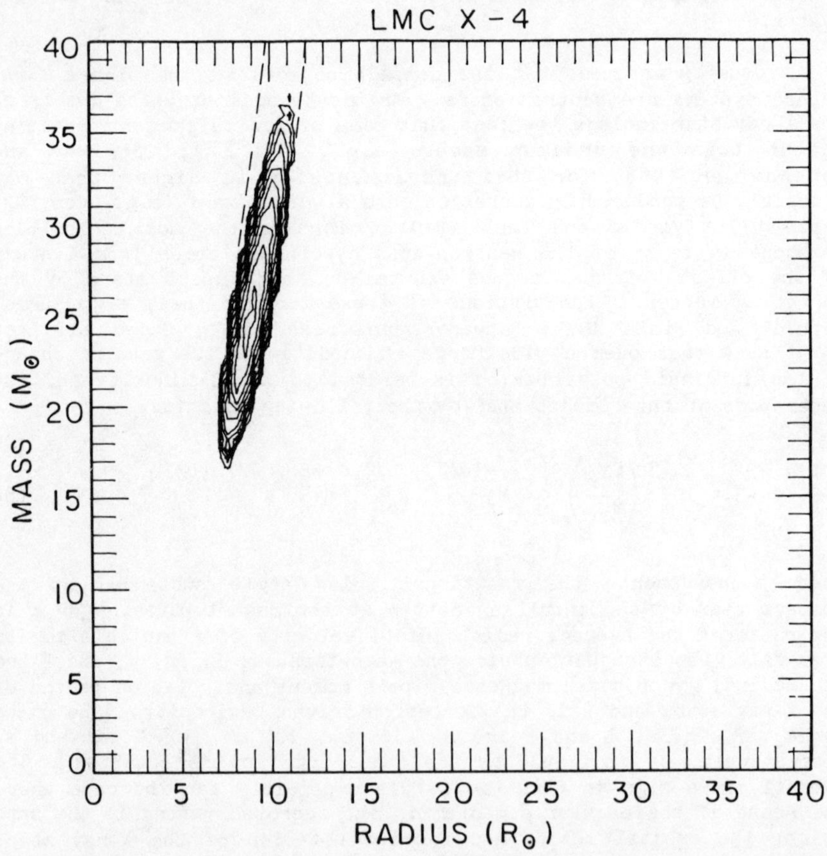

Figure 6: Empirically determined error contours for the companion star of LMC X-4. The input data and calculations are the same as for the systems shown in Figure 5, with the exception that the orbit of the X-ray star is inferred from optical measurements. The dashed curve represents the extension of the outer contour for the case where the velocity inferred from the He II line ($\sim$490 km s^{-1}; Hutchings et al. 1978) is regarded as only a lower limit to the velocity of the X-ray star. The range of M_c and R_c in this plot is the same as in Figure 5.

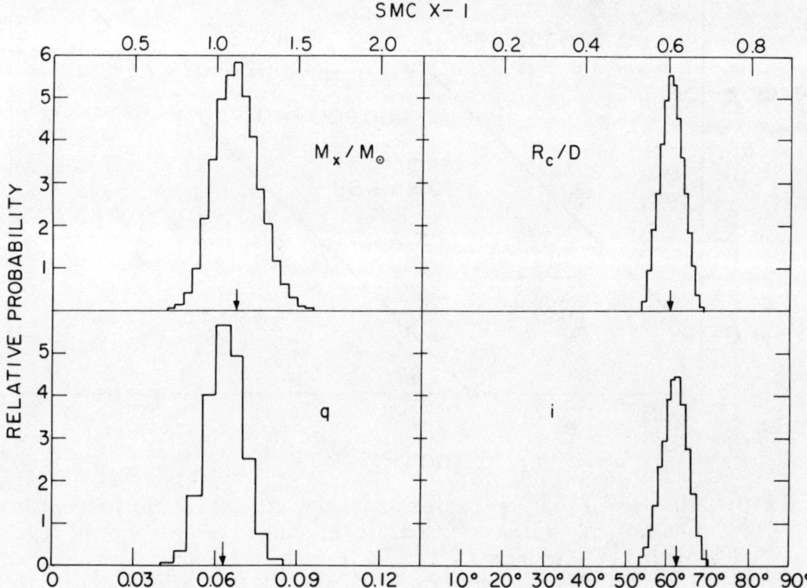

Figure 7:   Several of the binary system parameters and their uncertainties, for the SMC X-1 system, as determined by the Monte Carlo analysis technique described in the text. The quantities M_x, R_c/D, q and i are the neutron star mass, radius of the companion (Sk 160) divided by the orbital separation, the mass ratio, M_x/M_c, and the orbital inclination angle, respectively.

In these cases, we have adopted the average value of $\dot{P}$ during the
longest interval with generally decreasing pulse period as a
representative measure of the accretion torques during times when
equation (6) is most likely to be valid (see, e.g., Ghosh and Lamb
1979). A plot of $-\dot{P}/P$ vs the quantity $PL^{6/7}$ is shown in Figure 8 for
ten X-ray pulsars. The solid line is the relation expected from

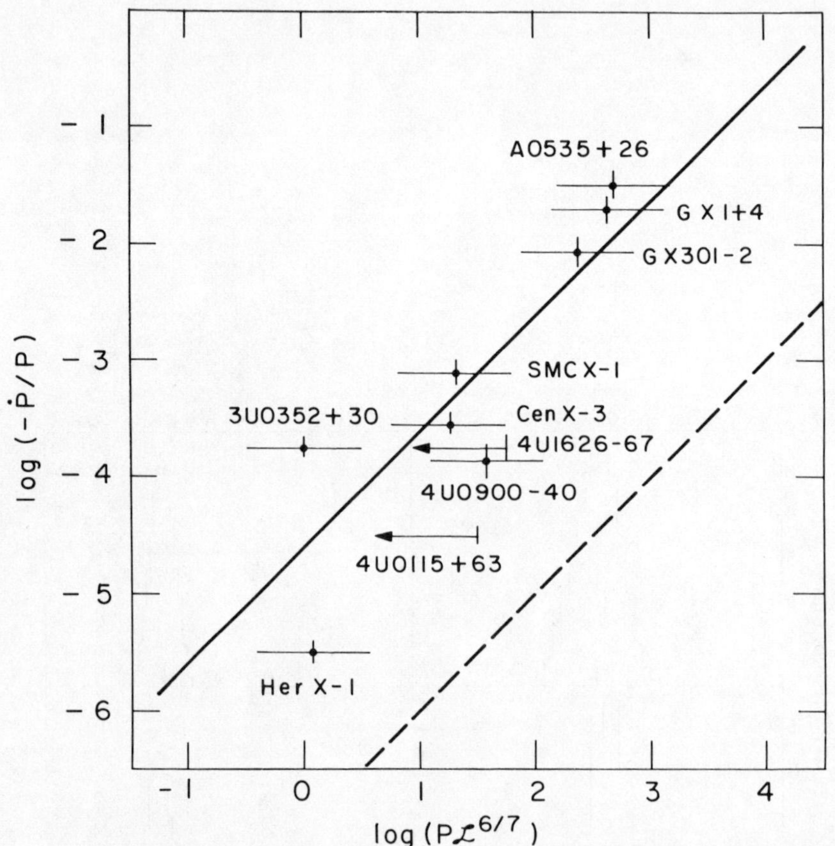

Figure 8: The empirical relation between the fractional rate of
change of pulse period, $\dot{P}/P$, and the parameter (pulse
period x luminosity $^{6/7}$) for ten binary x-ray pulsars
(adapted from Rappaport and Joss 1977b; Joss and Rappaport
1980). The units of $\dot{P}/P$, P and L are yr^{-1}, s and 10^{37} erg
s^{-1}, respectively. The solid line is the best fit to a
power law with logarithmic slope 1.0; the data point for
Her X-1 and the limits for 4U1626-67 and 4U0115+63 were
excluded from the fit. The intercept at log $(\dot{P}/P)$ = -4.6
is in good agreement with the expected value if the X-ray
emitting objects are accreting neutron stars (see
Rappaport and Joss 1977b). The dashed line is the
expected relation for $\sim 1M_\odot$ white dwarfs.

equation (6) if the X-ray stars are neutron stars with commonly accepted values of mass, radius, and magnetic moment. The dashed line, lying nearly two orders of magnitude below most of the data points, is the expected relation for white dwarfs.

With the knowledge that the X-ray stars in binary X-ray pulsar systems are neutron stars, we would next like to investigate some of the basic properties of these objects, e.g., masses, radii, surface magnetic fields, internal structure, and so forth. Detailed studies of pulse arrival times can yield information about the internal structure of neutron stars (see, e.g., Lamb, Pines and Shaham 1978), studies of energy spectra as a function of pulse phase may provide information on magnetic field strengths (see, e.g., Trümper et al. 1978, Wheaton et al. 1979), and measurements of the orbits of X-ray pulsars and their companions yield information on neutron star masses (see, e.g., Rappaport and Joss 1977c; Avni 1977; Bahcall 1978b). The analysis described earlier in this section directly yields the mass of the neutron star and its uncertainties.

The measured masses for neutron stars in the four binary systems SMC X-1, Cen X-3, 4U0900-40 and 4U1538-52 are shown in Figure 9. Also included are the masses of Her X-1, analyzed in a somewhat different but related manner (see above), and the binary radio pulsar PSR1913+16 (Taylor et al. 1979), added for completeness. These six are the only systems for which reliable measurements of the masses of neutron stars have been made. Not included are the estimates of masses of the compact objects in such systems as LMC X-4 (Hutchings et al. 1978) and Cyg X-2 (Cowley et al. 1979), where we cannot be certain that the compact objects are neutron stars due to the lack of detectable pulsations. Note that the masses given in Figure 9 are all consistent, within the uncertainties, with a mass of 1.4 ± 0.2 $M_\odot$ (shaded region, Figure 9). This is the range of masses that might be expected for neutron stars if they are formed in supernova events resulting from the collapse of the degenerate cores of highly evolved stars (Arnett and Schramm 1973; Iben 1974 and references therein). Furthermore, all of the masses are consistent with neutron-stellar models based on conventional nuclear and high-energy physics (Arnett and Bowers 1977 and references therein).

3.   THREE BINARY X-RAY PULSAR SYSTEMS

a)   4U1626-67

4U1626-67 is a 7.68 s X-ray pulsar with a hard energy spectrum and a pulse profile that varies dramatically with energy (Rappaport et al. 1977). The celestial position of 4U1626-67 was determined by Bradt et al. (1977), and the source was subsequently identified with a faint blue star ($V \simeq 18.7$, $B-V \simeq 0.0$, $U-B \simeq -1.2$) by McClintock et al. (1977). The pulse period for this source is observed to be decreasing at a rate $\dot{P}/P \simeq -1.8 \times 10^{-4}$ yr^{-1} which, from the theory of accretion torques (discussed in §2), implies an X-ray luminosity of $\sim 7 \times 10^{36}$ erg s^{-1}. This luminosity, together with the measured X-ray flux of $\sim 2 \times 10^{-9}$ ergs cm^{-2} s^{-1}, implies a distance of $\sim 3-12$ kpc

(Joss, Avni and Rappaport 1978). The absolute magnitude of McClintock's star is therefore $M_v \gtrsim 3$. This is consistent with a system containing a late-type dwarf companion with negligible intrinsic optical luminosity, wherein most of the light results from reprocessed X-radiation. Giant, supergiant, or Be stellar companions are apparently ruled out.

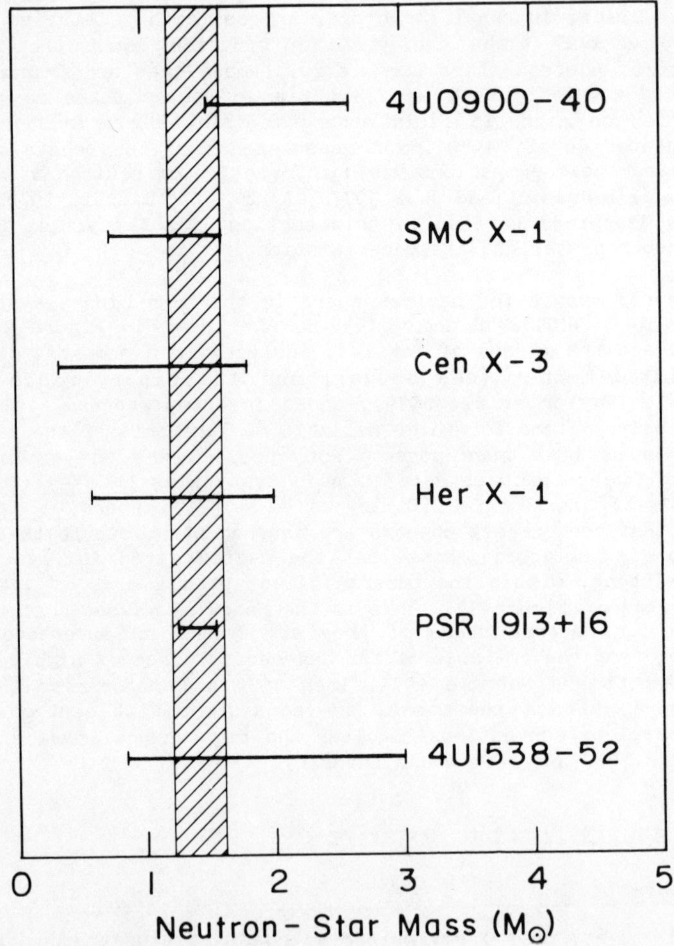

Figure 9: Empirical knowledge of neutron-star masses (from Joss and Rappaport 1980). Five of the neutron-star masses are derived from observations of X-ray binaries (see text). PSR1913+16 is a binary radio pulsar (Taylor et al. 1979) and is added for completeness. The hatched region represents the range of neutron-star masses that might be expected on the basis of current theoretical scenarios for neutron-star formation (see, e.g., Iben 1974).

Severe constraints have been placed on the size of the orbit of the X-ray star. The lack of Doppler shifts in the pulse arrival times has yielded limits on $a_x \sin i$ of 0.1-0.2 lt-sec for orbital periods in the range 1 hr $\lesssim P_{orb} \lesssim$ 20 d (Rappaport et al. 1977; Joss, Avni and Rappaport 1978; Pravdo et al. 1979); these constraints are summarized in Figure 10. With these orbital constraints, reasonable mass functions can be found only for very short orbital periods ($P \lesssim 1/3$ day) or very long orbital periods ($P_{orb} \gtrsim$ 200 days). The lack of a giant or Be star companion tends to rule against very long orbital periods because a smaller star, which would be far from filling its critical potential lobe and which should have a relatively low mass-loss rate, would very probably be unable to supply sufficient matter to the neutron star.

Quasi-periodic oscillations, on a time scale of $\sim$1000 s, in the X-ray intensity of 4U1626-67 were reported by Joss, Avni and Rappaport (1978). A subsequent observation of 4U1626-67 with the SAS-3 satellite (Li et al. 1980) again contained these "oscillations." The X-ray intensity during one 18-hour segment of the pointed observation is shown in Figure 11. The count rate is dominated by rather symmetric, sharply peaked, flare-like events which have a characteristic recurrence time of $\sim$1000 s. The power spectrum from this observation is shown in Figure 12, and the corresponding autocorrelation function is shown in Figure 13. The power spectrum demonstrates that there is a preferred time scale of $\sim$1000 s, but that the flare-like events are not strictly periodic. The autocorrelation function shows that the characteristic width of a flare is $\sim$200 s (FWHM), and also indicates the low quality factor of the oscillations.

To better understand the "clock" mechanism underlying the flaring activity, we have simulated counting rate data from 4U1626-67 with a variety of models for the temporal characteristics of the flares. The three basic models considered were: (a) a clock with white noise errors in phase; (b) a clock with random walk errors in phase; and (c) random arrival times. For white noise phase errors, the arrival time of the nth flare is given by:

$$t_n = nP_o \pm \delta R,$$

where P_o is the period of the underlying clock, δ is a constant , and R is a linearly distributed random number between 0 and 1. The random walk in phase is described by:

$$t_n = t_{n-1} + P_o \pm \delta R.$$

The random arrival-time model for the flares is generated from

$$t_n = t_{n-1} - P_o \ln R.$$

The power spectra of typical simulated data trains for the various models are shown in Figure 14. Note that for the case of the random-arrival-time model, the power spectra continue to rise towards decreasing frequencies. In the other two models, for moderately

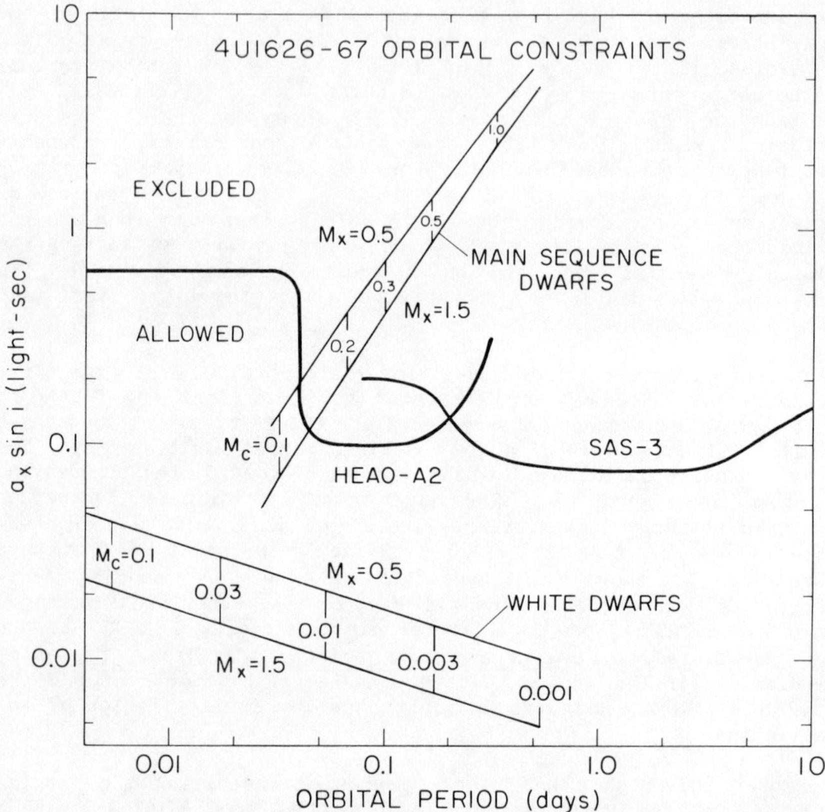

Figure 10: Orbital constraints for the 4U1626-67 system. The upper limits for $a_x \sin i$ (heavy curve) are derived from SAS-3 data (Li et al. 1980) and HEAO-1 data (Pravdo et al. 1979). Under the assumptions that 4U1626-67 is a compact star of mass M_x, in a circular orbit (with $\sin i = 1$) about a companion that corotates with the orbit and fills its critical Roche lobe, each point in the $a_x \sin i$-P_{orb} plane uniquely corresponds to a mass and radius for the companion star (Joss, Avni and Rappaport 1978). For these assumptions, the loci of points for main sequence dwarfs are shown for a range of possible masses for the X-ray star; a simple mass-radius relation , $M_c/M_\odot \simeq R_c/R_\odot$ for the companion, was used. The loci of points for degenerate dwarf companions are also shown for a range of possible masses for the X-ray star. The mass-radius relation for degenerate dwarfs, with mean molecular weight per electron $\mu_e = 2$, was taken from Paczyński (1967). Very low mass main sequence dwarfs as well as degenerate dwarf companions are consistent with the limits on $a_x \sin i$. For values of $\sin i < 1$ the loci of companion star parameters are shifted down by a factor of $\sin i$.

large values of δ (0.4 P_o $\lesssim$ δ $\lesssim$ 0.8 P_o), the frequency of the underlying clock is masked, although the power spectrum still exhibits a preferred time scale; in particular, it decays towards low frequencies. We conclude from this study that while it is difficult to make a precise statement about the statistical character of the flares´ arrival times, it is apparent that they are not random.

The ~1000 s oscillations could be the result of quasi-periodic (a) mass transfer, e.g., caused by pulsations of the companion star, (b) episodes of accretion from a disk, or (c) obscuration of X-rays by matter orbiting near the outer edge of an accretion disk. The first of these possibilities is unlikely because the ~200 s rise-time of the flares would probably not be preserved if the mass accretion is mediated by a disk. It is unclear whether mechanism (b), quasi-periodic episodes of accretion from a disk, can account for either the timescale or the symmetric rise and fall of the flares. It is also difficult to explain the sharply peaked flares by means of obscuration by orbiting matter (mechanism c), and, furthermore, the available spectral data are inadequate to determine whether the

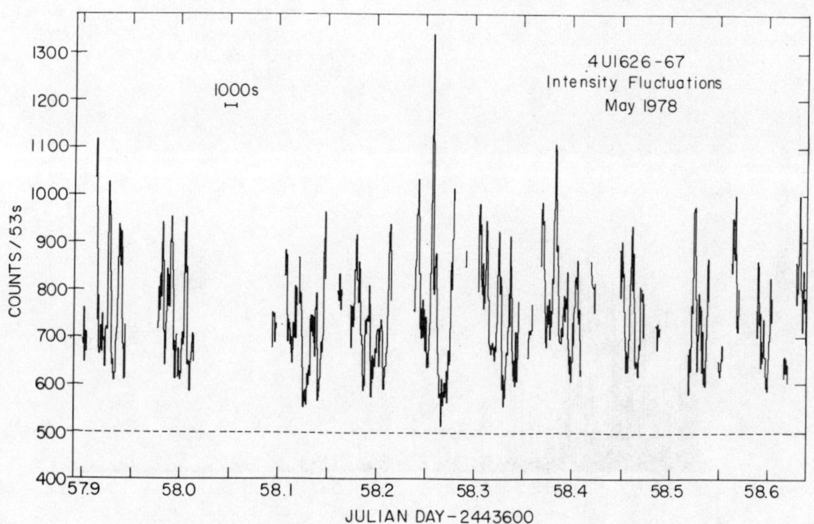

Figure 11: SAS-3 counting rate data from 4U1626-67 in the 1.5-12 keV energy band (from Li et al. 1980). The dashed line indicates the non-source background level. The large flares in intensity have a characteristic recurrence time of ~1000 s. Gaps in the data train are times when the source was occulted by the earth, the satellite was in the South Atlantic Anomaly, or data were lost in transmission.

amplitudes of the flares are highly energy dependent (cf. Joss, Avni and Rappaport 1978). If the ~1000 s oscillations could be physically linked to the orbital motion (i.e., mechanism c), our understanding of this system would be greatly enhanced. At present, however, this connection has not been made.

Although we have not measured orbital parameters for 4U1626-67, we can present a model for the system that explains many of its observed properties. Consider a main sequence dwarf with $M_c \simeq 0.1\ M_\odot$ in orbit about a neutron star with $M_x \simeq 1\ M_\odot$ (Figure 15). This type of system would be consistent with the stringent limits on $a_x \sin i$ shown in Figure 10. Such a system may have evolved from a cataclysmic variable, wherein the matter accreting onto the white dwarf eventually drives it over the Chandrasekhar limiting mass, to form a neutron star. This evolutionary scenario has been discussed by Gursky (1976), van den Heuvel (1977) and Joss and Rappaport (1979).

Mass transfer in the binary system pictured in Figure 15 could be driven by the decay of the orbit due to gravitational radiation. The timescale, $T \sim P_{orb}/(dP_{orb}/dt)$, for orbital decay can be calculated from the Einstein formula for gravitational quadrupole radiation

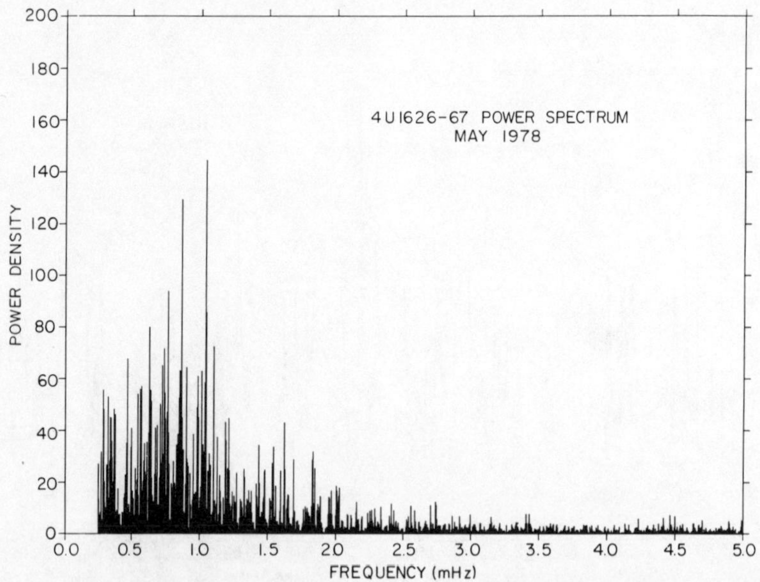

Figure 12: Power density spectrum for 4U1626-67 computed from the X-ray counting rate data, a portion of which is shown in Figure 11 (from Li et al. 1980). Note the enhanced power density for frequencies near 1 mHz (~1000 s period). The plot is terminated for frequencies below ~0.25 mHz because of excessive noise near the ~95 min orbital period of the satellite.

(Paczyński 1967):

$$T = 1 \times 10^7 (M_x + M_c)^{1/3} M_x^{-1} M_c^{-1} P_{orb}^{8/3} \text{ yrs} \quad , \tag{7}$$

where all masses are in units of solar masses and P_{orb} is now in units of hours. Under the assumptions that no mass is lost from the system during mass transfer, and that the main sequence dwarf continues to fill its Roche lobe as the orbit decays, the mass transfer rate can be computed analytically (Paczyński 1967; Faulkner 1971). For a main sequence dwarf that follows the mass-radius

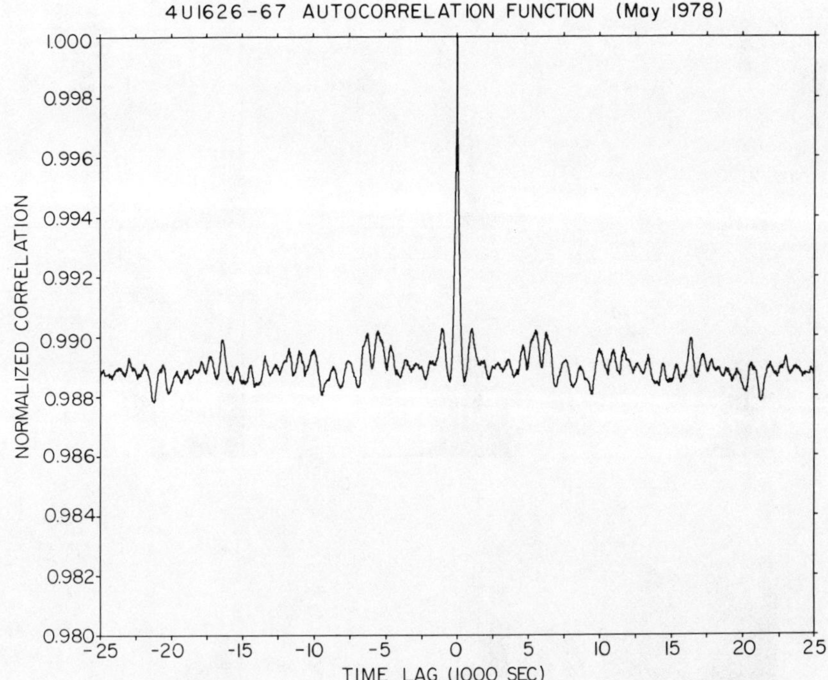

Figure 13: Autocorrelation function for 4U1626-67 computed from the X-ray counting rate data, a portion of which is shown in Figure 11 (from Li et al. 1980). The characteristic width (FWHM) of the flares inferred from the central peak in the autocorrelation function is ~200 s. The data are nearly uncorrelated for time lags $\gtrsim$ 1000 s. The smaller correlation features, spaced by ~1000 s, are due to the finite number (~100) of large flares in the data. The small enhancement near ~6000 s is due to earth occultations of the source.

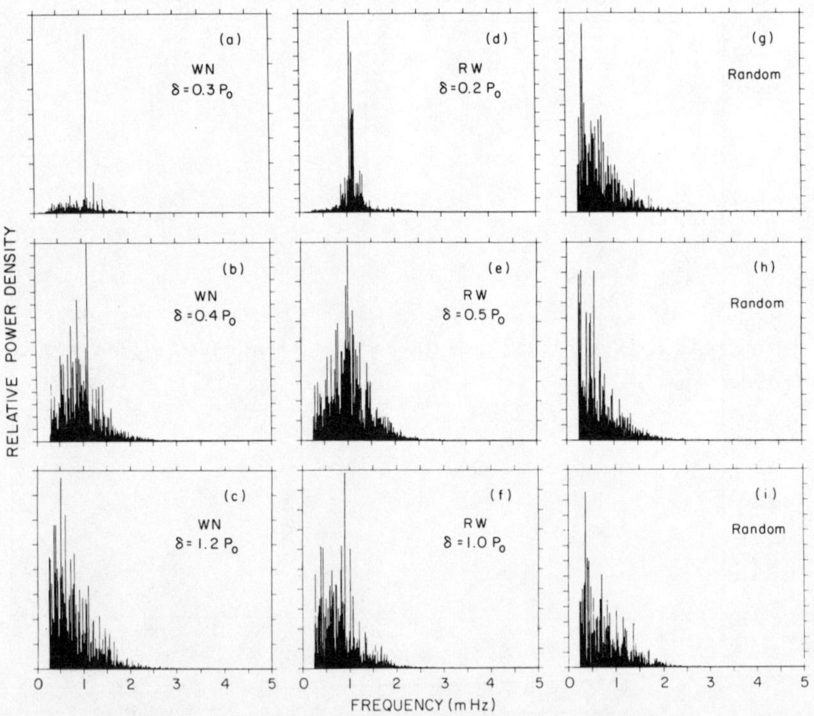

Figure 14: Power spectra for simulated data trains of 4U1626-67. Models for the temporal characteristics of the flare arrival times included: i) white noise (WN) errors in phase [(a)-(c)] , ii) random walk (RW) in phase [(d)-(f)] , and iii) random arrival times [(g)-(i)] (see text for details). The simulated power spectra are terminated for frequencies below 0.25 mHz to match the power spectrum from 4U1626-67 shown in Figure 12.

relation, $M_c/M_\odot \simeq R_c/R_\odot$, the mass accretion rate is:

$$\dot{M} \simeq 7 \times 10^{15} (1-\mu)^2 \mu^{-2/3} (4-7\mu)^{-1} \text{ g sec}^{-1} \quad , \qquad (8)$$

where $\mu \equiv M_c/(M_x + M_c)$.

For the system parameters sketched in Figure 15, $\mu \simeq 0.09$ and $\dot{M} \simeq 10^{16}$ g s^{-1}, which corresponds to an X-ray luminosity of $\sim 10^{36}$ ergs s^{-1}. This is marginally consistent with the estimated luminosity of $\sim 7 \times 10^{36}$ ergs s^{-1}, which is uncertain by perhaps a factor of 5. Because the gravitational timescale in this system ($\sim 10^8$ yrs) is smaller than the Kelvin time of the low-mass main sequence dwarf ($\sim 10^9$ yrs), it is unlikely that the assumed mass-radius relation continues to be appropriate as matter is stripped from the dwarf, and hence equation (8) may have to be substantially modified. It is also possible that the mass transfer is driven by a self-excited stellar

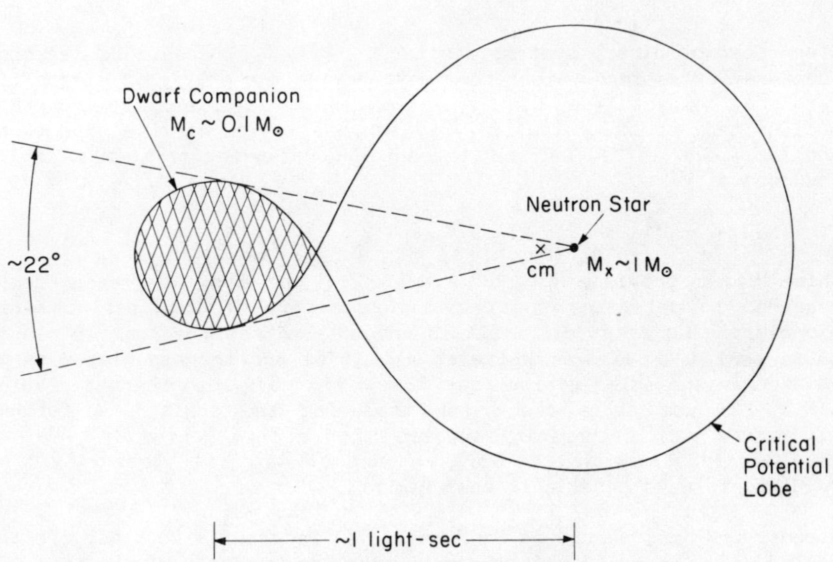

Figure 15: Compact-binary model for 4U1626-67 (adapted from Joss and Rappaport 1979). The figure is drawn to scale for illustrative values of M_c and M_x. *(Reproduced by kind permission from Astronomy & Astrophysics, Vol 71, p217, 1979).*

182

wind (Faulkner 1974; Basko et al. 1977) and/or by the evolution of
the late-type dwarf companion (Faulkner 1974).

Several other properties of 4U1626-67 can be understood in terms of
this compact binary model. The binary separation in Figure 15 is
about 1 lt-sec, and the distance of the neutron star from the center
of mass is $\leq$0.1 lt-sec. This is consistent with the stringent limits
placed on $a_x \sin i$ (Figure 10). The eclipse probability for the
inverted mass ratio in this system (Joss and Rappaport 1979) is given
by

$$P_{ecl} \simeq 0.46 \ (1+q)^{-1/3} \simeq 0.2 \ . \qquad (9)$$

Thus, the lack of observed eclipses in this system is not surprising.
Finally we note that the ratio of the X-ray luminosity of 4U1626-67
to the intrinsic optical luminosity of the companion star would be
$\geq 10^5$. Therefore, if any significant fraction of the X-ray flux is
reprocessed into optical light, this would dominate the optical
emission of the binary system. Direct evidence for such reprocessing
is provided by the discovery of optical pulsations at the 7.68 s
X-ray period (Ilovaisky et al. 1978). No periodic variations in the
optical pulse period, corresponding to orbital motion, have been
found and the reprocessing thus probably takes place in an accretion
disk rather than on the surface of the companion star (Grindlay et
al. 1978; McClintock et al. 1980). The latter process would produce
an optical light curve similar to that seen in Her X-1 (see, e.g.,
Bahcall, Joss, and Avni 1974).

Other compact binary systems similar to 4U1626-67 should be searched
for X-ray pulsations, since they may provide the only good diagnostic
probes of compact X-ray binary systems. This is especially important
because many of the galactic X-ray sources may be of this type (Joss
and Rappaport 1979), and our present understanding of them is still
inadequate.

b) GX301-2 (4U1223-62)/ WRA 977

This system provides a good example of the type of measurements
required to determine the orbital parameters of long period X-ray
binaries. The first difficulties encountered are the long intrinsic
pulse period ($\sim$696 sec; White et al. 1976) and the associated rapid
changes in the pulse period (see §2 and Fig. 8); both characteristics
limit its use as a clock for measuring the orbit. A further
difficulty with determining the orbit of a long period ($P_{orb} \gtrsim 10$
days) X-ray binary is the large amount of satellite observing time
required to make a reliable measurement.

Recently, White, Mason and Sanford (1978) reported a 40.8 day orbital
period for GX301-2 based on two X-ray observations of duration $\sim$20
and $\sim$70 days, respectively, and separated by $\sim$140 days. In contrast,
optical observers (Hammerschlag-Hensberge et al. 1976; Pakull 1978,
1979) have found evidence for 22.6 day ellipsoidal light variations
in studies of the companion star, WRA 977 (Vidal 1973; Dower et al.
1978).

Motivated by these findings, we observed GX301-2 for 30 days with the Y-axis detectors of SAS-3 (Lewin et al. 1976). For each of 109 satellite orbits of data we obtained a pulse "arrival time" (see, e.g., Rappaport et al. 1978). The arrival times were then fitted with a function representing a general eccentric Keplerian orbit, plus a term that represents a constant rate of change in intrinsic pulse period. For each trial orbital period greater than ~23 d, good fits were obtained and stringent constraints on the orbital parameters were set. These orbital constraints, superposed on contours of constant mass function in the $a_x \sin i$-P_{orb} plane, are shown in Figure 16. Note that the orbital period suggested by the

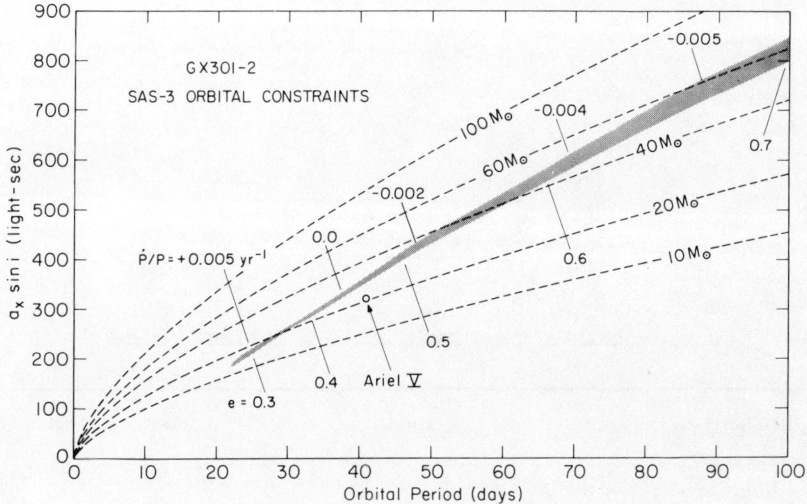

Figure 16: Constraints on the orbital elements of GX301-2 determined from the SAS-3 data (from Kelley, Rappaport and Petre 1980). The shaded region represents the range of allowed values of $a_x \sin i$ (95% confidence) as a function of trial orbital period, P_{orb}. For various trial orbital periods, the best-fit values of the eccentricity, e, and the fractional rate of change in the pulse period, $\dot{P}/P$, are indicated. The dashed curves are contours of constant mass function. Also shown is the orbital solution of White, Mason, and Sanford (1978).

184

Ariel V group is consistent with our results, while the shorter
period suggested by the optical observations is only marginally
consistent.

In an attempt to determine the orbital parameters of the system
uniquely, we have combined the SAS-3 data with the Ariel V pulse
arrival time data (White 1979). The combined data set was fitted
with a general orbit and a variety of $\dot{P}$ and $\ddot{P}$ terms for the various
data segments [see Kelley, Rappaport and Petre (1980) for details of
the analysis]. We find that the best-fitting orbits have P_{orb} of
32.5 and 35.0 days, respectively. However, we regard the 32.5 day
orbit as less probable because the fits for this orbit require a
large and positive value of $\dot{P}/P$ (+2% yr^{-1}) during the first segment
of Ariel V data. For both orbital solutions, the basic parameters of
the system are very similar, and the conclusions drawn from the
analysis will be almost independent of which orbit is correct.

The 35 day orbit has an eccentricity of 0.44 and a semi-major axis
that is ~2/3 of an AU. The Doppler delays expected for this orbit
are compared to the pulse arrival times in Figure 17. A schematic
for the GX301-2/WRA 977 binary system (35-day orbit), is shown in
Figure 18; the orbital parameters are listed in Table 2.

<u>Table 2</u>

<u>Tentative Orbital Parameters for the GX301-2 Binary System</u>*

Projected Semi-Major Axis	$a_x \sin i$	304 ± 5 lt-sec
Orbital Period	P_{orb}	34.99 ± 0.02 days
Mass Function	$f(M)$	$25.2\ M_0$
Eccentricity	e	0.44 ± 0.02
Longitude of Periastron	ω	$-46^{\circ} \pm 5^{\circ}$
Time of Periastron Passage	τ	JD $2,443,906.7 \pm 0.3$

* Solution A of Kelley, Rappaport and Petre (1980). Uncertainties are
  approximate single parameter 95% confidence limits.

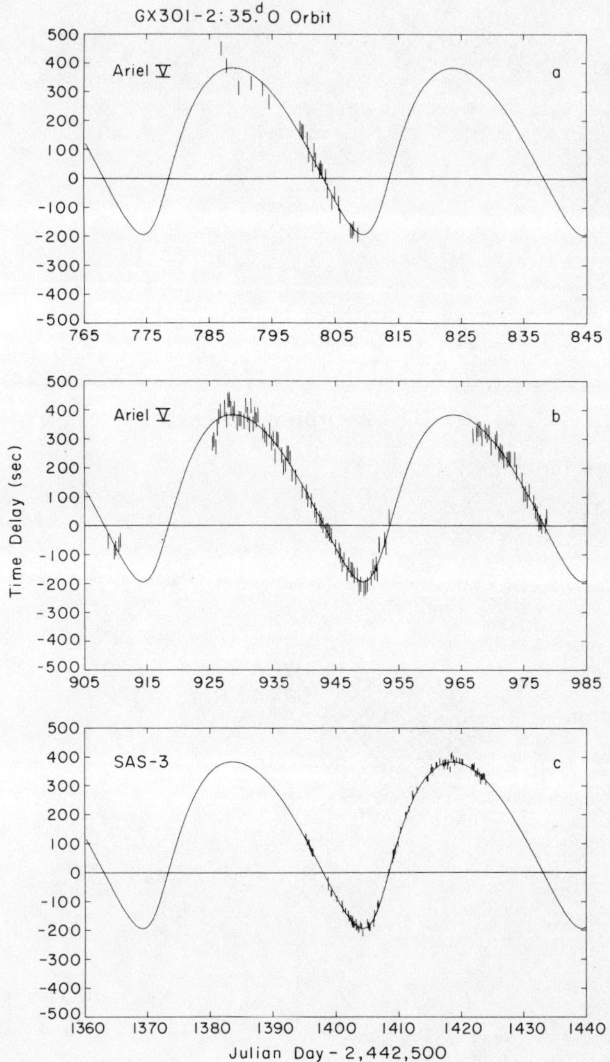

Figure 17: Doppler delays for GX301-2 (from Kelley, Rappaport and Petre 1980). The solid curves represent the delays predicted by the orbital parameters in Table 2. The vertical bars are the measured delays after a second-order polynomial has been subtracted from the data. The length of each bar represents ±1σ uncertainties in the arrival times: ~10 s for the SAS-3 data and ~26 s for the Ariel V data (see Kelley, Rappaport and Petre 1980; White, Mason and Sanford 1978).

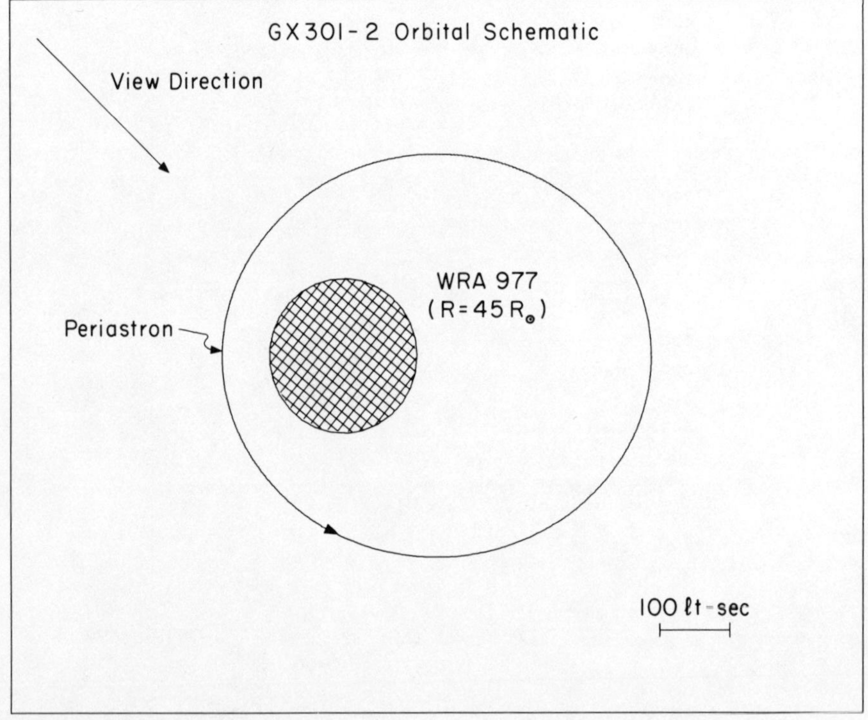

Figure 18: Orbital schematic for the GX301-2/WRA 977 system, derived from parameters in Table 2 (from Kelley, Rappaport, and Petre 1980). An illustrative radius of ~45 R$_\odot$ has been adopted for the companion star WRA 977 (Hammerschlag-Hensberge et al. 1976) which is placed at the center of mass of the system. For orbital inclination angles, i, less than 90°, the size of the orbit scales as 1/sin i.

GX301-2 has the largest orbit and greatest eccentricity of any of the seven X-ray binaries with measured orbits (see, e.g., Rappaport et al. 1978). Despite the large orbit and the large value of the eccentricity, GX301-2 is not classified as a transient source (cf. 4U0115+63; Rappaport et al. 1978), and in fact is usually bright in X-rays. The source is variable and, on occasion, flares up to ~5 times its normal intensity (White, Mason and Sanford 1978; Kelley and Bradt 1979). There is some evidence that these "flaring" episodes coincide with periastron passages.

We suggest two possible mechanisms for mass transfer in the GX301-2/WRA 977 system. Both are slight variants of the conventionally accepted pictures.

In the first scenario, mass transfer takes place via a stellar wind (see, e.g., Davidson and Ostriker 1973). Because the mass capture rate by the neutron star depends on both the stellar wind density and its speed relative to the neutron star, there should be significant variations throughout the cycle of the eccentric orbit (see, e.g., Okuda and Sakashita 1977). In particular, if the captured matter is accreted onto the surface of the neutron star in a time short compared to P_{orb}, then one expects an asymmetric X-ray light curve that can reach maximum intensity at any part of the orbit, depending on the parameters of the orbit and the stellar wind (see Figure 9 of Kelley, Rappaport and Petre 1980). Both the SAS-3 and Ariel V data show qualitative evidence for such temporal behavior with a maximum in intensity near periastron passage (White, Mason and Sanford 1978; Kelley, Rappaport and Petre 1980).

An alternative mechanism involves mass transfer through the inner Lagrange point of a critical potential surface whose size and shape varies with orbital phase. We have used the potential given by Avni (1976) together with the measured orbital parameters to compute the size of the critical potential lobe as a function of orbital phase. These calculations, carried out for a range of possible rotation rates for WRA 977 (Kelley, Rappaport and Petre 1980), indicate that the size of the critical lobe can vary from ~50 $R_{\odot}$/sin i at periastron passage to ~140 $R_{\odot}$/sin i at apastron. Thus, the companion star WRA 977 (Hammerschlag-Hensberge et al. 1976) may fill or nearly fill its critical potential lobe at periastron. If the transferred matter is stored in an accretion disk for a time $\geq P_{orb}$, then the source could appear variable but should not resemble the X-ray transients.

c) 4U0900-40/ HD77581

Almost all of our knowledge of stellar interiors is derived from theoretical studies. At present, very few observations can be performed to test any theoretical predictions. Two direct measurements are, however, possible: the measurement of the neutrino flux from the sun (Davis 1978) and the apsidal motion test for binary stellar systems (Russell 1928). In the former case, one attempts to gain information about the thermal and chemical structure of the solar interior, while in the latter, a one-parameter measure of the mass distribution within a star (in a binary system) may be obtained.

Reliable measurements of apsidal motion, which results from the tidally and rotationally induced gravitational quadrupole moments of the stars, have been made for about fourteen close binary systems (Kopal 1965; see Stothers 1974 for more recent references), wherein both stars are nondegenerate and are usually on the main sequence. The rate at which the longitude of periastron of an eccentric orbit advances can then be related to the "apsidal motion constant" k, which is calculated for various stellar model mass distributions under standard theoretical assumptions (see Schwarzschild 1958). In general, the measured values of k are smaller than the standard calculated values by factors of ~2. In all cases, however, the interpretation is complicated by the fact that both stars of the binary system contribute to the apsidal motion.

Binary X-ray pulsars, on the other hand, provide a potentially less ambiguous means of carrying out these measurements. In such systems one of the stars is a neutron star, which acts essentially as a point mass and has a completely negligible apsidal motion constant compared to that of the companion star. Of the well studied X-ray binaries, Her X-1 (Fechner and Joss 1977), Cen X-3 (Fabbiano and Schreier 1977) and SMC X-1 (Primini, Rappaport and Joss 1977) have highly circular orbits, and 4U1538-52 is not sufficiently well measured to signif- icantly constrain the eccentricity (Becker et al. 1977; Davison, Watson, and Pye 1977). Only 4U0900-40 (Rappaport, Joss and McClintock 1976), GX301-2 (White et al. 1978; Kelley et al. 1979), and 4U0115+63 (Rappaport et al. 1978) are appreciably eccentric. The latter is a transient source that was detected in X-rays on only two occasions separated by ~7 years. The orbit of GX301-2 has only recently been measured and the observational baseline is too short to allow a determination of the rate of apsidal motion, $\dot{\omega}$. 4U0900-40 is therefore the best system known in which one can attempt a measure- ment of $\dot{\omega}$.

The orbit of 4U0900-40 was first determined in July 1975 (Rappaport, Joss and McClintock 1976) and found to have a small but detectable eccentricity, e = 0.12 $\pm$ 0.02 (1σ), with longitude of periastron ω = 146° $\pm$ 11° (1σ). Subsequent measurements of the orbit by Ögelman et al. (1977) and Becker et al. (1978) confirmed the eccentricity but had neither the precision nor the temporal baseline to attempt a measurement of apsidal motion.

The pulse phase of 4U0900-40 was again tracked for one orbital cycle with SAS-3 during November 1978 (Rappaport, Joss and Stothers 1980). The Doppler delays derived from these new observations of 4U0900-40 are shown in Figure 19. Figure 19a displays a fit to a circular orbit (i.e., a fit with e set equal to zero); the residual delays, after subtracting a best-fitting circular orbit, show strong evidence for a periodic variation with half the orbital period ($P_{orb}/2$) as one would expect if the orbit is eccentric. The best-fitting eccentric orbit is shown in Figure 19b. The residuals from this fit are significantly reduced and appear more randomly distributed throughout the orbital cycle. The orbital parameters derived from the new

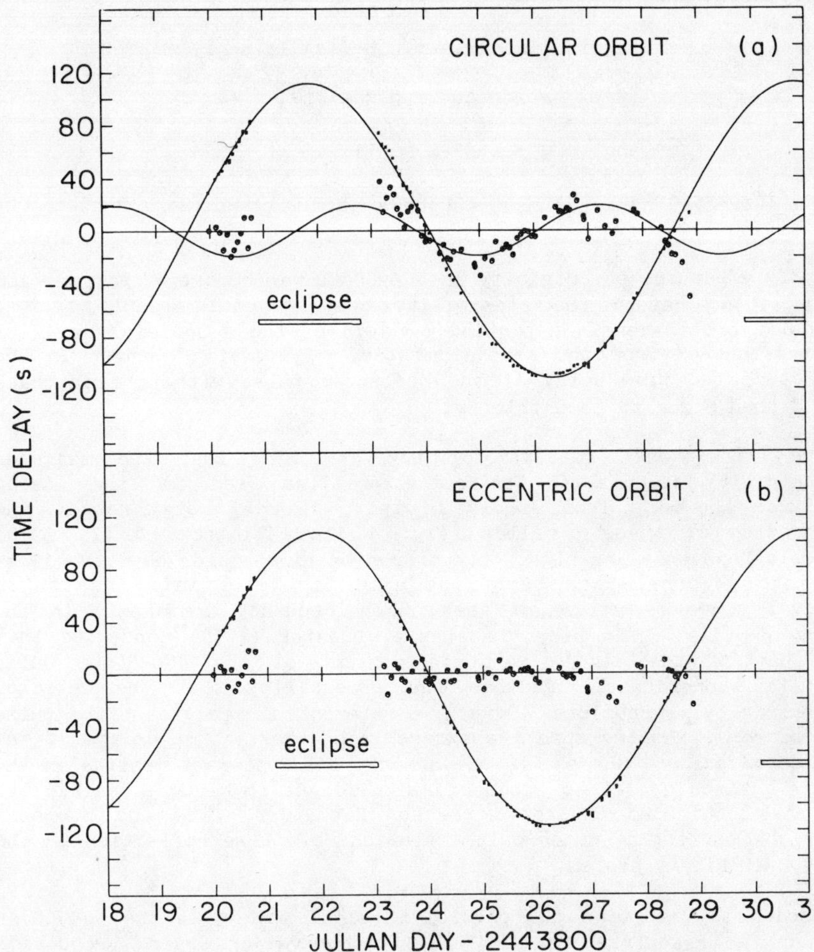

Figure 19: Doppler corrections to the SAS-3 1978 November 4U0900-40
pulse arrival times as a function of 8.9649 d orbital
phase for (a) the best-fit circular orbit and (b) the
best-fit eccentric orbit (from Rappaport, Joss and
Stothers 1980). The vertical bars are the measured delays;
the height of each bar represents ±1σ error limits. The
large-amplitude solid curves are the best-fit orbits; the
circles are the residuals, multiplied by a factor of 5.
The small-amplitude sinusoidal curve in (a), with period
$P_{orb}/2$, indicates the systematic trend in the residuals of
the circular orbital fit. The indicated eclipse intervals
are derived from the best-fit orbital parameters and the
representative eclipse duration of 1.90 d that was
observed by Forman et al. (1973).

190

observations are:

$$a_x \sin i = 112.3 \pm 0.8 \; (1\sigma) \; \text{lt-sec},$$

$$e = 0.095 \pm 0.005 \; (1\sigma)$$

$$\text{and} \quad \omega = 158^{\circ} \pm 5^{\circ} \; (1\sigma).$$

Note, in particular, that the apparent eccentricity has a statistical significance of ~20σ.

Shortly after the eccentricity in 4U0900-40 was reported, Milgrom and Avni (1976) raised the possibility that it could result from a circular orbit wherein a portion ($\lesssim$ 10%) of the detected X-ray flux is "reflected" from the companion star. Systematic changes in the amplitude and time delay of the reflected pulse with orbital phase could then mimic an eccentricity.

There are several reasons for now believing that the measured eccentricity is real. The pulse profiles used for the timing analyses are made from X-rays in the 3-12 keV energy range. An inspection of these profiles (Fig. 1 of McClintock et al. 1976) reveals that they are highly structured on time scales down to ~15 s. The cross-correlation technique used for determining the pulse arrival times makes use of these high-frequency components in the pulse profile. In a recent analysis, Chester (1979) concluded that the use of the higher harmonic content of the 4U0900-40 pulse profiles considerably reduces the probability that the detected eccentricity is spurious. When the relevant time scale of the pulse structure is shorter than the excess light-travel-time delays of the reflected pulses, one should not observe the systematic shifts in the arrival times needed to produce a significant spurious eccentricity. We have confirmed Chester's result (Rappaport, Joss and Stothers 1980) by performing numerical simulations of pulse reflection in the 4U0900-40 binary system.

In order to determine the possible apsidal motion in the 4U0900-40 system, we reanalyzed the 1975 SAS-3 observations using production data and performed an orbital fit to the joint 1975/1978 data set (Rappaport, Joss and Stothers 1980). We find an improved orbital period of 8.9649 d $\pm$ 0.0002 d (1σ) but obtain only an upper limit to the rate of advance of the longitude of periastron: $\dot{\omega} < 3^{\circ}_{.}8 \; \text{yr}^{-1}$ (97% confidence).

For a binary system where the orbital and stellar parameters are known, limits on $\dot{\omega}$ can be used to constrain the allowed values of k through the relation

$$\dot{\omega} \simeq \left(\frac{2\pi k}{P_{\text{orb}}}\right)\left(\frac{R_c}{D}\right)^5 \left\{15qg(e) + \Omega^2(1+q)h(e)\right\} \tag{10}$$

(Cowling 1938; Sterne 1939). The factors g(e) and h(e) in this

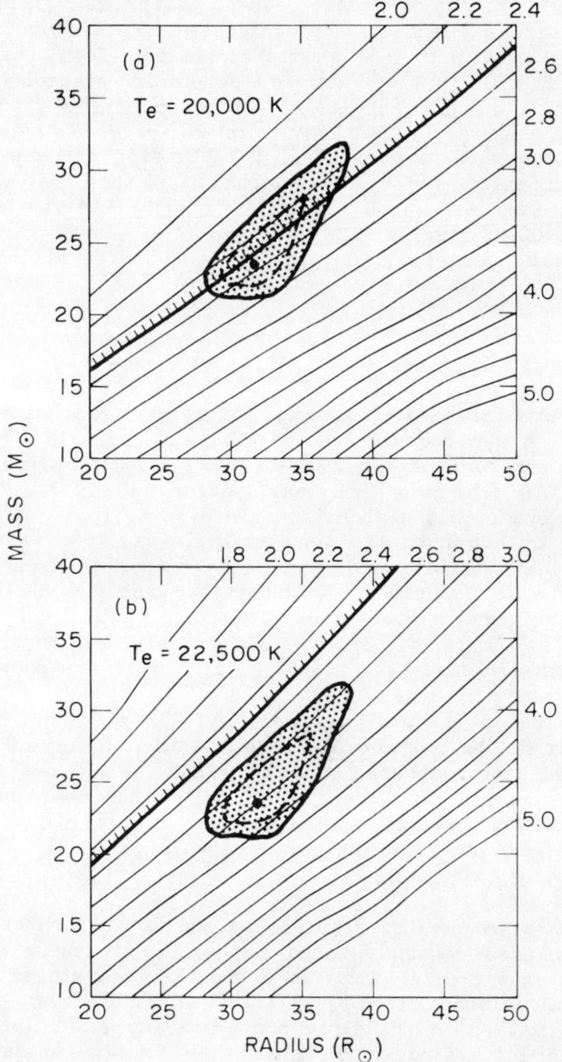

Figure 20: Contours of apsidal motion constant, k, in the mass/radius plane, calculated for two assumed values of the effective temperature, T_e.   (a) T_e = 20,000 K; (b) T_e = 22,500 K (from Rappaport, Joss, and Stothers 1980). The numbers to the top and to the right of each figure denote the value of -log k for the various contours.   The shaded zone represents the error region for the companion star HD 77581 in the 4U0900-40 binary system (see also Figure 5). The heavy curve is the SAS-3 measured limit on k and separates the allowed (lower) from the excluded (upper) region.

192

expression are slightly greater than unity and incorporate the contributions to $\dot{\omega}$ from the non-infinitesimal eccentricity of the orbit. (The contribution to apsidal motion from the rotational distortion of the primary was neglected in Rappaport, Joss and Stothers (1980), but in fact its effect can be significant. However, this omission does not affect any of their results or conclusions.) The contribution to $\dot{\omega}$ from general relativistic effects (see, e.g., Landau and Lifshitz 1962) is negligible in this system. We have evaluated the quantity $(R_c/D)^5(15qg + \Omega^2(1+q)h)$ and its uncertainties for the 4U0900-40 system by means of the Monte Carlo analysis described in §2. Our limit on apsidal motion then yields a limit on log k ($\lesssim -2.5$) that is a direct measure of the structure of the companion star HD 77581.

We have calculated values of k for a wide range of stellar models (Rappaport, Joss and Stothers 1980). Companion masses in the range 10-40 $M_\odot$, radii of 20-50 $R_\odot$, and effective temperatures of T_e = 17,000-28,000 K were considered. The value usually quoted for the effective temperature of the B0.5 Ib supergiant, HD 77581, is 22,500 (+5000,-2500) K (Morgan, Code and Whitford 1955; Avni and Bahcall 1975b) . However, only combinations leading to luminosities that are greater than or equal to the zero-age main-sequence values and less than the Eddington limit were used. The method of generation of the stellar models is described by Rappaport, Joss and Stothers (1980). The results of these calculations are shown in Figure 20 as contours of constant log k in the M_c- R_c plane; only the results for T_e = 20,000 K (Fig. 20a) and T_e = 22,500 K (Fig. 20b) are shown.

We have evaluated M_c and R_c for HD 77581 and the corresponding uncertainties by the methods discussed in §2. The results are shown superposed on the contours of k in Figure 20. The shaded error region for M_c and R_c and the inner region enclosed by the dashed curve contain 98% and 70% of the Monte Carlo events, respectively. The filled circle denotes the most probable parameter values (i.e., M_c = 24 $M_\odot$ and R_c = 32 $R_\odot$).

The heavy curve in each of Figures 20a and 20b separates the allowed from the excluded regions, based on our limit for k (i.e., log k $\lesssim -2.5$). The experimental limit on k and the theoretical calculations of k are inconsistent for T_e $\leq$ 18,000 K (not shown in Fig. 20). Thus, these results in principle constrain the effective temperature of HD 77581 (cf. Nandy and Schmidt 1975, who suggest that the effective temperature of the B0 Ia star ϵ Ori could be as low as ~18,000 K). For T_e = 20,000 K, the lack of observed aspidal motion is just consistent with theoretical expectations. For T_e $\gtrsim$ 23,000 K, aspidal motion should not have been observed.

We anticipate that another measurement of $\dot{\omega}$, three or more years in the future, should be able to detect the apsidal motion of 4U0900-40. This will provide a unique measure of the internal mass distribution of an evolved supergiant. Such a measurement will, in turn, yield a direct check on theoretical stellar models and on our understanding of the evolution of close binary stellar systems.

The author is grateful to P.C. Joss, T. Chester, J. Faulkner, R. Kelley, D. Lamb, F. Li, J. van Paradijs, R. Webbink, and N. White for helpful discussions. I also acknowledge Ms. T. Welch for her able assistance in the preparation of the manuscript and M. Brodheim and J. Pelz for help carrying out several of the calculations involved in this work.

REFERENCES

Arnett, W.D., and Bowers, R.L., 1977. Astrophys. J. (Suppl.), 33, 415.
Arnett, W.D., and Schramm, D.N., 1973. Astrophys. J., 184, L47.
Avni, Y., 1976. Astrophys. J., 209, 574.
Avni, Y., 1977. Talk presented at the 16th General Assembly of the
      IAU, in "Highlights of Astronomy", 4, in press.
Avni, Y., 1978. In "Physics and Astrophysics of Neutron Stars and
      Black Holes", eds. R. Giacconi and R. Ruffini, Amsterdam: North
      Holland, p. 43.
Avni, Y., and Bahcall, J.N., 1975a. Astrophys. J., 197, 675.
Avni, Y., and Bahcall, J.N., 1975b. Astrophys. J. (Letters), 202, L131.
Avni, Y., and Bahcall, J.N., 1976. X-ray Binaries, NASA SP-389), p. 615.
Baan, W.A., and Treves, A., 1973. Astron. Astrophys., 22, 421.
Bahcall, J., 1978a. In "Physics and Astrophysics of Neutron Stars and
      Black Holes,"ed. R. Giacconi and R. Ruffini, Amsterdam: North Holland.
Bahcall, J., 1978b. Ann Rev. of Astron. Astrophys., 16, 241.
Bahcall, J.N., Joss, P.C., and Avni, Y., 1974. Astrophys. J., 191, 211.
Bahcall, J.N., and Chester, T.J., 1977. Astrophys. J. (Letters), 215,
      L21.
Basko, M.M., Hatchett, S., McCray, R., and Sunyaev, R.A., 1977. Astro-
      phys. J., 215, 276.
Becker, R.H., Swank, J.H., Boldt, E.A., Holt, S.S., Pravdo, S.H., Saba,
      J.R., and Serlemitsos, P.J., 1977. Astrophys. J. (Letters), 216,
      L11.
Becker, R.H., Rothschild, R.E., Boldt, E.A., Holt, S.S., Pravdo, S.H.,
      Serlemitsos, P.J., and Swank, J.H., 1978. Astrophys. J., 221,
      912.
Bradt, H.V., Apparao, K.M.V., Dower, R., Doxsey, R.E., Jernigan, J.G.,
      and Markert, T.H., 1977. Nature, 269, 496.
Bradt, H.V., Doxsey, R.E., and Jernigan, J.G., 1979. In "X-ray Astronomy",
      (COSPAR), eds. W.A. Baity, and L.E. Peterson, Pergamon Press, Oxford.
Chester, T.J., 1979. Astrophys. J., 229, 1085.
Chevalier, C., and Ilovaisky, S.A., 1977. Astron. Astrophys., 59, L9.
Conti, P.S., 1978. Astron. Astrophys., 63, 225.
Cooke, B.A., Rickets, M.J., Maccacaro, T., Pye, J.P., Elvis, M., Watson,
      M.G., Griffiths, R.E., Pounds, K.A., McHardy, I., Maccagni, D.,
      Seward, F.D., Page, C.G., and Turner, M.J.L., 1978. MNRAS, 182,
      489.
Cowley, A.P., Crampton, D., and Hutchings, J.B., 1979. Astrophys. J.,
      231, 539.
Cowling, T.G., 1938. MNRAS, 98, 734.
Crampton, D., Hutchings, J.B., and Cowley, A.P., 1978. Astrophys. J.
      (Letters), 225, L63.
Davidson, K., and Ostriker, J.P., 1973. Astrophys. J., 179, 585.

194

Davies, R.E., Fabian, A.C., and Pringle, J.E., 1979. MNRAS, 186, 779.

Davis, R. Jr., 1978. Proc. Brookhaven Solar Neutrino Conf., Vol. 1, p1.

Davison, P.J.N., Watson, M.G., and Pye, J.P., 1977. MNRAS, 181, 73P.

Bower, R.G., Apparao, K.M., Bradt, H.V., Doxsey, R.E., Jernigan, J.G., and Kulik, J., 1978. Nature, 273, 364.

Fabbiano, G., and Schreier, E.J., 1977. Astrophys. J., 214, 235.

Faulkner, J., 1971. Astrophys. J. (Letters), 170, L99.

Faulkner, J., 1974. In "IAU Symp", No. 66, eds. R.J. Tayler, (Dordrecht: Reidel), p. 155.

Fechner, W.B., and Joss, P.C., 1977. Astrophys. J. (Letters), 213, L57.

Forman, W., Jones, C., Tananbaum, H., Gursky, H., Kellogg, E., and Giacconi, R., 1973. Astrophys. J. (Letters), 182, L103.

Forman, W., Jones, C., Cominsky, L., Julien, P., Murray, S., Peters, G., Tananbaum, H., and Giacconi, R., 1978. Astrophys. J. (suppl.), 38, 357.

Ghosh, P., Lamb, F.K., and Pethick, C.J., 1977. Astrophys. J., 217, 578.

Ghosh, P., and Lamb, F.K., 1978. Astrophys. J. (Letters), 223, L83.

Ghosh, P., and Lamb, F.K., 1979. Astrophys. J., in press.

Grindlay, J.E., 1978. Astrophys. J., 225, 1001.

Gursky, H., 1976. In "IAU Symp.", No. 73, eds. P. Eggleton, S. Mitton, and J. Whelan, (Dordrecht: Reidel), p. 19.

Hammerschlag-Hensberge, G., Zuiderwijk, E.J., van den Heuvel, E.P.J., and Hensberge, H., 1976. Astron. Astrophys., 49, 321.

Hutchings, J.B., 1976. In "X-ray Binaries", NASA SP-389, p. 531.

Hutchings, J.B., Crampton, D., Cowley, A.P., and Osmer, P.S., 1977. Astrophys. J., 217, 186.

Hutchings, J.B., Crampton, D., and Cowley, A.P., 1978. Astrophys. J., 225, 548.

Hutchings, J.B., Cowley, A.P., Crampton, D., van Paradijs, J., and White, N.E., 1979. Astrophys. J., 229, 1079.

Iben, I., 1974. Ann. Rev. Astron. Astrophys., 12, 215.

Ilovaisky, S.A., Motch, Ch., and Chevalier, C., 1978. Astron. Astrophys., 70, L19.

Ives, J.C., Sanford, P.W., and Bell Burnell, S.J., 1975. Nature, 254, 578.

Johnston, M., Bradt, H., Doxsey, R., Gursky, H., Schwartz, D., and Schwarz, J., 1978. Astrophys. J. (Letters), 223, L71.

Joss, P.C., Avni, Y., and Ra-paport, S., 1978. Astrophys. J., 221, 645.

Joss, P.C., Fechner, W.B., Forman, W., and Jones, C., 1978. Astrophys. J., 225, 994.

Joss, P.C., and Rappaport, S., 1979. Astron. Astrophys., 71, 217.

Katz, J.I., 1977. Astrophys. J., 215, 265.

Kelley, R., Rappaport, S., and Petre, R., 1979. Submitted to Astrophys. J.

Kelley, R., and Bradt, H.V., 1979. Preprint.

Kopal, Z., 1965. Adv. Astron. and Astrophys., 3, 89.

Kylafis, N.K., and Lamb, D.Q., 1979. Astrophys. J. (Letters), 228, L105.

Lamb, F.K., Pethick, C.J., and Pines, D., 1973. Astrophys. J., 184, 271.

Lamb, F.K., Pines, D., and Shaham, J., 1978. Astrophys. J., 224, 969.

Landau, L.D., and Lifshitz, E.M., 1962. "The Classical Theory of Fields", Pergamon Press, Oxford, p. 374.

Lecar, M., Wheeler, J.C., McKee, C.F., 1976. Astrophys. J., 205, 556.

Lewin, W.H.G., Li, F.K., Hoffman, J.A., Doty, J., Buff, J., Clark, G.W., and Rappaport, S., 1976. MNRAS, 177, 93P.

Li, F., Rappaport, S., and Epstein, A., 1978. Nature, 271, 37.

Li, F., Rappaport, S., Joss, P.C., McClintock, J.E., and Wright, E., 1979. In preparation.

Markert, T., Winkler, P.F., Laird, F.N., Clark, G.W., Hearn, D.R., Sprott, G.F., Li, F.K., Bradt, H., Lewin, W.H.G., and Schnopper, H.W., 1979. Astrophys. J. (Suppl.), 39, 573.

Mason, K.O., 1977. MNRAS, 178, 81P.

McClintock, J.E., Rappaport, S., Joss, P.C., Bradt, H., Buff, J., Clark, G.W., Hearn, D., Lewin, W.H.G., Matilsky, T., Mayer, W., Primini, F., 1976. Astrophys. J. (Letters), 206, L99.

McClintock, J.E., Canizares, C.R., Bradt, H.V., Doxsey, R.E., Jernigan, J.G., and Hiltner, W.A., 1977. Nature, 270, 320.

McClintock, J.E., Canizares, C.R., Li, F.K., and Grindlay, J.E., 1979. Preprint.

Middleditch, J., and Nelson, J., 1976. Astrophys. J., 208, 567.

Milgrom, M., and Avni, Y., 1976. Astron. Astrophys., 52, 157.

Morgan, W.W., Code, A.D., and Whitford, A.E., 1955. Astrophys. J. (Suppl.), 2, 41.

Nandy, K., and Schmidt, E.G., 1975. Astrophys. J., 198, 119.

Ogelman, H., Beuermann, K.P., Kanbach, G., Mayer-Hasselwander, H.A., Capozzi, D., Fiordilino, E., and Molteni, D., 1977. Astron. Astrophys., **58**, 385.

Okuda, T., and Sakashita, S., 1977. Astrophys. and Space Sci., 47, 385.

Ostriker, J.P., and Bodenheimer, P., 1968. Astrophys. J., 151, 1089.

Paczynski, B., 1967. Acta Astronomica, 17, 287.

Paczynski, B., 1971. Ann. Rev. Astron. Astrophys., 9, 183.

Pakull, M., 1978. IAU Circ. No. 3317.

Pakull, M., 1979. Private communication.

Parsignault, D., and Grindlay, J., 1978. Astrophys. J., 225, 970.

Petterson, J.A., 1978. Astrophys. J., 224, 625.

Plavec, M., 1968. Adv. in Astron. Astrophys., 6, 201.

Pounds, K.A., Cooke, B.A., Ricketts, M.J., Turner, M.J., and Elvis, M., 1975. MNRAS, 172, 473.

Pounds, K.A., 1976. Talk presented to the Meeting of the American Astronomical Society (HEAD), Cambridge, MA, January 1976.

Pravdo, S.H., White, N.E., BOldt, E.A., Holt, S.S., Serlemitsos, P.J., Swank, J.H., and Szymkowiak, A.E., 1979. Astrophys. J., in press.

Primini, F., Rappaport, S., Joss, P.C., Clark, G.W., Lewin, W., Li, F., Mayer, W., and McClintock, J., 1976. Astrophys. J. (Letters), 210, L71.

Primini, F., Rappaport, S., and Joss, P.C., 1977. Astrophys. J., 217, 543.

Pringle, J.E., and Rees, M.J., 1972. Astron. Astrophys., 21, 1.

Rappaport, S., Joss, P.C., and McClintock, J.E., 1976. Astrophys. J. (Letters), 206, L103.

Rappaport, S., and Joss, P.C., 1977a. Nature, 226, 123.

Rappaport, S., and Joss, P.C., 1977b. Nature, 266, 683.

Rappaport, S., and Joss, P.C., 1977c. Ann. N.Y. Acad. Sci., 302, 460.

Rappaport, S., Markert, T., Li, F.K., Clark, G.W., Jernigan, J.G.,
     and McClintock, J.E., 1977. Astrophys. J. (Letters), 217, L29.

Rappaport, S., Clark, G.W., Cominsky, L., Joss, P.C., and Li, F.K.,
     1978. Astrophys. J. (Letters), 224, Ll.

Rappaport, S., and Li, F.K., 1979. Unpublished SAS-3 data.

Rappaport, S., Joss, P.C., and Stothers, R., 1979. Astrophys. J.,
     in press.

Rappaport, S., and Joss, P.C., 1979. In preparation.

Russell, H.N., 1928. MNRAS, 88, 641.

Schreier, E.J., Levinson, R., Gursky, H., Kellogg, E., Tananbaum, H.,
     and Giacconi, R., 1972a. Astrophys. J. (Letters), 172, L79.

Schreier, E.J., Giacconi, R., Gursky, H., Kellogg, E., and Tananbaum,
     H., 1972b. Astrophys. J. (Letters), 178, L71.

Schwarzschild, M., 1958. "Structure and Evolution of the Stars",
     Princeton University Press.

Sterne, T.E., 1939. MNRAS, 99, 451.

Stothers, R., 1974. Astrophys. J., 194, 651.

Tananbaum, H., Gursky, H., Kellogg, E.M., Levinson, R., Schreier, E.,
     and Giacconi, R., 1972. Astrophys. J. (Letters), 174, L143.

Taylor, J., Fowler, L.A., and McCulloch, P.M., 1979. Nature, 277,
     437.

Trumper, J., Pietsch, W., Reppin, C., Voges, W., Staubert, R., and
     Kendziorra, E., 1978. Astrophys. J. (Letters), 219, L105.

Ulmer, M.P., 1976. Astrophys. J., 204, 548.

van den Heuvel, E.P.J., 1975. Astrophys. J. (Letters), 198, L109.

van den Heuvel, E.P.J., 1976. In "Structure and Evolution of Close
     Binary Systems", ed. P. Eggleton (Dordrecht: Reidel), p. 35.

van den Heuvel, E.P.J., 1977. Ann. N.Y. Acad. Sci., 302, 14.

van Paradijs, J.A., Zuidersijk, E.J., Takens, R.J., Hammerschlag-
     Hensberge, G., van den Heuvel, E.P.J., and de Loore, C., 1977.
     Astron. Astrophys. Suppl., 30, 195.

van Paradijs, J.A., Hammerschlag-Hensberge, G., and Zuiderwijk, E.J.,
     1978. Astron. Astrophys. Suppl., 31, 189.

Vidal, N.V., 1973. Astrophys. J. (Letters), 186, L81.

Watson, M.G., and Griffiths, R.E., 1977. MNRAS, 178, 513.

Wheaton, W.A., et al., 1979. Nature, 282, 240.

White, N.E., 1978. Nature, 271, 38.

White, N.E., 1979. Private communication.

White, N.E., Mason, K.O., Huckle, H.E., Charles, P.A., and Sanford,
     P.W., 1976. Astrophys. J. (Letters), 209, L119.

White, N.E., Mason, K.O., and Sanford, P.W., 1978. MNRAS, 184, 67P.

White, N.E., and Pravdo, S.H., 1979. Preprint.

Young pulsars in binary systems

L. Maraschi and A. Treves

Instituto di Fisica dell'Universita and
Laboratorio di Fisica Cosmica e Tecnologie Relative del CNR,
Milano, Italy.

1.   Introduction

2.   Observational appearance

3.   Three candidates

1.   INTRODUCTION

Many X-ray sources in the Galaxy are binary systems where a black
hole or a neutron star accretes matter from a companion star. The
evolutionary history is fairly well understood in the case of massive
systems (Van den Heuvel, 1978). For low mass X-ray binaries various
possibilities have been suggested -

   a)   the progenitors are massive systems which undergo
         substantial mass loss during their first stage of
         evolution (Gursky, 1978),

   b)   the progenitors are U-Gem type systems where the
         collapse to a neutron star occurs due to accretion
         of the white dwarf above the Chandrasekhar limit,

   c)   formation by capture has been considered especially
         in relation to globular cluster X-ray sources (Clark,
         1975).

Except for case c) the accretion phase should therefore be preceded
by a phase when the binary contained a young collapsed object giving
rise to strong activity through rotational energy loss. (In the
following we shall consistently use the word pulsar only for objects
in this phase.) The problem of a rapidly spinning black hole surrounded by plasma, though possibly relevant, has not been explored yet.
The case of a young neutron star is more tractable, and we shall make
use of the observations and models of isolated pulsars in order to
infer the observational consequences of the presence of a fast
rotating neutron star in a binary system.

The most direct indication of a recent collapse is the presence of a supernova shell associated with the binary system. However active pulsars would also be found in binaries not associated with a supernova shell. In fact, as we discuss in the next section, detectable emission from the pulsar may last $\sim 10^5$y while the possibility of unambiguous association is limited to the early phases of the expansion ($\sim 10^4$y), and this time may be even shorter for the collapse of a white dwarf, which would give rise presumably to a less massive shell.

In the end we consider three cases where pulsars in binaries may have been detected already.

2.   OBSERVATIONAL APPEARANCE

An isolated magnetized rotating neutron star releases energy in the form of particles and waves at a rate given by

$$\frac{dW}{dt} \simeq \frac{2}{3} \frac{\omega^4}{c^3} a^6 B_o^2 \tag{1}$$

where ω is the spin rate, a the radius and B_o the magnetic field of a neutron star.

In fig. 1 the energy loss and period are plotted as a function of the pulsar age. It is important to realise that only a fraction of the rotational energy loss is in the form of pulsed emission and, for an isolated pulsar, the rest goes undetected. We recall however the case of the wisp activity near the Crab pulsar which may represent an example of the interaction of the "relativistic wind" emitted by the pulsar with the surrounding matter (Rees and Gunn, 1974). Something analoguous should happen in a binary.

If the energy density of the relativistic wind at the accretion radius, $r_A = \frac{GM}{V^2}$, is larger than the local gas pressure, accretion is inhibited and a cavity is formed around the pulsar (Illarionov and Sunyaev, 1975). Assuming spherical symmetry both for the mass loss and for the pulsar wind, this requires a pulsar luminosity

$$L \gtrsim 10^{31} \dot{M}_{-10} A_{12}^{-2} V_8^{-3} \; \text{erg s}^{-1} \tag{2}$$

where $\dot{M}_{-10}$ is the mass loss rate, in units of $10^{-10} M_\odot y^{-1}$, of the normal star, A_{12} is the separation in units of 10^{12} cm and V_8 is the gas velocity in units of 10^8 cm s^{-1}.

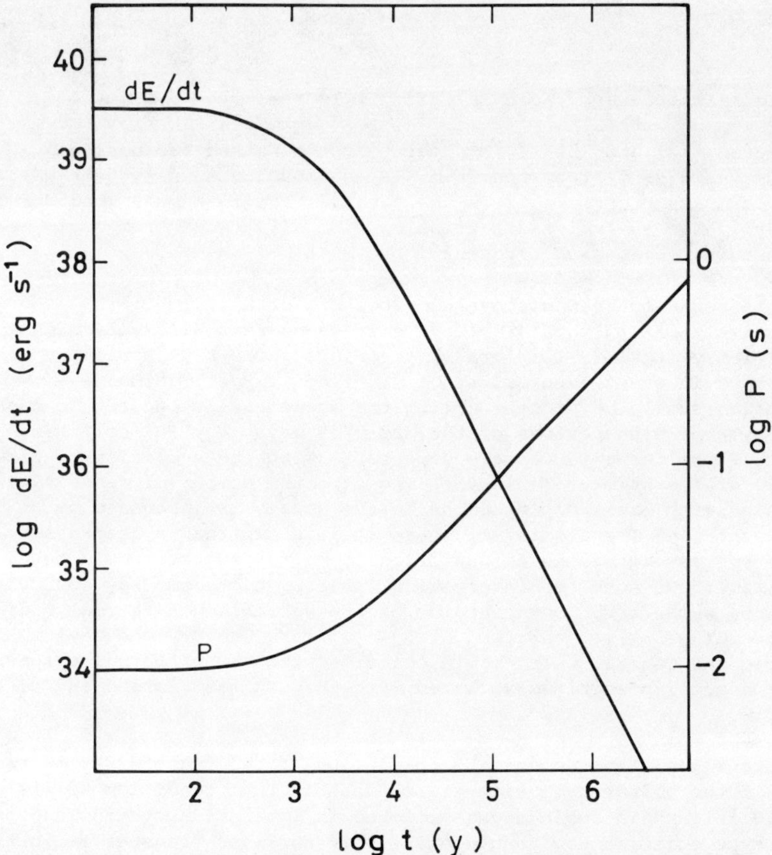

Figure 1:   The energy loss and spin period of the pulsar as a function of its age.

A substantial fraction of the rotational energy loss should then be radiated at or behind the shock which surrounds the cavity. For gas velocities of the order of 10^7-10^8 cm s^{-1} temperatures of 10^6-10^8°K are expected. Moreover non thermal radiation is likely to occur due to magnetic field compression and possibly local acceleration of particles.

During the cavity stage the mechanism of pulsed emission should be unaffected. The radio pulses however are easily absorbed. In fact the optical depth to freefree absorption is greater than 1 for a mass loss rate

$$\dot{M} \gtrsim 3.6 \times 10^{-10} \, A_{12}^{3/2} \lambda_{100}^{-1} \qquad (3)$$

where λ is the wavelength in units of 10^2 cm.

Comparing (2) and (3) it is clear that the pulsed radio signal, which has a very steep spectrum, is quenched even in systems where the low mass transfer allows the cavity stage to have a long duration (10^6y). Non thermal radio emission with a flat spectrum would still be detectable at high frequency.

It has recently been discovered (Buccheri et al., 1980) that some relatively old pulsars ($\tau = 10^5$y) emit gamma-rays with very high efficiency ($L_\gamma \simeq 10^{33} - 10^{34}$ erg s^{-1}). Since gamma-rays would not be absorbed in winds even orders of magnitude larger, gamma-ray emission should be present during the whole cavity phase. In massive systems with periods of the order of days ($A \simeq 10^{12}$cm) this should last for a time scale comparable with the evolutionary time-scale of the primary ($\simeq 10^6$y). The expected number of gamma-rays pulsars with massive companions in the Galaxy is discussed by Maraschi and Treves (1979), where it is shown that a search for such systems among the unidentified gamma-ray sources could be fruitful. We note the interesting cases of two gamma-ray sources, CG 135, whose error box contains of the variable non thermal radio-source GT 0236+61 identified with an O star (Gregory and Taylor 1978), and CG 78+1 which could coincide with a non thermal compact radiosource, possibly associated with the SNR W66 (Cordes and Dickey) 1979),.

If one takes into account the anisotropy of the transferred matter and of the pulsar radiation it is possible that condition (2) is valid in certain regions and violated in others. Accretion and pulsar type emission may then coexist. If the mass transfer is sufficiently high and anisotropic, this may happen also when the pulsar luminosity is large, $\simeq 10^{37}$ erg s^{-1}. This is not expected to occur in massive systems because of the early evolutionary stage of the primary after the SN explosion. In low mass systems the high mass transfer may result from Roche lobe overflow, possibly induced by the activity of the pulsar itself.

3. THREE CANDIDATES

We now examine three cases where we suspect a young pulsar in a binary. Two of them are contained in a list of variable radio sources associated with SN shells compiled by Ryle et al., (1978). It is possible that other objects in this list are indeed of the type of interest here.

Table 1

	Cyg X-3	Cir X-1	SS 433
Binary period	4.8^h	16.6^d	13.3^d
Other periods	17^d	-	164^d
Mass of the primary	$1 \, M_\odot$	-	$2 \, M_\odot$?
Separation	10^{11} cm	-	10^{12} cm ?
Optical, IR luminosity*	$10^{35}-10^{36}$	3×10^{37}	10^{37}
X-ray luminosity*	10^{38}	10^{38}	$10^{34} - 10^{35}$
Radio emission	Variable with strong outbursts	Variable with periodic flares	Variable
Gamma-ray emission	possibly present at 35 MeV, 10^{12} eV	Associated with CG 321-1 ?	Unknown
Supernova shell	Absent	Present	Present

* erg/s

1) Cyg X-3

The proposal that the source could contain a young pulsar was made by
Basko, Sunyaev and Titarchuk (1974), Lamb et al., (1977), Bignami, Maraschi
and Treves (1977) and Milgrom and Pines (1978). The observations of
this source were well summarized by the Cygnus X-3 panel in the GSFC
Symposium on X-ray Binaries (Boldt & Kondo, 1976). Its distinctive
features are the large radio outbursts, which imply the capability
of accelerating high energy particles and the gamma-ray emission at
10^8 MeV, Lamb (1977) (not confirmed by COS B) and 10^{12} eV (Vladimirskii et al., 1975). The total luminosity of the source is 10^{38}
erg s^{-1}, mostly in X-rays, which requires an age of 10^3-10^4y (see
Fig. 1). No supernova shell is observed around the source. However,
if the system derives from a U Gem type binary, as proposed by
Davidson and Ostriker (1974), this difficulty may not be crucial.
The age of the system could also be increased if one attributes to
the pulsar only the nonthermal radiation and the X-ray luminosity,
which contains the maximum power, is interpreted as due to accretion.

2) Cir X-1

A full description can be found in the paper by John Whelan and colleagues (1977). We wish to emphasize the following characteristics. The source is active in the radio band, and near to a supernova shell, in fact it belongs to the list of Ryle et al (1978). Hard X-ray emission has been observed with Ariel V (Coe et al., 1977) and a COS B gamma-ray source lies within 2° (IAU Symposium, Bangalore, 1979), though no claim for association has been made as yet. The orbital period is 17^d and there is evidence of a large eccentricity in the orbit and of an associated apsidal motion (S.S. Holt, private communication). Current models of the source involve accretion onto a neutron star or black hole (Whelan et al., 1977). The estimated age of the SN is $\simeq 10^5 y$, therefore a pulsar model has difficulty in accounting for the total (X-ray) luminosity and a composite model is to be preferred. In particular, due to the highly eccentric orbit, the ratio between the accretion luminosity and the luminosity in the pulsar type emission (radio and hard X-rays) may be variable.

3) SS433

This very peculiar object is also contained in the Ryle et al., (1978) list. It coincides with a weak X-ray source. Its most unusual characteristic is the presence of two satellites to the Hα line, one blueshifted and one redshifted, corresponding to velocities of the order of $10^{-1}C$ (Margon et al., 1979; Liebert et al al., 1979 and Mammano, 1980). The shifts are sinusoidally modulated with a period of 164^d. Radial velocity and photometric observations indicate a binary period of 13^d (Crampton, 1979).

No comprehensive model of the source has been proposed as yet. The presence of large velocities is attributed either to the internal rim of an accretion disk around a black hole (Terlevich & Pringle, 1979) or to two opposite jets of matter (Fabian and Rees 1979). The 164^d modulation requires then either the disk or the jets to precess. The jets could conceivably be produced by supercritical disk accretion or by the interaction of a rapidly spinning pulsar with matter transferred from a companion.

In the context of the present discussion the last possibility is particularly interesting. A distinctive feature of this type of model is that the total power involved in the jets should be less than 10^{38} erg s^{-1}, otherwise the required spin (P $\sim$ 30 ms) would be excessive even for a young neutron star. While the radiated luminosity satisfies this requirement (see Table 1) the kinetic energy output could be substantially higher. Simple arguments show however that one can satisfy all the observational constraints with a reasonable (10^{37}-10^{38} erg/s) kinetic energy output, provided the density is 10^{15}cm^{-3} and the geometry of the radiating region is elongated with cross section A$\simeq 10^{16}$cm^2 and length l $\simeq 10^{13}$cm. Therefore the pulsar hypothesis is at least consistent.

DISCUSSION

We have discussed some observational characteristics of pulsars in
binary systems and specifically the cases of three sources. While
the general picture seems persuasive, for each of the specific
cases alternative models exist which are entirely viable. One
consequence of dealing with three young systems should be considered.
In fact if the pulsar phase (10^3-10^4y) were followed by a much
longer X-ray active accretion phase (10^{6-7}y) the implied number of
X-ray binaries would be too high.

The expected duration of the X-ray stage in binaries depends
strongly on the mass of the companion varying from 10^4y for massive
systems (10-20M_O) to 10^8y for low mass system (1 M_O). Unfortunately
due to high extinction in all three cases the spectral type of the
companion is unknown. In Cir X-1 it could be massive, in SS433 a
value of 2 M_O is derived from radial velocity measurements, in Cyg
X-3 a low mass is implied, if 4.8^h is the binary period. It is
therefore impossible at present to say how serious the difficulty
is. In any case it would affect also models which interpret in
terms of accretion the present X-ray activity of SS433 and Cir X-1,
whose age can be derived from the association with the supernova
independently of models. In the pulsar hypothesis the difficulty
could be evaded if the strong mass transfer presently assumed in
these systems is induced by the pulsar itself. After the pulsar
phase the systems would become nearly detached binaries and these
later stages could give rise to the low mass X-ray transients.

REFERENCES

Basko, M.M., Sunyaev, R.A., and Titarchuck, I.G., 1974. Astron.
     Astrophys., 31, 249.
Bignami, G.F., Maraschi, L., and Treves, A., 1977. Astr. Astrophys.,
     55, 155.
Boldt, E., and Kondo, Y., 1976. Proc. GSFC Symp. on X-ray Binaries,
     NASA SP 395.
Buccheri, R., et al.,1979. Astron. Astrophys., in press.
Caravane Collaboration, 1979. Paper presented at the COSPAR/IAU Symp.
     Bangalore.
Clark, G., 1975. Astrophys. J. (Letters), 199, L143.
Coe, M.J., Engel, A.R., and Quenby, J.J.,   . Nature, 262, 563.
Cordes, J.M., and Dickey, J.M., 1979. Nature, 281, 24.
Crampton, D., Cowley, A.P., and Hutchings, J.B., 1979. IAU Circ. 3388.
Davidson, A., and Ostriker, J.P., 1974. Astrophys. J., 189, 331.
Fabian, A.C., and Rees, M.J., 1979. MNRAS, 187, 13P.
Gregory, P.C., and Taylor, A.R., 1978. Nature, 272, 704.
Gursky, H, 1978. In "Structure and Evolution of Close Binary Systems",
     eds. P. Essleton, S. Mitton and J. Whelan, IAU Symp. No. 73.
Holt, S.S., private communication.
Illarionov, A.F., and Sunyaev, R,A., 1975. Astr. Astrophys., 39, 185.
Lamb, R., et al., 1977. Astrophys. J., 212, L63.

204

Liebert, J., et al., 1979. Nature, 279, 384.
Mammano, A., Ciatti, F., and Vittone, A., 1979. In press.
Maraschi, L., and Treves, A., 1979. Nature, 279, 401.
Margon, B., et al., 1979. Astrophys. J. (Letters), 230, L41.
Milgrom, M., and Pines, D., 1978. Astrophys. J., 220, 272.
Rees, M.J., Gunn, J.E., 1974. MNRAS, 167, 1.
Ryle, M., et al., 1978. Nature, 276, 571.
Terlevich, R.J., and Pringle, J.E., 1979. Nature, 278, 719.
Van den Heuvel, E.P.J., 1978. In "Physics and Astrophysics of Neutron
     Stars and Black Holes", eds. R. Giacconi and R. Ruffini, North
     Holland Pub. Co., Amsterdam.
Vladimirskii, B.M., et al., 1975. Soc. Astr. Letters, 1, 57.
Whelan, J.A.J., et al., 1977. MNRAS, 181, 259.

The pulsating X-ray source GX1+4 (4U 1728-24)

E. Kendziorra and
R. Staubert

Astronomisches Institut der Universitat Tubingen

and

C. Reppin, W. Pietsch, W. Voges and J. Trumper

Max-Planck-Institut fur Physik und Astrophysik,
Institut fur Extraterrestrische Physik,
W. Germany.

1. Abstract
2. Introduction
3. Observations
4. The Energy Spectrum
5. The Pulsations

1.   ABSTRACT

GX 1+4 (4U 1728-24) was observed at hard X-rays (>22 keV) during a
balloon flight on 1978 November 22-24. The photon number spectrum is
measured up to $\sim$ 150 keV and found to be consistent with a thermal
bremsstrahlung spectrum with kT = (41 $\pm$ 8) keV or a power law with
spectral index α = 2.6 $\pm$ 0.2. A comparison with low energy data
favours the thermal bremsstrahlung spectrum. The pulsed fraction of
the periodically modulated source flux was (52 $\pm$ 8) percent in the
22 - 80 keV interval, significantly higher than in previous measurements. We present evidence that the true pulsational period is
(227.5 $\pm$ 0.5) s rather than half this value.

2.   INTRODUCTION

GX 1+4 was found as a hard X-ray source during a balloon observation
in October 1970 (Lewin et al. 1971) and was later identified by
Uhuru with the low energy source 4U 1728-24. It is the dominant X-ray
source in the galactic centre region. The first observation indicated
a possible periodic modulation with the approximate frequency of one
cycle per 2.3 minutes. Since then it has been confirmed by a number
of observations that GX 1+4 is a slow X-ray pulsar (see Table 1). No
eclipses and no Doppler-shifts of the pulsational period have been
observed and only indirect arguments may be used to identify this

TABLE 1

DATA ON THE PULSATION AND THE ENERGY SPECTRUM OF THE X-RAY FLUX FROM GX 1+4 (4U 1728-24)

INSTRUMENT	TIME OF OBSERVATION	ENERGY RANGE (keV)	PULSE PERIOD (SEC)	DEGREE OF MODULATION (fpf) 1)	ENERGY FLUX SPECTRUM α	kT (keV)	INTEGRAL FLUX (erg/cm^2 s)	REFERENCE
MIT Balloon	1970 Oct.15-16	15-70	$135 \pm 4^{2)}$		1.4 ± 0.7	$28 \pm 12^{4)}$	18-50 keV: $\sim 2.4 \times 10^{-9}$	Lewin et al. 1971.
MIT Balloon	1972 April 5	15-60	No Pulses	Upper Limit of 0.25 (2σ)	1.6 ± 0.2	$18 \pm 2^{4)}$	20-50 keV: factor 3 higher than 1970.	Ricker et al. 1976.
Copernicus	1972 Sept. 11	3 - 8	$258.9 \pm 0.24^{3)}$	0.12, 0.10	-	-	-	White et al. 1975.
	1972 Sept. 18	3 - 8	257.2 ± 1.2	-	-	-	-	and
	1973 March 25	3 - 8	263.4 ± 8.4	-	-	-	-	White et al. 1976.
SAS - 3	1975 Dec. 25-27	3 - 20	122.607 ± 0.006	-	-	-	-	Doty 1976.
OSO - 8	1975 Sept. 16 1975 Sept. 17	2 - 20	122.46 ± 0.03	0.09 high) 0.14 low)	0.3	}>35	$\sim 8.8 \times 10^{-10}$ $\sim 5.8 \times 10^{-10}$ (2-6 keV)	Becker et al. 1976.
AIT/MPI Balloon	1978 Nov. 22-23	18-130	227.5 ± 0.5	0.52 ± 0.08 (22-80 keV)	1.6 ± 0.2	41 ± 8	1.1×10^{-9} (20-50 keV)	Kendziorra et al. 1979. (This work).

1) The values given are the 'fractional pulsed flux'; values in the literature defined differently have been recalculated under the assumption of a sine wave type modulation, e.g. (max-mean)/mean = fpf, or (max-min)/mean = 2 fpf, or (max-min)/min = 2 fpf/(1-fpf).

2) The originally published value was 2.3 min; Becker et al. 1976 quoted (135 $\pm$ 8) s (2σ error) as private communication of W.H.G. Lewin.

3) Because of experimental limitations the period search was done in the range 2.5-50 min (Nyquist frequency 2.9 min), so the periods given could be an alias of several smaller periods, e.g. half the given values.

4) This kT value results from a simple exponential fit, other values given are from a thermal bremsstrahlung law.

source with a binary system. One of these is the observed rapid spin up t/τ which Becker et al. (1976) estimated to be of the order -0.026 per year. Another characteristic of this source is non-periodic time variability on time scales of years (Ricker et al., 1976), months (White et al., 1976), days (Becker et al., 1976) and minutes (Lewin et al., 1971).

In this paper we present new results on:

1) the energy spectrum,
2) intensity changes from pulse to pulse,
3) variability over long time scales, and
4) the periodic pulsation.

Regarding this latter point we present evidence that the true period is 227.5 s, rather than half this value as is generally assumed. The pulsed fraction of the flux is found to be high when compared with previous observations of the source.

3.   OBSERVATIONS

The large area AIT/MPI balloon X-ray detector (Reppin et al., 1978) was flown successfully on 1978 November 22 - 24 from Alice Springs, Australia. During the 54 hours of 'float' in an average atmospheric depth of 3.5 g/cm^2 several galactic and extragalactic X-ray sources could be observed. In particular, we concentrated on the pulsating X-ray sources using accumulated periods of pointing from 0.5 to 3.5 hours.

The detector system consists of four phoswich units with 3 mm thick NaI crystals and 50 mm thick CsI shielding crystals. A graded shield (W, Sn, Cu) honeycomb collimator defines a hexagonal field of view with $\sim$ 3 degrees (full width at half maximum). The total collecting area is 766 cm^2. For each event the energy (64 channels for the range 10 - 200 keV), arrival time (with 6.4 ms resolution and the pulse rise time are recorded. Background reduction is achieved by a passive graded shield (Pb, Sn, Cu), an active plastic scintillator (for anticoincidence) and a rise time discrimination system which vetoes all events with an energy loss above a defined threshold in the CsI shield crystal.

The typical background rates for different energy bands are:

| | Background |
Energy Range	(Counts/cm^2 sec keV x 10^3)
20 - 30 keV	0.59
30 - 50 keV	0.31
50 - 100 keV	0.30
100 - 130 keV	0.19

Energy calibration is performed every 30 min during the flight using 109_{Cd} radioactive sources. The detector system is mounted equatorially on a platform which is magnetometer stabilized to typically 0.1 degrees. The observation was aided by a Hewlet Packard 1000 computer system which was interlocked with the observing instrument by data and command telemetry. The observing programme was stored in the HP 1000 computer memory and performed automatically. An on-board microprocessor controlled the telescope pointing direction. The average pointing and tracking accuracy was better than 0.1 degree, which, for the 3° FWHM of the collimator, corresponds to a geometrical efficiency of better than 96%. A total time of about two hours was spent for the observation of GX 1+4. The pointings were made on November 22 for 19 and 14 minutes with an effective collecting area of 190 cm^2 through an atmospheric depth of 4.0 g/cm^2 (the first observation) and on November 23 for 10 and 14 minutes with 766 cm^2 through 4.5 g/cm^2 (the second observation). An equal time was always used for the corresponding background measurement.

4. THE ENERGY SPECTRUM

In Figure 1 we show our photon number spectrum from the second observation. We note that this is the first measurement which

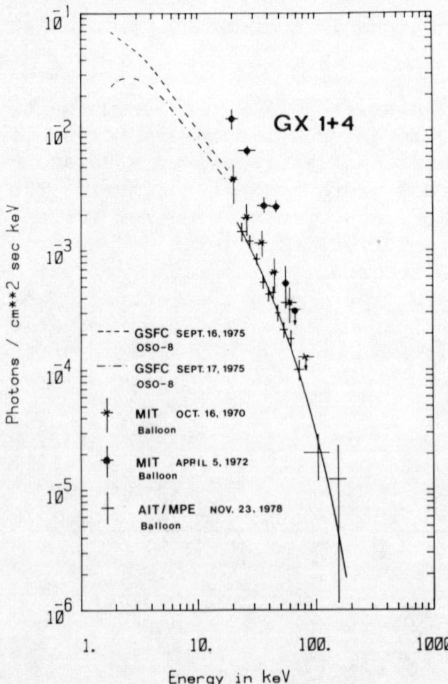

Figure 1: Photon number spectra of GX 1+4. Error bars are $\pm$ 1σ statistical errors. (For references see text or Table 1).

extends beyond 70 keV. Our data can be well fitted by a thermal bremsstrahlung law:

$$I \text{ (Photon intensity)} \propto E^{-1.4} \exp(-E/kT)$$

where $kT = (41 \pm 8)$ keV and the $\chi^2/_{d.o.f.} = 1.0$ for 9 degrees of freedom or (slightly less well) by a power law:

$$I_{photon} \propto E^{-\alpha}$$

where $\alpha = 2.6 \pm 0.2$ ($\chi^2/_{d.o.f.} = 1.7$ for 9 d.o.f.).

The first observation gives a very similar spectrum. This spectrum is consistent with the first balloon observation by the MIT group (Lewin et al. 1971) but is somewhat harder than their second balloon observation (Ricker et al. 1976). A comparison with the low energy OSO-8 data (Becker et al. 1976) supports a thermal bremsstrahlung shape and not the power law. Apparently GX 1+4 was in a low intensity state during our observation. The integral energy flux was 1.1×10^{-9} erg/cm^2 s between 22 and 50 keV.

5.   THE PULSATIONS

The analysis of our data clearly establishes a pulsational period of 113.75 $\pm$ 0.25 sec. However, below we present evidence that the true period may be twice this value. In Figure 2 we have plotted our measurements in three different energy bands as light curves with a period of 227.5 sec.

If we model the source flux with the sum of a steady component and a modulated component, we can define the fractional pulsed flux (fpf) as the ratio of the modulated component to the total flux from the source. Over our energy range we find values which are independent of energy within error limits, and which average (52 $\pm$ 8) percent for the 22 - 80 keV range. This has to be compared with 9 to 14 percent measured between 2 and 20 keV (see Table 1), and a 2σ upper limit of 25 percent measured at balloon energies (Ricker et al. 1976).

In order to make the comparison possible the published values which are defined differently have been recalculated assuming a sine wave modulation. We note that there is at least one other pulsar showing a higher degree of modulation at higher X-ray energies: Vela X-1 (4U 0900-40) has a pronounced increase of the fractional pulsed flux with energy, this was observed in the same balloon flight (Staubert et al. 1980). In these objects either the beaming mechanism works more effectively at higher X-ray energies or a "wash out" of the original pattern is less effective.

Now we turn to the question of the pulsational period. The first hint that this could be in question came from looking at the raw (count rate) data. Individual peaks could clearly be identified, and in spite of a fair amount of scattering, a systematic behaviour seemed to be present such that every second minimum appeared to be

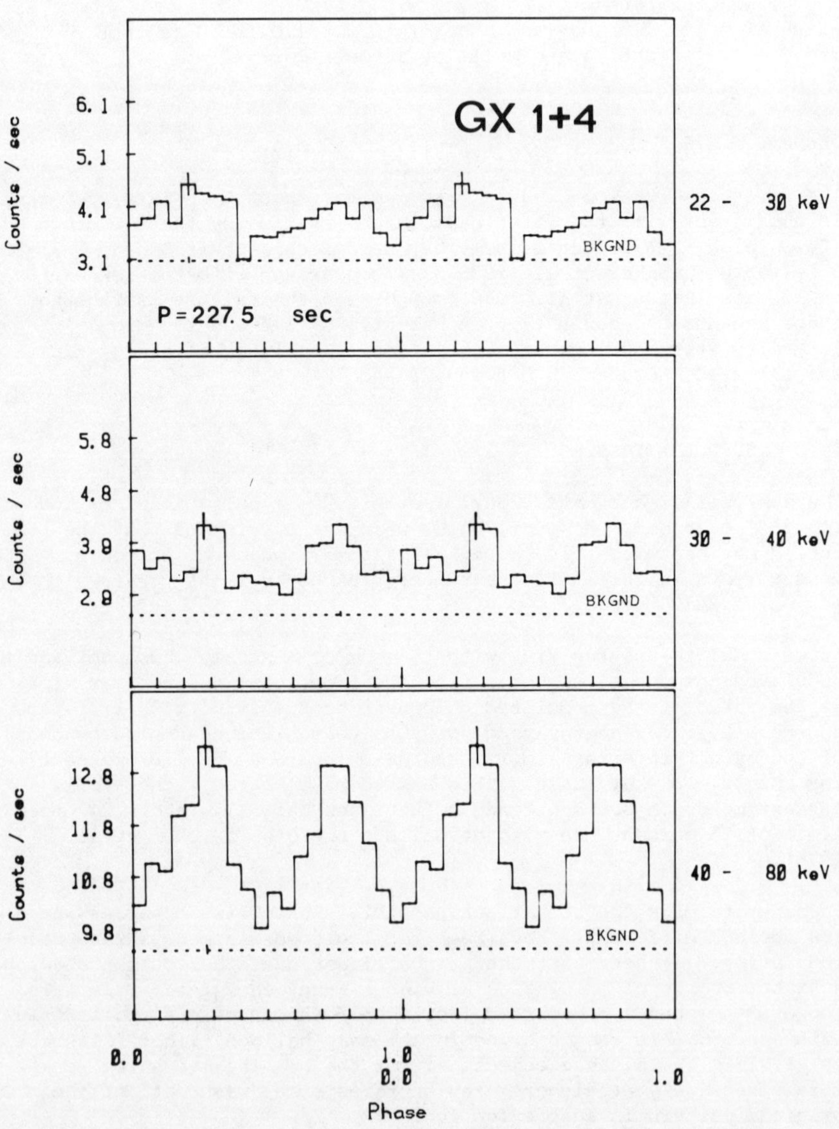

Figure 2: Pulsational light curves of GX 1+4 for three different
energy intervals assuming a double pulse structure. The
observed period is 227.5 ± 0.5 s. Error bars are ± 1σ
statistical errors.

deeper than those in between. If the true period is 227.5 s rather than 113.75 s, the light curve has a double peak structure (like several other X-ray pulsars) instead of a single peak. In general three possible signatures can identify a true double peak structure:

1) a separation of the two peaks other than 0.5 in phase and/or

2) a different shape of the two peaks, and/or

3) a difference in the energy spectrum for phase intervals corresponding to the two peaks.

The separation of the two peaks in GX 1+4 is consistent with a phase separation of 0.5. To search for a difference in the energy spectrum we have computed a hardness ratio N (26-34 keV)/N (50-70 keV). For the second observation the ratios are: 2.44 ± 0.46 for the first peak and 1.71 ± 0.35 for the second peak. This is no significant difference.

In Figure 3 the time averaged light curve (A) for the second observation is shown. Curve B is shifted by 0.5 in phase compared to A. The difference A-B then illustrates a difference in shape for the two peaks. The formal significance for the difference, assuming purely Poissonian fluctuations, is $\sim$ 3.5 standard deviations. This difference may be due to a real difference of the two peaks or may be the result of intensity fluctuations from the source.

Another method of searching for differences in shape of the light curve is to compare individual pulses rather than the time averaged ones. In Figure 4 the individual peaks from the second observation are shown. As a measure of the difference between any two peaks (k and l), the quantity

$$\chi^2_{k\,l} = \sum_{i=1}^{20} \frac{(F_{ik} - F_{il})^2}{\sigma_{ik}^2 + \sigma_{il}^2}$$

has been used. F_{ik} is the count rate in bin i of pulse k and σ_{ik} its corresponding standard error. This quantity χ^2_{kl} has been computed for all even/even-, odd/odd- and odd/even-combinations of peaks. The resulting distributions of χ^2 values for even/even- plus odd/odd-combinations on the one hand, and odd/even-combinations on the other hand are shown in Figure 5. The distributions are normalized to the same integral number. Also shown (as a dotted line), is the distribution expected when there is no difference in shape between peaks and when there are only Poissonian fluctuations. This distribution has been generated by Monte Carlo simulations. Two things are to be noted. Firstly, the observational data produce significantly higher χ^2 values. This is because of additional fluctuations due to real intensity changes. Secondly, the odd/even-distribution is shifted to higher χ^2 values, compared to the even/even- plus odd/odd-distribution, again indicating a difference in shape between even and

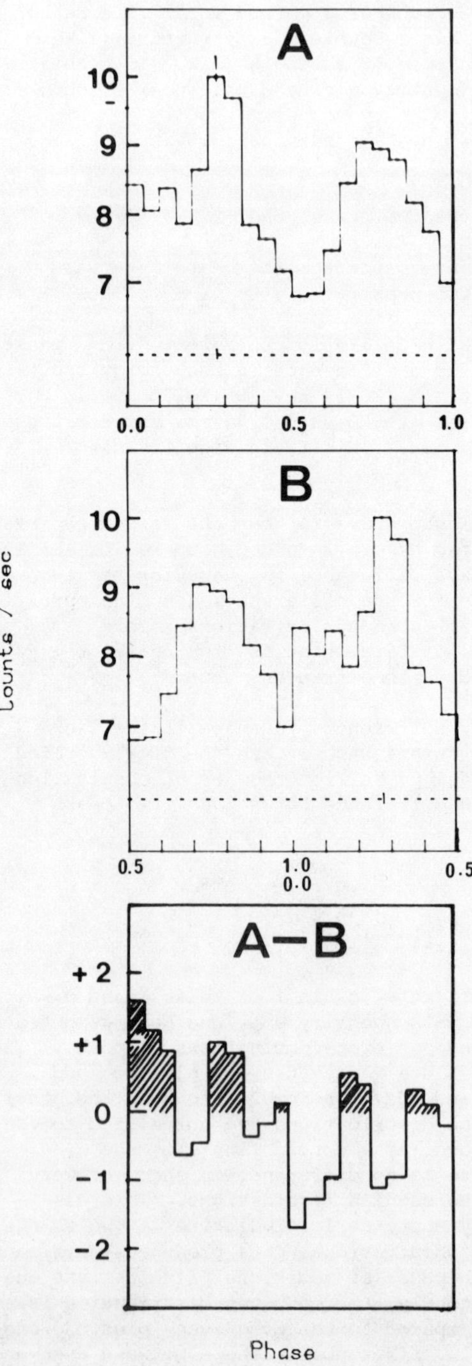

Figure 3A: Time averaged light curve in the 26 - 50 keV range.

Figure 3B: Time averaged light curve in the 26 - 50 keV range shifted by 0.5 in phase with respect to A.

Figure 3C: The difference (A-B) illustrating that the two peaks are not identical.

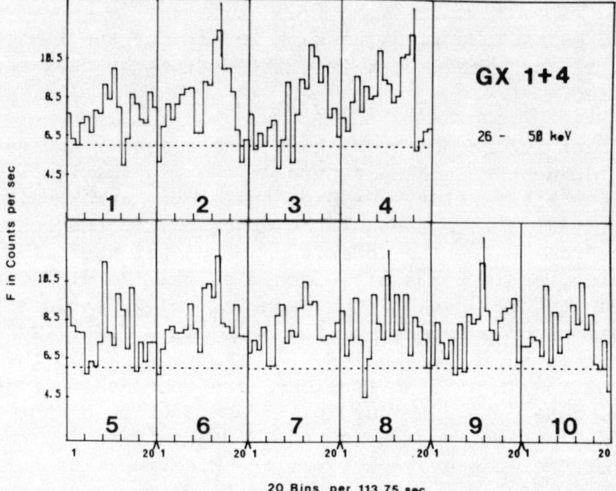

Figure 4:   Count rate versus time containing ten individual peaks.
The degree of inequality between any two peaks has been
tested using a χ^2 value as defined in the text.

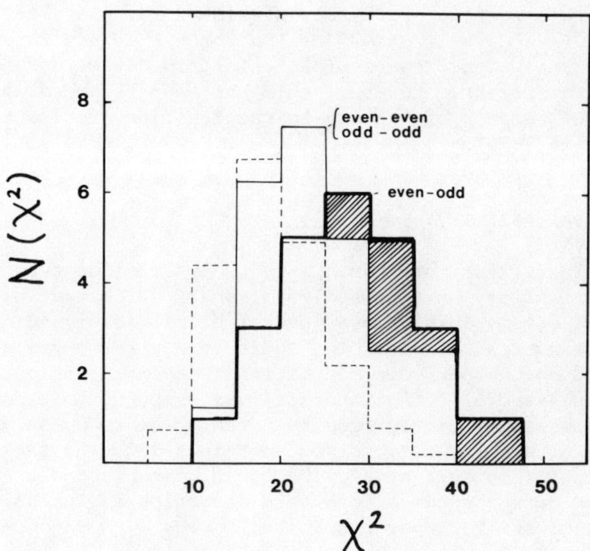

Figure 5:   Distributions of χ^2 values measuring the degree of inequality between individual pulses. Light line: even/even- and
odd/odd- combinations (pulse numbers as in Figure 4). Heavy
line: even/odd-combinations. Dotted line: theoretically
expected distribution for Poisson fluctuations only and no
inequality in pulse shapes.

odd numbered peaks. We find the same result for the first observation. Also, if any single peak is omitted from the data the result is not altered.

How significant is the observed shift in the mean χ^2 values? Because of the non-Poissonian nature of the fluctuations, we cannot simply apply χ^2 statistics. We have therefore performed Monte Carlo simulations, taking into account the observed fluctuations in the intensity of the peak. The underlying model was that of a single peak pulsation, or in terms of a double pulse: no difference in shape between the two peaks. The experimental situations for the two observations have been simulated 5000 times each. These sets of simulated pulses have been analysed in the same way as the observed pulses, resulting in χ^2 distributions like those shown in Figure 5 for the real data. As a measure of the separation of the even/even-plus odd/odd-distribution and the even/odd-distribution, we take the separation of the mean values. From the distributions for this measure we can infer the probability of finding, in one experiment, a separation as observed, or greater, due only to fluctuations. These individual probabilities are: for observation 1, 2×10^{-4} and for observations 2, 9×10^{-3}. The combined probability suggests that the model of a single peak pulsation is not correct.

We would like to add however, that more observations are necessary to confirm this result, preferably at high X-ray energies. An analysis of five days of SAS-3 data in the 8 - 19 keV range has yielded a formal significance of 3.5 standard deviations for the two peaks being different. However, this has been regarded as insufficient evidence because of the non-Poissonian type of fluctuations present (Lewin 1979).

6.   TIME VARIABILITY

Like many binary X-ray sources, GX 1+4 shows various types of non-periodic variability on time scales of years (Ricker et al. 1976), months (White et al. 1976), days (Becker et al. 1976) and down to minutes (Lewin et al. 1971). The short term erratic variability superimposed on the periodic modulation is apparent in our data (see Figures 4 and 5). The intensity varies by up to a factor of 1.3. This is significantly less than the pulse to pulse variability of up to a factor of two, observed in Vela X-1 during the same balloon flight (Staubert et al. 1980). A comparison between the spectra in Figure 5 shows a long term variation in intensity of up to a factor of six for energies above 20 keV.

The pulsed fraction is also variable for energies above 20 keV. We note, that for the three high energy observations available (see Table 1 and Figure 1), the pulsed fraction is anticorrelated with the intensity: we observed the lowest intensity and the highest pulsed flux of (52 ± 8) percent. Ricker et al. (1976) report an intensity six times higher with no measurable modulation (2σ upper limit of 25 %), whereas the original detection of the pulsation was

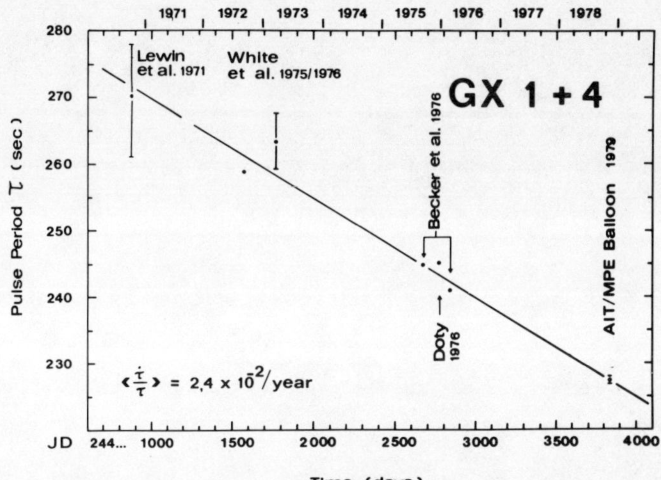

$\langle \frac{\dot{\tau}}{\tau} \rangle = 2.4 \times 10^{-2}/\text{year}$

Figure 6: Pulsational period as a function of time (data from Table 1 assuming a double peak lightcurve). The mean spin up rate is $-2.4 \times 10^{-2}/\text{year}$.

made at an intermediate intensity (no pulsed fraction was quoted) (Lewin et al. 1971).

Finally, we note that the regular decrease in pulsational period can be described by a relative spin up rate of $\dot{\tau}/\tau = -2.4 \times 10^{-2}/\text{year}$. In Figure 6 we have plotted all pulse periods from Table 1 at the times of observation (assuming a double peak light curve). The spin up rate is the highest observed in X-ray pulsars.

REFERENCES

Becker, R.H., Boldt, E.A., Holt, S.S., Pravdo, S.H., Rothschild, R.E., Serlemitsos, P.J., and Swank, J.H., 1976. Ap. J. (Letters), 207, L167.

Doty, J., 1976. IAU Circ. 2910.

Lewin, W.H.G., Ricker, G.R., and McClintock, J.E., 1971. Ap. J. (Letters), 169, L17.

Reppin, C., Pietsch, W., Trumper, J., Voges, W., Kendziorra, E., and Staubert, R., 1978. in "European Sounding Rocket-, Balloon- and Related Research", ESA SP-135, 293.

Ricker, G.R., Gerassimenko, M., McClintock, J.E., Ryckmann, S.G., and Lewin, W.H.G., 1976. Ap. J., 207, 333.

Staubert, R., Kendziorra, E., Pietsch, W., Reppin, C., Trumper, J., and Voges, W., 1980. Submitted to Ap. J.

White, N.E., Huckle, H.E., Mason, K.O., Charles, P.A., Pollard, G., and Culhane, J.L., 1975. IAU Circ. 2870.

White, N.E., Mason, K.O., Huckle, H.E., Charles, P.A., and Sanford,
    P.W., 1976. Ap. J. (Letters), 209, L119.

Twisted accretion disks

J.A. Petterson

Department of Physics,
University of Illinois at Urbana-Champaign,
Urbana, IL 61801,
USA.

1.   INTRODUCTION

Accretion disks which occur in X-ray binary systems are expected to
lie in the binary plane. This is natural, because the mass transfer
between the two stars generally takes place in the orbital plane of
the system. Nevertheless, it is interesting to ask what happens if
a disk forms in some other plane, unnatural as it may seem. The
compact object around which the disk lies is not likely to oppose
such a situation, because its gravitational field is almost per-
fectly spherically symmetrical. However, the tidal forces from the
companion star now acquire a vertical component, causing the disk to
precess with a frequency ω which depends on the masses and separation of the stars, and also strongly on the disk radius ($\omega \sim -r^{3/2}$).
The negative sign of ω indicates that this precession is retrograde,
i.e. opposite to the direction of rotation of the binary system, if
we assume that the binary motion of the stars, and the orbital
motion of the matter in the disk are in the same sense.

Accretion disks are obviously not solid bodies, and so there is no
reason why they should precess as if they were. Rings at different
radii in the disk try to precess at different rates, each in accor-
dance with its own value of ω. On the other hand, individual rings
are not completely independent of one another; they are coupled by
viscosity. Their differential precession soon leads to a winding up
of the disk, which creates stronger and stronger viscous forces to
oppose the differential precession: <u>viscous disks resist being
wound up indefinitely.</u>

If the viscosity is successful in stopping the winding up, the
result is <u>a twisted disk</u> with stationary shape, precessing with a
single period (which is some weighted average over the precession

periods of the individual rings). How far such a disk is wound up
obviously depends on the strength of the viscosity. The situation
is similar to the way the deformation of an elastic body by a given
stress tensor depends on the elasticity coefficients of the body. A
large viscosity (expressed in the usual dimensionless parameter α,
the ratio of viscous stress to pressure in the disk) allows only a
small twist; a small viscosity α allows a large twist. It should
be clear, however, that regardless of the value of α, <u>any tilted
disk must necessarily be twisted to some extent</u> (Fig. 1).

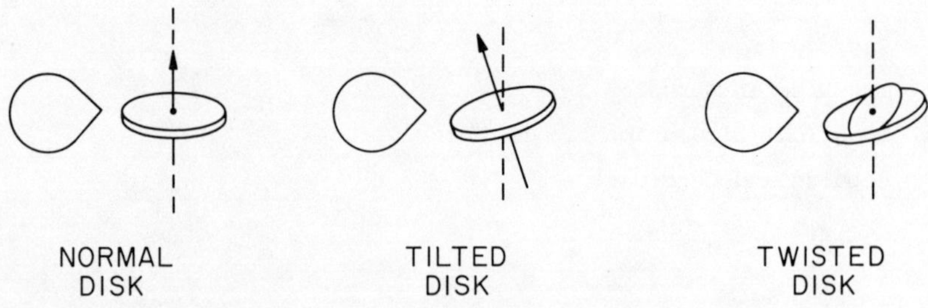

Figure 1: Three variations of the theme "accretion disk".

Two effects must now be mentioned which can change both the structure of a twisted disk and its precession rate. The first is the
<u>slow radial drift</u>, occurring in any disk which is actually accreting.
That this drift affects both twisting and precession rate is best
illustrated by the extreme case in which the inward drift-time is
much shorter than the precession period of the disk. In that case
all disk rings must lie nearly in the same plane as the outer ring
into which the new material is fed; they simply have no time to
precess away from it. The disk is therefore nearly flat (untwisted),
and its precession rate is that of the outer ring. The second
factor influencing both the structure and the precession rate of a
disk is the <u>angular momentum L</u> of the material that is fed into it
at the outer edge. The "slaved disk model" (Roberts 1974), in which
the outer ring slavishly follows the orientation of the equatorial
plane of a (slowly precessing) companion star because it constantly
receives matter out of this equatorial plane, illustrates this.

The equilibrium structure and precession rate a twisted disk
achieves when all the above mentioned effects are taken into account
in a specific model, can best be determined with a numerical
approach. However, before we plunge into this, let us ask whether a
situation as bizarre as that of a disk that does not lie in the
binary plane, could possibly ever occur in nature.

2. THE DISK IN HERCULES X-1

Perhaps to one's surprise there is indeed a system, Hercules X-1,
for which a twisted accretion disk model appears to be relevant.
Figure 2 is a sketch of this system, a low mass X-ray binary with a
1.7 day orbital period. The figure contains a rapidly rotating
neutron star (P = 1.24 sec), and an artist's conception of the
twisted disk, which is thought to be precessing with a 35 day
period, and is probably responsible for the 35 day variations seen
in the X-ray and optical light emission from the system (Katz 1973,
Roberts 1974, Petterson 1975, Gerend and Boynton 1976).

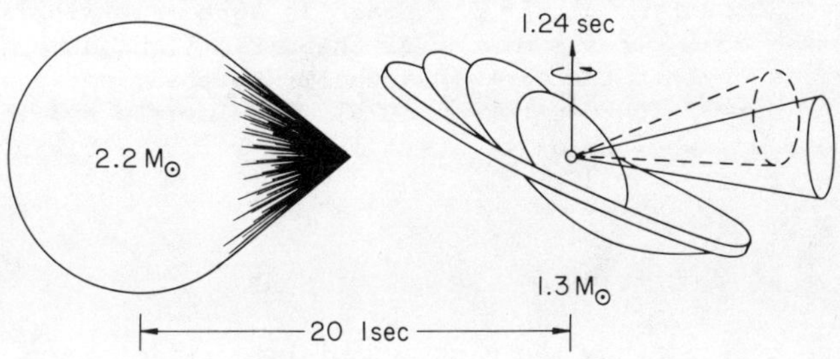

Figure 2 :   The X-ray binary Her X-1.

The most compelling evidence for the presence of a <u>precessing disk</u>
in Her X-1 is the occurrence of optical light variations at frequencies which are integer combinations of the orbital frequency
($\omega_{1.7}$) and the sum frequency ($\omega_{1.7} + \omega_{35}$). Deeter et al. (1976)
found the following frequencies (in order of decreasing amplitude):

$$\omega_{1.7},$$

$$\omega_{1.7} + \omega_{35},$$

$$3\omega_{1.7} + 2\omega_{35} = \omega_{1.7} + 2(\omega_{1.7} + \omega_{35}),$$

$$2\omega_{1.7} + \omega_{35} = \omega_{1.7} + (\omega_{1.7} + \omega_{35}),$$

$$2\omega_{1.7},$$

$$2\omega_{1.7} + 2\omega_{35} = 2(\omega_{1.7} + \omega_{35}),$$

$$4\omega_{1.7}, \dots$$

The prominent occurrence of the sum frequency $\omega_{1.7} + \omega_{35}$ strongly
suggests the presence of a retrogressively rotating structure which
modulates the (X-ray) illuminating source. A progressively rotating
structure would have produced the difference frequency $\omega_{1.7} - \omega_{35}$,

which is conspicuously absent.

Other pieces of evidence for a precessing disk are the observed 35 day variation in the depth of the optical secondary minimum (at orbital phase 0.5, when the disk obscures part of the companion star; Gerend and Boynton 1976) and the variation in the disk's direct contribution to the optical light (best discerned near X-ray eclipse). These observations are most easily explainable by a disk which systematically changes its projected surface area, as a precessing tilted disk would. Furthermore, comparisons between the X-ray spectra from the 35 day "on" and "off" states indicate that the X-ray "offs" are probably caused by cold material obscuring the X-ray source (Becker et al. 1977), again in agreement with a precessing disk model. The asymmetry of the X-ray lightcurve gives a direct indication that the disk is actually twisted, not just tilted out of the binary plane (Petterson 1975). It largely remains a mystery, however, why the disk material ever left the binary plane in the first place.

3.   NUMERICAL SIMULATION OF THE HERCULES DISK

Avoiding the question how the disk material left the binary plane, but assuming that it somehow did, we attempted to numerically determine what equilibrium structure the disk would achieve in the system Hercules X-1, and what its precession rate would be, making some assumptions about the complicating effects mentioned above. This work was a collaboration between D. Merritt and the author (Merritt and Petterson 1979).

We characterized each ring in the disk by a radius r, an inclination angle β, and a longitude γ for its line of nodes, and formulated the time-dependent "twist equations" which govern the evolution of the twist structure of a "standard" thin accretion disk (Shakura and Sunyaev 1973). The equations have the following symbolic form

$$\frac{\partial \beta}{\partial t} = \left[\begin{array}{c} \text{vertical viscous} \\ \text{interaction} \end{array} \right] + \left[\begin{array}{c} \text{radial} \\ \text{drift} \end{array} \right] \qquad (1)$$

$$\frac{\partial \gamma}{\partial t} = \left[\quad " \quad \right] + \left[\quad " \quad \right] + \omega \qquad (2)$$

The "vertical viscous interaction" term takes account of the tendency of neighbouring rings to align themselves with each other through viscosity, the radial drift is determined from a standard α disk model, and ω is the above mentioned forced precession due to the companion star.

We assumed that the outer ring of the disk always maintains a constant inclination angle β_0 and a constant value for the derivative $\left[\frac{\partial \gamma}{\partial r}\right]$. The particular value of β_0 is unimportant, because the equations (1) and (2) are linear in β, and therefore scale. However, the lack of specific knowledge about the L of matter newly added to the disk prevented us from assigning a definite value to $\left[\frac{\partial \gamma}{\partial r}\right]$, and we therefore treated this boundary condition as a free parameter. The assumed constancy of β_0 and $\frac{\partial \gamma}{\partial r}$ implies a restriction on the mechanism that causes the transferred matter to leave the binary plane (see Merritt and Petterson, 1979), but does not require this mechanism to be a precession of the companion star. The freedom to give $\frac{\partial \gamma}{\partial r}$ different values was hoped to provide some flexibility here.

To judge how well the numerical runs reproduced the characteristics of the Hercules disk, we formulated the following observational constraints:

a)   The precession period τ_e of the disk at equilibrium is
     approximately 35 days.

b)   The line of nodes of the outer rings must at
     equilibrium vary over an angle $\nabla \gamma_e \sim$ 1-2 radians,
     to reproduce an X-ray cycle which is "on" for
     1/3 and "off" for 2/3 of the time.

c)   The outermost ring must be trailing the rest of
     the disk in its precessional motion, to make the
     X-ray "turn on" sharp, and the "turn off" gradual.

We time-evolved an arbitrary initial configuration of rings which extended out to a radius of 2×10^{11} cm (80% of the Roche lobe radius), until it achieved a stationary structure. This would typically occur after only a few radial drift times. Runs were made with different values for the boundary parameter $\frac{\partial \gamma}{\partial r}$ and the viscosity parameter α. The results are summarised in Fig. 3. Each run is characterised by a point in the τ_e, $\nabla \gamma_e$ plane, indicating the precession period and the "twist angle" of its final equilibrium configuration. Runs with the same value for α but different values for $\frac{\partial \gamma}{\partial r}$ are connected by lines. The observational constraints appear to severely restrict the acceptable values of the two parameters. Only disks with a trailing outermost ring are represented in Fig. 3.

4.   RESULTS AND CONCLUSIONS

Before we draw our striking conclusion about the value of the viscosity parameter α required in our model disk for Her X-1, let us clearly state our assumptions. We assumed:

a)   the <u>binary parameters</u> of the Her X-1 system, as specified
     by Middleditch and Nelson (1976): Roche geometry with
     $M_{\text{X-ray}} = 1.3\ M_\odot$ and $M_{\text{companion}} = 2.2\ M_\odot$.

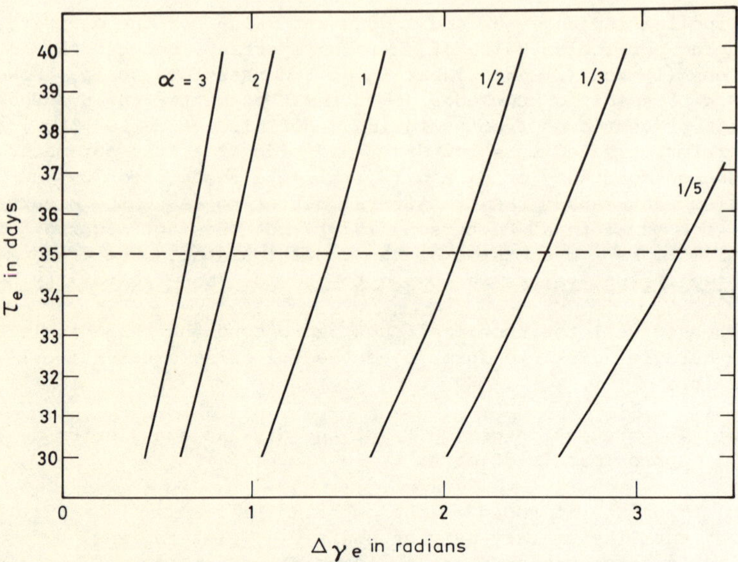

Figure 3: Precession periods (τ_e) and "twist angles" ($\Delta\gamma_e$) of a set of equilibrium disks with different α (viscosity parameter) and $\frac{\partial\gamma}{\partial r}$ (boundary condition). Points with the same α but different $\frac{\partial\gamma}{\partial r}$ are connected by lines.

b) a <u>disk radius</u> of 2×10^{11} cm, consistent with a tidal cut-off of the disk. A large value, but practically demanded by the observed very deep secondary optical minimum.

c) the standard optically thick <u>α-accretion disk model</u> of Shakura and Sunyaev (1973).

d) that the 35 day on-off <u>X-ray cycle is caused by variable obscuration</u> of the X-ray source, by matter from the outer parts of a tilted and precessing accretion disk.

e) that the <u>boundary condition</u> of constant β_o and $\frac{\partial\gamma}{\partial r}$ at the outer disk rim does not contradict the workings of the (unknown) mechanism that causes matter to leave the binary plane.

f) <u>an isotropic disk viscosity.</u>

g) that <u>X-ray heating</u> on the outer parts of the disk may be neglected.

Even if the first four assumptions are reasonable, the fifth is
quite uncertain, the sixth completely ad hoc, and the last one
probably somewhat unrealistic. We know that the use of an anisotropic viscosity affects our results in a simple way, broadening
the range of acceptable values for α (Merritt and Petterson, 1979).
However, we do not know how the X-ray heating affects our results.
Earlier time-independent calculations (Petterson 1977) indicated
that the effect may be important.

If we nevertheless make the above seven assumptions boldly, we find
that our precessing twisted disk model satisfies the stated three
observational constraints only for values of α in the range

$$1/2 < \alpha < 2 \qquad\qquad (3)$$

We hope to have made it clear enough to the reader that the uncertainties in the various assumptions require this result to be
treated with great caution. In particular, the narrow limits on α
stated in (3), reflect the restrictions set by <u>our</u> interpretation
of the observations on <u>our</u> disk model for Her X-1, not on disk
models in general! Nevertheless, we find it remarkable that the
observations <u>can</u> be fit by the model, that the range for α is so
narrow and that it lies so close to the value that is often quoted
as "canonical" ($\alpha=1$).

Supported by NSF Grant No. PH7 78-04404.

REFERENCES

Becker, R.M., Boldt, E.A., Holt, S.S., Pravdo, S.H., Rothschild, R.E.,
    Serlemitsos, P.J., Smith, B.W., and Swank, J.H., 1977. Ap.J.,
    214, 879.
Deeter, J.E., Crosa, L., Gerend, D., and Boynton, P.E., 1976. Ap.J.,
    206, 861.
Gerend, D., and Boynton, P.E., 1976. Ap.J., 209, 562.
Katz, J.I., 1973. Nature Phys. Sci., 246, 87.
Merritt, D., and Petterson, J.A., 1979. preprint.
Middleditch, J., and Nelson, J., 1976. Ap.J., 208, 567.
Petterson, J.A., 1975. Ap. J., (Letters), 201, L61.
Petterson, J.A., 1977. Ap. J., 218, 783.
Shakura, N.K., and Sunyaev, R.A., 1973. Astr. Ap., 24, 337.

The 1975 May transition of Cygnus X-1

Lucio Chiappetti

Mullard Space Science Laboratory <u>and</u> Instituto di Fisica dell'
University College London,   Universita and Laboratory
Holmbury St. Mary, Dorking,   di Fisica Cosmica e Tecnologie,
Surrey, UK.   Relative del CNR, Milan, Italy.

We have known, since 1971, that the powerful X-ray source Cyg X-1
exhibits a bi-modal behaviour, sometimes in an active (HI) state
(characterised by a steep complex spectrum) but usually in a quiet
(LO) state (with a photon spectral index $\alpha \simeq 1.6$). A transition from
the active to quiet state was observed in that year; other transitions
occurred in 1975 April, May and September and 1976 February (Heise et
al., 1975; Sanford et al., 1975; Holt et al., 1975, 1976).

The MSSL instrument (experiment C) on the Ariel 5 satellite observed
Cyg X-1 during May 1975 (May 3rd to May 15th), for part of the HI
state and the subsequent gradual recovery to the quiet state. These
data, the only known spectral observations of moderate energy resolution made during a transition, are presented here.

The spectrum (see Figure 1) has at least two components. The low
energy part is well approximated by a blackbody with kT $\sim$ 0.3 keV
and (in the higher energy region) by a power law. The ratio of the
luminosities of these two components agrees with the predictions of
inverse Compton models (e.g. Sunyaev and Titarchuk, 1979; Liang, 1980).
During the transition the spectral index of the power law component
flattens from $\alpha \sim 2.8$ to $\alpha \sim 2.0$. A bump in the spectrum at $\sim$ 3 keV
is possibly explained by a cut-off in the power law, but the data
are also consistent with a continuation of the power law to the low
energy region. If the cut-off hypothesis is accepted, the transition
between HI and LO states is both the result of an increase of the
blackbody temperature and a flattening of the power law (see Figure
2a). We note that the temperature increases before α drops. In the
case of no assumed cut-off to the power law the temperature remains
constant throughout the transition. Note also that the enhanced
flux of the power law component, without spectral changes, is responsible for flares in the HI state. A very weak emission feature
($<5\sigma$ above the continuum) is present between 6 and 7 keV. Due to
its faintness, it has not been possible to determine its energy with
sufficient precision to identify it with the 6.7 keV iron line.

An estimate of the electron temperature and the optical depth for
Thomson scattering, by the electron cloud as it reprocesses the soft
flux, has been attempted on the basis of recent theoretical calculations (Sunyaev and Titarchuk, 1979). In the HI state we find
$kT_e \gtrsim 5$ keV, $\tau_0 \sim$ 5 or $kT_e \lesssim 1$ keV $\tau_0 \sim$ 12 respectively with and
without a cut-off in the power law at 3 keV. In the LO state,

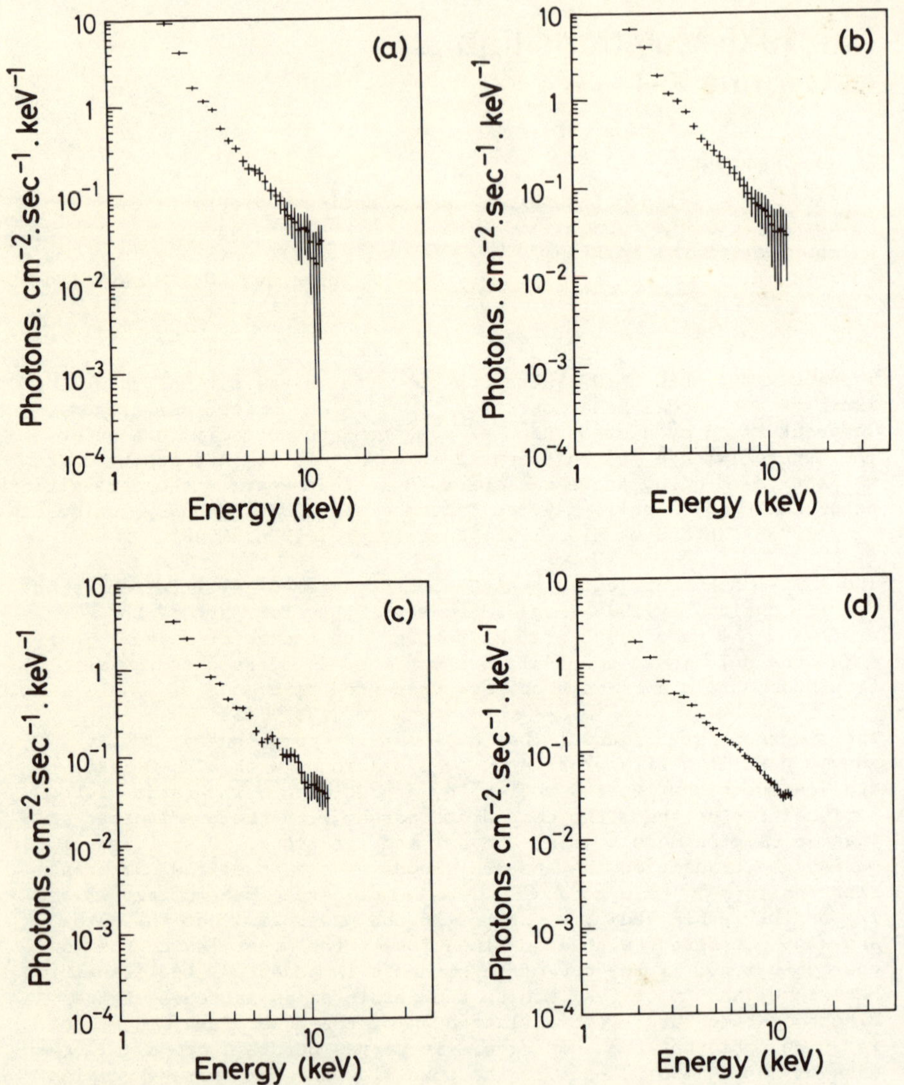

Figure 1: Some representative spectra of Cyg X-1, restored by the method of Blissett and Cruise (1979) (see also Blissett, these proceedings).

a) orbits 3064-3073 (combined)
b) orbit 3119
c) orbit 3153
d) orbits 3172-3219 (combined)

The variation between HI and LO state (from Figs. 1a to 1d) is clearly visible.

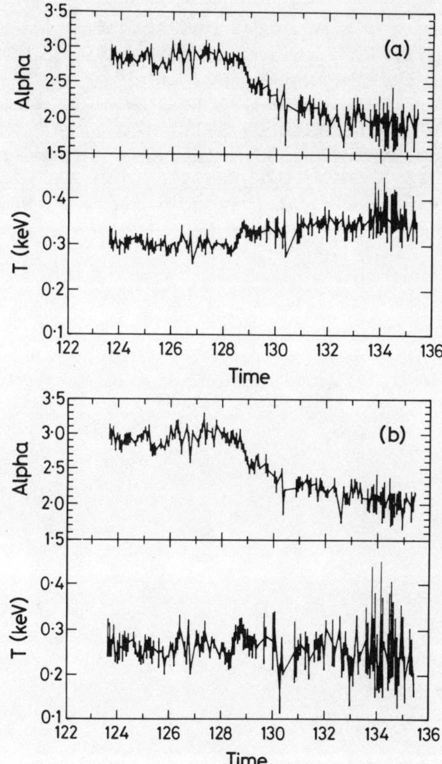

Figure 2:   Photon spectral index and blackbody temperature versus
time (days 1975), as obtained by a fit with cut-off at
3 keV for the power law component (Fig. 2a) and without
cut-off (Fig. 2b). Errors shown are 3σ.

Sunyaev and Trümper (1979) find kT_e = 27 keV, $\tau_0 \sim 5$. In our last
observation the recovery to the LO state was not yet complete
($\alpha \sim 2.0 \gtrsim 1.6$, the usual LO state value) the flux was also greater
and the recovery was completed within the next month (Eyles et al.,
1975). Since the formula which ties kT_e to the ratio of hard and
soft luminosities has a singular behaviour around $\alpha = 2$, it was not
possible to compute kT_e and τ_0 for the data from the less intense
state. However a hint of an increase in kT_e is evident at the start
of the transition.

This work, a more detailed version of which has been submitted to
Monthly Notices of the Royal Astronomical Society, was in part supported by the Foundation Boncompagni-Ludovisi-Bildt of Stockholm.
Professor R.L.F. Boyd, CBE, FRS is thanked for his hospitality at
the MSSL - also my co-workers.

REFERENCES

Blissett, R.J., Cruise, A.M., 1979. MNRAS, 186, 45.
Eyles, C.J., Skinner, G.K., Wilmore, A. P., 1975. MNRAS, 173, 63P.
Heise, J., et al., 1975. Nature, 262, 196.
Holt, S.S., et al., 1975. Nature, 265, 108.
Holt, S.S., et al., 1976. Nature, 261, 214.
Liang, E.P.T., 1980. Nature, 283, 642.
Sanford, P.W., et al., 1975, Nature, 256, 109.
Sunyaev, R.A., Titarchuck, L.G., 1979. preprint.
Sunyaev, R.A., Trümper, J., 1979, Nature, 279, 506.

X-ray emission from white dwarfs

N.D. Kylafis and D.Q. Lamb

Department of Physics
University of Illinois at Urbana-Champaign
Urbana, Illinois 61801,
USA.

Magnetic white dwarfs comprise an important class of galactic X-ray
sources. At present it is not known whether any of the galactic
X-ray sources are nonmagnetic or weakly magnetic white dwarfs. In
order to help find out we have carried out a comprehensive study of
X-ray emission from accreting nonmagnetic and weakly magnetic white
dwarfs. Various aspects of accretion onto such stars have been
studied by several authors (see e.g. Hoshi 1973; Aizu 1973; De
Gregoria 1974; Fabian et al. 1976; and Katz 1977). Our calculations
span the entire range of accretion rates and stellar masses, and
include the important, but previously unexplored, regime at moderate
and high accretion rates. A brief discussion of our calculations
and results has already appeared (Kylafis and Lamb 1979a), while an
extended description of our work is in preparation (Kylafis and Lamb
1979b). This study dramatically altered our understanding of the
X-ray spectral properties of accreting white dwarfs. Our results
show that it may be possible to determine the nature of some X-ray
sources and their physical properties (such as mass and intrinsic
luminosity) from their X-ray spectra and from correlations between
their X-ray luminosity and spectral temperature. The most important
conclusions of our study are the following:

a)   White dwarfs undergoing spherical accretion can produce
     a maximum hard X-ray luminosity of 4×10^{36} erg s^{-1}.
     This luminosity occurs for an accretion rate more than
     an order of magnitude larger, than earlier estimates
     suggested (DeGregoria 1974; Fabian et al, 1976).

b)   Contrary to previous reports (Hoshi 1973; Aizu 1973;
     Fabian et al, 1976; Katz 1977) even high-mass white
     dwarfs can produce low-temperature ($\lesssim 10$ keV) X-ray
     spectra.

c)   At moderate and high accretion rates, the spectra
     have a power-law high energy tail.

d)   At moderate and high accretion rates the spectral
     temperature varies dramatically and the star exhibits
     a pronounced correlation between X-ray spectral
     temperature and luminosity. Such a correlation is

not expected for nonmagnetic neutron stars, and may
be an important signature of white dwarf sources.

e) All spectra have an intense UV or soft X-ray component that is not expected for neutron stars.
Again, detection of such a component would be
strong evidence that the observed source is a white
dwarf.

f) Iron line emission is expected from white dwarfs
accreting at low rates. At moderate rates the line
is broadened by scattering, and becomes unobserved
at higher rates.

g) White dwarfs in which the magnetic field is sufficiently strong to channel the flow of accreting
matter, but not so strong that cyclotron emission
dominates bremsstrahlung, behave essentially as
nonmagnetic white dwarfs (Fabrian et al. 1976;
Katz 1977). In such a case our results may still
be used, together with knowledge of the overall
geometry, to determine the character of the emitted
spectrum.

Unlike the case of strongly magnetic white dwarfs, as yet no X-ray
source has been unequivocally identified as a nonmagnetic or weakly
magnetic white dwarf. Cyg X-2 shows a correlation between spectral
temperature and luminosity very similar to that predicted by our
calculations for a $\sim 0.4\ M_\odot$ white dwarf (Branduardi et al. 1979).
However, spectroscopic observations made by Cowley et al. (1979) do
not support this interpretation. Sco X-1, and many other sources
like it, also exhibit correlations between spectral temperature and
luminosity (Mason et al. 1976; Parsignault and Grindlay 1978).
Although, as a class, these sources cannot be white dwarfs, some of
them might be. In such a case, comparison of theory and observation will allow their masses, intrinsic luminosities and distances
to be determined from their X-ray spectra alone.

REFERENCES

Aizu, K., 1973. Progr. Theoret. Phys., 49, 1184.
Branduardi, G., Kylafis, N.D., Lamb, D.Q., and Mason, K.O., 1979. in
     preparation.
Cowley, A.P., Crampton, D., and Hutchings, J.B., 1979. Ap. J., 231,
     539.
DeGregoria, A.J., 1974. Ap. J., 189, 555.
Fabian, A.C., Pringle, J.E., and Rees, M.J., 1976. MNRAS, 175, 43.
Hoshi, R., 1973. Progr. Theoret. Phys., 49, 776.
Katz, J.I., 1977. Ap. J., 215, 265.
Kylafis, N.D., and Lamb, D.Q., 1979a. Ap. J. (Letters), 228, L105.
Mylafis, N.D., and Lamb, D.Q., 1979b. In preparation.
Mason, K.O., Charles, P. A., White, N.E., Culhane, J.L., Sanford, P.
     W., and Strong, K.T., 1976. MNRAS, 177, 513.
Parsignault, D.R., and Grindlay, J.E., 1978. Ap. J., 225, 970.

Line emission from neutron star surfaces

Joachim Trümper

Max-Planck-Institut für Physik und Astrophysik
Institut für Extraterrestrische Physik,
München, Germany.

1. Introduction

2. The Spectra of Pulsating X-Ray Sources

3. Nuclear Gamma Ray Lines and Electron-Positron Annihilation

4. Cyclotron Lines

5. Atomic and Ionic Lines from the Hot Polar Plasma

6. Concluding Remarks

1.   INTRODUCTION

Neutron stars are fascinating in many respects, but one of the most
exciting aspects is that they represent macroscopic matter in its
most condensed, but still observable form. Indeed, at the surface of
a neutron star, gravitational and magnetic fields assume their most
extreme values. Gravity is of the order of 10^{13} g, leading to free
fall velocities of a few tenths of the velocity of light. In the
superstrong magnetic fields, which are of the order of 10^{12} Gauss,
quantum effects govern the motion of free charged particles, and the
structures of atoms is controlled by Lorentz rather than by Coulomb
forces, resulting in small atomic volumina and large densities of
condensed matter (cf. Rudermann 1974). The latter are of the order
of 2×10^3 gcm^{-3} $Z_{26}^{2/5}$ $B_{12}^{6/5}$ (Flowers et al. 1977).

Our empirical knowledge of these extreme conditions, coming from the
observation of radiation from neutron star surfaces is still rather
limited. But neutron stars were discovered only twelve years ago and
since then the subject has undergone rapid development. Today,
neutron stars are found in probably four different astrophysical
situations - radio pulsars, pulsating binary X-ray sources, X-ray
bursters and point-like X-ray objects in supernova remnants.

In this talk I shall concentrate mainly on the pulsating X-ray sources
which have proven to be a rich source of information on neutron star
physics. The fascinating results of timing studies have been
summarized by Saul Rappaport at this meeting (page 159, this volume).

At the same time, broad band X-ray spectroscopy has been very important since it relates to source luminosities and temperatures.

As in any other branch of astronomy, line spectroscopy is of particular diagnostic value, since it provides the most specific information. In the following I want to discuss the potential of line spectroscopy of neutron stars and review a few results which have been obtained so far. In doing so, I concentrate on the surface emission from neutron stars, leaving alone the interesting phenomena which occur in their surroundings, namely in stellar winds, accretion disks, Alfven shells or at companion surfaces. Some of these topics have been reviewed by Dick McCray at this meeting.

2.   THE SPECTRA OF PULSATING X-RAY SOURCES

In general pulsating X-ray sources show quite hard power law spectra with a spectral break at 10 to 25 keV, beyond which the intensity drops rather rapidly. Most of the X-ray luminosity is found in the spectral region just below the cutoff energy.

It is generally accepted that the bulk of the observed X-ray luminosity originates at the polar caps to which the accreted material is channeled by the magnetic field and where its kinetic energy is converted into thermal energy (cf. Fig. 1). This emission region is quite small, the extent in height above the neutron star surface is probably not larger than $\sim$ 1 km, and in fact, it may be much smaller. Therefore, line spectroscopy should tell us about the physical conditions in a narrow radius band near the neutron star surface.

The principal processes which can give rise to X-ray or gamma ray lines are

- nuclear de-excitation & $e^{+}e^{-}$ annihilation

- cyclotron transitions of free electrons in the strong magnetic field, and

- atomic transitions, e.g. from highly ionized iron.

By observations of such lines one can get information on the surface gravitational and magnetic fields as well as on physical properties of the emitting plasma. In the following we shall discuss the three processes separately.

3.   NUCLEAR GAMMA RAY LINES AND ELECTRON-POSITRON ANNIHILATION

Excited nuclei may be produced by nuclear collisions of the infalling semirelativisitic particles (10-100 MeV/nucleon) with matter at the su face of neutron stars, which can emit gamma ray lines and positrons (Ramaty et al. 1973; Higdon et al. 1977). Positrons in return will give rise to annihilation gamma quanta. The strongest nuclear lines are expected at 4.4 MeV from ^{12}C and at 6.129 MeV from ^{16}O (Ramaty 1978).

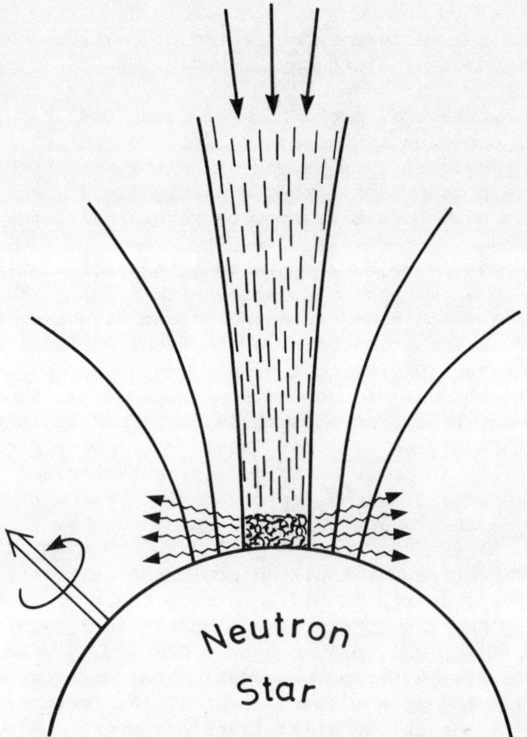

Figure 1:   Schematic picture of the accretion flow on to a magnetized
neutron star.

Of course, any line emitted from the surface of a neutron star will
be subjected to the gravitational redshift, which depends on mass M
and radius R of the neutron star

$$z = - \frac{\Delta E}{E} = \left[1 - \frac{2GM}{Rc^2} \right]^{-1/2} - 1$$

z is expected to be in the range 0.1 - 0.4 for standard neutron stars
(Borner, 1973). The observation of gamma ray lines offers the exciting possibility to measure M/R and to determine R, if the mass is
known from double star dynamics. Such a measurement would provide an
extremely important test of the equation of state of matter of ultrahigh densities.

The effects of superstrong magnetic fields on the nuclear energy
levels should be very small since the potential energy of a nuclear
magnetic dipole in a magnetic field is of the order of

$$\Delta E = \mu_B \, B = 20 \text{ eV } B_{12}$$

where μ_B is the nuclear magneton. Therefore, any magnetic shift or broadening of gamma ray lines may be neglected.

If there is a substantial production of positrons at the surface of neutron stars, one expects to observe the characteristic annihilation radiation. In contrast to nuclear lines the resulting spectrum is affected by the presence of strong magnetic fields, and the corresponding effects have been discussed by several authors.

Wunner and Herold (1979) have shown that the annihilation of bound pairs results in a two-photon emission with a sharp line at 511 keV, which has a transition rate depending on the magnetic field strength. In the case of annihilation of free electrons and positrons, one has in general a mixture of two-photon and single-photon final stages, whereby the branching ratio depends on the magnetic field strength (Wunner 1979, Daugherty and Bussard 1979). The single photon decay becomes dominant only at $B \gtrsim 10^{13}$ Gauss. The two-photon decay, which is important at lower energies, is characterized by a width of the 511 keV peak which depends on the angle between the photons and the direction of the magnetic field.

Tentative evidence for a redshifted positron annihilation line at 400 keV from the Crab nebula had been reported by Leventhal et al. 1977. However, the result was of low statistical significance (Cherry et al. 1980) and could not be confirmed by other measurements (Ling et al. 1977). A second observation has been reported by Jacobsen et al. (1978), who found several line features, including one at 413 keV in an unidentified transient event. Also this measurement was of rather low statististical significance and must remain in doubt (Cherry et al. 1980). The first real case seems to be the unusual March 5, 1979, gamma ray burst event (reported by Shakura at this conference, Mazets et al. 1979), which shows a strong spectral feature of 430 keV. If it is a redshifted annihilation line from a neutron star of 1.3 $M_\odot$, the corresponding R is 15 km, which is consistent with neutron star dimensions. The occurence of a damped 8 sec oscillation after the initial gamma ray burst suggests that one is indeed dealing with a rotating magnetic neutron star.

4.   CYCLOTRON LINES

In the superstrong magnetic fields of neutron stars electrons can move freely only along the field lines while their transverse energies are quantized in units of the cyclotron frequency

$$E_n = n\ \hbar\omega_B = n\ \hbar\ \frac{eB}{m_e c} = 11.6 \text{ keV}\ B_{12}.$$

This is a nonrelativistic expression. The relativistic quantum mechanical treatment leads to eigen value solutions of the Dirac equation with energies given by

$$E_{js} = m_e c^2 \left[\left\{ 1 + \left(\frac{P_z}{m_e c} \right)^2 + (2j + s + 1) \left(\frac{B}{B_{cr}} \right) \right\}^{1/2} - 1 \right]$$

where $j = 0,1,2...$ and $s = \pm 1$ are angular momentum and spin quantum numbers, respectively, and

$$B_{cr} = \frac{m_e^2 c^2}{eh} = 4.414 \times 10^{13} \text{ G}$$

is the critical magnetic field strength, P_z is the electron momentum parallel to the magnetic field lines. The expression shows that spin-flip transitions ($\Delta s = 2$) have the same energy as angular momentum transitions ($\Delta j = 1$), which is not exactly true but holds as long as $(g - 2)$ effects can be neglected.*

If the temperature in the hot radiating plasma at the neutron star pole is high enough, excitation of the lower Landau levels may occur and the corresponding Landau transitions may give rise to discrete spectral features (Gnedin and Sunyaev, 1974). In addition, the Thomson scattering cross section as well as free-free emission and absorption processes become resonant at the cyclotron resonance frequencies, at least for the extraordinary photons. Figure 2 shows as an example the Thomson cross sections in a strong magnetic field for different photon directions with respect to the magnetic field (Ventura, 1979 and this volume page 241ff). In view of the strength of the resonances one may expect that they manifest themselves in the X-ray spectrum of the hot plasma, if the magnetic field is sufficiently uniform over the emission region.

The first observational evidence in this context was obtained in 1976 by a MPI/AIT balloon observation of Her X-1 in which spectral structures were discovered in the pulsed hard X-ray spectrum (Trumper et al., 1977, 1978; Kendziorra et al., 1977). As shown in Figure 3, there is a peak at 58 keV. seen with a statistical significant of 5.4σ and one at 110 keV seen at 3.2σ. If one interprets the features as cyclotron emission lines, the corresponding magnetic field strength is $\sim 5.3 \times 10^{12}$ Gauss. If the resonances are seen in absorption one has $\sim 4 \times 10^{14}$ Gauss. These values have to be corrected for the unknown gravitational redshift, which is $\sim 10 - 40\%$ (Borner, 1973).

These spectral structures show up both in the pulsed spectrum obtained (obtained by subtracting the interpulse spectrum from the pulse spectrum) and in the spectrum measured during the pulse phase. This is very important since it excludes the possibility that the features are an artifact introduced by subtracting spectra.

The reality of the first peak has been confirmed by further balloon observations in 1977 of the MPI/AIT group with large statistical significance ($\sim 15\sigma$), while only upper limits on the feature at 110 keV could be obtained, which are substantially lower than the flux measured in 1976 (Voges et al., 1979). At the same time the total flux of Her X-1 was somewhat lower than in the earlier measurements,

*g is the spin g factor of the electron (=2.00232).

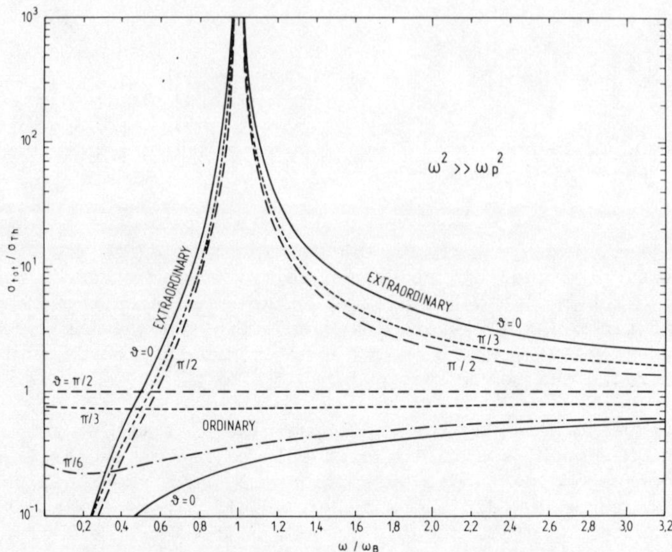

Figure 2: Frequency dependence of the cross sections near the fundamental cyclotron resonance at various angles of propagation. The ordinary mode photon cross sections remain unaffected by the resonance at all angles of propagation. Effects at the higher harmonic frequencies are neglected. *(Reproduced by permission from the American Physical Society, Ventura, J., Phys. Rev. D19, 1684).*

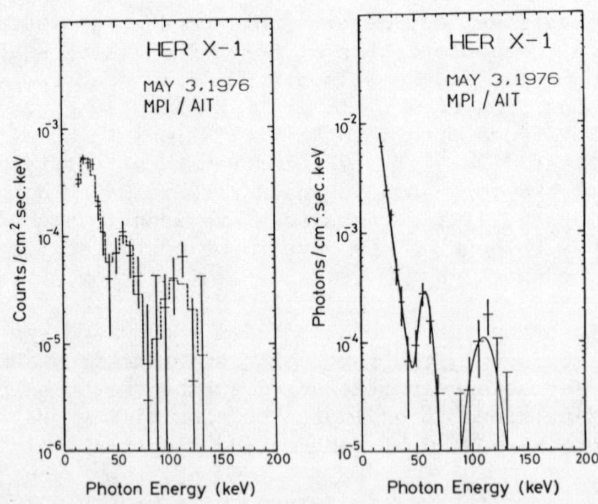

Figure 3: Hard X-ray spectrum of Hercules X-1 obtained during the pulse phase of the 1.24 sec pulsations. The left diagram shows raw count rate spectra. The other diagram shows the deconvolved spectrum, assuming spectral lines at 58 keV and 110 keV (from Trumper, 1979).

and the absence of the second harmonic feature may be due to a
lower source intensity and temperature.

Pointed observations with the HEAO-1 /A-4 experiment have confirmed
the existence of the spectral feature at $\sim$ 58 keV and shown that it
is stable throughout the on-state of the 35 day cycle of Her X-1
(Gruber et al. 1979).

An important property of any line feature is its width, which in the
case of cyclotron lines, may result from several effects. While
collisional and radiative broadening are negligible, Doppler
broadening plays a role. Its effect on the line width depends on the
viewing angle with respect to the magnetic field because the free
motion of the electrons is one-dimensional (Trumper et al. 1977). If
the emission region extends in height (Δr) over the neutron star
surface, line broadening will occur due to differential gravitational
redshift and variations of cyclotron frequency. The latter effect is
the dominant one and leads to $\Delta E/E = \Delta B/B \stackrel{\sim}{\sim} 3 \Delta r/r$ for a magnetic
dipole field.

The observations indicate a line width of $\Delta E/E \stackrel{\sim}{\sim} 0.2$ to 0.3 (Voges
et al. 1979, Gruber et al. 1979, unpublished MPI/AIT data), viz
$\Delta r/r \lesssim 0.1$. Hence the radial extent of the emission region must be
smaller than $\sim$ 1 km for a neutron star radius of $\sim$ 10 km.

In the meantime, two other sources with possible cyclotron line
features have been reported. Wheaton et al. (1979) find an absorption type feature at $\sim$ 20 keV in the pulsed spectrum of the 3.6 sec
pulsar 4U0115+63 while the pulse phase spectra of 4U1626-67 (7.8 sec
period) show a rather broad bump around 20 keV (Pravdo et al. 1979).
Interpreted in terms of cyclotron resonance the corresponding magnetic fields would be $\sim 2 \times 10^{12}$ Gauss, however, the cases are less
clear than Her X-1: In 4U0115+63 the feature has been observed only
in the pulsed flux while the spectral enhancement observed in
4U1626-67 at certain pulsational phases is rather broad and may be
due to continuum emission processes.

The Her X-1 observations have stimulated an extensive theoretical
discussion on the spectral formation in strongly magnetized plasmas
(Weaver 1978, Meszaros 1979, Yahel 1979 a, b; Bussard 1979, Bonazzola, Heyverts and Puget 1979, Melrose and Zheleznyakov 1979, Langer,
McCray and Baan 1980, Meszaros, Nagel and Ventura 1980, Nagel 1980 a,
b; Wassermann and Salpeter 1980, and others). As a result the
photon-electron cross sections under strong magnetic field conditions
are now known in some detail and the first radiative transfer calculations have been performed using more or less idealized models.
The talks of Joseph Ventura and Steve Langer at this meeting address
themselves to some of the theoretical problems involved.

5.   ATOMIC AND IONIC LINES FROM THE HOT POLAR PLASMA

The properties of atoms and ions - binding energies, energies of

238

excited states, cross sections, transition rates etc. - are strongly
modified in superstrong magnetic fields (Canuto and Ventua 1977). A
comprehensive account of the electromagnetic transitions of the
hydrogen atom in strong magnetic fields has been given recently by
Wunner and Ruder (1979). In view of the high temperatures, which can
be inferred from the continuum spectra of accreting neutron stars and
which are in the range $kT \sim 10$ to 30 keV, one may expect to observe
line emission from highly ionized heavy atoms. Burdyuzha and Pavlov-
Verevkin (1980), who have calculated the Lyman-α quantum energies of
hydrogen-like ions from helium to iron in superstrong magnetic
fields, find a general shift towards higher energies compared with
the non-magnetic case. For example, the Lyman-α iron line at 6.9
keV is shifted to an energy of ~ 15 keV in a 5×10^{12} Gauss field.

'Iron lines' have been found around 6.9 keV in the spectra of
several binary X-ray sources including Her X-1 (Pravdo 1979). These
lines are usually wide and sometimes shifted or double. They have
been interpreted in terms of iron fluorescence in matter surrounding
the neutron star, e.g. an Alfven shell or stellar companions surface.
The fact, that the Her X-1 'iron lines' are observed between 5.5 and
8.2 keV (Boldt 1977), and not at ~ 15 keV, supports the view, that
they do not originate in the polar plasma, but in low magnetic field
regions. We conclude that the detection of ionic lines from strongly
magnetic polar plasma remains a task for future observations.

6.   CONCLUDING REMARKS

A number of fundamental questions can be answered by observations of
line emission from neutron star surfaces.

The measurements of the neutron star gravitational field by detecting
redshifted nuclear lines relates to the problem of neutron star
structure and behaviour of matter at ultrahigh densities. Such
measurements may provide an integral test of high energy elementary
particle physics in domains which are not accessible by man-made
machines.

Cyclotron line spectroscopy may represent the only accurate method
of determining neutron star magnetic fields, whose knowledge is
interesting with regard to the origin and decay of neutron star
magnetic fields. No less exciting are the various aspects of plasma,
atomic and solid state physics in superstrong magnetic fields.

In this context, we mention also the possibility to observe yet un-
tested effects of quantum electrodynamics, in particular vacuum
polarization which should become important when the magnetic fields
come close enough to the critical value $B_C = 4.414 \times 10^{13}$ G (Mészáros and Ventura 1979, Mészáros 1980).

Finally, the spectroscopy of atomic lines may not only give insight
into the chemical composition of the hot polar plasma, but also

provide information on both the gravitational and magnetic field. Of
course, the ultimate dream would be the simultaneous observation of
nuclear, cyclotron and atomic lines from a single object.

REFERENCES

Borner, G., 1973. Springer Tracts in Modern Physics, 69, p. 1.
Boldt, E.A., 1977. Ann. N.Y. Acad. Sci., 302, 329.
Bonazzola, S., Heyverts, J., and Puget, J.L., 1979. Astr.Ap., 78, 53.
Burdyuzha, V.V., and Pavlov-Verevkin, V.B., 1980. preprint 505,
    Space Research Institute, Moscow.
Bussard, R.W., 1979. preprint.
Canuto, V., and Ventura, J., 1977. Fund. of Cosmic Phys., 2, 203.
Cherry, M.L., Chupp, E.L., Dunphy, P.P., Forrest, D.J., and Ryan, J.
    M., 1980. preprint.
Daugherty, J., and Bussard, R.W., 1979. preprint.
Flowers, E.G., Lee, J.F., Ruderman, M.A., Sutherland, P.G., Hille-
    brandt, W., and Muller, E., 1977. Ap.J., 215, 291.
Gnedin, Yu.N., and Sunyaev, R.A., 1974. Astron. Astrophys., 36, 379.
Gruber, D.E., et al., 1979. preprint.
Higdon, J.C., and Lingenfelter, R.E., 1977. Ap.J. (Letters), 215, L53.
Jacobsen, A.S., Ling, J.C., Mahoney, W.A., and Willett, J.B., 1978.
    Gamma Ray Spectroscopy in Astrophys. NASA, Tech. Memor. 79619.
Kendziorra, E., Staubert, R., Pietsch, W., Reppin, C., Sacco, B.,
    and Trümper, J., 1977. Ap.J. (Letters), 217, L93.
Langer, S.H., McCray, R., and Baan, W., 1980. Ap.J. (in press).
Leventhal, M., MacCallum, C., and Watts, A., 1977. Ap.J., 216, 491.
Ling, J.C., Mahoney, W.A., Willett, J.B., and Jacobsen, A.S., 1977.
    Nature, 270, 36.
Mazets, E.P., Golenetskii, S.V., Il'inskii, V.N., Aptekar', R.L., and
    Guryan, Yu.A., 1979. Nature 282, 587.
Melrose, D.B., and Zheleznyakov, V.V., 1979, to be published in
    Astron. Astrophys.
Mészáros, P., 1979. Astron. Astrophys., 63, L19.
Mészáros, P. and Ventura, J., 1979. Phys. Rev., D19, 3565.
Mészáros, P., 1980. J. of Magnetism and Magnetic Mat., 15-18, 1551.
Mészáros, P., Nagel, W., and Ventura, J., 1980. preprint.
Nagel, W., 1980a, Ap.J., 236, 904.
Nagel, W., 1980b, Thesis, Techn. Univers. Munich.
Pravdo, S.H., 1979. Workshop on Compact Galactic X-ray Sources,
    eds. F.K. Lamb and D. Pines, p 194.
Pravdo, S.H., White, N.E., Boldt, E.A., Holt, S.S., Serlemitsos, P.J.,
    Swank, J.H., and Szymkowiak, A.E., 1979. Ap. J., 231, 912.
Ramaty, R., Borner, G., and Cohen, J.M., 1973. Ap.J., 181, 891.
Ramaty, R., 1978. Gamma Ray Spectroscopy in Astrophys., NASA, Tech.
    Memor., 79619.
Rudermann, M.A., 1974. IAU Symp. No. 53, D. Reidel Pub. Co, Holland.
Trümper, J., Pietsch, W., Reppin, C., Sacco, B., Kendziorra, E., and
    Staubert, R., 1977. Ann. N.Y. Acad. Sci., 302, 538.
Trümper, J., Pietsch, W., Reppin, C., Voges, W., Staubert, R., and
    Kendziorra, E., 1978. Ap.J. (Letters), 217, L105.

240

Ventura, J., 1979. Phys. Rev., D19, 1684.
Voges, W., Pietsch, W., Reppin, C., Trümper, J., Kendziorra, E.,
    and Staubert, R., 1979. IAU/COSPAR Symp. on X-ray Astron.,
    Innsbruck.
Wassermann, J., and Salpeter, E., 1980. preprint.
Weaver, R.P., 1978. Nature, 274, 571.
Wheaton, W.A., et al., 1979. Nature, 282, 240.
Wunner, G., 1979. Phys. Rev. (Lett), 42, 79.
Wunner, G., and Herold, H., 1979. Astrophys. Space Sci., 63, 503.
Wunner, G., and Ruder, H., 1979, to be published in Ap.J.
Yahel, R.Z., 1979a. Astr. Ap., 78, 136.
Yahel, R.Z., 1979b. Ap.J. (Letters), 229, L73.

Radiative opacities and transfer in magnetic X-ray pulsars

J. Ventura

Max-Planck-Institut fur Physik and Astrophysik
Institut fur Extraterrestrische Physik,
Germany.

1.   INTRODUCTION

The recent discovery of cyclotron line features in the spectra of
two X-ray pulsars (Trümper 1979, and this volume) offers us, for the
first time, the possibility to determine the magnetic field of
neutron stars. As is natural this development has motivated a new
surge of theoretical activity aimed at a better understanding of the
physical conditions in the polar cap region and the lower accretion
column of magnetic X-ray pulsars.

This is the region where most of the infalling matter's energy is
converted into radiation. The pulsar's strong magnetic field along
with the relatively high accretion rates ($\sim 10^{17}$ gs^{-1}) conspire to
produce fairly dense accretion funnels, whose transverse optical
depth (in free fall) exceeds one (e.g. Davidson 1973). One is then
confronted with the problem of a hot spot (the deceleration region)
whose radiation escapes after several scatterings in the lower
accretion column. The properties of this very anisotropic magnetic
filter should therefore be reflected in the observed radiation to a
lesser or greater extent.

Early calculations of the Thomson scattering coefficients (Canuto et
al. 1971; Gnedin et al. 1973) have shown a strong directional anisotropy of these coefficients as well as strong frequency and polarization dependence. At low frequencies relative to the electron gyrofrequency ω_H, the extraordinary wave (mode 1) is found to have a much
smaller opacity than the ordinary wave (mode 2), i.e.

$$< \sigma_1 > /< \sigma_2 > \simeq (\omega/\omega_H)^2 \tag{1}$$

where $< \sigma_i >$ denotes the opacity averaged over the photon's direction.

A medium with this property is a natural polarizer. A slab of such plasma, which is optically thick to the ordinary wave, can be transparent in the extraordinary. An unpolarized beam incident on one side of the slab should then give highly polarized transmitted radiation cf. Fig. 1. The first polarization measurements of the radiation from Her X-1 and Cen X-3 have indeed yielded polarized radiation at the frequencies 2.6 keV and 5.6 keV (Silver et al. 1979).

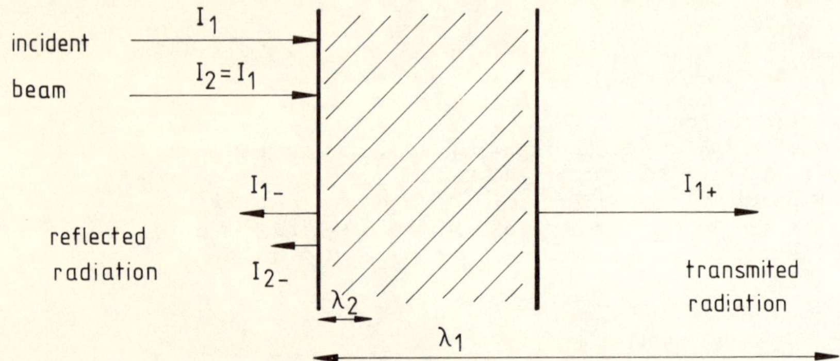

Figure 1: Reflection and transmission of two polarizations from a semitransparent slab. Absorption is assumed insignificant ($\kappa_i \ll \sigma_i, i = 1,2$), while the scattering coefficients are assumed to obey the relation $\sigma_2 \gg \sigma_1 \approx \sigma_{21} = \sigma_{12}$. The slab is transparent to mode 1 but opaque to mode 2 ($\lambda_i = \sigma_i^{-1}$). Incident mode 2 radiation scatters within a thin surface shell contributing (via mode conversion) almost pure mode 1 to the transmitted intensity.

Similarly the strong anisotropy in the cross section ought to give rise to beaming effects, which should also depend on the photon frequency. Strong beaming effects are indeed implied by the multiple-peak structure in the measured pulse profiles of most X-ray pulsars. In fact the complexity of the pulse profiles tends to diminish at high frequency (Lamb 1977; Kanno 1980) just as the anisotropy in the cross section would imply.

Finally, one expects the resonant structure of these cross sections, at the electron gyrofrequency and its multiples to be reflected into observable spectral line features at the frequency

$$\hbar\omega_H \sim 11 \, B_{12} \, keV$$

and its multiples (Gnedin and Sunyaev 1974; Daugherty and Ventura 1977, 1978).

Recent theoretical work in Garching and Munich has extended the previous classical cold plasma results of Canuto et al. (1971) and of Gnedin et al. (1973) to include propagation and polarization effects appropriate for X-ray frequencies. We have found that the polarization properties of the magnetic vacuum are a dominant ingredient in determining the radiative opacities and the transfer of radiation in the lower accretion column. The magnetic vacuum should determine further the polarization properties of the radiation observed at infinity. I will present in the following new simplified analytic expressions for the scattering and bremsstrahlung opacities, which allow a better intuitive understanding of the anisotropic and polarizing properties of these media than was previously available. These expressions are also generalizable, and it is therefore now possible to include vacuum polarization and other propagation effects in the medium. I will discuss finally results of simple model spectra radiated by a homogeneous self-radiating magnetic medium when the above effects are included.

2.   SCATTERING AND BREMSSTRAHLUNG OPACITIES IN A PLASMA

Propagation in a magnetized 'cold' plasma at high frequencies occurs in two normal modes, designated usually as ordinary (mode 1) and extraordinary (mode 2) waves. In an optically thick medium these modes propagate independently if their relative phase shift is large within one photon mean-free path, i.e.

$$\omega c^{-1} \left| n_1 - n_2 \right| \gg (\sigma_1 + \sigma_2)$$

where σ_i is the scattering opacity coefficient of mode 1 and n_i its refractive index (Gnedin and Pavlov 1974). The four transfer equations for the Stokes parameters may then be replaced by two equations describing the transfer of the normal modes. These two equations are coupled in general through the mode-exchange scattering (processes $1 \rightarrow 2$, $2 \rightarrow 1$).

The simple (single particle) description of propagation in a plasma predicts that the bremsstrahlung absorption coefficient for a wave of mode i is simply related to the scattering coefficient for that mode. Propagation of a wave in a medium is accompanied by an induced polarization current density

$$j_\alpha^i \propto \Pi_{\alpha\beta} \, e_\beta^i \tag{2}$$

where e_α^i is the electric vector of mode i and $\Pi_{\alpha\beta}$ is the medium susceptibility tensor. This current suffers radiation damping as well as collisional damping. The first of these losses is simply related to the total scattering coefficient σ_i (θ,ω) while the latter relates to the bremsstrahlung coefficient $\kappa_i(\theta,\omega)$. Here is the angle between the direction of propagation and the external magnetic field and ω is the wave frequency. The detailed evaluation of these coefficients, in a weakly dispersive medium, gives the expressions (Ventura 1979; Nagel 1980).

244

$$\sigma_T^{-1} \, \sigma_i(\theta,\omega) = \frac{\omega^2 \left|\hat{e}_+^i\right|^2}{(\omega+\omega_H)^2} + \frac{\omega^2 \left|\hat{e}_-^i\right|^2}{(\omega-\omega_H)^2+\gamma^2} + \left|\hat{e}_z^i\right|^2 \qquad (3)$$

$$\sigma_T^{-1} \, \kappa_i(\theta,\omega) = \eta g_\perp \left[\frac{\omega^2 \left|\hat{e}_+^i\right|^2}{(\omega+\omega_H)^2} + \frac{\left|\hat{e}_-^i\right|^2}{(\omega-\omega_H)^2+\gamma^2} \right] + \eta g_{\shortparallel} \left|\hat{e}_z^i\right|^2 \qquad (4)$$

$$\eta = \frac{3\pi}{2} \frac{Ze^2}{\hbar v_T} \frac{c^3}{\omega^3} n_e \, (1-e^{-\hbar\omega/kT}) \qquad (5)$$

$\hat{e}_\pm^i$ are the cyclic components of the normal mode vectors. $\sigma_T = (8\pi/3)(e^2/mc^2)^2 \, n_e$ is the Thomson scattering coefficient. ω_H is the electron gyrofrequency, γ the damping frequency, n_e the electron number density, T the electron temperature characterizing the motion along the magnetic field, v_T the electron thermal velocity. The coefficients $g_{\shortparallel}$, $g_\perp$ are Gaunt factors, which vary smoothly with the frequency. Finally eq (5) assumes a Maxwellian of temperature T for the velocity distribution of electrons along the magnetic field.

The right hand side of equation (3) is a representation of the matrix product $\left|\Pi_{\alpha\beta} \, e_\beta^i\right|^2$ and thus gives the relation between the power radiated and the induced polarization current density of eq. (2). The differential cross section for the scattering of a photon of polarization $\hat{e}^i(\underline{k})$ into the state $\hat{e}^j(\underline{k}')$ is given by a similar expression (Ventura 1979).

$$\frac{d\sigma_{ij}}{d\Omega} = n_e r_o^2 \frac{\omega^2}{\omega_p^2} \left|\hat{e}_\alpha^{j\,*} \, \Pi_{\alpha\beta} \, \hat{e}_\beta^i\right|^2$$

$$= r_o^2 \left| \frac{\omega \hat{e}_+^{j\,*} \hat{e}_+^i}{\omega+\omega_H} + \frac{\omega \, \hat{e}^{j\,*} - \hat{e}^i}{\omega-\omega_H} + \hat{e}_z^{j\,*} \hat{e}_z^i \right|^2 \qquad (6)$$

where $r_o = e^2/mc^2$. The explicit expressions, given above, follow from the diagonal representation of the tensor $(\omega^2/\omega_p^2) \, \Pi_{\alpha\beta}$ in a cyclic coordinate frame with its z axis along the magnetic field $\underline{B}$. The eigenvalues of this tensor in the cold plasma approximation, are:-

$$\Pi_\pm = \omega \, (\omega \pm \omega_H)^{-1}, \quad \Pi_z = 1.$$

The frequency dependence of eqs. (3), (4) and (6) is characterized by a resonant behaviour near the electron gyro-frequency, while at low frequencies (relative to ω_H) these equations are dominated by the $\hat{e}_z$ term. We thus find at low frequencies ($\omega \ll \omega_H$):

$$\frac{d\sigma_{ij}}{d\Omega} \simeq n_e r_o^2 \left| \hat{e}_z^j (\underline{k}') \; \hat{e}_z^i (\underline{k}) \right|^2 + O(\omega/\omega_H) \tag{7}$$

$$\sigma_i \simeq \sigma_T \left| e_z^i (\underline{k}) \right|^2 + O(\omega^2/\omega^2_H) \tag{8}$$

The interpretation of these results is straight-forward. An electron responds as a free particle to photons with their e-vector along the magnetic field and the corresponding cross section is σ_T. On the other hand the electron 'appears' bound to photons having their e-vector perpendicular to the external field. It is instructive to compare this to the Rayleigh cross section $\sigma_R = \sigma_T \, \omega^4 (\omega^2 - \omega_o^2)^{-2}$ for the scattering of light from bound atomic electrons. As is well known, the low frequency dependence of this scattering process is responsible for the blue colour of the terrestrial sky. We see from this comparison that the magnetic medium displays a combined Rayleigh-Thomson behaviour. This behaviour is anisotropic, with a dependence on the angles of scattering and angle of illumination due to the directional dependence of the medium normal modes.

It is worth noting that eq (6) gives a simple relation between the T-matrix for photon scattering and the susceptibility tensor Π, which is valid in the single particle limit. This relationship as well as eqns. (3) and (4) can be recovered from detailed quantum electro-dynamic calculations (Hamada, 1975; Herold, 1979; Nagel, 1980a) in their non-relativistic limit, if one uses the dipole approximation in evaluating the matrix elements of radiative transitions between the Landau levels of the electrons.

3. PROPAGATION EFFECTS OF THE MEDIUM

At high frequencies ($\omega^2 >> \omega_p^2 \equiv 4\pi \, n_e e^2/m$) the medium is only weakly dispersive and the normal modes are transverse to a very good approximation. In a magnetized electron plasma the modes are generally elliptical and given the vectors (coordinate frame x', y', z' with $\hat{z}' \equiv \hat{k}$, $\hat{y}' \propto \hat{B} \times \hat{k}$),

$$e^1 \equiv (1, i\alpha , 0), \; e^2 \equiv (i\alpha, 1, 0) \tag{9}$$

For real α these correspond to mutually orthogonal polarization ellipses. α is then a measure of the normal modes ellipticity, and is a function of the propagation angle θ.

From Figure 2a we see that at low frequencies, $\omega << \omega_H$, the cold plasma modes tend to be linearly polarized i.e. $\alpha \approx 0$ at almost all angles of propagation. The components of the normal modes to be used in eqns. (7) and (8) are obtained from (9) by a coordinate rotation over the angle θ giving e.g. for the z component

$$\hat{e}_z^1 \simeq - \alpha \sin\theta, \; e_z^2 \simeq - \sin \theta, \tag{10}$$

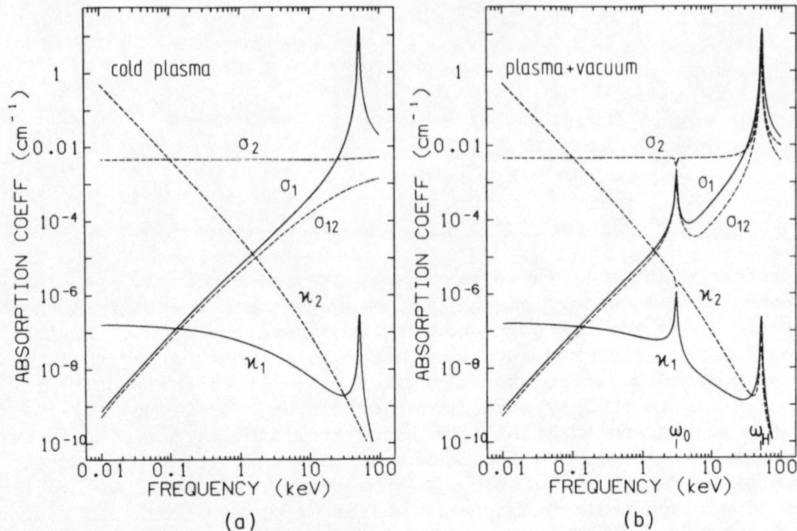

Figure 2: Frequency dependence of the angle-averaged scattering and absorption coefficients. σ_{ij} denotes the coefficient for the mode exchange process $i \rightarrow j$, while σ_i stands for the total scattering opacity of mode i, $\sigma_i = \sigma_{ii} + \sigma_{i \neq j}$. The calculation corresponds to a plasma density $n_e = 10^{22} cm^{-3}$, a temperature $T = 10$ keV, and a magnetic field $B = 0.1 B_{cr}$.

The total cross section of eq. (8) is thus $\sigma_2 = \sigma_T \sin^2\theta$ for the ordinary mode while for the extraordinary mode, it has a more complicated dependence on θ and rises to a maximum. Detailed studies of the anisotropy of these cross sections have been given recently by Borner and Meszaros (1979) and by Ventura (1979). The low frequency maximum in the cross section (σ_1 (θ)) corresponding to a minimum in the effective mean free path $\frac{1}{2}(\sigma_1^{-1} + \sigma_2^{-1})$ has been exploited recently by Kanno (1980) to interpret the multiple peak structure observed in the pulse profiles of some X-ray pulsars at low frequencies.

The cold plasma approximation has been traditionally used to describe propagation effects in the pulsar magnetosphere. With a quantum electrodynamical approach this amounts to a long-wavelength and low temperature limit of the theory (Canuto and Ventura 1977; Melrose and Stoneham 1977 and references therein). The approximations involved become questionable, however, in the X-ray range of frequencies. Strong departures of the absorption coefficients from the classical theory have been reported near the electron gyro-frequency (Daugherty and Ventura 1977, 1978; Herold 1979). The structure of the normal modes is also significantly affected by thermal effects near the gyrofrequency, the parameter α becoming complex (Nagel 1980a; Kirk 1980).

4.   MAGNETIC VACUUM EFFECTS

Strong departures from the cold plasma behaviour are also predicted
at high frequencies, due to vacuum polarization effects induced by
the presence of the external magnetic field (Meszaros and Ventura
1978, 1979; Gnedin et al. 1978, 1979).

As is well known, in the absence of matter the vacuum itself is
expected to display a birefringent behaviour (e.g. Erber 1961; Adler
1971). The normal modes of propagation are plane polarized in
directions parallel and perpendicular to the Bk-plane with corresponding refraction indices

$$n_{\parallel,\perp} = 1 + (4\,\delta,\ 7\,\delta)\ \sin^2\theta, \tag{11}$$

$$\delta \equiv \frac{1}{45\Pi} \equiv \frac{e^2}{\hbar c}\ \left(\frac{B}{B_{cr}}\right)^2 \simeq 0.5 \times 10^{-4}\ \left(\frac{B}{B_{cr}}\right)^2 \tag{12}$$

where $B_{cr} = m_e^2 c^3/e\hbar = 4.4 \times 10^{13}$ G is the critical electrodynamic
field. In the presence of a material component, vacuum polarization
thus contributes terms of order δ in the medium susceptibility compared to an analogous effect of order

$$\omega_p^2/\omega^2 \simeq 1.4 \times 10^{-7}\ n_{20}\ (\hbar\omega/keV)^{-2} \tag{13}$$

which is the electron plasma component. We therefore see (in the keV
range of frequencies), the propagation and polarization properties
of the combined vacuum + plasma is dominated by the vacuum component.
This can have implications both on the observable X-ray polarization
properties of pulsars and on the resultant spectrum as we shall see.

Both propagation and absorption characteristics are most severely
affected within a range of frequency and density which marks the
transition from the vacuum dominated (high frequency, low density) to
the plasma dominated (low frequency, high density) regime. The effect
on the normal mode ellipticity is seen in Fig. 3b at a given frequency
$\omega = (1/2)\omega_H$. Starting with the plasma dominated ellipticity at high
density ($n_e > 10^{24}$ cm^{-3}) measured here in units of

$$\frac{n_e}{B^2/4\Pi mc^2} = \frac{\omega p^2}{\omega_H^2} \simeq 10^{-9}\ n_{20}\ B_{12}^{-2}, \tag{14}$$

we see that the structure of the normal modes changes from elliptical
to circular polarization, as we pass the critical density at which

$$\omega_p^2/\omega_H^2 = 3\sigma\ (\omega^2/\omega_H^2)\ (1-\omega^2/\omega_H^2) \simeq (9/32) \times 10^{-4}\ \left(\frac{B}{B_{cr}}\right)^2, \tag{15}$$

and then again to linearly polarization at low densities $n_e < 10^{20}$
cm^{-3} where the limit is dominated by the vacuum characteristics.

These changes to the normal modes are reflected in changes of the

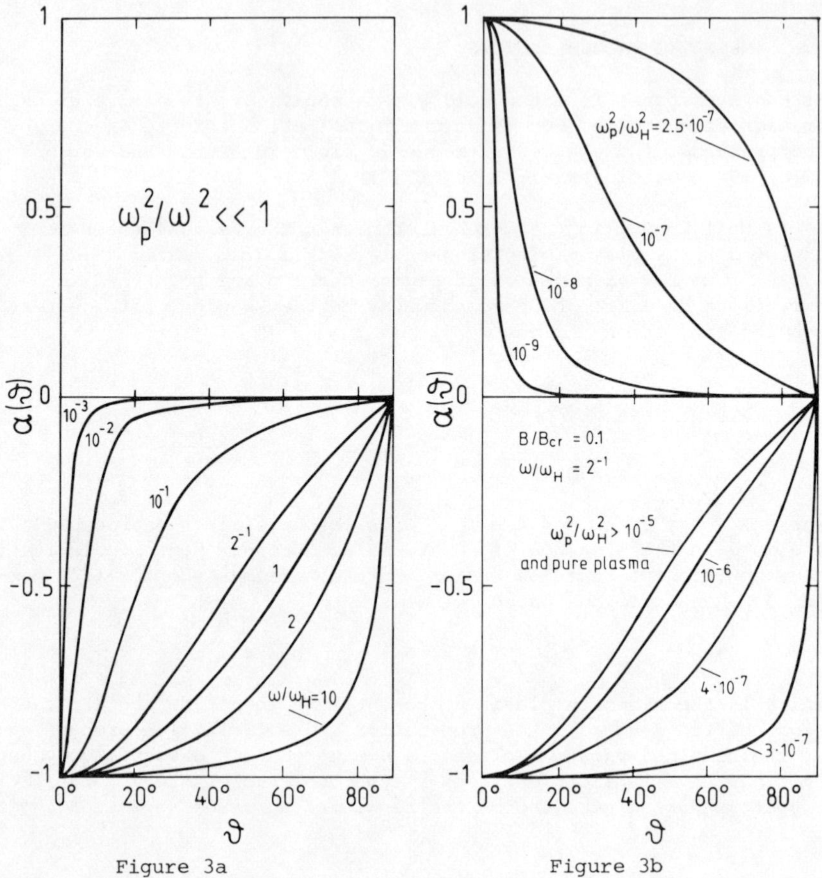

Figure 3a   Figure 3b

The ellipticity parameter $\alpha(\theta)$ of the e-vector normal modes varies
with the angle of propagation θ between the values $|\alpha| = 1$ (circular
polarization) and $\alpha = 0$ (linear). *(Reproduced by permission from the
American Physical Society, Ventura, J., Phys. Rev. D19, 3565, 1979).*

a) for the cold plasma normal modes at various frequencies
   and
b) for a plasma + vacuum medium at the frequency $\omega = 2^{-1}\omega_H$ but
   for changing value of the plasma density parametized here
   through the ratio ω_p^2/ω_H^2. For the accretion column of a
   typical pulsar of $n_e \sim 10^{20}$ cm^{-3}, $B \sim 0.1\,B_{cr}$, $\omega_p^2/\omega_H^2 < 10^{-9}$.

corresponding opacity coefficients in both the Thomson and bremsstrahlung processes. Large scale changes of the opacity coefficients are also found near the electron gyrofrequency, where the
ordinary wave of the cold plasma now acquires a resonant behaviour
of order δ^{-1}.

These effects are seen in the opacities angle-averaged shown in Figure 2 b (Ventura, Nagel and Meszaros 1979). Strong effects are also found on the directional anisotropy of the medium near the gyrofrequency (Meszaros and Ventura 1979) and near the critical frequency

$$\hbar\omega_o \simeq 3 \; n_{20}^{1/2} \; (B/B_{cr})^{-1} \; keV \tag{16}$$

where the transition from plasma dominated to vacuum dominated regions occurs (Pavlov and Shibanov 1979, Hamada 1980).

5.   SPECTRA EMITTED FROM SELFRADIATING CYLINDER

The microphysics, discussed in the previous sections has been incorporated into calculations of the spectrum emitted at the transverse surface of a hot homogeneous cylinder of magnetized plasma (Nagel 1980; Ventura, Nagel and Meszaros 1979). We make use of the angle-averaged cross sections shown in Figure 2 and concentrate on the resultant spectra. We have assumed that the electron velocities along the magnetic field obey a Maxwellian distribution of temperature T. However, the transverse distribution, i.e. the occupation of the Landau levels is not given by the same temperature. Under the conditions assumed here, the rate of Coulomb excitation or deexcitation is negligible compared to the radiative transition rates. The occupation of the Landau levels is thus controlled by the local density of photons having energies near $\hbar\omega_H$.

Photons are produced thermally by the bremsstrahlung process with a volume emissivity determined from Kirchhoff's law, where:

$$j_i = B_\omega(T) \; \kappa_i \tag{17}$$

where κ_i is the bremsstrahlung absorption coefficient and B_ω is the Planck specific intensity. Once produced, photons diffuse outwards and are absorbed after $\sim (\sigma+\kappa)/\kappa$ scattering events in the average, or escape from the side of the cylinder. This gives rise to the radiated intensities shown in Figure 4.

As in the nonmagnetic case the calculated intensities are way below the black body values in the range of frequencies where $\sigma_i \gg \kappa_i$ (radiation dominated).

It turns out from such calculations, however, that the coupling of the two modes has a strong influence on the emitted spectrum. The calculated emergent spectra (Fig. 4) thus differ qualitatively and quantitatively from the corresponding spectra emitted by nonmagnetic plasmas (e.g. Felten and Rees 1972).

We thus find that in the range of frequencies above 2 keV, where $\sigma_{12} \gg \kappa_1, \kappa_2$ (strong coupling case) the overall intensity radiated is limited by the quantity

$$I \simeq R^{1/2} \epsilon^{1/2} B_\omega \tag{18}$$

where $\varepsilon \simeq \kappa_1/\sigma_1 \simeq \kappa_2/\sigma_2$ and $R = \sigma_>/\sigma_<$ is the ratio of the larger to the smaller of the total scattering coefficients σ_1 and σ_2. This quantity is only reached in the fully self-absorbed limit of 'infinite' optical depth and can greatly exceed the corresponding $\varepsilon^{1/2} B_\omega$ limit predicted for the radiation from non-magnetic plasmas. The nonmagnetic case would actually fall close to the lower of the two partial polarization curves shown in Figure 4.

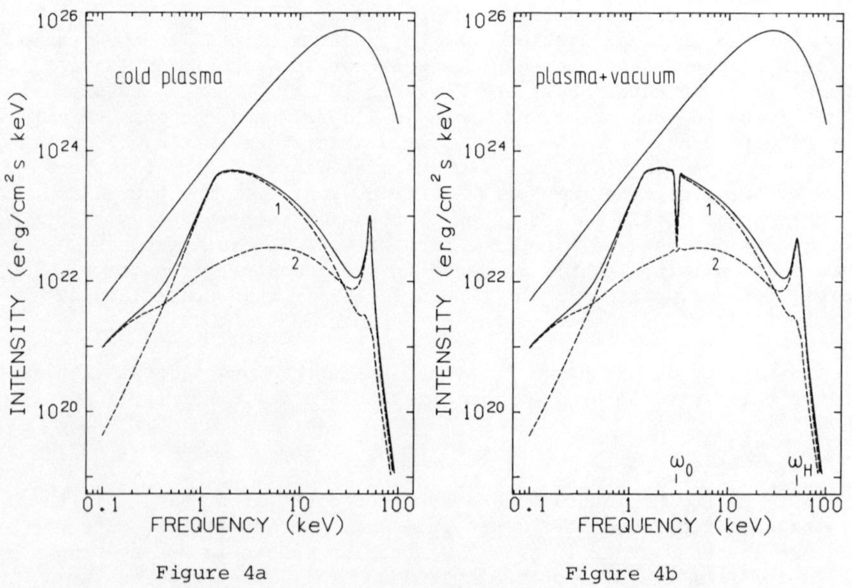

Figure 4a   Figure 4b

Simplified model spectra from a self-radiating homogeneous cylinder of transverse dimension $d = 10^5$ cm and characterized by the same plasma parameters as in Figure 2. Emission and absorption lines are found even at infinite optical depth. (a) pure plasma (b) plasma + vacuum case. The spectral feature at $\omega_0 \sim 3$ keV marks the change of regime from plasma dominated at $\omega << \omega_0$ to vacuum dominated at high frequencies. The top full line represents the 10 keV black body spectrum. The lower full line gives the total intensity emitted by the slab sum of the two partial polarizations (broken lines).

Another new feature predicted by eq (18) and confirmed by the detailed calculation is the possibility of spectral line features from isothermal and optically thick media whenever the quantity $R^{1/2}$

undergoes rapid changes. This is a totally new mechanism for producing spectral line features in the radiation from polarizing media, when the strong coupling condition exists for the two polarizations. It was first obtained by Werner Nagel by solving the coupled diffusion equations for the transfer of two polarizations (Nagel 1980).

As a result of this mechanism we see strong spectral line features at the frequencies ω_O and ω_H reflecting corresponding sharp features in the cross sections of Figure 2. A more detailed discussion of this new approach to the transfer of polarized radiation has been given recently by Meszaros, Nagel and Ventura (1980).

A possible observation of the vacuum feature at ω_O in the X-ray frequencies cannot be ruled out at present. Except for confirming the role of vacuum polarization in the pulsar magnetosphere such an observation together with an additional feature at ω_H would determine the electron plasma density of the accretion column. The shape of the feature is insensitive to the temperature but will be influenced strongly by density gradients and comptonization effects which were neglected in the above simple models. A very conspicuous characteristic which is likely to be preserved in a more sophisticated treatment is the prediction that the plane of linear polarization changes by 90° at ω_O (Ventura, Nagel and Meszaros 1979).

I should stress that the spectra shown above represent primitive model calculations. Improvements have to be made before we can draw definitive conclusions on the nature of the observed X-ray sources. A severe restriction is the neglect of incoherence in the scattering process, which is thus assumed to conserve the photon frequency. In a dominated scattering (thick atmosphere) incoherence should bring large scale modification of the above spectra (e.g. Felten and Rees 1972; Pozdnyakov et al. 1979; Bonazzola, Heyvaertz and Puget 1979). Furthermore, we have neglected directional anisotropy, density and temperature gradients such as could occur by the presence of a possible hydro dynamical shock, and the effects of possible deviations from thermodynamic equilibrium (cf. Langer and McCray 1980).

6.   DISCUSSION

Though primitive, the above calculations represent the first comprehensive attempt to include details of the polarizing properties of magnetic media. They show most clearly the qualitative aspects of the effects to be expected in more sophisticated treatments. These effects are so remarkable that we conclude that the detailed frequency, polarization and directional dependence of the cross sections cannot be neglected in interpreting the observational details of the pulsating X-ray sources.

Detailed quantitative studies, currently in progress at the Max-Planck-Institut and elsewhere, can take advantage of the advances in the physics of high magnetic fields reported above to resolve some

of the current riddles presented by the X-ray pulsators. The next
several months may thus bring a firm interpretation of the observed
break at 25 keV and of the cyclotron line feature at $\sim$ 58 keV in the
Her X-1 spectrum (Trümper et al. 1978). Similar features were
reported recently in the spectrum of 4U0115+63 (Wheaton et al. 1979).

REFERENCES

Adler, S.L., 1971. Ann. Phys. (N.Y.), 67, 599.

Borner, G., Meszaros, P., 1979. Plasma Phys., 21, 357.

Canuto, V., Lodenquai, J., Ruderman, M., 1971. Phys. Rev. D., 3,
     2303.

Canuto, V., Ventura, J., 1977. Fund. Cosmic Phys., 2, 203,

Daugherty, J., Ventura, J., 1977. Astron. Astrophys., 61, 723.

Daugherty, J., 1978. Phys. Rev. D., 18, 1053.

Davison, K., 1973. Nature Phys. Sci., 246, 1.

Erber, T., 1961. Nature, 190, 25.

Felten, J.E., Rees, M.J., 1972. Astron. Astrophys., 17, 226.

Gnedin, Yu., Sunyaev, R., 1973. Sov. Phys. JETP, 38, 51.

Gnedin, Yu., 1974, Astron. Astrophys., 36, 379.

Gnedin, Yu., Pavlov, G., 1974. Sov. Phys. JETP, 38, 903.

Gnedin, Yu., Pavlov, G., Shibanov, Yu., 1978. JETP Lett., 27, 305.

Gnedin, Yu., 1978. Sov. Astron. Lett., 4, 117.

Hamada, T., 1975. Publ. Astron. Soc. Jpn, 27, 275.

Hamada, T., 1980. Publ. Astron. Soc. Jpn, to be published.

Herold, H., 1979. Phys. Rev. D., 19, 2868.

Kanno, S., 1980. Publ. Astron. Soc., Jpn., to be published.

Kirk, J., 1980, Plasma Phys., submitted.

Lamb, F.K., 1977. Ann. N.Y. Acad. Sci., 302, 482.

Langer, S., McCray, R., 1980, Astrophys. J., submitted.

Melrose, D.B., Stoneham, R., 1977. Proc. Astron. Soc. Australia, 3,
     120.

Meszaros, P., Ventura, J., 1978. Phys. Rev. Lett., 41, 1544.

Meszaros, P., 1979. Phys. Rev., D., 19, 3565.

Meszaros, P., Nagel, W., Ventura, J., 1980. Ap. J., to be published.

Nagel, W., 1980. Astrophys. J., to appear.

Nagel, W., 1980a. Doctoral dissertation, Universitat Munchen.

Pavlov, G., Shibanov, Yu., 1979. Zh. E. Teor. Fix., 76, 1457.

Silver, E., Weisskopf, M., Kestenbaum, H., Long, K., Novick, R.,
     Wolf, R., 1979. Astrophys. J., 232, 248.

Trümper, J., 1979. In "X-ray Astronomy", (COSPAR), eds. W.A. Baity
     and L.E. Peterson, Pergamon, Oxford.

Trümper, J., Pietsch, W., Reppin, C., Voges, W., Staubert, R., Kend-
     ziorra, E., 1978. Astrophys. J., 219, L105.

Ventura, J., 1979. Phys. Rev. D., 19, 1684.

Ventura, J., Nagel, W., Meszaros, P., 1979. Astrophys. J. (Letters),
     233, L125.

Wheaton, Wm. A., et al., 1979. Nature, 282, 240.

The influence of collective effects on the deceleration of an ion in an accretion column

J. G. Kirk

Max-Planck-Institut fur Physik und Astrophysik,
Munich.

Calculations of the rate of energy deposition by a fast ion falling
through the plasma above a magnetised neutron star are important in
the physics of X-ray pulsars, since such effects may play a decisive
role in the question of the formation of the shock front. Treatments
of spherically symmetric accretion (Alme and Wilson, 1973) used the
formula for deceleration by Coulomb collisions in the absence of a
magnetic field. This approach requires modification when a magnetised
neutron star is considered. Basko and Sunyaev (1975) and Pavlov and
Yakovlev (1976) calculated the relevant cross-sections, and estimated
that the stopping length of an ion is increased by an order of
magnitude over the zero-field case. However, this conclusion no longer
holds when the collective effects in the plasma are included. The most
interesting of these is the scattering of ions on ion-sound fluctuations.
The importance of this effect arises from the nature of the ion-electron
cross-sections themselves. These are particularly small whilst the
ion is moving in the direction of the magnetic field, so that during
this time the otherwise negligibly small effect of these fluctuations
plays the major role. That such a scattering can occur at all is seen
only when full account is taken of the orbit of an ion in the magnetic
field. Even a modest level of ion-sound turbulence suffices to
deflect the ion, and severely reduce the stopping length. Numerical
investigations of this effect are under way, and should reveal the
momentum and energy deposition rates as well as the rate of collisional excitation of electrons to higher Landau levels.

REFERENCES

Alme, M.L., and Wilson, J.R., 1973. Astrophys. J., 186, 1015.
Basko, M.M., and Sunyaev, R.A., 1975. Sov. Phys. JETP, 41, 52.
Pavlov, G.G., and Yakovlev, D.G., 1976. Sov. Phys. JETP, 43, 389.

PART 3
Soft X-ray & UV measurements of stars and nebulae

Stellar winds and the evidence for coronae around early type stars

Joseph P. Cassinelli

Washburn Observatory,
Wisconsin, USA

1.   Introduction to the Theory of Stellar Winds

2.   Stellar Winds of Early Type Stars

3.   The Oxygen VI Problem and Stellar Coronae

1.   INTRODUCTION

Mass loss by a more or less steady expansion of the outer atmospheric layers is now known to occur across the Hertzsprung-
Russell diagram. Figure 1 shows stars on the H-R diagram for
which mass loss rates have been estimated either from studies of
line profiles or the infrared free-free continua. The size of the
dot represents the magnitude of the mass loss. The largest
correspond to about 10^{-5} $M_\odot$/yr. For comparison, the solar wind
results in a mass loss of 10^{-14} $M_\odot$/yr.

Note that the mass loss rate tends to be especially large for stars
of large luminosity. The hatched lines mark the Eddington instability limits for stars of 1 and 50 solar masses. Stars cannot exist
with luminosities above their Eddington limit because the outward
radiation force due to electron scattering opacity in the interior
would exceed the inward force of gravity. The concentration of
stars with high mass loss rates near the Eddington limits suggests
that radiation pressure plays an important role in the mass loss
process.

How do we know that stars are losing mass? The most convincing
evidence is the presence of "P Cygni" line profiles in the stellar
spectrum. A "P Cygni" line profile has an emission component and a
shortward displaced absorption component. The "absorption" is
produced because photospheric light is scattered out of our line of
sight by the outwardly moving wind material. Figure 2 shows the
exciting discovery of Morton (1967) of P Cygni lines in the rocket
ultraviolet spectra of OB supergiants. The shortward displaced
absorption components in the lines of Si IV and C IV indicated that
the stars are losing mass at a speed of about 2500 km/sec and with
a mass loss rate larger than 10^{-6} $M_\odot$/yr.

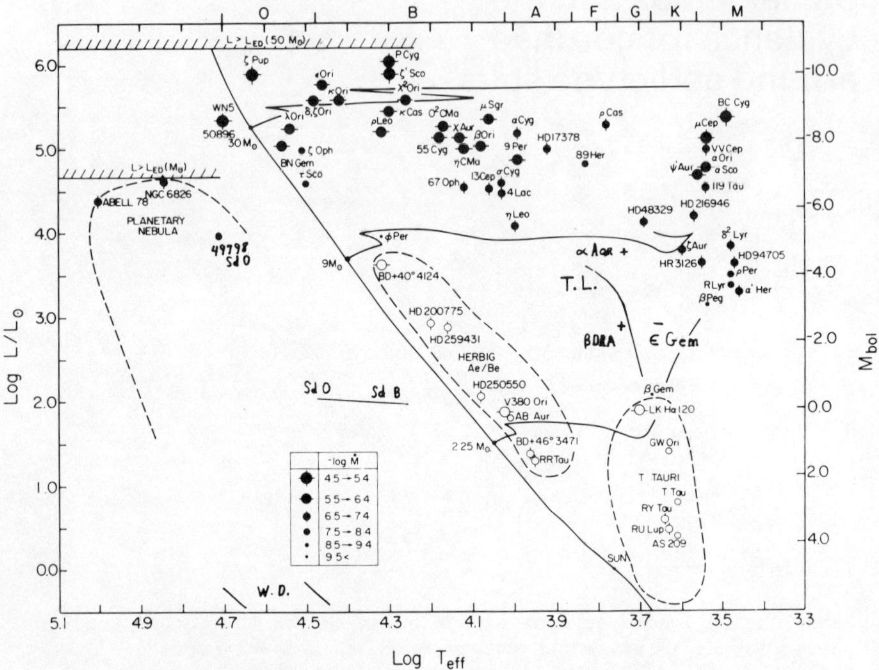

Figure 1: Shows many of the stars in the H-R diagram for which mass loss rates have been estimated. The size of the symbol indicates the magnitude of the mass loss in solar masses/year. The Eddington instability limits for stars of 1 $M_\odot$ and 50 $M_\odot$ are indicated by the boundaries of the cross-hatched areas.

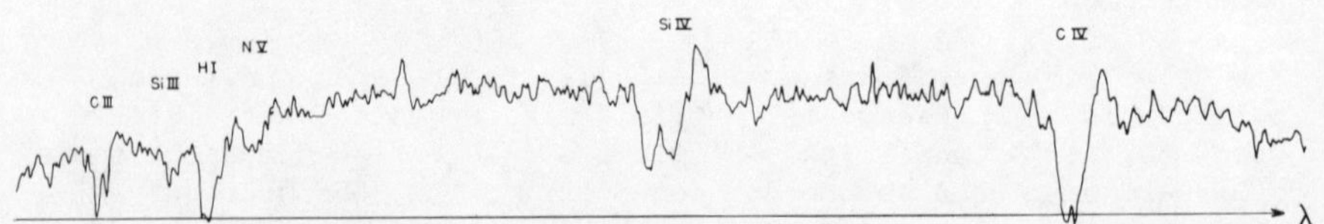

Figure 2: Shows Morton's (1967) early observations of the ultra-violet spectrum of ζ Ori O9.5 la. Wavelengths increase to the right from 1140 to 1630 Å. Shown are the P Cygni profiles of C IV and Si IV and shortward displaced absorption lines of C III, N V, and Si III.

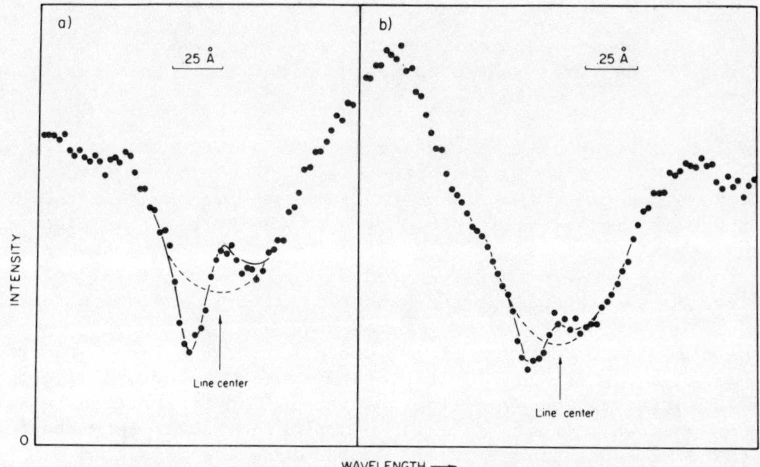

Figure 3: Circumstellar MnI $\lambda4030.8$ Å and $\lambda4033.1$ Å lines in α Ori. These have narrow P Cygni profiles superimposed on the background photospheric line. (Bernat and Lambert, 1976a).

The two component P Cygni lines are also seen in late stars. Figure 3 shows narrow P Cygni lines of Mn I superposed on the broad photospheric line in α Orionis. Such lines also indicate that the star is losing mass at a large rate, but in contrast with the hot stars the flow velocity is very small - about 30 km/sec.

2.   WINDS OF EARLY TYPE STARS

In this lecture, we will focus attention on the fast winds of the Of stars and OB supergiants. The theory that is usually used to describe these winds is the "line driven wind theory" of Lucy and Solomon (1970) and Castor, Abbott and Klein (1975).

Lucy and Solomon rejected the possibility that the winds could be produced by a coronal mechanism like that in the Sun because one sees the evidence of the expansion in lines of low stages of ionization such as C III and N IV. If winds with speeds of near 3000 km/sec were to be produced by the Parker solar wind mechanism, the coronae would have temperatures near 50 million degrees and of course negligible C III and N IV could exist.

Let us consider the process by which the radiation field could drive a steady outflow. The outward radiative acceleration (g_R) caused by opacity in the outer atmosphere is given by

$$g_R = \frac{1}{\rho}\frac{dP_R}{dr} = \frac{4\pi}{c}\int \kappa_\nu H_\nu \, d\nu$$

where κ_ν is the monochromatic opacity and $4\pi\ H_\nu$ is the flux at frequency ν, dP_R/dr is the pressure gradient and ρ the density at radius R.

For g_R to be large it is necessary to have a large flux at frequencies where κ_ν is large. For hot stars, H_ν tends to be large in the ultraviolet, because the Planck function at T_{eff} peaks in the UV, and there is also significant opacity in the UV resulting from many strong resonance lines. We know from diffusion theory that deep in the star $H_\nu \propto 1/\kappa_\nu$ so that the product $\kappa_\nu\ H_\nu$ may be small and the interior is stable against the potentially large radiation forces.

In the atmosphere however it is possible to get a very large product of $\kappa_\nu\ H_\nu$. For example if the atmosphere has even a slight outward expansion motion, the line opacity is Doppler shifted to a frequency at which there is a large continuum flux. Lucy and Solomon found that one line C IV λ1550 could drive a mass loss of $\dot{M} = 10^{-9}\ M_\odot/yr$ at speeds of 2000 km/sec. While the terminal speeds that were predicted are about right, the mass loss rate is too small. Castor, Abbott and Klein (1975) therefore extended the theory to account for the effects of very many lines: weak lines, strong lines, and overlapping lines. We will quote their work using the abbreviation CAK.

Consider the acceleration process in somewhat greater detail: a photon emitted from the photosphere at a frequency (ν^*) somewhat shortward of a line, travels outward and is intercepted at a shell, at which it is now resonant with the line opacity. It will scatter many times within the shell and eventually leave at some angle θ relative to the radial direction and frequency smaller than it had initially, i.e.,

$$\Delta\nu = -\nu^* \frac{v}{c}(1-\mu)$$

Inside the "resonant shell", the opacity in the line is κ_ℓ' and the flux may be approximated by $H_\nu = H_c^*\ \beta$, the continuum flux at the star is H_c^* and β is a "penetration probability".

$$H_\nu = H_c^*(1-e^{-\tau_s})\tau_s$$

with τ_s = the optical depth of the shell,

$$\tau_s = \kappa_\ell'\ \rho\ \Delta v_D/(dv/dr)$$

Δv_D is the Doppler width of the line in velocity units and dv/dr is the velocity gradient of the flow at the position of the resonant shell. Consider the limiting cases of very strong line opacity and very weak line opacity in the shell.

a)   Strong lines $\tau \gg 1$, $\beta = \dfrac{1}{\tau}$, $g_\ell = \dfrac{4\pi}{c} \dfrac{\rho}{\Delta v_D} \dfrac{dv}{dr} H_c^{*}$,

b)   Weak lines $\tau \gg 1$, $\beta = 1$, $g_\ell = \dfrac{4\pi}{c} \kappa_\ell \rho H_c^{*}$.

Note that the line opacity cancels out of the acceleration for strong lines. Therefore, the total force on the strong lines is simply proportional to the number of strong lines. That is, the well studied C IV line is no more important than that of some unfamiliar far UV line. Abbott (1977) has found that lines from the 20 most abundant elements can contribute to the strong line summation in the winds of Of. This means that it is necessary to consider huge line lists when calculating the total acceleration.

Note also that the acceleration due to strong lines is proportional to $(dv/dr)^1$ and that on weak lines to $(dv/dr)^0$. Thus, for a mixture of strong and weak lines the total acceleration will depend on some intermediate power of dv/dr. CAK let

$$g_R \propto \left(\frac{dv}{dr}\right)^\alpha$$

where $\alpha = 1$ if all the lines are strong and non-overlapping; $\alpha = 0$ if all the lines are weak, and $0 < \alpha < 1$ if there is a mixture of weak, strong and/or overlapping strong lines. Now that the radiative acceleration is non-linear the solution is somewhat more complicated than that of the more familiar Parker theory.

There are several important predictions of the theory. For example, if we assume that every photon from the photosphere is scattered by one resonant shell, we get an estimate of the maximum mass loss rate expected by equating photon and mass loss momenta; $L/c = \dot{M}_{max} v_\infty$,

$$\dot{M}_{max} = \frac{L}{v_\infty c}$$

This single scatter maximum mass loss rate is near 10^{-5} $M_\odot/yr$ for early type stars like ζ Pup. Other predictions are discussed in CAK. In summary, the theory predicts:

a) $\dot{M} < 10^{-5}$ $M_\odot/yr$;

b) $\dot{M} \propto L^{1.1}$;

c) $v_\infty \approx 2$ to $4 \times v_{esc}$;

d) $T_e(wind) \approx 0.8\, T_{eff}$.

The terminal velocity v_∞ is expected to be proportional to v_{esc} through a homology argument (Abbott, 1978). There is no need for mechanical energy to drive the flow and thus (d) is a statement that

262

the gas should be near its radiative equilibrium temperature (Klein
and Castor, =978). On most counts the theory holds up very well.
The maximum mass loss rates seldom exceed 10^{-5}. Perhaps a few Wolf
Rayet stars have $\dot{M}$ of order 2×10^{-5} $M_\odot$/yr but slight excesses above
$L/v_\infty c$ could perhaps be explained by the effects of multiple scattering. The infrared survey of mass loss rates by Barlow and Cohen (1977)
led to a good fit to relation (b). Abbott (1978) found that $v_\infty = 3$ v_{esc}
fit very well the data of 20 early-type stars after accounting for the
reduction of v_{esc} caused by electron scattering opacity. Only in prediction (d) is there a major flaw in the picture. This is known as
the "O VI problem" and is discussed in great detail by Cassinelli,
Castor and Lamers (1978).

3.   THE O VI PROBLEM AND STELLAR CORONAE

Figure 4 shows a Copernicus spectrum of the O VI line in ζ Pup. A
fractional abundance of O VI/O total of about 10^{-4} is necessary to
explain this line. There are several ways in which this degree of
ionization can be produced.

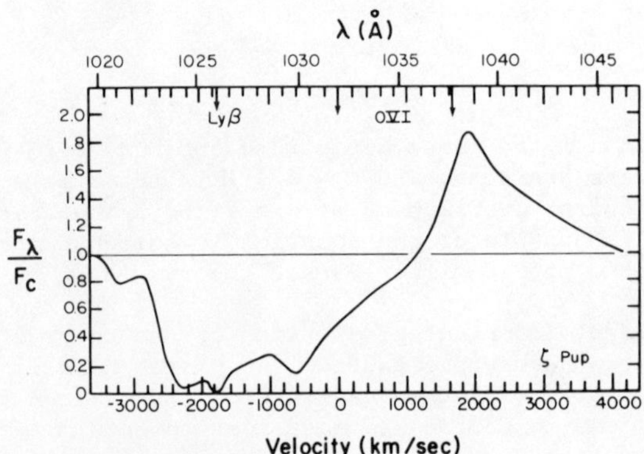

Figure 4: The P Cygni profile of the O VI 1031.945 - 1037.619 lines
in the Copernicus satellite spectrum of ζ Pup (O4f) Morton (1976).
The arrows indicate the rest wavelengths of the lines. The velocity
scale at the bottom gives the Doppler displacement from the 1031.8
component. *(Reproduced, with permission, from the Annual Review of
Astronomy & Astrophysics, Volume 17. Copyright 1979 by Annual Reviews
Inc.)*

Lamers and Morton (1976) suggested that the ion is produced by collisional ionisation in the wind and found that it is necessary for
the wind to have a temperature of 2×10^5 K over the whole extent of
the wind. This is known as the "warm wind model". It does explain
the overall ionisation balance seen in the ultraviolet of ζ Pup and
if the temperature is reduced to about 80,000 K it explains the ionisation anomalies (N V and C IV) seen in the winds of B supergiants.
It has one major problem, however; an unreasonably large input of
mechanical energy is required to sustain the elevated temperature of
2×10^5 K. A gas at that temperature radiates very effectively (Cox

and Tucker, 1969) and if it is to be maintained by the deposition of say acoustic flux, the star's acoustic luminosity must be about 10% the radiative luminosity.

An alternative explanation is the "hybrid" corona plus cool wind model. In this model it is assumed that the anomalously high stages of ionisation are produced in the wind by absorption of hard radiation that originated in a corona. Hearn (1975) noted that the densities in the winds of OB supergiants are so large $\sim 10^{11}$ cm^{-3} that if there is a coronal zone at the base of the flow the outflowing matter would like to cool rapidly because of radiative recombination radiation. Cassinelli, Olson and Stalio (1978) and Cassinelli and Olson (1979) have investigated the observational consequences of such a model.

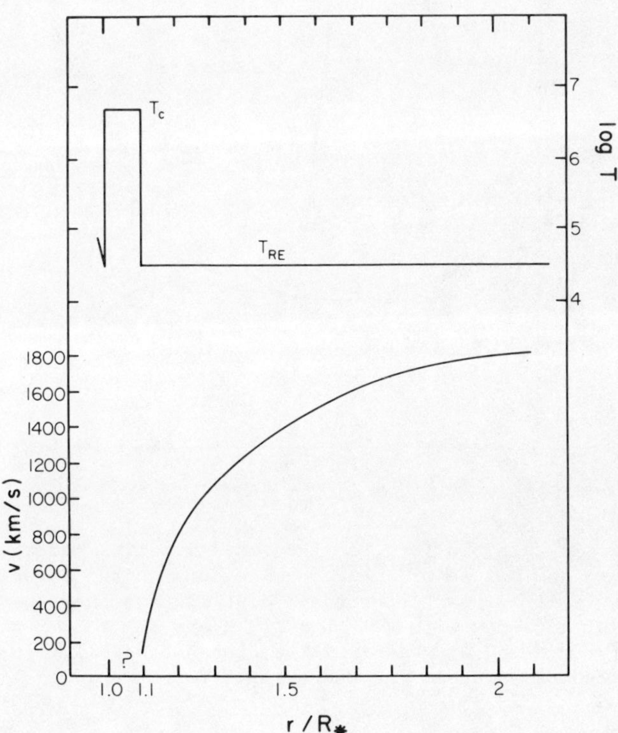

Figure 5: Shows a schematic run of temperature and velocity in a corona plus cool wind model. The anomalous ionisation stages are formed in the cool wind region as a result of X-rays emitted from the Coronal Zone. *(Reproduced by courtesy of the International Astronomical Union.)*

The energy distribution at the base of the cool wind for a model of
ζ Pup is shown in Figure 6. The star is assumed to radiate as a
black-body of 42,000 K at frequencies lower than the He II n = 1
edge, and in absence of a corona there would be negligible flux
beyond the O^{+4} to O^{+5} edge at 0.11 keV. The corona is assumed to
have a temperature of about 5 x 10^6 K. The emission measure
$EM_C = N_C^2$ Vol determines the magnitude of the coronal radiation.

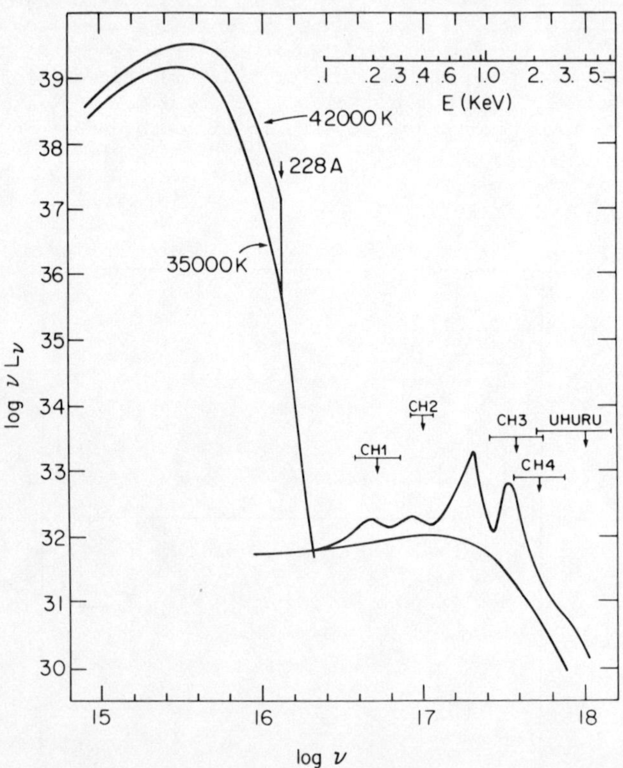

Figure 6: The assumed form for the radiation field emergent from the
photosphere and corona of ζ Pup and incident on the lower boundary of
the cool stellar wind. The coronal distribution shown here is for
the case of a corona with T_C = 5 x 10^6 K and EM_C = 10^{56} cm^{-3}. Upper
limits of the X-ray flux derived from the ANS satellite's Channel 1
to 4 are indicated as is the Uhuru limit for ζ Pup.

The distribution shown in Figure 6 corresponds to EM_C = 10^{56} cm^{-3},
which would be sufficient to provide enough O VI if the wind were
optically thin. The winds of Of stars, Wolf Rayet stars and OB
supergiants are far from being optically thin however. The run of
optical depth vs. frequency for the wind of ζ Pup is shown in Figure
7. Some important ionisation edges are indicated, such as the K

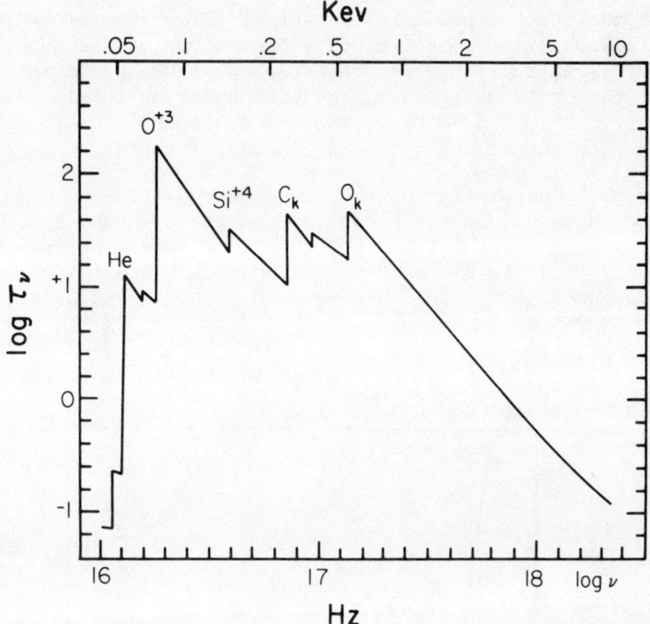

Figure 7:   Shown is the total optical depth of the wind of ζ Pup at frequencies beyond the He II edge.

shell edges of carbon and oxygen. The opacity beyond the oxygen K edge falls as $\nu^{-2.6}$ and the wind does not become optically thin until about 1 - 2 keV.

Because of the large optical thickness the ionisation balance in the wind depends both on the diffuse radiation field and on the attenuated coronal flux.

The corona-plus-cool-wind model makes two important predictions that can be tested by observations. The first concerns the production of the ionisation anomalies of the Of stars and OB supergiants. The dominant ionisation stages of any element are determined by the stellar and diffuse radiation fields. Thus we can determine the most abundant ions in the wind as a function of the stellar effective temperature. The anomalous ionisation states (such as O VI or N V) are produced in trace amounts (10^{-4}) by the absorption of coronal X-rays. For C, N and O the absorption of an X-ray by K shell photoionisation essentially always leads to the net ejection of two electrons instead of one, because of the Auger effect (Weisheit, 1974). Thus if oxygen is primarily in the form of O IV in the stellar wind we should see O VI in the spectrum. As we look to later spectral type the O VI should disappear at the spectral range for which the dominant ionisation is shifting from O IV to O III. Similarly, N V is produced with an "anomalously" high abundance as long as N III is the dominant stage and C IV is produced from C II.

Table 1 summarises the predicted ranges of effective temperatures for which various ions should produce "anomalously" strong spectral lines. For example, without the Auger process we should not see O VI in any Of star or OB supergiant, but with Auger it should be produced down to $T_{eff} \approx 30,000 \pm 3000$ K. Table 2 shows the comparison of the prediction with the observations and excellent agreement is found.

The second major prediction of the hybrid corona-plus-wind model is that X-rays should be observable by HEAO-B (now HEAO-2). This is shown in Figure 8 for the ζ Pup model. There should be essentially no soft X-ray flux because of the wind opacity and only X-rays with energies beyond about 1 keV can penetrate the flow and be detected.

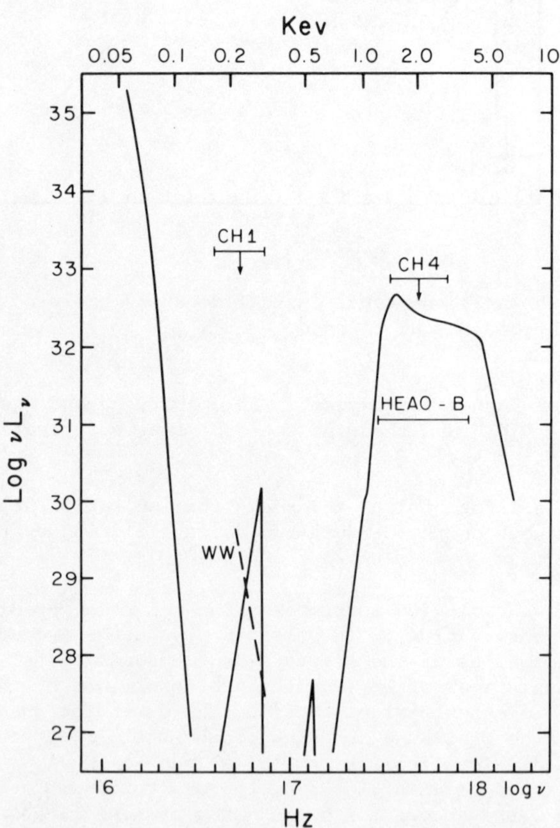

Figure 8:   The emergent X-ray spectrum predicted for ζ Pup, after accounting for attenuation by the cool wind. These results correspond to the case $T_C \simeq 5 \times 10^6$ K, $\beta M_C = 10^{58}$. Indicated by "WW" is the emergent flux expected from the warm wind model. The ANS channel 1 and 4 upper limits are shown as is the limit of sensitivity of the IPC device on the HEAO-2 satellite.

TABLE 1 - ULTRAVIOLET RESONANCE LINES LIKELY TO BE ENHANCED BY AUGER
IONIZATION

LINE	$(\overset{\circ}{A})$	f	Abundance N_{El}/N_H	T_{eff} range normally expected†	T_{eff} range ion expected via Auger process†	Parent ion	K-shell edge (keV)
C IV	1548.2 50.8	0.194 0.097	3.7×10^{-4}	25 - 65	9 - 20	C^{+1}	.296
N V	1238.8 1242.8	0.152 0.076	1.1×10^{-4}	$T_{eff} > 39$	20 - 37	N^{+2}	.432
O VI	1031.9 1037.6	0.130 0.065	6.8×10^{-4}	$T_{eff} > 54$	30 - 50	O^{+3}	.595
Si IV	1393.8 1402.8	0.528 0.262	3.5×10^{-5}	20 - 45	8 - 16	Si^{+1}	1.88
P V	1118.0 1280.0	0.495 0.245	2.7×10^{-7}	33 - 70	14 - 29	P^{+2}	2.24
S IV	1062.7 1073.0	0.038 0.037	1.6×10^{-5}	19 - 50	8 - 19	S^{+1}	2.54
S VI	933.4 944.5	0.426 0.210	1.6×10^{-5}	$T_{eff} > 37$	25 - 44	S^{+3}	2.58

$†\ (10^3\ K)$

TABLE 2 - RESONANCE LINES SEEN FROM HIGH STAGES OF IONIZATION IN OB
SUPERGIANTS

STAR		T_{eff}	M $10^{-6} M_\odot/yr$	S IV $\lambda 1065$	Si IV $\lambda 1400$	C IV $\lambda 1550$	P V $\lambda 1120$	N V $\lambda 1240$	O VI $\lambda 1035$	S VI $\lambda 940$
ζ Pup	O4f	42000	7.0	+	+	+	+	+	+!	+
α Cam ζ Ori	O9.5 Ia	30000	2.5	+	+	+	+	+!	+!	
ε Ori	BO Ia	24800 28800	4.3	+	+	+		+!	+!	
κ Ori	BO.5 Ia	26400	2.5	+	+			+!	+! ***	***
ρ Leo	B1 Iab	21000	1.0		+			+!	-	
θ Ara	B2 Ib	18000						+! ***	-	
o²CMa	B3 Ia	15500	2.0	-	+!	+!	***	-	-	
η CMa	B5 Ia	13300	.4	-	(+!)	(-) ***			-	-
β Ori	B8 Ia	11600	1.0	- ***	+! ***					

+ indicates that the line shows mass loss effects; P Cygni profile or
  displaced absorption ( ) indicates an uncertainty
- indicates that the line shows no mass loss effects
  blank indicates no information
+! indicates that this line is not expected in a cool radiative equilibrium wind, yet is seen.
*** above which is the region where Auger enhancement is expected.

The flux distribution shown in the figure could account for the observed abundance of O VI and it falls below the ANS channels 1 and 4 upper limits. The predicted X-rays have in fact now been seen by the HEAO-2 satellite, with X-ray luminosities of about $10^{32} - 10^{33}$ ergs/sec. (See Grindlay's paper in this symposium).

In future observations it will be important to (a) derive information on the variation of coronal temperature and emission measure on spectral type, (b) search for coronal line emission using the solid state spectrometer (SSS), (c) look for time dependence of the X-rays and correlation with ultraviolet line strengths.

On the theoretical side, there is now increased importance in investigating the various possible mechanisms by which coronal regions can be produced. Several possible mechanisms have been suggested. Thomas (1976) has argued that coronae should exist about these stars as a result of sub-photospheric storage modes (non-radial pulsations, rotation, etc.). Hearn (1972) has proposed that the strong radiation fields of OB supergiants will lead to the amplification of acoustic waves which could heat the flow. The coronae around hot stars promise to be the focus of exciting theoretical and observational research in the 1980's.

REFERENCES

Abbott, D.C., 1978. Ap.J., 225, 893.
Abbott, D.C., 1977, unpublished Ph.D. Thesis, Univ. of Colorado.
Barlow, M.J., Cohen, M., 1977. Ap.J., 213, 737.
Cassinelli, J.P., Castor, J.I., and Lamers, H.J.G.L.M., 1978. PASP, 90, October 1978.
Cassinelli, J.P., Olson, G.L., and Stalio, R., 1978. Ap.J., 220, 573.
Cassinelli, J.P., and Olson, G.L., 1979. Ap.J., 229, 304.
Castor, J.I., Abbott, D.C., and Klein, R.I., 1975. Ap.J., 195, 157.
Castor, J.I., Abbott, D.C., Klein, R.I., 1976, In Physique des Mouvements dans les Atmospheres Stellaires, eds. R. Cayrel, M. Steinberg, Paris: Cent. Nat. Rech. Sci.
Cox, D.P., Tucker, W.A., 1969. Ap.J., 157, 1157.
Hearn, A.G., 1972, Astr. Ap., 19, 417.
Hearn, A.G., 1975, Astron. Astrophys., 40, 277.
Klein, R.I., and Castor, J.I., 1978, Ap.J., 220, 902.
Lamers, H., and Morton, D.C., 1976. Ap. J. (Suppl.), 32, 715.
Morton, D.C., 1967. Ap. J., 147, 1017.
Morton, D.C., 1969, in Mass Loss from Stars, ed. M. Hack, p. 36.
Thomas, R.N., 1978. in Proceedings of Iau Symp., No. 83, Mass Loss and Evolution of O-Type Stars, in press.
Weisheit, J.C., 1974. Ap.J., 190, 735.

Ultraviolet spectroscopy
of galactic X-ray sources

A. K. Dupree

Harvard-Smithsonian Center for Astrophysics

1.   Introduction

2.   High Luminosity X-ray Sources

3.   Late-Type Stars

1.   INTRODUCTION

The ultraviolet spectral region provides a unique probe of stellar
atmospheres by directly sampling plasma with energies from 1 to
~ 50 eV; moreover the presence of non-equilibrium conditions can
produce emission in strong resonance lines such as O I ($\lambda 1304$),
C IV ($\lambda 1550$) and N V ($\lambda 1240$) that can be used to infer a wide variety
of physical conditions and phenomena in X-ray sources. In many
cases the ultraviolet measurements provide the necessary continuity
bridging the X-ray and optical spectral regions. The spectroscopy
of selected sources has provided new insights into the structure of
massive stellar winds and high luminosity binary systems. Although
new results from the Einstein Observatory have shown the ubiquitous
nature of X-ray emission from all stars at some level, we shall
focus on two classes – the high luminosity binaries with a compact
object and the binary systems composed of late-type stars. In this
paper, we first discuss the ultraviolet spectroscopy of four such
systems; 4U 0900-40 (HD 77581), Cygnus X-1 (HDE 226868), 4U 1700-37
(HD 153919) and 1653-40 (HD 152667 = V861 Sco). Additionally the
study of late-type binaries such as the RS CVn and W UMa systems has
revealed substantial X-ray and ultraviolet fluxes in contrast to
single stars of the same type. These late-type binary systems
possessing enhanced coronae are also discussed.

Reports in this volume by Hartmann treat the ultraviolet spectroscopy of globular clusters and the source HZ Herculis (p149ff).
Supernovae remnants and the spectroscopy of AM Her type systems are
discussed here by Raymond (p405ff).

Many substantial advances in our understanding of these sources are emerging from new observations made possible by the International Ultraviolet Explorer satellite launched in January 1978 (Boggess et al. 1978a,b). The 45-cm diameter telescope and echelle spectrograph allow the spectrum to be measured from $\lambda\lambda 1170$-3200 at both high ($\lambda/\Delta\lambda$ about 10^4) and low (6 to 7Å) dispersions. Most importantly X-ray binary systems can be measured spectroscopically for the first time - many of them in the high dispersion mode. The multiplexing nature of the echelle spectrographs permits simultaneous wavelength coverage - an important consideration when studying time varying phenomena present in most of these sources.

2.   HIGH LUMINOSITY X-RAY SOURCES

a)   HD 77581 = Vela X-1

Several binary systems containing supergiants of early spectral type have been monitored spectroscopically with IUE over their orbital period. The system HD 77581 corresponding to the X-ray source 4U 1700-37 (Vela X-1) is particularly interesting. The binary system contains a B0.5Ib primary star and a pulsar that is orbiting the primary with a period of 8.96 days. The X-ray source is apparently formed by accretion of a massive stellar wind from the primary by the compact object. The ultraviolet spectrum (see Fig. 1) is dominated by the primary and is typical of an early B supergiant as is found in the surveys made by the Princeton scanning spectrometer on the earlier Copernicus satellite (Snow and Jenkins 1977). The broad minimum centered near $\lambda 2200$ in the spectrum of HD 77581 is caused by interstellar absorption. The star is highly reddened with E(B-V)=0.7. Strong resonance lines of Si IV ($\lambda 1400$), C IV ($\lambda 1550$), He II ($\lambda 1640$) and N IV ($\lambda 1750$) are clearly present, and the asymmetries arising from mass outflow can be noted even in low dispersion spectra. Strong modulation of the line profiles is found with orbital phase. High dispersion spectra show the profiles particularly well and distinct changes can be seen (see Fig. 2 and 3). The lines are typical P Cygni profiles formed in a star with an extended atmosphere. The emission component on the long wavelength side of the transition arises from scattering of the photospheric continuum by ions in the extended atmosphere, whereas the broad absorption at shorter wavelengths occurs in the expanding material lying in front of the stellar disk. The shape and maximum extent of the absorption component can give the terminal velocity and structure of the massive stellar wind. Inspection of Figures 2 and 3 shows strong changes occur in the blue absorption feature at phases 0.44 and 0.52 near inferior conjunction when the compact object is between us and the supergiant primary. The short wavelength edge of the absorption feature moves to longer wavelengths implying a decrease in the maximum expansion velocity of the Si IV and C IV ions ions from about 1700 km s^{-1} (phase 0.0) to about 850 km s^{-1} (phase 0.5). At phase 0.52 there is a suggestion in the Si IV profile and clear presence in the C IV profile of a "high-velocity" absorption

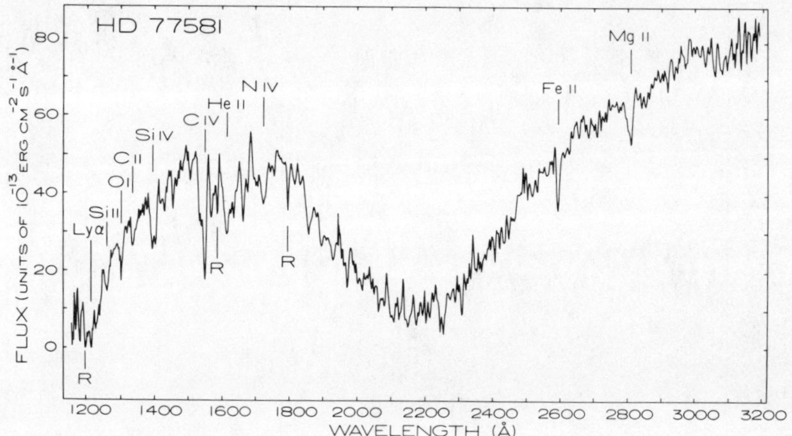

Figure 1: The observed ultraviolet spectrum of HD 77581 = Vela X-1
as measured by the International Ultraviolet Explorer
satellite (Dupree et al. 1980). Prominent line features
are noted. R denotes a fiducial mark (reseau).

feature. Additionally the emission components are weak near phase
0.0, and stronger near phase 0.5. Such behaviour can be understood
qualitatively as the effect of an X-ray source immersed in a stellar
wind as suggested by Hatchett and McCray (1977). An X-ray source
would be expected to ionise a surrounding volume in the stellar wind,
and destroy the ions producing the line feature.

Near X-ray eclipse, the part of the wind producing the blue-shifted
P Cygni absorption would be unaffected by the X-rays and the absorption part of the profile should approximately represent the undisturbed stellar wind. The wind on the far side of the star, as seen
from Earth, produces the P Cygni emission; and at X-ray eclipse
(orbital phase 0.04) this gas is ionized by the X-ray source,
reducing the emission. When the X-ray source is in front, near
phase 0.5, the gas which produces the P Cygni absorption is now
ionized to a higher state than before, resulting in a reduction of
the edge velocity. The emission component is increased because the
gas on the far side of the star has returned to its normal state.

The geometrical structure to be expected can be evaluated by using
theoretical calculations of the ionization (Hatchett and McCray 1977;

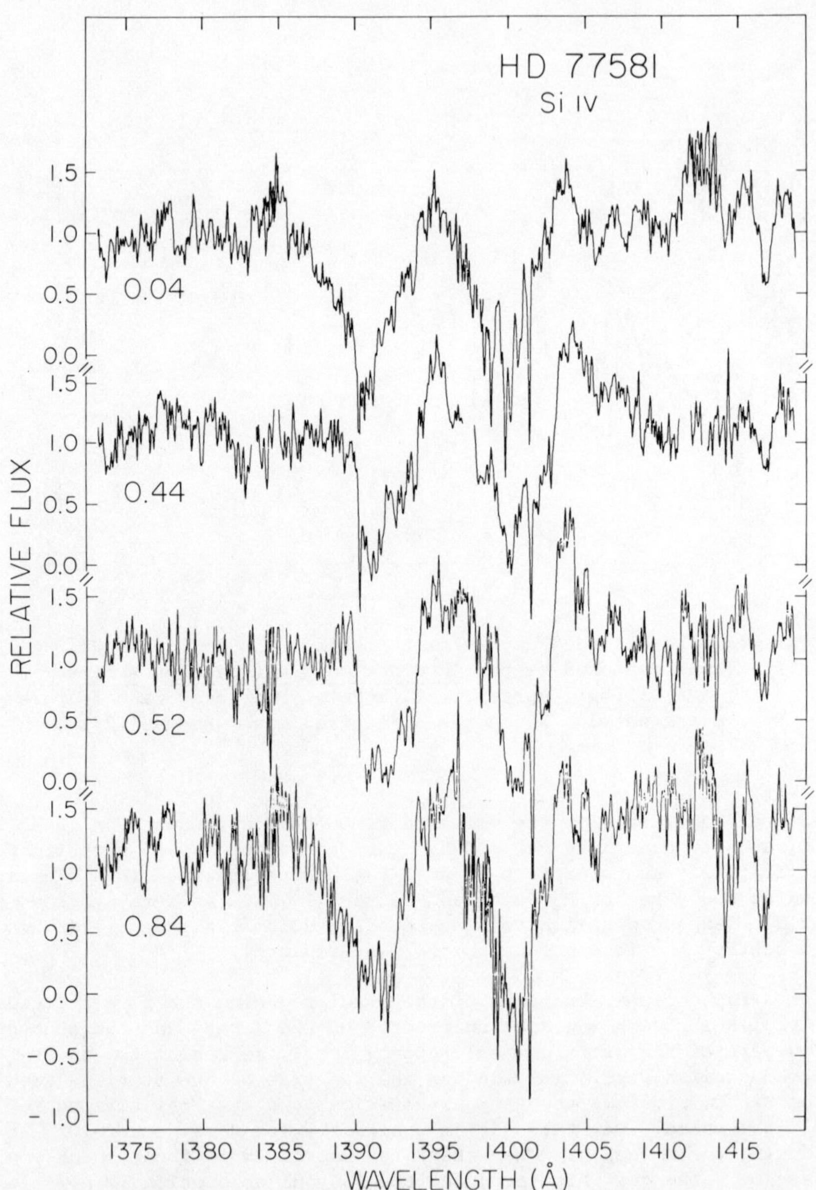

Figure 2: The spectral region near λ1400 in HD 77581 showing the Si
IV stellar profile (Dupree et al. 1980). At phase 0.04,
when the pulsar is eclipsed by the primary star, the profile is like those found in other early type stars. However, as the pulsar crosses in front of the stellar disk
(phase 0.52) substantial changes occur in the line profile.

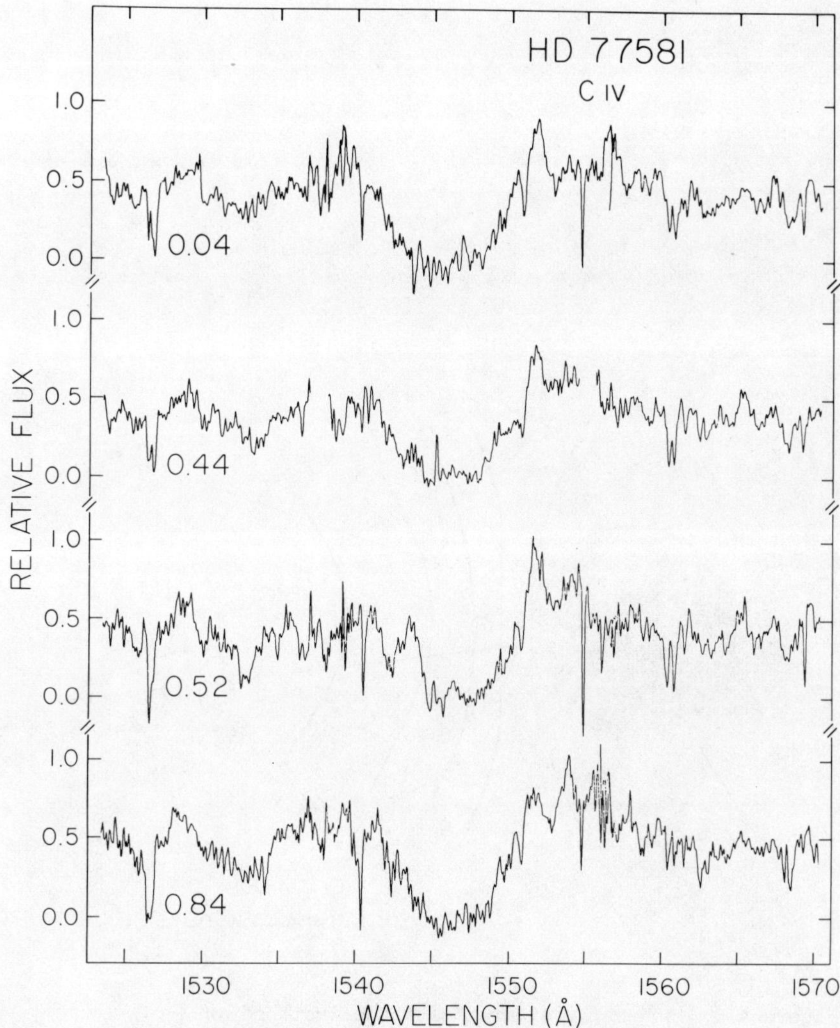

Figure 3: The spectral region in HD 77581 near λ1550 showing the changes in the C IV profile with orbital phase. From Dupree et al. (1980).

Kallman and McCray 1979). Dupree et al. (1980) find (see Fig. 4)
the volume where C IV is ionized to be roughly a sphere lying along
a line joining the primary star and compact object and centered at
a greater distance than the X-ray source from the primary star. We
see from inspection of Figure 5 that photoionization of C IV over
such a volume would naturally create decreased absorption corresponding in wind velocity to $v/v_\infty \sim -0.7$ and in addition produce high
velocity absorption features.

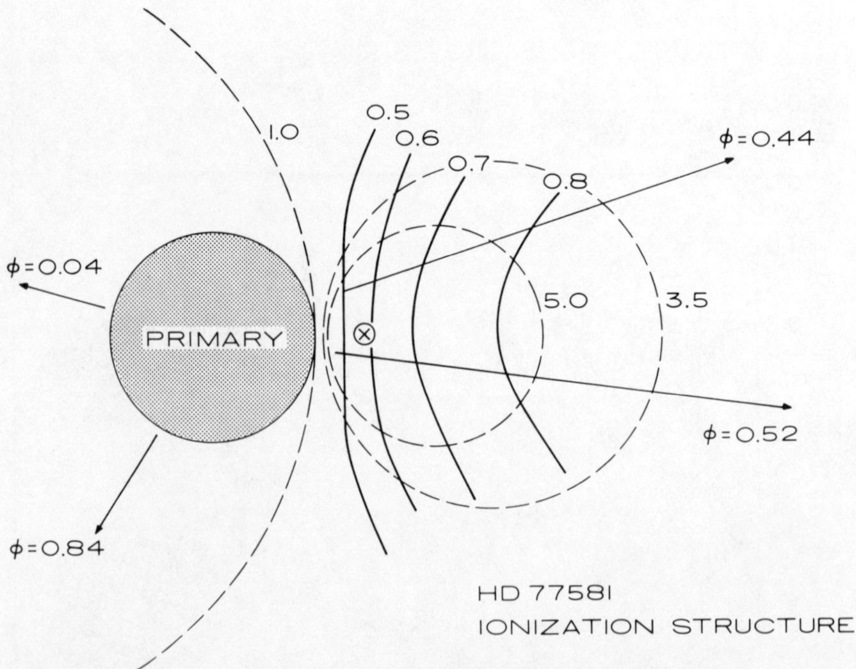

Figure 4: A schematic diagram of the effects of an X-ray source
sited in an expanding stellar atmosphere. The solid
lines denote various values of v/v_∞. The broken lines
indicate the loci of values of q. Within the circles,
it is assumed that C IV is absent.

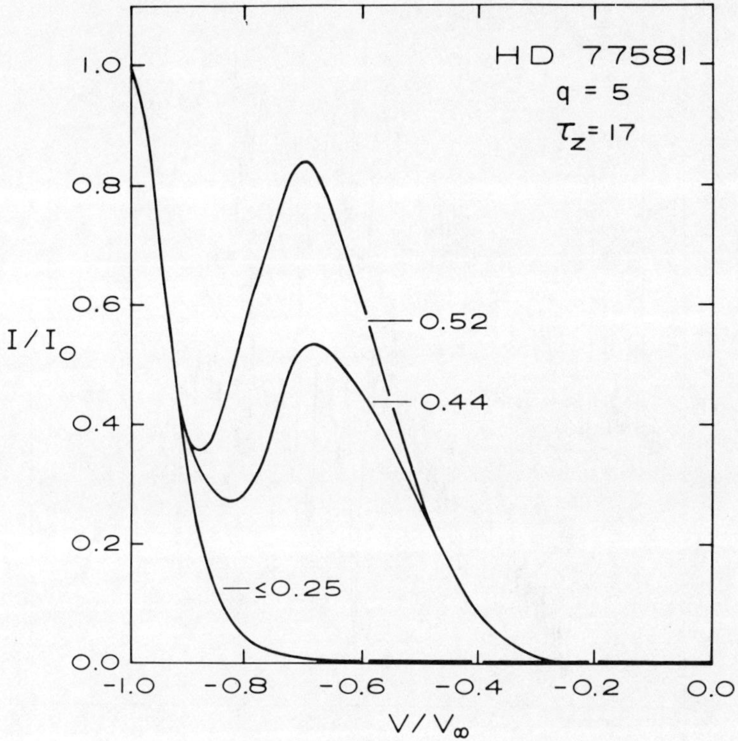

Figure 5: The blue wing of a line profile calculated by assuming
that the scattering ion is absent within the volume
defined by q = 5 (see previous figure). Various phases
are marked. Note the decrease in edge velocity of the
line with phase and the appearance of a high velocity
absorption feature near phase 0.44 and 0.52. *(Reproduced
by courtesy of the International Astronomical Union)*.

b)   HDE 226868 = Cygnus X-1

Other luminous supergiants have been studied with IUE and some show a
similar effect. The BO star, HDE 226868, identified with Cygnus X-1
has a spectrum similar to that of HD 77581, (see Fig. 6) and the
early observations revealed changes in the line profile with orbital
phase (Dupree et al., 1978). This star is too reddened to be easily
accessible at high dispersion but the changes in the profiles of
Si IV, for instance, are reminiscent of those found in HD 77581. The
Si IV line clearly weakens at inferior conjunction. More extensive
ultraviolet observations by an international consortium have borne
out these effects (Treves et al., 1980).

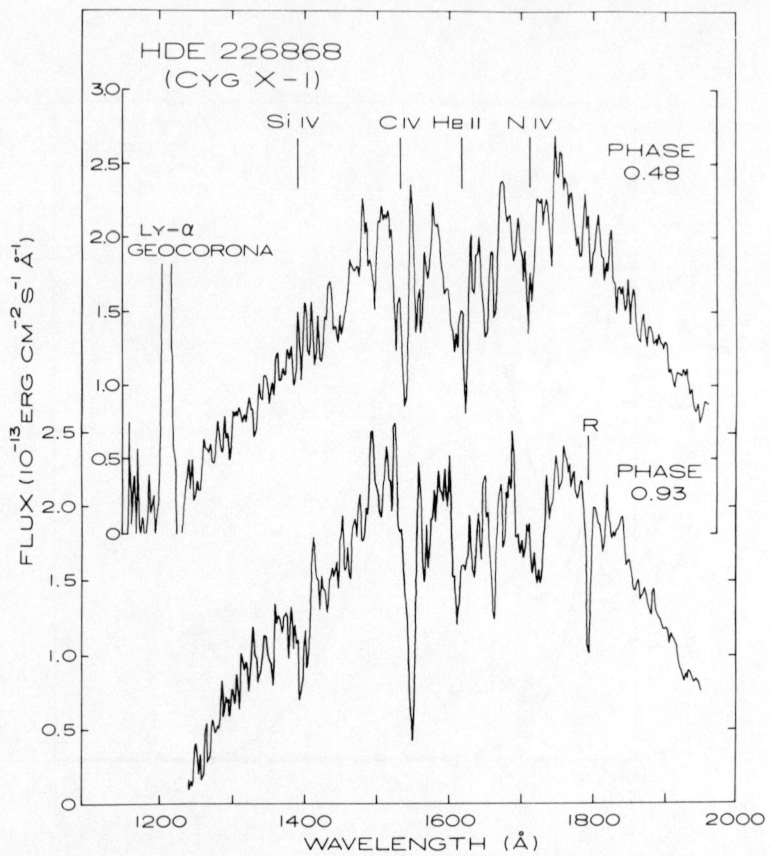

Figure 6:   Low dispersion ultraviolet spectra of Cygnus X-1 at 2
phases (Dupree et al. 1979). *(Reprinted by permission
from Nature, Vol. 275, No. 5679, p400, Copyright (c) 1979
Macmillan Journals Limited)*

c)   HD 153919 = 4U 1700-37

Another highly luminous X-ray source 4U 1700-37 is identified with
the O6f star HD 153919. Early theoretical calculations by McCray and
collaborators (McCray and Hatchett 1975) suggested that this object
should show the spectral effects of an X-ray source imbedded deep
within the stellar wind. The first high dispersion spectra (see
Fig. 7) revealed the presence of a massive wind, and a deep extended
absorption component on the C IV profile. No apparent phase effect
was found in these transitions (Dupree et al. 1978). The atmosphere
of this early-type supergiant contains a substantially higher column
density of C IV in its wind, and any modification of the ion abundance is not great enough to be spectroscopically apparent. A higher
wind density also restricts the volume that is affected by X-rays,
and can make it difficult if not impossible to observe. A less
saturated line is necessary to probe the atmosphere.

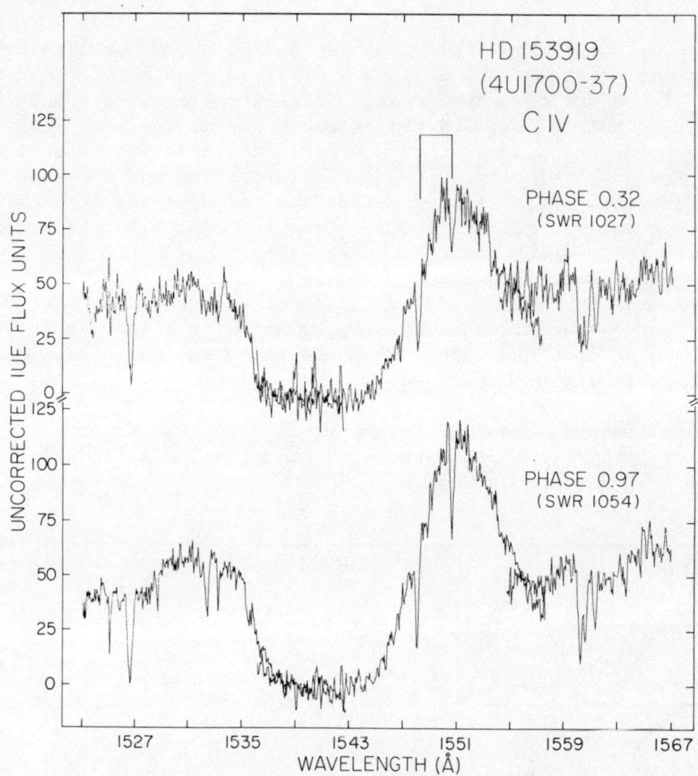

Figure 7:   The C IV region in HD 153919 = 4U 1700-37 at two orbital
            phases (Dupree et al., 1979). Note the strong narrow
            interstellar lines marked near their rest wavelengths.
            *(Reprinted by permission from Nature, Vol. 275, No. 5679,
            Copyright (c) 1979, Macmillan Journals Limited).*

   d)   HD 152667 = 1653-40

Another massive binary system, HD 152667 (spectral type B0 I) was
suggested as the optical counterpart to a weak X-ray source detected
by the Copernicus satellite (Polidan et al., 1978). Because of the
occurrence of a unique UV flare observed near an X-ray turn-on, by
both the UV and X-ray instruments on Copernicus, the likelihood of
the identification seemed strengthened. Subsequently, a 38 second
period X-ray pulsar was found by White and Pravdo (1979) 30' away
from HD 152667, and later observations with the Einstein Observatory
(HEAO-B) failed to confirm X-rays from the star at phases away from
eclipse (Holt, 1979). It appears that this source has a low luminosity and may be typical of a corona on a normal single early-type
star.

The ultraviolet spectroscopy of this object (Hutchings and Dupree,

1980) shows complex multiple components. Figure 8 contains high
dispersion spectra of the Si IV and C IV lines as a function of
phase. There are no marked changes with phase as were found in
HD 77581 or HDE 226868, and two observations at the same phase
(0.57, not shown) suggest that non-periodic variations are significant. The edge velocity of the profiles corresponds to -1430 km s^{-1}
which is within the normal range for stellar type and luminosity
(Hutchings and van Rudloff 1980). However, many lines or components
show relatively sharp absorptions at velocities in the -400 to -600
km s^{-1} range (Hutchings and Dupree 1980) that may result from an
enhanced density region such as gas streams surrounding the system.
In the source there is occasionally evidence of a hole in the wind
at high velocities associated with the secondary and enhanced
absorption at phases 0.51.

Thus X-ray sources imbedded in the winds of hot stars offer a
unique opportunity to probe the wind structure, mass loss and its
variation in a luminous star.

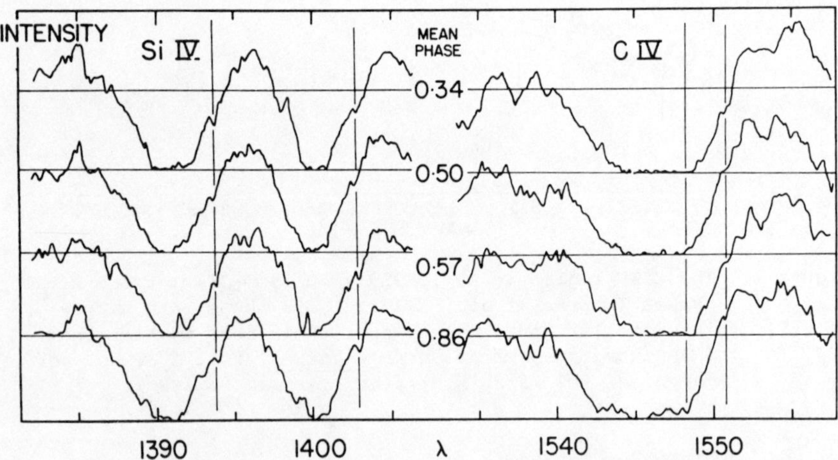

Figure 8: Ultraviolet spectra of the Si IV and C IV lines in HD
           152667 = OAO 1653-40 (Hutchings and Dupree 1980).

3.   LATE-TYPE STARS

We focus on the spectroscopy of cool stars that are in binary
systems and then overview the ultraviolet emission from single late-
type stars. The RS CVn systems and stars of W Ursae Majoris type
have been found to be soft X-ray sources with HEAO-1 and Einstein.
The X-ray properties of RS CVn sources are treated by Charles in
these proceedings. The Einstein Observatory has been able to survey
single late-type stars, and find for the most part that all stars
have X-ray emission to some degree (Vaiana et al. 1979).

Ultraviolet spectroscopy of late-type systems was first possible on
a large scale with the advent of IUE. Prior to this only a few
strong resonance emission lines were detected with the Copernicus
satellite (Bernat and Lambert 1976; Dupree 1975; Evans et al. 1975;
McClintock et al. 1975).

a) RS CVn and W UMa types

Binary systems containing late-type giants or dwarf members are
typically strong sources of ultraviolet emission. The RS CVn stars
are late-type giant and subgiant binaries whose surface activity has
been known from optical studies to be peculiar, in that large spots
cover their surfaces (Eaton and Hall 1979). Ultraviolet observations with Copernicus and IUE of members of this class such as λ
And, HR 1099, and HR 4665, have shown that these stars are powerful
emitters in transition-region lines formed at temperature T $\sim 10^5$K,
with much higher surface fluxes than the quiet sun. RS CVn binaries
have also been shown to be strong soft X-ray sources (Walter,
Charles, and Bowyer 1978). This combination of high chromospheric
and coronal activity with spot activity makes it plausible to
interpret these phenomena as extreme analogues of solar activity.

The W UMa stars are late-type, dwarf binaries which are orbiting so
closely together that they may share a common convective envelope
(Lucy 1976; Flannery 1976; Shu, Lubow and Anderson 1976). Two W
UMa systems have been observed with IUE: VW Cep (G5) and 44 Boo
(G2). The system VW Cep was recently discovered to be an X-ray
source by the NRL experiment on HEAO-A (Carroll et al. 1980). These
stars are contact binary systems that eclipse with periods of less
than one day. Both stars have ultraviolet spectra that are very
similar. The short wavelength spectra are shown in Figure 9. High
temperature emission lines, N V, C IV, and Si IV are apparent, as is
He II λ1640.

In Figure 10, the stellar surface fluxes in the binary systems are
compared to that of the quiet sun, assuming that the chromospheric
activity is uniformly distributed across the stellar surface. The
fluxes for λ And and HR 1099 were taken from Linsky et al. (1978).
The surface fluxes show a general progression of increasing enhancement with increasing temperature much like that found for a solar
active region (Dupree et al. 1973) and for active dwarfs (Hartmann
et al. 1979). However, the fluxes of the W UMa stars are an order
of magnitude larger than typical solar active regions. The He II
λ1640 line is strongly enhanced, which may reflect photoionization
by increasing coronal emission measures (Hartmann et al. 1979).

There appears to be a dependence of the surface flux upon orbital
period. The W UMa stars 44 Boo and VW Cep have periods $0\overset{d}{.}7$; the RS
CVn stars have longer periods: HR 1099 ($2\overset{d}{.}8$), λ And ($30\overset{d}{.}5$), HR 4665
($64\overset{d}{.}4$). The results in Figure 10 suggest that the flux is enhanced
with decreasing orbital period. Because of tidal synchronization,
the rotational periods should be close to the orbital periods in the
shorter-period systems; this is verified by the spot light curve

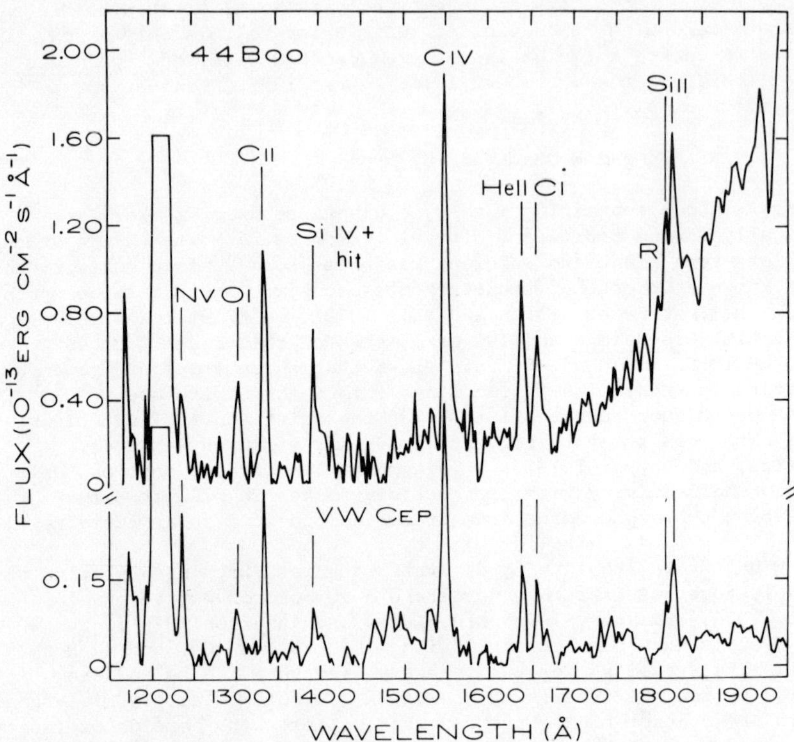

Figure 9: Ultraviolet spectra of two late-type binaries of the W Ursae Majoris class. R denotes a fiducial mark (reseau).

periods in RS CVn stars. Thus the behaviour of emission strength with orbital period may be viewed as an extension of the general increase of chromospheric flux with increasing rotation rate originally found for dwarfs (Kraft 1967; Skumanich 1972). Observations of such a general relationship encompassing stars of very different interior structure provide important constraints on theories of chromospheric and coronal heating.

b) Coronae in Single Stars

The presence of hot plasma with temperatures about 2×10^5 K in a stellar atmosphere can be inferred from the appearance of the C IV doublet ($\lambda 1550$). Figure 11 contains a compilation of results from many IUE spectra (Brown, Jordan and Wilson 1979; Carpenter and Wing 1979; Dupree et al. 1979; Hartmann et al. 1979; Linsky and Haisch 1979; Wing 1978). It is apparent that the dwarf stars (luminosity class V) consistently show evidence for C IV, but the

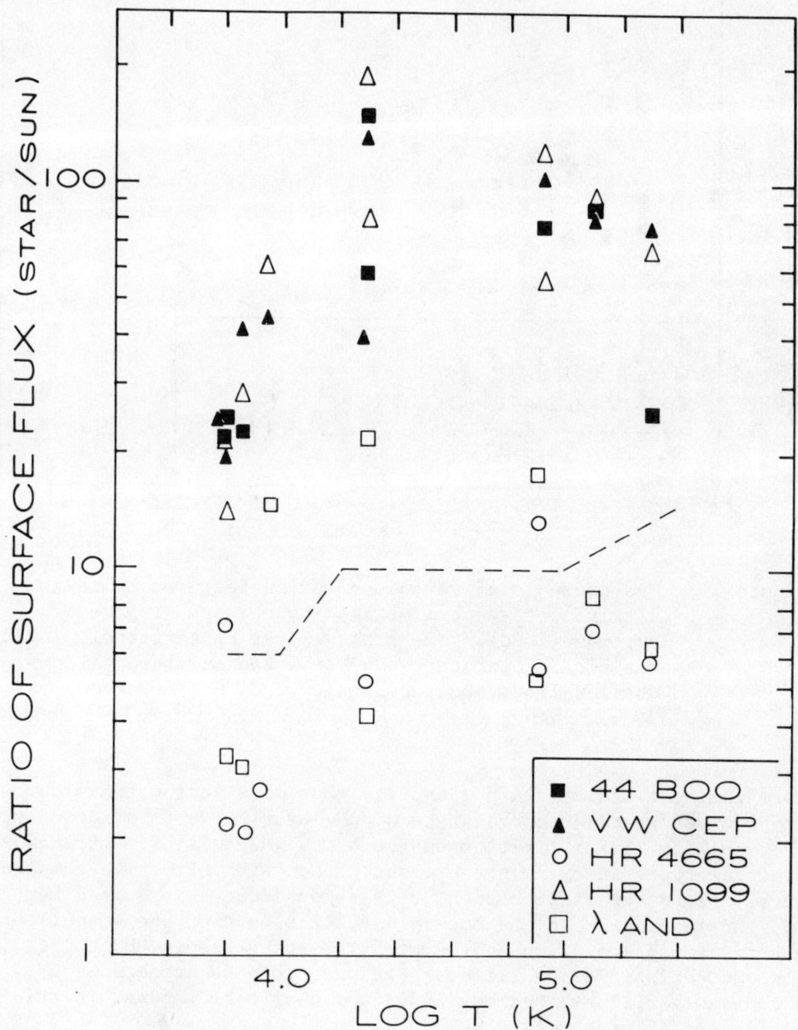

Figure 10: Flux ratio of stellar emission lines to those in the Sun as a function of temperature of formation (Dupree et al. 1979). These stars are all X-ray emitting binary systems. Those of shortest orbital period (VW Cep, 44 Boo, and HR 1099) show the highest flux enhancement. The broken line is a typical value for a solar active region.

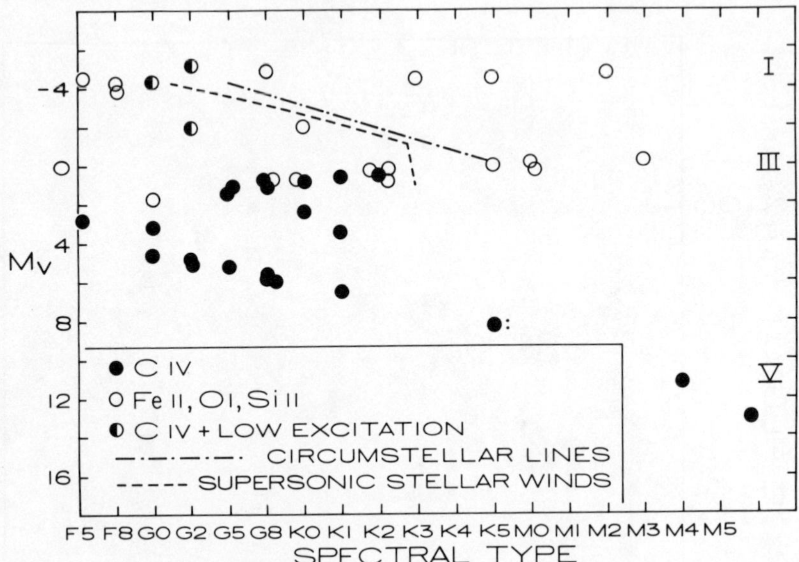

Figure 11: The presence of various emission features in the
ultraviolet spectra of late-type stars (Dupree and
Hartmann 1980). The broken lines approximately indicate
the high temperature edge of a region where circumstellar lines and supersonic winds are observed.
*(Reproduced by courtesy of the International Astronomical
Union)*

more luminous stars (class I and II) show more varied behaviour.
The coolest stars (spectral type M and later) give no evidence of
C IV as Wing (1978) first noted for γ Cru (M3.4 IIIb). At G and K
spectral types there is not a clear change with effective temperature. Along the giant branch, both types occur at spectral type
G8, for instance; in the supergiant stars we find the so-called
hybrid atmosphere discussed below. The stars showing both C IV and
low excitation species lie near the onset of the appearance of
circumstellar lines (Reimers 1977) and near the "supersonic transition locus" proposed by Mullan (1978).

Linsky and Haisch (1979) suggested that there were two types of
atmospheres - solar and non-solar, that the division was sharp
between these types at spectral type K0 III, and they speculated
that the presence of a massive stellar wind would suppress a corona
and eliminate high temperature species. The addition of more data
shows that the situation is more complicated. The "division"
between hot and cool atmospheres is not at all sharp. Cool atmos-

pheres dominate at a later spectral type than shown by Linsky and Haisch, and hybrid atmospheres exist that exhibit both high temperature lines and substantial mass loss. There is other, more subtle behaviour in the line ratios of C II, C IV, and O I emphasized by Brown, Jordan, and Wilson (1979) that support a smooth change in atmospheric structure with decreasing effective temperature. Generally speaking however, stars with large mass loss rates show the coolest emission features and apparently have weak or nonexistent coronae.

We can infer the presence of still hotter material at $T \sim 10^6$K, from the flux in the He II $\lambda1640$ transition. In solar active regions this transition can be enhanced by recombination following photoionization (Raymond, Noyes, and Stopa 1979). In active dwarf stars the line flux confirms the direct measurement of soft X-rays (Hartmann et al. 1979). The presence of the $\lambda1640$ transition in the supergiant spectra suggests that a hot corona may well exist on these stars; detection could be difficult if soft X-ray emission is absorbed by a substantial extended atmosphere. Whereas dwarf stars show He II $\lambda1640$ (Dupree et al. 1979, Hartmann et al. 1979), it is interesting that the spectrum of the T Tauri star RU Lupi has very weak, if any $\lambda1640$ emission, (Gahm et al. 1979) suggesting that relatively little coronal material is present.

X-ray measurements from the Einstein Observatory (HEAO-2) could of course give direct information on the highest temperature plasma in these stars, but the data are in an early stage of assessment. The data are consistent with the picture suggested by the ultraviolet results. It is not yet known whether cool supergiants have substantial coronae; the upper limits on coronal emission from very luminous stars can be lowered when exposure times are longer (Rosner 1979). At present, many of the objects in the stellar surveys are faint and have not been well studied optically (Vaiana et al. 1979). Their luminosity classes, activity, or possible binary nature are not known in most cases. The X-ray observations should shortly provide important advances in understanding the high temperature plasma in the atmospheres of cool stars.

These observations are beginning to provide some constraints for theories of coronal heating and structure.

Figure 12 shows a range of surface fluxes that are found in late-type single stars. The fluxes in binary systems are substantially larger, ranging up to 100 times the solar surface flux in some high excitation lines. The radiative losses in a wide variety of single stars vary by a factor of ~ 10 in Mg II (Basri and Linsky 1979). The ultraviolet emission is independent of stellar effective temperature for active dwarfs, and in general appears to be more dependent on $\log g$ than on T_e. And, in several cases, stars of the same spectral type and gravity exhibit quite different ultraviolet surface fluxes. To date no single theory appears to be successful at reproducing these results.

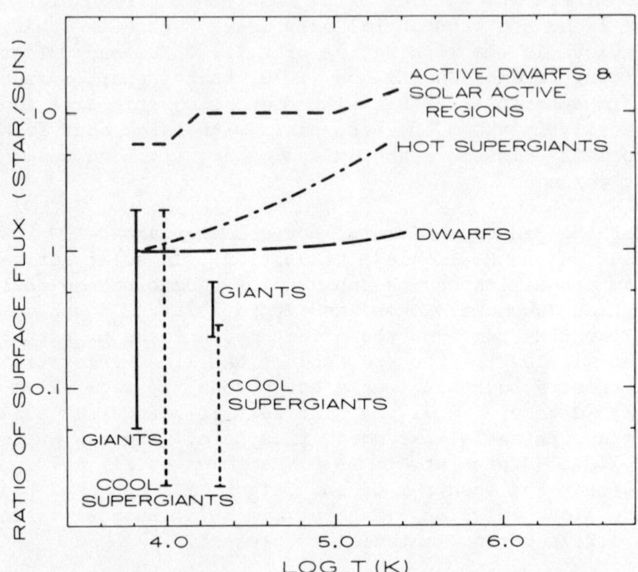

Figure 12:   Behaviour of the ratio of stellar: solar surface flux
              for chromospheric and coronal emission lines in single
              stars of various types. *(Reproduced by courtesy of the
              International Astronomical Union).*

I am grateful to L. Hartmann and J. C. Raymond for many discussions
on these topics.

REFERENCES

Basri, G.B., and Linsky, J.L., 1979. Ap. J., in press.
Bernat, A.P., and Lambert, D.L., 1976. Ap. J., 204, 830.
Boggess, A et al 1978a. Nature, 275, 356.
Boggess, A., et al., 1978b. Nature, 275, 361.
Brown, A., Jordan, C., and Wilson, R., 1979. in Proc. Sym. "The First
      Year of IUE, (Ed. A. Willis), University Col. Lond., p232.
Carpenter, K.G., and Wing, R.F., 1979. Bull. A.A.S., 11, 419.
Carroll, R.S., et al., 1980. Ap. J. (Letters), in press.
Dupree, A.K., 1975. Ap. J. (Letters), 200, L27.
Dupree, A.K., Huber, M.C.E., Noyes, R.W., Parkinson, W.J., Reeves,
      E.M., and Withbroe, G.L., 1973. Ap. J., 182, 321.
Dupree, A.K., et al., 1978. Nature, 275, 400.

Dupree, A.K., Black, J.H., Davis, R.J., Hartmann, L., and Raymond, J.
    C., 1979. in Proc. Symp "The First Year of IUE", (ed. A. Willis),
    University College London, p. 217.
Dupree, A.K., et al., 1980, Ap. J., in press.
Dupree, A.K., and Hartmann, L., 1980. IAU Symp. No. 51, "Turbulence
    in Stellar Atmospheres", in press.
Eaton, J., and Hall, D.S., 1979. Ap. J., 227, 907.
Evans, R., Jordan, C., and Wilson, R., 1975. MNRAS, 172, 585.
Flannery, B.P., 1976. Ap. J., 205, 217.
Gahm, G.F., Fredga, K., Liseau, R., and Dravins, D., 1979. Astron.
    Astrophys., 73, L4.
Hartmann, L., Davis, R., Dupree, A.K., Raymond, J., Schmidtke, P.C.,
    and Wing, R.F., 1979. Ap. J. (Letters), 233, L69.
Hatchett, S., and McCray, R., 1977. Ap. J., 211, 552.
Holt, S.S., 1979. personal communication.
Hutchings, J.B., and Dupree, A.K., 1980, Ap. J., in press.
Hutchings, J.B., and van Rudloff, I.R., 1980, Ap. J. in press.
Kallman, T., and McCray, R., 1979. private communication.
Kraft, R.P., 1967. Ap. J., 150, 551.
Linsky, J.L., et al., 1978. Nature, 275, 389.
Linsky, J.L., and Haisch, B., 1979. Ap. J., (Letters), 229, L27.
Lucy, L.B., 1976. Ap. J., 205, 208.
McClintock, W., Henry, R.C., Moos, H.W., and Linsky, J.L., 1975. Ap.
    J., 202, 733.
McCray, R., and Hatchett, S., 1975. Ap. J., 199, 196.
Mullan, D.J., 1978. Ap. J., 226, 151.
Polidan, R.S., Pollard, G.S.G., Sanford, P.W., and Locke, M.C., 1978.
    Nature, 275, 296.
Raymond, J.C., Noyes, R.W., and Stopa, M.P., 1979. Solar Physics,
    61, 271.
Reimers, D., 1977. Astron. Astrophys., 57, 395.
Rosner, R., 1979. Private Communication.
Shu, F.H., Lubow, S.H., and Anderson, L., 1976. Ap. J., 209, 536.
Skumanich, A., 1972. Ap. J., 171, 565.
Snow T.P., and Jenkins, E.B., Ap. J. (Suppl.), 33, 269.
Treves, A., et al., 1980. Ap. J., in press.
Vaiana, G., Forman, W., Giacconi, R., Gorenstein, P., Pye, J., Rosner,
    R., Seward, F., and Topka, K., 1979. Bull. A.A.S., 11, 446.
Walter, F., Charles, P., and Bowyer, S., 1978. Ap. J. (Letters), 225,
    L119.
White, N.E., and Pravdo, S.H., 1979. Ap. J. (Letters), 233, L121.
Wing, R.F., 1978. in High Resolution Spectrometry, (ed. M. Hack),
    Proc. of the 4th International Colloquium on Astrophysics, p. 683.

Features of the soft
X-ray background

J.A.M. Bleeker,

Cosmic Ray Working Group,
Huygens Laboratory,
Leiden, The Netherlands.

1.   Introduction

2.   Formation of Hot Interstellar Plasma

3.   Main Features Derived from the Observational Picture

4.   Conclusion

1.   INTRODUCTION

The soft X-ray background (photon energies < 1 keV) shows a very
patchy brightness distribution quite unlike the diffuse background
at higher energies, which is highly isotropic and predominantly of
extragalactic origin. Brightness distributions of soft X-rays have
now become available for almost the entire sky with a typical
spatial resolution of several square degrees (e.g. de Korte et al.
1976, Sanders et al. 1977). These data provide us with a number of
observational facts which point at a galactic origin for the bulk of
the soft X-ray background, i.e.:

- Absence of shadow effects caused by cold galactic gas and/or
  members of the Local Group (Small and Large Magallanic
  Clouds, Andromeda).

- Finite emission from the galactic plane at 0.25 keV, which
  should be non-existent if the emission originated from
  outside the gas disk due to the severe interstellar absorption at those energies.

- A number of energy dependent large scale features.

- The correlation of soft X-ray brightness with columnar
  hydrogen is consistent with what is expected from simple
  photo-electric absorption.

Figure 1 shows a soft X-ray brightness map of the southern galactic
hemisphere in the $\frac{1}{4}$-keV pass band compiled by the Wisconsin group
together with a map of columnar cold gas (N_H) obtained from 21 cm
radio data (Sanders et al. 1977). Correlation between low-N_H regions
and soft X-ray enhancements seems to exist, the reverse however does
not hold.

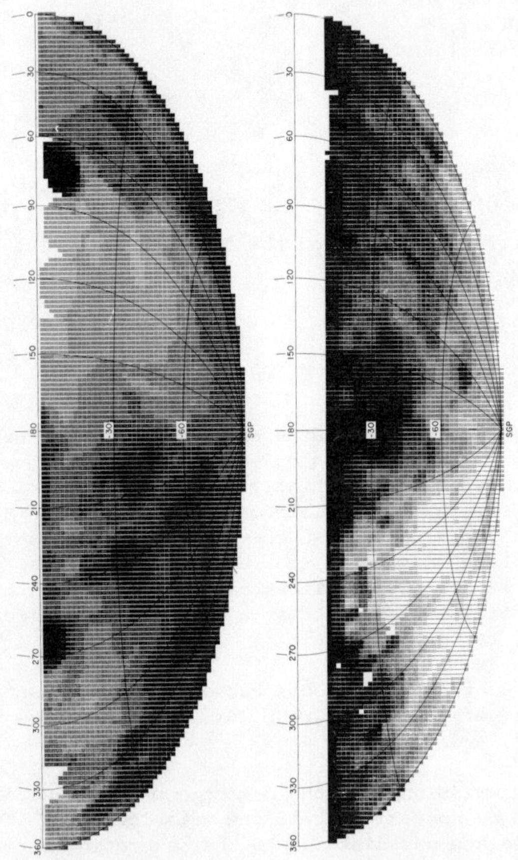

Figure 1: Upper map: brightness distribution of the southern galactic hemisphere in the ¼ keV band, the shading is proportional to the X-ray intensity. The two extremely
intense features are created by the Cygnus Loop and Vela
supernova remnants.

Lower map: distribution of the hydrogen column density.

(From Sanders et al., 1977).

A review of the possible origin mechanisms for a soft galactic X-ray background (see e.g. Tanaka and Bleeker 1977) shows that neither the unresolved contribution from discrete stellar sources nor diffuse non-thermal emission processes can account for the observed flux levels.

Concerning the stellar contribution, a soft X-ray luminosity averaged over stars of all types of $<q_{star}> \simeq 3 \times 10^{29}$ erg sec^{-1} would be required to account for the observed level of the background, assuming a space density of 7×10^{-2} stars pc^{-3} in the solar vicinity (Allen 1973). Upper limits placed on the soft X-ray luminosity of several nearby stars by ANS and Skylab are appreciably below this average value, the Einstein Observatory has now started to detect numerous main sequence stars at a luminosity level of $\simeq 10^{27}$ erg sec^{-1} which is far below the required value. It is also unlikely that a particular population of stars which are relatively luminous in soft X-rays (say 10^{31}-10^{32} erg sec^{-1}) makes a major contribution, since measurements on the small scale fluctuations of the background provide lower limits to the required number density which are close to or even exceed the number density of stars of all known types.

Concerning non-thermal processes, synchroton emission of energetic electrons ($\simeq 10^{13}$ eV) in an interstellar field of $\simeq 3$ microgauss falls short by about two orders of magnitude, inverse Compton effect on starlight or blackbody photons is inadequate by almost four orders of magnitude, scaling of the MeV-electron density in this case is limited by a potential bremsstrahlung gamma-ray excess. Non-thermal bremsstrahlung of low energy electrons cannot significantly contribute since this would result in an ionization rate per hydrogen atom of 10^{-11} sec^{-1}, which largely exceeds the limits set on heating of the interstellar gas.

The conclusion is therefore inescapable that the bulk of the galactic soft X-ray background is thermal in origin and originates from hot plasma regions ($\simeq 10^{6}$K) in interstellar space. The available spectral data are consistent with this interpretation, especially the recent detection of the emission line from helium-like oxygen at 570 eV (O VII) has confirmed the thermal nature (Inoue et al. 1978). Since the soft X-ray spectrum varies over the sky, the emitting plasma seems to contain several temperature components. Another tracer of the presence of tenuous hot plasma has been found in the detection by Copernicus of, most probably interstellar, O VI absorption lines in the spectra of numerous hot bright stars (Jenkins 1978). As we shall see later, the O VI absorption is probably caused by plasma regions of a somewhat lower temperature ($\simeq 3 \times 10^{5}$K) than the X-ray emitting regions.

Let us in the following paragraph dwell shortly on possible formation and maintenance of hot plasma regions or "coronal" gas in interstellar space. I shall then turn to a brief description of what constraints can be derived from the available observational data on the properties of the hot gas like filling factor, thermal pressure, spatial and temperature distribution, in order to discriminate between theories on origin, evolution and physical state.

2.   FORMATION OF HOT INTERSTELLAR PLASMA

The possible formation and maintenance of large regions of hot plasma in the interstellar medium (ISM) has been discussed by several authors. A scenario, originally proposed by Cox and Smith, is that supernova (SN) explosions occurring at the presently estimated rate could create and maintain a mesh of interconnected plasma regions containing hot gas at a temperature of $\simeq 10^6$K, the filling factor of these regions can become quite large and they in fact constitute a 'hot phase' component of the global ISM.

Cox and Smith (1974) considered SN explosions in a uniform medium. Assuming a rate of one SN/40 yrs per galaxy, a mean life of 2 million years and a diameter up to 80 pc, one finds a filling factor $f \simeq 0.1$ for isolated SN-cavities. The probability that the cavities interact is however appreciable ($\simeq 0.3$) and this may result, as Smith showed with a Monte Carlo simulation, in interconnected plasma regions at $\langle T \rangle \simeq 10^6$K with f as large as 0.5 and average density $\langle n \rangle \simeq 10^{-2}$ cm^{-3}. SN-explosions in an inhomogeneous substrate (cloudy medium), which may better comply with the actual case, was considered by McKee and Ostriker (1977). They showed that the well-accepted two-phase-model of the ISM, comprising cold clouds (T<100K) embedded in a warm intercloud medium (T $\simeq 10^4$K), cannot remain stable if the SN-rate is anywhere near the value mentioned above, the two-phase ISM will self-destruct on a time scale of less than 10^7 years. Imposing pressure equilibrium, they arrive at a three component model for the interstellar medium featuring the following main constituents:

Clouds of cold gas : $\langle n_c \rangle \simeq 40$, $\langle T_c \rangle \simeq 80$ K, $f_c \simeq 0.02$-0.04

Cloud coronae   : $\langle n_{cc} \rangle \simeq 0.1$, $\langle T_{cc} \rangle \simeq 10^4$K, $f_{cc} \simeq 0.2$

Hot intercloud medium:$\langle n_{ic} \rangle \simeq 3 \times 10^{-3}$,$\langle T_{ic} \rangle \simeq 5.10^5$K, $f_{ic} \simeq 0.7$-0.8

This model therefore predicts that the interstellar volume is mainly filled with hot 'coronal' gas with a temperature close to that required for the soft X-ray production.

The existence of hot plasma spheres around early type stars with strong stellar winds (so called 'stellar bubbles') was suggested by Castor et al. (1975). The idea was primarily put forward to suggest a possible circumstellar origin for the observed O VI absorption lines. The shocks produced by the strong wind can heat the circumstellar gas to T $\simeq 10^6$K, however the space density of stars possessing these winds would yield a very small filling factor ($\simeq 10^{-4}$). With more data becoming available, no positive indication for the

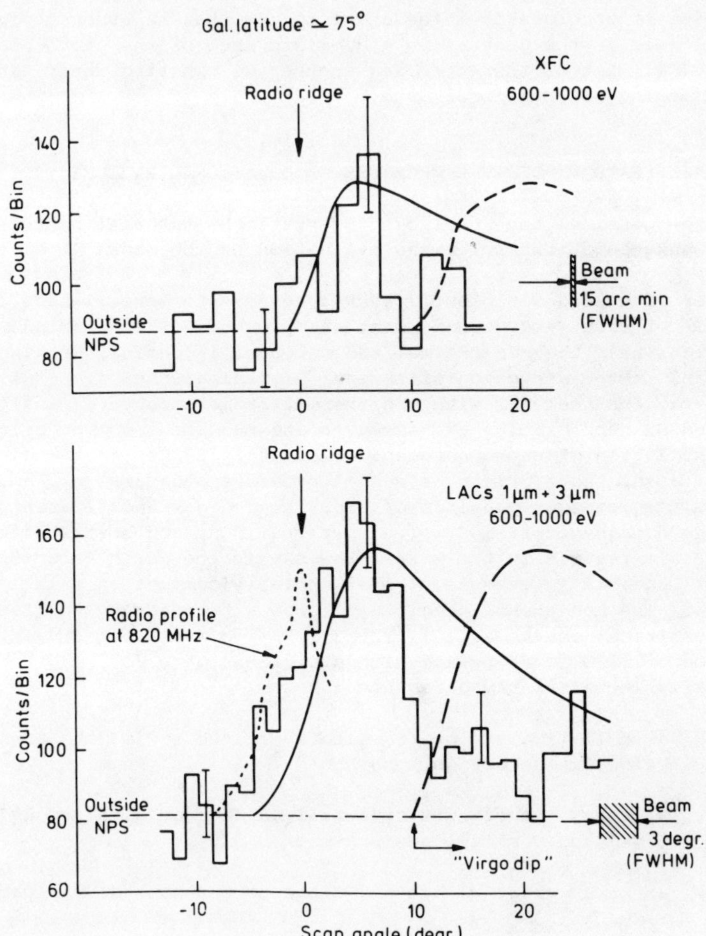

Figure 2: The X-ray brightness profiles of the North Polar Spur ridge at b = 75° in the energy range 0.6 - 1.0 keV measured with two instruments of different angular resolution (15 arc min, 3 degrees respectively). The radio ridge profile at 820 MHz is indicated together with the brightness profiles of a supernova remnant in two different stages of evolution. Solid line: adiabatic phase, dashed line: snow-plough phase. (From Davelaar and Bleeker, 1979).

existence of these bubbles has yet been found. The O VI data do not relate in any way to the properties of the associated stars, indicating an interstellar rather than a circumstellar origin. The first observations with the Einstein Observatory have also not identified the existence of these bubbles, although the expected soft X-ray flux should be far above the detection threshhold. It seems justified to conclude from the above models that supernovae play a crucial role in the generation and maintenance of the 'hot' phase of the ISM. Let us therefore see whether we can find their fingerprints in the observational data.

3.   MAIN FEATURES DERIVED FROM THE OBSERVATIONAL PICTURE

If we now consider the available observational material on the soft X-ray background, the following evaluation can be made:

A number of extended regions (several degrees to several tens of degrees) of soft X-ray enhancement are present. These include detections near the non-thermal radio loops, in Gemini, Eridanus and Lupus and in the direction of several HI-minima. Associations of soft X-ray enhancements with non-thermal radio structures and extended Hα-nebulosities are observed and suggest a common origin. The association of such phenomena immediately points in the direction of a supernova origin, since it provides a natural explanation for the simultaneous presence of shock heated gas and filaments and compressed magnetic fields. The most prominent and best studied large scale feature in the soft X-ray sky is the North Polar Spur (Loop I) and, as an example, I shall briefly comment on this feature. The non-thermal radio data show a fragmentary shell with the geometrical centre at $l = (329 \pm 1.5)^{\circ}$, $b = (17.5 \pm 3)^{\circ}$, at a distance of (130 ± 75) pc and with a diameter of (230 ± 135) pc. The available soft X-ray data show:

a)   a shell-like structure, also where the radio data are
     below the detection threshhold

b)   a plasma temperature of 3 million degrees with strong
     O VII and O VIII lines present

c)   an X-ray ridge within and close to the non-thermal radio-
     ridge

d)   the width of the X-ray ridge correlates with the width of
     the radio-ridge in a relative manner.

All these data suggest that the North Polar Spur (NPS) is a supernova remnant of phenomenal scale. It has been suggested that because of its size the NPS must be a very old remnant already in the snow-plough (radiative) phase of its evolution. Support for this was also derived from the presence of a dense sheet of neutral gas near the position of the shock front, which is expected to form when radiative cooling has commenced. Figure 2 shows observed

X-ray profiles of the NPS in the 0.6 - 1 keV band (Davelaar and Bleeker, 1979). Computed brightness profiles are also shown based on a model for SN-evolution developed by Mansfield and Salpeter (1974); thick line for the adiabatic stage, dashed line for the snow-plough phase. The latter is much too far behind the shock-front (=radio ridge) compared to the observed position, indicating that the NPS is probably still in its adiabatic phase. The extraordinary scale should then arise from the low ambient gas density at the time the explosion took place. Applying the model of Mansfield and Salpeter to the NPS emerging SN-parameters are: age $\simeq 7.10^4$ yrs, diameter $\simeq 160$ pc, explosion energy $\simeq 3 \times 10^{51}$ ergs, ambient gas density $\simeq 10^{-2}$ cm^{-3}. The presence of the dense sheet remains to be explained. Weaver (1978) actually suggests that the neutral sheet does not belong to the NPS but is swept-up interstellar gas created by intense stellar winds from numerous O and B stars in the Sco-Cen association. This also creates a low density cavity of large radius in which the NPS explosion could have taken place, consistent with the low ambient gas density inferred from the X-ray data. Alternatively, the observed radio and X-ray structures could also be the result of several SN-explosions in the Sco-Cen association which cavities have interacted and amalgamated into the observed large scale feature. Obviously many more observations are needed to uncover the proper scenario.

The hot spots observed in Gemini, Eridanus and the Lupus Loop show plasma temperatures very similar to the NPS and also suggest a supernova origin. Thus, supernovae undoubtedly play an important role in the formation and reheating of plasma in interstellar space.

Let us next consider sky regions which do not contain any pronounced enhancements. The dependence of the observed X-ray spectra on the hydrogen column density N_H in various directions has been used to evaluate simple emission models (Sanders et al. 1977, Hayakawa et al, 1978).

If the emission is assumed to arise from hot gas interspersed in cool matter, the spectral hardness ratio (H/L), defined as the observed flux at higher energies (e.g. 0.4 - 0.8 keV band) divided by the observed flux at lower energies (e.g. 0.1 - 0.4 keV band), should vary strongly if N_H ranges from $5.10^{19} - 2.10^{21}$ atoms cm^{-2}. This is caused by the steep dependence of the absorption cross-section on energy ($\sim E^{-3}$). Investigation of this hardness ratio as a function of N_H for several sky areas shows only very small dependence (factor <2), much smaller than the expected factor >5 (Hayakawa et al. 1978, Matsuoka 1979).

If we, on the other hand, assume the emission to come from a localized region attenuated by cold gas, the spectral data are only consistent with very small N_H-values ($\leqslant 5.10^{19}$ atoms cm^{-2}), Sanders et al. (1977) and Hayakawa et al. (1978) have therefore suggested that the background mainly arises from a local hot gas region in which the solar system is embedded. The X-ray data indicate that the temperature of this hot gas region is in the range 0.7 - 1.5 x

294

10^6K (O VII line). The observed intensity increase towards the
north galactic pole can be explained by assuming an asymmetric
position of the sun inside the gas region; Sanders et al. have
interpreted the intensity variation as a displacement effect. Other
evidence for a region of low gas density near the solar system
comes from the observed Lα-absorption in nearby stars and data on
resonantly scattered solar λ584-radiation.

What can we say about the properties of this local hot region? For
an isothermal approximation we can express the soft X-ray intensity
as

$$I_{sx} = \frac{\Lambda}{4\pi} \, n_e^2 \, R_L \tag{1}$$

in which $\Lambda \equiv \Lambda$ (E, T) represents the volume emissivity of the X-ray
emitting plasma and $n_e^2 R_L$ the emission measure (EM). At T = 10^6K,
the observed intensity in the 0.1 - 0.4 keV band requires EM $\simeq 10^{-2}$
cm^{-6} pc. If we impose pressure equilibrium with the two-phase ISM
(p/k $\simeq 2.10^3$ cm^{-3}K), the extent of the local plasma region becomes

$$R_L = \frac{4T^2}{p/k^2} \, EM \; \rightarrow \; R_L \simeq 10 \text{ kpc } ! \tag{2}$$

This value is of course highly excessive and we need to look for
constraints to be imposed on R_L. An estimate of R_{Lmax} can be
derived from the O VI absorption measured in the spectra of 72 hot
stars (Jenkins 1978). The local region should not violate the
observed O VI column density. If we impose from the data $N_{O\,VI}$ <
10^{13} cm^{-2}, we obtain at T = 10^6K a value $n_e R_L \lesssim 1.2$ cm^{-3} pc. This
yields, with the EM-value above, $R_{Lmax} \simeq 140$ pc, which gives rise to
$(p/k)_{min} \simeq 1.7 \; 10^4$ cm^{-3}K. This value exceeds by an order of magnitude the value derived for the two-phase medium, and is still a
factor of 2-3 higher than that obtained for the three-phase model of
McKee and Ostriker. There is however no a-priori reason why
pressure equilibrium should hold, the ISM as a whole will most
probably be in a continuous dynamical state (see e.g. McCray and
Snow 1979). Incidentally, the overpressure problem always emerges,
independent of the model assumption (e.g. Shapiro and Field 1976
for an alternative model).

Of course the local region will not have to be isothermal. The
observed properties are for instance in rough agreement with an
interpretation in terms of an old supernova remnant which exploded
about 2.10^5 years ago at about 20 pc from the earth. This remnant
should have entered the snow-plough phase with a shock temperature
$\simeq 10^6$K and encompasses the solar system. The chance of having a
supernova go off within about 20 pc from the earth is small (about
1 in 10^8 years), but cannot be excluded. In this evolutionary
stage the remnant would generate a copious flux of EUV-emission
caused by cooling of the outer layer. In fact large fluxes of EUV
background are observed (Bowyer and Stern 1979), indicating the
presence of temperatures in the range 1 - 4.10^5K.

What about the filling factor of the X-ray emitting plasma? One
may approximate the contribution from more distant regions by hot
plasma interspersed in neutral gas, and again use the variation of
the spectral hardness ratio as a tool to determine the filling
factor f. If the local region has a temperature T and electron
density n_e, and the more distant regions are assumed to be at
temperature T' and electron density n_e', the maximum variation in
the hardness ratio can be approximated by:

$$\delta(H/L) = 1 + f \ \frac{\Lambda'_H}{\Lambda_H} \ \left(\frac{n_e'}{n_e}\right)^2 \ \frac{D_H}{R_L} \ , \qquad (3)$$

with Λ_H, Λ'_H the volume emissivities in the high band for plasma
temperatures T and T' respectively, D_H unity optical depth for H-
band photons ($\simeq 2$ kpc) and R_L the extent of the local region
($\leqslant 140$ pc). For many sky regions, except the X-ray enhancements,
one observes $\delta(H/L) < 2$, hence

$$f \leqslant 7.10^{-2} \ \frac{\Lambda_H}{\Lambda'_H} \ \left(\frac{n_e}{n_e'}\right)^2 \qquad (4)$$

The contribution from more distant regions having T', n_e'-values
similar to the local region is apparently quite small ($f < 7\%$).
The allowed filling factor increases however rapidly for regions with
$T' < T$ and/or $n_e < n_e'$. It is very likely that these lower temperature regions do exist and in fact make the major contribution to the
O VI absorption. Several stars in the observations of Jenkins
(1978) exhibit an O VI-column density which appreciably exceeds the
value which can be accommodated for by the local X-ray emitting
plasma, some of them by an order of magnitude. To raise the contribution to the O VI column, plasma regions with temperatures near
3.10^5K have to be invoked. The O VI-ion is most abundant at these
temperatures and (4) allows still a considerable filling factor,
since the contribution to the soft X-ray flux can remain minor.
McKee and Ostriker have suggested that the O VI could be contained
within the conductive interface between dense clouds and their hot
intercloud medium which would be responsible for the bulk of the
soft X-rays. The latter explanation does not seem to meet however
the observational constraints on required temperature and filling
factor.

4. CONCLUSION

The soft X-ray background data show us the hottest phase of the ISM.
The solar system is most probably embedded in a hot plasma region
with a temperature of $\simeq 10^6$K, which could be the cavity of an old
SNR which has already entered the radiative phase. The data furthermore indicate that within 1-2 kpc from the sun, except for the
enhancements, not many more of these 'bubbles' with similar
temperature and electron density exist. In addition to this, several
hotter regions exist (soft X-ray enhancements) at a temperature of
about 3.10^6K. These are probably all SNR's or amalgamates of SNR's

in the adiabatic stage of evolution, the North Polar Spur being
the best studied example. The O VI data also refer to a hot
component of the ISM, this component is however at a lower
temperature ($\simeq 3.10^5$K) and a different filling factor than the X-ray
emitting plasma. Three main temperature regimes of the hot ISM have
therefore already been identified and much more observational data
is needed to assess the picture in any detail. Spectroscopic observations and a full sky survey in the EUV/soft X-ray band with
typical spatial resolution of a few arc minutes is certainly
required. This allows the hot gas to be mapped over a large regime
of temperatures, which is necessary to make real progress in dis-
entangling the complex structure of the ISM.

REFERENCES

Allen, C.W., 1973. Astrophysical Quantities (3rd ed.; London:
     University of London, The Athlone Press), page 30.
Castor, J., McCray, R., and Weaver, R., 1975. Ap. J. (Letters), 200,
     L107.
Cox, D.P., and Smith, B.W., 1974. Ap. J. (Letters), 189, L105.
Davelaar, J., and Bleeker, J.A.M., 1979. Preprint.
Kayakawa, S., Kato, T., Nagase, F., Tanaka, Y., and Yamashita, K.,
     1978. Astron. Astrophys., 62, 21.
Inoue, H., Koyama, K., Matsuoka, M., Ohashi, T., Tanaka, Y., and
     Tsunemi, H., 1978. Proc. Cospar/IAU Symp., Innsbruck, Austria.
Jenkins, E.B., 1978. Ap. J., 219, 845.
de Korte, P.A.J., Bleeker, J.A.M., Deerenberg, A.J.M., Hawakawa, S.,
     Yamashita, K., and Tanaka, Y., 1976. Astron. Astrophys., 48,
     235.
Mansfield, V.N., and Salpeter, E.E., 1974. Ap. J., 190, 305.
Matsuoka, M., 1979. Preprint.
McCray, R., and Snow, T.P., 1979. Ann. Rev. of Astronomy and Astro-
     physics, Vol. 17.
McKee, C.F., and Ostriker, J.P., 1977. Ap. J., 218, 148.
Sanders, W.T., Kraushaar, W.L., Nousek, J.A., and Fried, P.M., 1977.
     Ap. J. (Letters), 217, L87.
Shapiro, P.R., and Field, G.B., 1976. Ap. J., 205, 762.
Stern, R., and Bowyer, S., 1979. Ap. J., to be published.
Weaver, H., 1978. paper presented at IAU symp. No. 84.

Ultraviolet observations
of the Cygnus loop with IUE

J.C. Raymond

Harvard-Smithsonian Centre for Astrophysics

and

R.S. Wolff,
Bell Laboratories, New Jersey.

We have obtained a series of low resolution spectra of supernova remnants with the IUE satellite (Boggess et al. 1978). Here we discuss observations of a filament on the western edge of the Cygnus Loop; position 3 of Miller (1974).

Table 1 gives intensities of the UV lines corrected for a reddening E(B-V)= .08 (Parker 1967) using the Bless and Savage (1972) extinction curve. Two models of plane parallel, steady-flow shocks from Raymond (1979) for a gas of cosmic abundances and shock velocities of 120 and 140 km s^{-1} are included for comparison. The models of Cox (1972) or Shull and McKee (1979) lead to the same conclusions.

The ratios of the highest temperature lines, N V and O V, with the lower temperature lines indicate a shock of $\sim$130 km s^{-1}, higher than the 90 km s^{-1}, inferred by Raymond (1979) from the optical spectrum of Miller (1974). It is similar to the velocity obtained by Benvenuti, D'Odorico and Dopita (1979) for Miller's position 1 on the eastern edge of the Cygnus Loop.

The discrepancy may be attributed to a departure from steady flow which probably accounts for the other anomalously strong [O III] filaments (Parker 1964; R. Fesen, private communication; J. Percival, private communication) as well. The hydrogen recombination zone (T $\sim$ 8 x 10^{3}K) predicted by the steady flow models seems to be nearly absent.

The departure from steady flow can be interpreted in several ways;

    a)   thermal instability of the post-shock cooling (McCray, Stein and Kafatos 1975);

    b)   a shock in a cloud reached very recently ($\tau \sim$ 200 years) by a shock in the intercloud medium (e.g. McKee and Cowie 1975);

    c)   presently occurring cold shell formation as the SNR evolves.

The coherence of the instability (over $\sim 10^{10}$ cm) is surprising if we interpret the departure from steady flow as the result of thermal instability in the cooling after passage of a shock. In the other interpretations (b and c) the fast shock $\sim 10^{18}$ cm <u>ahead</u> of the bright filaments is surprising (Raymond et al., 1979).

The second outstanding feature, apparent from Table 1, is the weakness of C IV λ1550 compared with the models. Benvenuti, D'Odorico and Copita (1979) attribute this to carbon depletion onto grains. The higher C III λ1909 intensity relative to the models then requires the progressive destruction of carbon bearing grains present in the cooling gas. An alternative explanation is that the observed filament is a sheet of gas seen edge-on. Scattering in resonance lines will then reduce their strength relative to inter-combination lines. This hypothesis is supported by the lines λ1335/λ2325 from Carbon II.

<u>Table 1 - Relative UV Line Fluxes for Miller's Position 3</u>

Lambda (Å)	Ion	Iλ	Model F 120 km s^{-1}	G 140
1240	N V	19	6	52
1335	C II	10*	202	100
1371	O V	6*	.1	11
1400	O IV		32	57
	S IV	73	8	11
	Si IV		55	22
1485	N IV	11	18	27
1550	C IV	108	890	1130
1640	He II	27	38	36
1666	O III	100	100	100
1745	N III	52**	29	29
1817	Si II	6*	4	1
1892	Si III	27	90	53
1909	C III	181	350	220
2325	C II	102	169	87
2420	Ne IV	37	18	22
2470	O II	21*	31	27
I(1666)		2.5 (-4)		

*   Estimated uncertainty greater than 30%

**   Particle noise contributes up to 50% of this value.

We gratefully acknowledge the assistance of the IUE Observatory
staff, particularly A. Holm, in the acquisition and reduction of
these data, and we thank T. Gull, R. Parker, and R. Kirshner for the
use of filter photographs of the Cygnus Loop. This work has been
supported by NASA Grant NSG 5370 to the Harvard College Observatory.

REFERENCES

Benvenuti, P., D'Odorico, S., and Dopita, M., 1979. Nature, 277, 99.
Bless, R., and Savage, B., 1972, Ap. J., 171, 293.
Boggess, et al., 1978. Nature, 275, 372.
Cox, D.P., 1972. Ap. J., 178, 143.
McCray, R., Stein, R.F., and Kafatos, M., 1975. Ap. J., 196, 565.
McKee, C.F., and Cowie, L.L., 1975. Ap. J., 195, 715.
Miller, J.S., 1974. Ap. J., 189, 239.
Parker, R.A.R., 1964. Ap. J., 139, 493.
Parker, R.A.R., 1967. Ap. J., 149, 363.
Raymond, J.C., 1979. Ap. J. (Suppl.), 39, 1.
Raymond, J.C., and Davis, M., Gull, T., and Parker, R.A.R., 1979.
    in preparation.
Shull, J.M., and McKee, C.R., 1979. Ap. J., 277, 131.

Galactic soft X-ray sources:
RS Cvn systems, flare stars and A stars

Philip Charles*

Space Sciences Laboratory,
University of California,
Berkeley, USA.

1.   Introduction

2.   X-ray Emission from RS CVn Systems
     a)   Spectral Observations
     b)   Variability
     c)   Models of Stellar Coronae

3.   X-ray Emission from Flare Stars
     a)   Observations
     b)   Models of Stellar Coronae

4.   X-ray Emission from A and B Stars
     a)   Observations
     b)   Implications

1.   INTRODUCTION

Since the last major review of soft X-ray sources by Gorenstein and
Tucker (1976), the field of soft X-ray astronomy has expanded
considerably, principally as a result of the HEAO-1 satellite. The
Low Energy Detectors (LEDs) of the HEAO-1 A-2 experiment[†] were
sensitive over the energy range 0.15-3 keV and conducted an all-sky
survey. Here I shall consider only galactic, point sources of soft
X-ray emission where the bulk of the emission is in the 0.1-1 keV
range (supernova remnants are amply discussed in the above review
and by Culhane (1977)). Such a restriction will provide a natural
bias to relatively local (< 1 kpc) sources of X-ray emission
because of the effects of interstellar absorption. At 0.5 keV, an
optical depth of unity is reached through an X-ray absorbing
column density of $N_x \sim 10^{21}$ cm^{-2} (Fireman 1974; see also Ryter et al.
1975) which is at a distance of $\sim$ 300 pc for $n_H \sim 1$ cm^{-3}, although
there is evidence for much lower densities in our immediate locale

[†]The HEAO-1 A-2 experiment is the result of a collaborative effort
led by E. Boldt (GSFC) and G. Garmire (CIT) with collaborators at
UCB, JPL, GSFC and CIT.

*Present address: Department of Astrophysics, Oxford University.

($\lesssim$ 100 pc; Bohlin et al. 1978) and in certain directions.   Nevertheless, this is a very important spectral region because objects
at temperatures $\sim$ 1-15 million degrees will emit most strongly at
these energies and, more importantly, will radiate the bulk of
their energy in emission lines if their atmospheres have close to
cosmic abundances (see e.g. Raymond and Smith 1977).   SNRs are the
perfect example of such astrophysical plasmas but, recently, active
stellar coronae, such as those found in RS CVn systems, have been
added to the list of examples.   Figure 1 lists the various classes
of soft X-ray sources as a function of their luminosity.   Since the
dwarf novae and magnetic white dwarfs are discussed elsewhere in
this volume, I shall concentrate on the HEAO-1 studies of active
stellar coronae.

As most of the observational material in this review was obtained
with the LEDs onboard the HEAO-1 satellite, a brief description is
in order (full details are contained in Rothschild et al. 1979).
The satellite (in an ecliptic orbit) has a spin period of 30
minutes.   All the experiments view perpendicular to the spin axis,
which is itself pointed at the sun.   A scan in ecliptic latitude
thus covers the whole sky in six months.   The narrow field of view,
($1^\circ.55$ x $3^\circ.03$ FWHM of LED 1), is used most often to avoid source
confusion.   LED 1 has a geometrical collecting area of 177 cm^2 and
an energy range of 0.15-3 keV in its normal gain setting.   Total
counting rates are read out every 1.28 s (equivalent to $0^\circ.26$ of scan
motion) while spectral (pulse height) data are accumulated in 10.24s
 intervals   ($2^\circ.05$ of scan motion) and read out in 32 pseudo-logarith-
mic channels.   The spacecraft aspect is determined to an accuracy of
$\sim$ $0^\circ.1$ for this experiment.

2.   RS CVn SYSTEMS

The most extensive reviews of these systems can be found in Hall
(1976), who defines them and the related long and short period
systems.   Basically, RS CVn systems are slightly evolved (Popper and
Ulrich 1977) binaries with periods in the range 1-14 days consisting
of a hotter component, generally late F to G V, and a cooler component KO IV the latter being responsible for strong CaII and Hα
emission.   Some are radio sources marked by strong flaring episodes
(e.g. Feldman et al. 1978; Gibson et al. 1975) and most seem to be
X-ray sources (see Table 1).   Perhaps the most fundamental property
of RS CVn systems is the "photometric wave" in the optical light
curve (of amplitude $\lesssim 0^m.3$), which, in the canonical model of Hall
(1972), is attributed to one hemisphere being dominated by starspots.
 Detailed models of these have been presented by Eaton and Hall
1979.

    a) Spectral Observations

A complete search of the HEAO-1 LED data for evidence of X-ray
emission from the 59 systems listed by Hall (1976), Eggen (1978) and

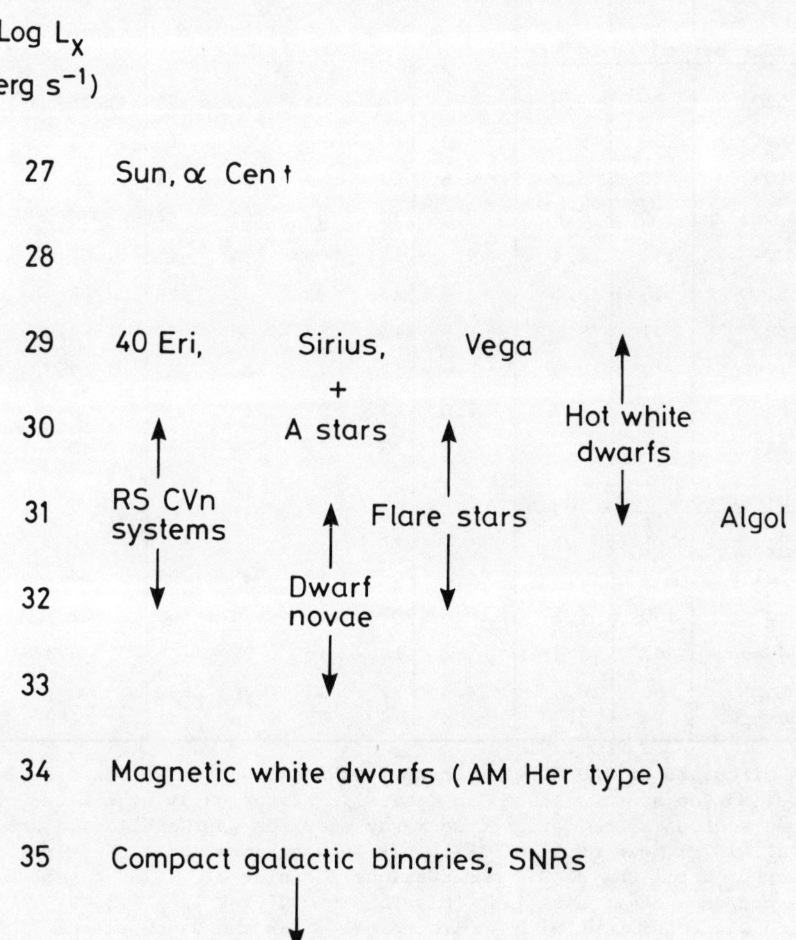

Figure 1 — Soft X-ray Sources

TABLE 1: RS CVn Systems Observed by HEAO-1

NAME	Dist pc	V_{max}	Period Days	α h	m	δ $^\circ$	'	L_x (0.2-2.8) KeV 10^{30} erg s^{-1}
Regular period (1-14^d):								
HD 5303	66	7.9		0	51	-74	55	14
UX Ari	50	6.5	6.4	03	24	28	33	21
HR 1099	33	5.9	2.8	03	34	00	26	12
SAO 015338	130	8.8	7.5	10	53	60	44	70
RS CVn	145	8.4	4.8	13	08	36	12	62
HR 5110	46	5.0	2.6	13	33	37	26	3
σ CrB	21	5.8	1.9	16	13	33	59	3.9
HD 155555	17	7.5		17	12	-66	53	3.2
AR Lac	50	6.9	2.0	22	07	45	30	15
HD 224085	26	7.3		23	52	28	06	40
Long Period (>14^d):								
α Aur (Capella)	14	0.1	104	05	13	45	57	4
σ Gem	150	4.2	19.6	07	40	29	00	21
HR 4665	47	5.4		12	13	72	45	8.4
HK Lac	150	6.5	24.4	22	03	46	59	100
λ And	24	3.9	20.5	23	35	46	11	1.9

the circulars of IAU Commission 42 (Working Group on RS CVn binaries)
revealed the sources listed in Table 1. Historically, the first of
these sources to be detected at X-ray energies was Capella (Catura
et al. 1975; Mewe et al. 1975). Little was known about the X-ray
spectrum until the HEAO-1 observations (Cash et al. 1978) revealed
the presence of intense line emission at 0.85 keV (see Fig. 2)
which was consistent with an isothermal solar abundance plasma
(Raymond and Smith, 1977; hereafter RS) at 10.5 million degrees.
There is evidence for higher temperature gas in the HEAO-2 observations (Holt et al. 1979). No model can fit the data which does not
include line emission. The principal contribution to the emission
feature visible is a complex of Fe XVII and XVIII (Rugge and Walker
1978).

However, it was not realized at that time that Capella was a member

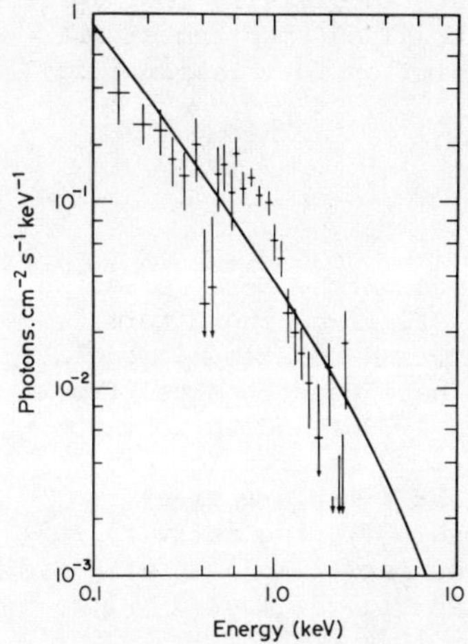
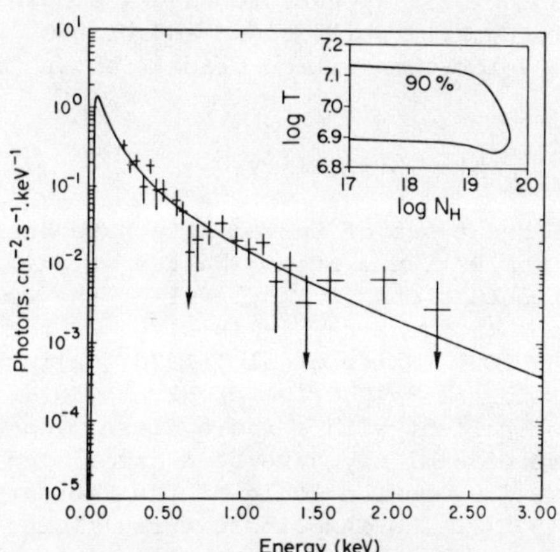

Figure 2: The LED X-ray spectrum of Capella (left) and UX Ari
(right) taken from Cash et al. (1978) and Walter et al.,
(1978a). In both cases the solid line represents the
best fit bremsstrahlung model with $T = 5.6 \times 10^7$ K
(Capella) and 10^7 K (UX Ari). The intense line emission
at 0.85 keV is clearly visible in Capella but notably
absent in UX Ari.

of the long period group of RS CVn systems, and so UX Ari was the
first regular period RS CVn system to be identified as an X-ray
source (Walter et al., 1978a), followed by HR 1099 and RS CVn (Walter
et al., 1978b) and the remainder of those listed in Table 1 (Walter
et al., 1978c, 1980).

The spectrum of UX Ari (also in Figure 2) is substantially different
from that of Capella in that there is no evidence for the iron line
emission with a 99% confidence upper limit of 0.03 solar. However
the temperature of 10^7 K is the same.

Because of the low flux from most of the RS CVn systems, source
confusion or poor sky coverage, we were unsuccessful in extracting
detailed spectra for any of the other sources. However, the LEDs do
provide broad energy band counting rates through the different energy
responses of the two layers of anodes in the detector. Assuming the
intrinsic source spectrum to be either bremsstrahlung or an RS plasma
these layer ratios can uniquely define the temperature and column
density. The spectral parameters of Capella and UX Ari are known
independently ($N_x \lesssim 10^{19}$ cm^{-2} in both cases) and can be used as a
reference. Walter et al. (1979) find that most have temperatures
consistent with $5.10^6 \lesssim T \lesssim 2.10^7$ K and only AR Lac requires a high

column density of $\sim 5.10^{21}$ cm^{-2} (which is consistent with the reddening in this direction of $A_v = 0.17$, Lacy 1979). Although the uncertainties in the count rates make it difficult to classify most systems, SAO O15338 appears to have a simple bremsstrahlung spectrum similar to UX Ari, while σ CrB and HD 5303 appear similar to emission from a solar composition plasma, as in Capella.

b) Variability

Three types of variability have been observed in the X-ray emission from RS CVn systems (Walter et al., 1980): flaring, short term fluctuations in the level of the luminosity, and rotational modulation of the X-ray luminosity. X-ray flares have been reported from HR 1099 (White et al., 1978; Walter et al., 1978b) and 2A 1052+606 = SAO O15338 (Charles et al., 1979). The HR 1099 observation was coincident with a radio flare, and the bright X-ray transient 4U 0336+01 may have been associated with radio flaring activity in HR 1099 on the scale of the February 1978 outburst (Feldman et al., 1978). The SAO O15338 X-ray flare (shown in Fig. 3) lasted ~ 1.5 days and reached a peak luminosity of 1.7×10^{32} erg s^{-1}.

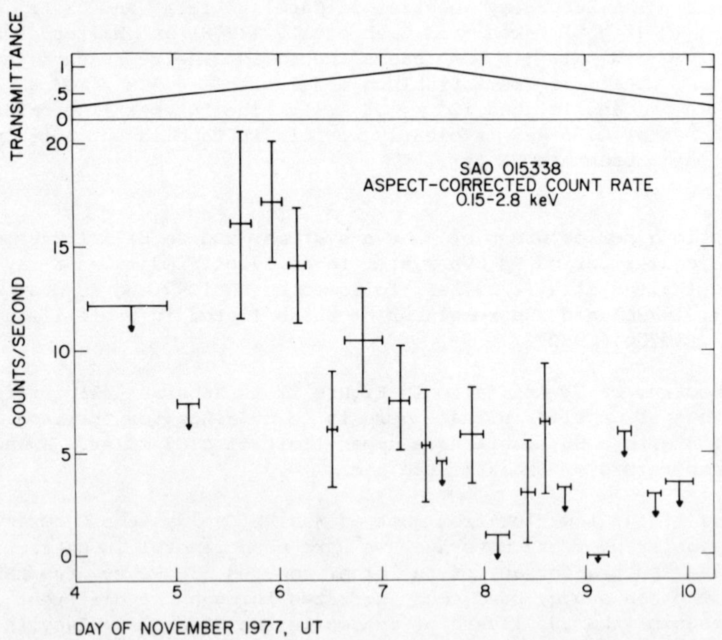

Figure 3: Aspect corrected 0.2-2.8 keV light curve of 2A 1052+606 (= SAO O15338) in November 1977 (Charles et al., 1979). The time of peak source transit was day 311 (collimator response plotted at the top) and the variability is clearly evident.

Short term fluctuations on a timescale of hours have been seen by
HEAO-1 in Capella; changes by up to a factor 2 occurred on consecutive scans separated by 30 minutes. This scale is similar to
the largest flares routinely observed from the Sun, as opposed to
the X-ray flares described above. Capella was scanned by HEAO-1 in
September 1977 and March 1978 (photometric wave phase 0.5 and 0.0
respectively) during which time its <u>average</u> X-ray luminosity did not
change, contrary to the speculation of Mullan (1976a).

During the February 1978 scans of UX Ari the source was not detected
between binary phases 0.98-0.15, but outside these phases was
present at the same intensity as six months earlier. From Landis et
al. (1978) and Weiler et al. (1978) these binary phases corresponded
to a minimum in the photometric wave, i.e. we are seeing rotational
modulation because the starspot-dominated hemisphere was pointed
away from us and the system was at maximum optical light. If the
X-ray emitting region is small compared to the size of the star,
then its scale height h is < (sec $(i-\theta)-1$)R where R is the stellar
radius, i is the inclination and θ is the maximum latitude. Taking
$i = 55^{\circ}$ (Popper and Ulrich 1977) and $\theta = 30^{\circ}$, then h < 0.1 R and,
from the occultation duration, the longitudinal extent of the X-ray
emitting region is $\leq 120^{\circ}$. Outside of the occultation there is no
evidence for any other variability in UX Ari. Similarly, no
variability was observed in HR 1099 ten days before the giant
February 1978 radio outburst, or in σ CrB.

    c)   Models of Stellar Coronae.

The observed range of coronal emission measures from all classes of
stars now extends over five orders of magnitude, from $\sim 10^{49}$ cm^{-3}
for the Sun to $\sim 6 \times 10^{54}$ cm^{-3} for the brightest RS CVn's, and
coronal temperatures now range from 10^{6} K to 10^{7} K. In addition,
these hot coronae are stable phenomena in that they persist for
long periods (at least five years in Capella). This immediately
dismisses the minimum flux model (Hearn 1975) which predicted a
temperature $\sim 5.10^{5}$ K for Capella, a factor of at least 20 below
that observed; and at $\sim 10^{7}$ K all cooling should occur via stellar
winds, so a stable corona could not exist, contrary to observations.
Mechanical models, where the corona is heated by acoustic shock
waves (e.g. de Loore and de Jager 1970; Landini and Fossi 1973),
also fail to describe these observations, predicting X-ray luminosities more than a factor 1000 below that seen.

Currently, the most attractive explanation of these active stellar
coronae lies in magnetic confinement of the gas (e.g. Rosner et al.
1978; Craig et al. 1978) as applied to solar coronal loops. This
has been scaled up to RS CVn levels by Walter et al. (1980). The
model combines N loops with parameters indicated in Figure 4.

Rosner et al. show that the temperature of a coronal loop, its
length L and gas pressure p are related by

$$T = 1.4 \times 10^{3} \, (pL)^{1/3}$$

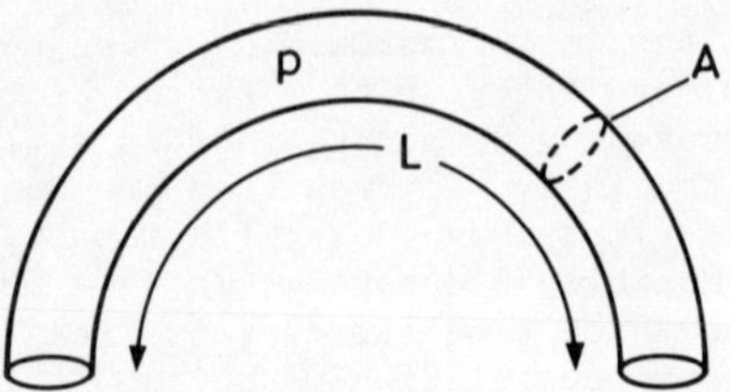

Figure 4: The basic parameters of the coronal loop model.

and then the superposition of N loops yields an emission measure of

$$E.M. = 1.07 \times 10^9 \left(\frac{T}{10^7}\right)^4 L^{-1} \left(\frac{R_*}{R_o}\right)^2 f \quad 10^{54} cm^{-3}$$

where the average cross-sectional area, A, of a loop times N has been expressed as a fraction f of the stellar surface area. The number of loops can be written as

$$N = 4\pi R_*^2 f (\alpha L)^{-2}$$

where the radius of a loop is expressed as αL. From these three expressions the parameters of interest are:

$$f = 340 \left(\frac{E.M.}{10^{54}}\right) \left(\frac{T}{10^7}\right)^{-1} \left(\frac{R_*}{R_o}\right)^{-2} p^{-1}$$

$$N\alpha^2 = 50 \left(\frac{E.M.}{10^{54}}\right) \left(\frac{T}{10^7}\right)^{-7} p$$

$$L = 3.10^{11} \left(\frac{T}{10^7}\right)^3 p^{-1}$$

which are evaluated in Table 2 for various systems.

For UX Ari, Weiler (1978) gives $p \sim 10$ dyne cm^{-2}, whence $L \sim 3.10^{10}$ cm (similar to the Sun). A large number of loops ($\sim 10^4$) is required and f is large (f can be > 1 because the average value of A is greater than the cross-section at the base of the loop) in agreement with optical models (e.g. Vogt 1978; Eaton and Hall 1979). In Capella, however, $p \sim 1.5$ dyne cm^{-2} (Haisch and Linsky 1976) giving $L \sim 2.10^{11}$ cm, $f \sim 0.1$ and a small number of loops ($\sim 10^2$). A few loops will therefore dominate and give rise to greater variability as in the solar corona, in contrast to the constancy of X-ray emission from UX Ari (outside the occultation). This scaling can also be applied to the lower luminosity systems 40 Eri (an intriguing triple system discussed by Cash et al. 1979) and ξ Boo (Walter et al. 1978c) giving small N in both cases, which predicts that their X-ray emission will be highly variable. For comparison,

Table 2 also includes the quiet Sun and two hypothetical cases: a
plage-covered Sun (Vaiana and Rosner, 1978) and a plage-covered
subgiant.

TABLE 2:   Coronal loop parameters for RS CVn systems

	$T(K)$	EM (cm^{-3})	p $(DYN\ CM^{-2})$	f	L (cm)	$N\alpha^2$
UX ARI	10^7	4×10^{54}	10	16	3×10^{10}	2000
CAPELLA	10^7	1.2×10^{53}	1.5	.12	2×10^{10}	9
ξ BOO A	10^7	2.3×10^{51}	30	.02	10^{10}	3
40 ERI A	10^7	4×10^{50}	30	.10	10^{10}	.6
SUN (quiet)	3×10^6	10^{50}	1	.11	10^{10}	20
SUN (covered with plages)	10^7	4×10^{52}	10	1.3	3×10^{10}	20
SUBGIANT (covered with plages)	10^7	4×10^{53}	10	1.3	3×10^{10}	200

The HEAO-1 observations interpreted within the framework above,
suggest that there is no fundamental difference between the activity
in RS CVn's and that in the Sun and that the observed difference is
merely a matter of scale. RS CVn's may just be an extreme case of
normal late-type stellar activity.

3.   X-RAY EMISSION FROM FLARE STARS

Until the HEAO-1 survey X-ray flares had only been seen from two of
the more active dMe stars UV Ceti and YZ Cmi (Heise et al., 1975;
see also the Gorenstein and Tucker 1976 review) and an UV flare had
been seen from Prox Cen (Haisch et al., 1977). These stars are cool
(surface temperature $\lesssim$ 3000 K), low mass ($\lesssim$ 0.5 $M_\odot$) and very faint
($\sim 10^{-4}$ $L_\odot$; or $M_V \sim$ 15). Their optical flaring characteristics
have been discussed by Kunkel (1975) and Gershberg (1975). However,
the observed L_x/L_{opt} ratios conflict with some of the theoretical
model predictions (e.g. Gurzadyan, 1966, 1971; Grindlay, 1971;
Mullan, 1976b; Kodaira, 1977) and therefore question our understanding of the flare emission mechanism.

310

a) Observations

Using the HEAO-1 LEDs, Kahn et al. (1979) searched for X-ray emission
during single scans across nearby dMe stars. They discovered two
flares from AT Mic and two from AD Leo. The AT Mic flares on 1977
October 25.62 and 27.62 were at the $\sim 9\sigma$ and $\sim 4\sigma$ level. The former
was strong enough to derive a spectrum which is shown in Figure 5.
This spectrum includes information from the Medium Energy Detector
(MED; 2-18 keV) which is co-aligned with LED 1 and saw the same
event. The best fit spectrum is a thermal model (kT = 2.7 keV, with
N_X fixed at 10^{19} cm^{-2}, which corresponds to the distance of 8.2 pc)
plus a line at 6.55 keV which is significant at the 2.5σ level.
Power law models are marginally inconsistent with the data. The two
AD Leo flares occurred on 1977 November 22.34 and 22.86 but, as they
were much weaker than AT Mic, no spectral information was available,
nor was this source detected by the MED. Assuming the same spectral
shape as AD Leo, the X-ray luminosities and emission measures of all
four flares are given in Table 3. There is weak evidence for ~ 5
sec variability in AT Mic during the 30 sec transit by LED 1 of the
strong flare. Because of the incomplete coverage of the flare, only
a lower limit of $\gtrsim 5.10^{32}$ erg can be set to the total energy of one
strong AT Mic flare.

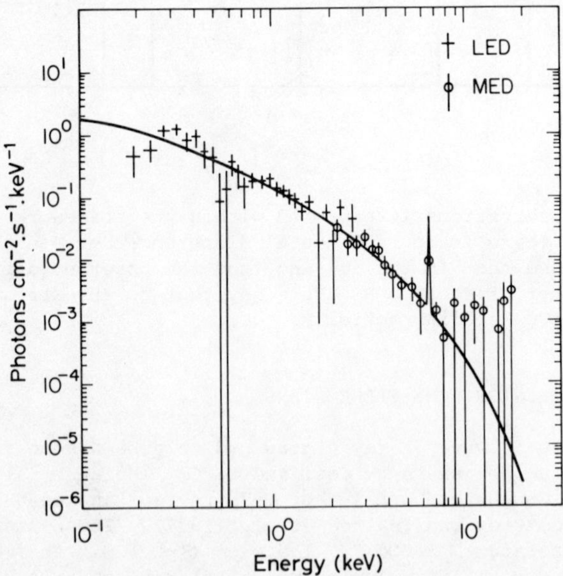

Figure 5: 0.2-18 keV spectrum of the strong AT Mic flare obtained
with LED 1 and the MED on 1977 October 25.62 (Kahn et al.
1979). The solid curve represents the best fit bremsstrahlung model of kT = 2.7 keV with a single emission line at
6.55 keV. *(Reprinted by permission from Nature, Vol. 282,
No. 5740, p691, Copyright (c) 1979, Macmillan Journals Ltd).*

TABLE 3: Observed and Predicted Parameters for Flare Stars

STAR	Spectral Type	L_x (erg s^{-1})	EM (cm^{-3})	L_x/L_{opt} (Observed)	L_x/L_{opt} (Predicted)[1]
AT Mic	dM4.5e	1.6×10^{31} 4.6×10^{30}	1.4×10^{54} 4.0×10^{53}	>4	<0.073[2]
AD Leo	dM3.5e	1.3×10^{30} 1.6×10^{30}	1.1×10^{53} 1.4×10^{53}	>10	<0.079
UV Cet[4]	dM6e	2.0×10^{30}	1.2×10^{53}	$\sim$0.04[6]	<0.029
YZ CMi[4]	dM4.5e	8×10^{30}	5×10^{53}	<0.03[6]	<0.066
Prox Cen (3,4)	dM5e	$0.7-1.8$ $\times 10^{30}$	4.6×10^{53}	> 2	<0.030
Sun[5]	G2 V	4×10^{27}	6×10^{49}	2	0.030

[1]Using the highest value in Mullan's (1976b) table for the quoted upper limit.

[2]The value for AT Mic is computed from the absolute magnitude using Mullan's prescription and the magnitude-radius relation given in Greenstein, Neugebauer, and Becklin (1970).

[3]L_x extrapolated from EUV measurements.

[4]Data from Haisch et al. (1977).

[5]Data from Colgate (1978) for the very large solar flare of 1972 August 4.

[6]Values based on actual simultaneous X-ray and optical observations.

    b) Models of Stellar X-Ray Flares.

Unfortunately, there were no simultaneous optical observations to give L_x/L_{opt} values directly for these two flare stars. However, Kahn et al. derive a lower limit to this parameter in the following manner. The empirical expression

$$R(M_u) = \exp\,(a(M_u - M_{u_o}))\ \ hr^{-1}$$

given by Kunkel (1970) relates the rate of occurrence of a flare with its absolute magnitude where a, M_{u_o} are constants given for these two stars by Kunkel (1975).   So in the total X-ray observing time of the star, T, one would expect T x $R(M_u)$ optical flares. If these are Poisson distributed then by setting the probability of seeing two flares to 1%, one can calculate the 99% confidence lower limit to M_u corresponding to the X-ray flares observed. The Haisch et al. (1977) expression

$$\log L_{opt} = 35.523 - 0.4 \, (M_u + B.C.) \; \text{erg s}^{-1}$$

then converts this to an upper limit to L_{opt} and hence a lower limit to L_x/L_{opt} which is given in Table 3.

It should be noted that $L_x/L_{opt} \ll 1$ for UV Ceti and YZ CMi flares but ≥ 1 for AT Mic, AD Leo and Prox Cen flares. The impulsive stage of a flare rapidly heats the upper chromosphere/corona. In Mullan's (1976b) model this then cools mainly by thermal conduction down the magnetic field lines to the chromosphere and photosphere, which is faster than the soft X-ray radiative cooling. Consequently, optical photospheric emission at the magnetic footpoints is enhanced and $L_x/L_{opt} \ll 1$ as in UV Ceti and YZ CMi (see Table 3). However, this model disagrees strongly with the AT Mic and AD Leo data.

Kahn et al. point out several complicating factors in Mullan's model that could reduce the disagreement. The emission measures quoted in Table 3 are very large and, for reasonable electron densities, require large emitting volumes of radius $\sim$ 0.1-0.2 that of a typical dMe star radius. Since the optical emission is enhanced at the magnetic footpoints, it is possible for the X-ray flare to be visible while the optical emission is reduced by foreshortening or not visible at all. In addition, the cross-section of the footpoint of the flare bubble or loop will be much smaller than the flare radius and so conductive heating along the field lines will be inhibited. Both these effects will result in less optical emission by conductive heating and instead optical flares will be produced by the inefficient radiative heating of the lower atmosphere which gives $L_x/L_{opt} > 1$.

Clearly future observations of flare stars should be simultaneous over _all_ wavelengths in order to advance these studies further.

4.   X-RAY EMISSION FROM A AND B STARS

The sources discussed above, particularly the RS CVn systems, have demonstrated the importance of magnetic fields in modelling stellar coronae. It would therefore be of great interest to determine the effect of spectral type on the intensity of a corona. The mechanical model (Landini and Fossi 1973) predicts that A stars should be at a minimum in coronal intensity (supported through observations of microturbulence). Yet Mullan (1976) has suggested that the convective theories are inaccurate and that Sirius A has a corona (HEAO-2 has shown Sirius A to be responsible for most of the X-rays from this system), and Vega (AOV) has also been observed (Topka et al. 1979).

a) Observations

Cash et al. (1979b) therefore decided to search the HEAO-1 LED data for evidence of X-ray emission from a list of eighteen nearby early A, late B and peculiar A stars. The results are summarized in Tables 4 and 5. Two stars from this list, π Ceti and ϕ Her, were

TABLE 4

Peculiar A Stars Searched for X-ray Emission

Star	HD	Spect.	V	r*(pc)	X-Ray 2σ Upper Limit counts/sec	X-Ray Upper Limit Log (ergs/sec)	Comments[+]
HR976	20210	A6m	6.24	53	1.32	29.94	SB, $5.^d5$ period; mag.; X-rays?
θ Aur	40312	AOp	2.69	56	2.19	30.21	Si; susp. mag.
κ Cnc	78316	B8p	5.24	70	1.64	30.29	SB, $6.^d4$ period; Hg(Mn); mag.
HR4072	89822	AOp	4.93	25	0.62	28.96	SB, $11.^d6$ period; Hg; mag.
ε UMa	112185	AOp	1.76	125	1.41	30.72	Cr; susp. mag. (var.)
3 Cen A	120709	B5IIIp	4.72	59	5.86	30.69	He-weak
α Dra	123299	AOp	3.64	91	3.83	30.88	SB,$51.^d4$ period; Si; susp. mag.
χ Lup	141556	Ap	3.94	39	0.70	29.41	
Ø Her§	145389	B9p	4.24	83	(2.42)	(30.60)	Hg-Mn; susp. mag.; X-rays
46 Dra	173524	AO Si	5.06	143	1.41	30.83	SB, $9.^d8$ period; Hg
β Scl	221507	B9p	4.37	47	1.33	29.85	Hg-Mn; mag.

* Distances are based on measured trigonometric parallaxes except for those which are underlined; these are crudely estimated from standard absolute magnitude calibrations.

+ Comments; SB refers to spectrum binary; notations of elements refer to those which are enhanced unless otherwise specified; remarks on the presence or suspected presence of a magnetic field are based on the work of Babcock (1958).

§ X-Ray source - numbers in parentheses are detections - see text.

detected at the 4.1 and 4.0 σ levels respectively (X-ray sources
H 0242-13 and H 1605-45) with no evidence for variability. Given
the size of the X-ray error boxes Cash et al., computed that the
probability of finding two or more of the eighteen target stars by
chance coincidence with other sources is less than 0.8%. No
alternative candidates for the X-ray sources were found. The most
interesting upper limit in Table 4 is that for the Am star HR 976
which was seen by ANS (den Boggende et al., 1978) about a factor of 2
above the upper limit. It should also be noted that most of the
upper limits are below the detection levels of the two stars. In
particular, an upper limit for α Leo (same spectral type as π Ceti)
a factor of 10 below the detection of π Ceti was derived, and
similarly for β Scl. Presumably π Ceti and ø Her are particularly
active members of this class.

TABLE 5:   Normal Stars Search for X-ray Emission

Star	HD	Spec.	V	r* (pc)	X-Ray 2σ Upper Limit counts/sec	X-ray Upper Limit Log (ergs/sec)
π Cet[+]	17081	B7V	4.23	70	(1.64)	(30.30)
η Leo	87737	AOIb	3.48	1250	4.14	33.19
α Leo	87901	B7V	1.36	26	1.41	29.36
α CrB	139006	AOV	2.23	23	2.11	29.41
ε Sgr	169002	B9IV	1.84	67	1.80	30.28
α Aql	187642	A7V	0.77	5	0.86	27.70
α PsA	216956	A3V	1.16	7	1.88	28.34

* Distances are from measured trigonometric parallaxes, except for
underlined values, which are estimated from adopted absolute
magnitudes.

[+] X-ray source - numbers in parentheses are detections - see text.

b)   Implications

ø Her is an Ap star with abundance peculiarities (it is a mercury-
manganese star), the principal model for which involves radiative
diffusion which acts selectively on certain ions to enhance their
abundances at the surface (Michaud, 1970). The diffusive velocities
are extremely slow ($\lesssim 1$ cm s^{-1}) and so any other mass motions, which
are implied by the presence of X-ray emissions, will neutralize this.
Hence the X-ray data appears to rule out this model. However, Cash
et al. point out that these results must be interpreted cautiously
since (a) only a handful of these objects are known as X-ray sources,

(b) the X-rays might be localized, away from the regions where the peculiar abundances occur (e.g. at the stable poles of these magnetic stars) and (c) note that π Ceti is apparently normal and so the peculiarities of the Ap star may be unrelated to the X-ray emission.

The results I have summarized here are the cumulative effort of the Berkeley HEAO-1 team consisting of Stu Bowyer, Webster Cash (now at LASP, Colorado), Steve Kahn, Keith Mason, Gail Reichert and Fred Walter, many of whom helped in the preparation of this presentation. In addition, Jeff Linsky contributed to the flare star work and Ted Snow to the Ap star search through the HEAO-1 Guest Investigator programme. This work was supported by NASA under contract CIT 44-466866.

REFERENCES

Babcock, H.W., 1958. Ap. J. (Suppl.), 3, 141.
Bohlin, R.C., Savage, B.D., and Drake, J.F., 1978. Ap. J., 224, 132.
Cash, W., Bowyer, S., Charles, P., Lampton, M., Garmire, G., and
     Riegler, G., 1978. Ap. J. (Letters), 223, L21.
Cash, W., Charles, P., Bowyer, S., Walter, F., Ayres, T., and Linsky,
     J., 1979a. Ap. J. (Letters), 231, L137.
Cash, W., Snow, T.P., and Charles, P., 1979b. Ap. J. (Letters), 232,
     L111.
Catura, R.C., Acton, L.W., and Johnson, H.M., 1975. Ap. J. (Letters),
     196, L47.
Charles, P., Walter, F., and Bowyer, S., 1979. Nature, 282, 691.
Colgate, S.A., 1978. Ap. J., 221, 1068.
Craig, I.J.D., McClymont, A.N., and Underhill, J.H., 1978. Astron.
     Ap., 70, 1.
Culhane, J.L., 1977. In "Supernovae", ed. D.N. Schramm, 29, Reidel,
     Dordrecht, Holland.
de Loore, C., and deJager, C., 1970. IAU Symp. 37, ed. L. Gratton,
     238, Reidel, Dordrecht, Holland.
den Boggende, A.J.F., Mewe, R., Heise, J., Brinkman, A.C., Gronenschild,
     E.H.B.M., and Schrijver, J., 1978. Astron. Ap., 67, L29.
Eaton, J.A., and Hall, D.S., 1979. Ap. J., 227, 907.
Eggen, O.J., 1978. IBVS, 1426.
Feldman, P.A., Taylor, A.R., Gregory, P.C., Seaquist, E.R., Balonek,
     T.J., and Cohen, N.L., 1978. A. J., 83, 1471.
Fireman, E.L., 1974. Ap. J., 187, 57.
Gershberg, R.E., 1975. In "Variable Stars and Stellar Evolution", 47.
Gibson, D.M., Hjellming, R.M., and Owen, F.N., 1975. Ap. J. (Letters),
     200, L99.
Gorenstein, P., and Tucker, W.H., 1976. Ann. Rev. Astron. Ap., 14,
     373.
Greenstein, J.L., Neugebauer, G., and Becklin, E.E., 1970. Ap. J.,
     161, 519.
Grindlay, J.E., 1970. Ap. J., 162, 187.
Gurzadyan, G.A., 1966. Dokl. Akad. Nank SSSR, 166, 53.
Gurzadyan, G.A., 1971. Astron. Ap., 13, 348.
Haisch, B.M., and Linsky, J.L., 1976. Ap. J. (Letters), 205, L39.
Haisch, B.M., Linsky, J.L., Lampton, M., Paresce, F., Margon, B.,
     and Stern, R., 1977. Ap. J. (Letters), 213, L119.

316

Hall, D.S., 1972. PASP, 84, 323.
Hall, D.S., 1976. IAU Colloquium No. 29, 287.
Hearn, A.G., 1975. Astron. Ap., 40, 355.
Heise, J., Brinkman, A.C., Schrijver, J., Mewe, R., Gronenschild, E.,
    den Boggende, A., and Grindlay, J., 1975. Ap. J. (Letters), 202,
    L73.
Holt, S.S., White, N.E., Becker, R.H., Boldt, E.A., Mushotzky, R.F.,
    Serlemitsos, P.J., and Smith, B.W., 1979. Ap. J., 234, L65.
Kahn, S.M., Linsky, J.L., Mason, K.O., Haisch, B.M., Bowyer, C.S.,
    White, N.E., and Pravdo, S.H., 1979. Ap. J.(Letters), 234, L107.
Kodaira, K., 1977. Astron. Ap., 61, 621.
Kunkel, W.E., 1970. Ap. J., 161, 503.
Kunkel, W.E., 1975. In Variable Stars and Stellar Evolution, 15.
Lacy, C.H., 1979. Ap. J., 228, 817.
Landini, M., and Fossi, B.C., 1973. Astron. Ap., 25, 9.
Landis, H., et al., 1978. A. J., 83, 176.
Mewe, R., Heise, J., Gronenschild, E.H. B.M., Brinkman, A.C., Schrijver,
    J., and den Boggende, A.J.F., 1975. Ap. J. (Letters), 202, L67.
Michaud, G., 1970. Ap. J., 160, 641.
Mullan, D.J., 1976a. Ap. J., 209, 171.
Mullan, D.J., 1976b. Ap. J., 207, 289.
Nugent, J., and Garmire, G., 1978. Ap. J. (Letters), 226, L83.
Popper, D.M., and Ulrich, R.K., 1977. Ap. J. (Letters), 212, L131.
Raymond, J.C., and Smith, B.W., 1977. Ap. J. (Suppl.), 35, 419.
Rosner, R., Tucker, W.H., and Vaiana, G.S., 1978. Ap. J., 220, 643.
Rothschild, R., et al., 1979. Space Sci. Inst., 4, 269.
Rugge, H.R., and Walker, A.B.C., 1978. Ap. J., 219, 1068.
Ryter, C., Cesarsky, C.J., Audouze, J., 1975. Ap. J., 198, 103.
Topka, K., Fabricant, D., Harnden, F.R., Gorenstein, P., and Rosner,
    R., 1979. Ap. J., 229, 661.
Vaiana, G.S., and Rosner, R., 1978. Ann. Rev. Astron. Ap., 16, 393.
Vogt, S., 1978. Ph.D. Thesis, University of Texas, Austin.
Walter, F., Charles, P., and Bowyer, S., 1978a. Ap. J. (Letters),
    225, L119.
Walter, F., Charles, P., and Bowyer, S., 1978b. Nature, 274, 569.
Walter, F., Charles, P., and Bowyer, S., 1978c. Astron. J., 83,
    1539.
Walter, F.M., Cash, W., Charles, P.A., and Bowyer, S., 1980. Ap. J.
    236, 212.
Weiler, E.J., 1978. A. J., 83, 795.
Weiler, E.J., et al., 1978. Ap. J., 225, 919.
White, N.E., Sanford, P.W., and Weiler, E.J., 1978. Nature, 274, 569.

PART 4
Globular clusters and burst sources

X-ray sources in globular clusters and burst sources

Walter H. G. Lewin and George W. Clark

Massachusetts Institute of Technology,
Cambridge, U.S.A.

Most of the 50 brightest X-ray sources which lie within 30 degrees of
the direction of the galactic center have the following characteristics:

Star-like (angular diameter less than several arc seconds)

Soft X-ray spectra (in comparison to the massive X-ray binaries
of Population I)

Variability on time scales of minutes to days

Luminosities greater than 10^{35} ergs/sec

No periodic pulsations

No eclipses.

About two dozen produce X-ray bursts. As a class they are often
referred to as the galactic bulge sources, a somewhat misleading
name since most of them do not lie within the strictly defined galactic bulge (Salpeter, 1973). Nevertheless, their location outside
the regions of current or recent active star formation identify them
as being members of an older stellar population. Eight high luminosity, variable, non-pulsing, non-eclipsing sources with soft spectra
lie in globular clusters, and of these at least five (probably six)
emit bursts. The similarities of the bulge and cluster sources lead
one to consider them together as a class of high luminosity old
population (II) X-ray sources.

What could these objects be? It is generally assumed that they, like
the high luminosity X-ray sources of Population I, are compact objects
(i.e. white dwarfs, neutron stars, or black holes of either stellar
or greater mass) which derive their X-ray power from the process of
accretion.

In this paper we will argue on the basis of the observed properties
that the majority of these sources are weakly magnetised neutron stars
in low-mass binary systems.

SOFT SPECTRA

Considering the uncertain state of knowledge about the physics of
accretion, the softness of the spectra does not tell us much about
the nature of the compact objects. However, it does suggest that the
area of the X-ray emitting regions of Population II sources is larger
than for the highly magnetised pulsating sources of Population I.

VARIABILITY

The observed variability in the persistent X-ray emission down to time
scales of minutes also does not tell us much either about the nature
of the compact objects.

HIGH LUMINOSITY

On the basis of the high luminosity ($> 10^{35}$ erg/sec) one can rule out
white dwarfs for the bulge sources as a class (see Kylafis and Lamb,
1979 and references therein).

ABSENCE OF PERIODIC PULSATIONS

More than twenty of these objects have been observed intensively in
search of periodic pulsations. None were found. Cominsky et al.,
(1980_a) recently analysed the persistent flux of 7 burst sources.
They found upper limits on the pulsed fractions of the order of 1-3%
for periods from $\sim$ 1.6 to 10^3 sec (Table 1). Spada, Rappaport and Li
(private communication) searched for periodicities in seven of the
brightest (non-bursting) sources near the galactic centre. They also
found none, with typical upper limits of 3% in the range $\sim$ 2 msec -
2 sec.

The absence of detected pulsations tells a little about the nature of
the compact objects, but on the basis of this information alone one
cannot distinguish between black holes and weakly magnetised neutron
stars.

Pulsations are expected from accreting neutron stars only if they
have magnetic fields sufficiently strong (B $\sim$ 10^{12} Gauss) to channel
the accretion flows, as is presumably the case among the short-lived
binary X-ray pulsars of Population I. Old neutron stars may have lost
their magnetic fields, their rotation and magnetic dipole axes may
have become coaligned (Flowers and Ruderman, 1977; Gunn and Ostriker
1970, and references therein) or differences in their formation may
have prevented the development of strong magnetic fields (Joss, 1979a).
In fact, the strong magnetic field which is presumably responsible for
the anisotropic X-ray emission of an X-ray pulsar may supress thermonuclear flashes in the accreted material (Joss, 1978; Joss and Li,
1980; Taam and Picklum, 1978). This may explain why Type I X-ray
burst sources never pulse and why X-ray pulsars never burst (Joss,
1979a). Thus the absence of pulsations is not a strong argument against
neutron stars.

TABLE 1 - Upper limits to periodicities in the persistent components
of Type I burst sources

This table is from Cominsky et al (1980a)

Source	Upper limits (90% confidence) to pulsed fraction* in %, for periods in the ranges listed.		
	2 msec-2 sec	1.6 sec-66 sec	66 sec-10^3 sec
1636-53		1.9	3.5
1659-29		3.5	9.5
1728-34		3.6	9.2
1735-44	8	1.9	13.0
1820-30	1.5**	1.5	1.5†
1837+05		1.1	6.1
1916-05		1.3	15

*   Fraction of total flux

†   Period range 66-200 sec

** F. K. Li private communiation.

Upper limits of $\sim$ 3% were found for seven bright sources (not
bursters) near the galactic centre in the range $\sim$ 2 m sec -
2 sec (see text).

ABSENCE OF ECLIPSES

No X-ray eclipses have been found among these objects. However, this
does not disprove the binary nature of these systems. On the one hand,
Joss and Rappaport (1979) have shown that an X-ray source powered by
accretion through Roche lobe overflow from a low-mass companion (less
than or about 0.3 solar masses) has a low probability (about 25%) of
being eclipsed by its companion. The eclipse probability varies as
the cube root of the mass of the companion and would therefore be less
than 10% for a companion of 0.02 solar masses. On the other hand,
Milgrom (1978) proposed a model in which an accretion disc casts an
X-ray shadow in the orbital plane, thereby excluding from our detection
those very systems which would otherwise show eclipses.

GLOBULAR CLUSTER X-RAY SOURCES

The similarities between the bulge and cluster sources favour the
assumption that they are basically the same kind of systems (Canizares, 1975). Therefore an examination of the circumstances of their
occurrences in globular clusters, based on knowledge of the clusters
themselves, may throw light on the nature of the Population II sources
in general.

The structure of a globular cluster can be broadly characterised by
three observables which are the core radius R_C, the dial radius R_t,
and the central brightness B_C (King, 1966). Peterson and King (1975)
summarised the data as of 1975, and several studies updating and
extending their work have been carried out since then with special
attention to the clusters containing X-ray sources (Bahcall and Hausman, 1977; Bahcall et al., 1977; Canizares et al., 1978; Kleinmann et
al., 1976; Peterson, 1976). The data show a striking correlation between the presence of an X-ray source and the quantity B_C/R_C, which is
a measure of the central density of luminous stars (Figure 1). We
found that the mean value of this measure for the seven X-ray clusters
(NGC 1851, 6440, 6441, 6624, 6712, 7078, Liller I) is so large that it
was not exceeded by the mean for seven clusters taken at random from
among the 72 for which data are available in 3000 trials, in spite of
the presence among the X-ray clusters of NGC 6712, which stands out
as peculiarly low in central density.

The expectation that the cluster sources would be found to lie close
to the cluster cores followed naturally from the recognition of the
correlation with high central density. This expectation was confirmed
by position measurements, carried out with the RMC detectors on SAS-3,
which showed that in five cases the positional error circles with
radii of 20 to 30 arc seconds (90% confidence) include the optical
centres of the clusters (Jernigan and Clark, 1979).

The evidence clearly points away from any explanation of the cluster
sources based on the evolution of primordial binary systems that might
be found in all clusters and toward an explanation based on the formation of the X-ray sources under the circumstances peculiar to the
dense cores of the most centrally condensed clusters. One likely circumstance is a high concentration of compact remnants of the short-
lived massive stars that were presumably present at the beginning,
i.e. neutron stars and black holes that were not ejected from the
clusters by the processes of their formation and subsequently settled
into the cores. The frequency of close encounters between such remnants and nuclear burning stars is certainly greater in a dense core.
Clark (1975) suggested that the cluster sources are neutron stars
that have acquired close nuclear burning companions by capture.
Fabian, Pringle and Rees, (1975) estimated that captures could occur
with sufficient frequency in typical dense cores through tidal diss-
ipation or orbital energy in two-body encounters, and Press and Teukolsky (1977) confirmed the viability of the mechanism by detailed
calculations. Hills (1975) examined binary formation through star
exchange interactions between neutron stars and primordial low-mass
binaries, though the apparently low occurrence frequency of binaries

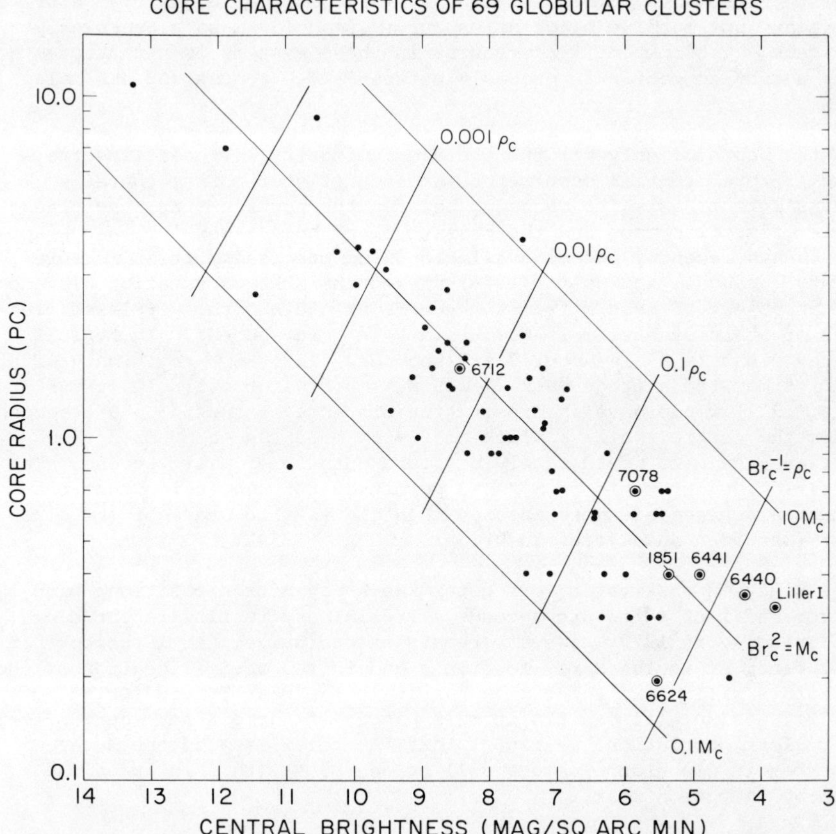

Figure 1: Log-log plot of the core radius r_c (in parsec), versus the central core brightness (B) for 69 globular clusters. The grid lines are loci of constant core mass ($M_c=Br_c^2$) and core density ($\rho_c=M_c/r_c^3$). The core densities have a range of $\sim 10^5$. The globular clusters that contain X-ray sources (except NGC6712) have relatively high core densities (see text). Not all X-ray clusters are indicated. This figure is from F. K. Li, Ph.D. thesis.

in clusters makes the importance of this process less certain. Capture through three body encounters in a dense sub-cluster may also be significant (Fall and Malkan, 1978).

It has been proposed that the deep potential well of a centrally concentrated cluster might retain gas lost by stars and that a black hole of sufficient mass, formed perhaps in a collapse of the dense core, could accrete this gas at the rate required to produce the observed X-ray luminosities (Bahcall and Ostriker, 1975; Silk and Arons, 1975). However, as we will discuss in this article, there exists persuasive evidence that the X-ray sources in globular clusters are not massive black holes but instead low-mass close-binary systems. The compact X-ray source in these systems is a neutron star, the nearby companion is probably between a few tenths and one solar mass.

On the basis of only (1) the average luminosity, (2) soft spectra, and (3) preferential occurrence in dense cluster cores, there is insufficient basis for choosing between the neutron star, the black hole of stellar mass and the massive black hole. However, a new approach has recently become available using precise position measurements. Bahcall and Wolff (1976) showed that the expectation distance of an object of mass M from the centre of a dynamically relaxed cluster of stars of mass m is $0.9\ R_C\ (m/M)^{1/2}$ arc seconds. In typical X-ray clusters R_C is about 7 arc seconds. Thus an X-ray binary with four times the average mass ($\sim 0.5\ M_\odot$) of cluster stars (Ostriker et al., 1972) would have an expectation distance of about 3.5 arc seconds from the centre, while a black hole with more than a hundred times the average mass would rarely be found outside of 1 arc second. The position measurements from SAS-3 permitted the conclusion that the cluster sources are more massive than the visible stars of the cluster (Jernigan and Clark, 1979).

The Einstein observatory can determine X-ray source positions with error radii of a few arc seconds. Preliminary results reported by J. Grindlay in 1979, however, remain inconclusive. It is hoped that a refinement in the X-ray locations and in the optical centres of the clusters will yield a decisive result in the near future (Grindlay, 1979). As we will explain in this paper, if these X-ray sources form a uniform population, we expect that the total mass of the X-ray sources in globular clusters will be near 2 $M_\odot$ (this includes the nearby companion star).

Another possibility to keep in mind about the low-mass binary hypothesis is that it implies a fair chance of finding two X-ray sources within the same cluster. Of the 20 known centrally condensed clusters, 6 contain an X-ray source (Figure 1; Table 2). Therefore the chance of finding two or more in a single centrally condensed cluster is $\sim (6/20)^2 = 0.09$. Obviously, the discovery of two in one would independently exclude the supermassive black hole hypothesis.

The anomaly in central density presented by the case of NGC 6712 is especially intriguing in the light of calculations by Shapiro (1977) which suggest that core collapse may be followed by puffing up and dissolution of the remaining cluster through injection of kinetic energy from interactions with a massive central body. Perhaps X-ray binaries and a massive black hole are produced in the processes of core collapse, and in NGC 6712 we are witnessing the aftermath as the cluster proceeds toward oblivion. Some consequences of this theme of evolution and death in globular clusters have been explored by Lightman, Press and Odenwald (1978).

TABLE 2

GLOBULAR CLUSTER X-RAY SOURCES

		R_c (arc sec)	Burst Source?
47 Tuc	1E002151-7221.5	29	No bursts observed
NGC 1851	4U1513-40	7	Yes
Terzan 2	{ 4U1722-30 { 1E172420-3045.6 }	7	Yes
Liller 1	MXB1730-335	7	Yes
NGC 6641	4U1746-37	9	Probably
NGC 6624	4U1820-30	6	Yes
NGC 6712	4U1850-08	49	Almost Certainly
NGC 7078	4U2131+11	10	No bursts observed
*Kron 3	4U0026-73		No bursts observed
*NGC 6440	MX1746-20		No bursts observed

*
These two associations are uncertain.

For detailed references and for accurate source locations, see
Table 4 (this paper) and Bradt et al., (1979).

X-RAY BURST SOURCES

So far we have only been able to reject white dwarfs as a class for
the bulge sources. We will now turn to the last characteristic, namely that many of the Population II sources produce X-ray bursts. We
believe that the wealth of information on bursts harvested over the
last few years provides strong evidence that these objects are neutron
stars.

X-ray bursts were discovered in 1975 independently by Grindlay et al.
(1976) and by Belian, Conner and Evans (1976a, 1976b). Grindlay et
al. were able to associate the two observed bursts with a known X-ray
source in the globular cluster NGC 6624; the Los Alamos group

(Belian, Conner and Evans, 1976a, 1976b, 1976c) could not make any association with a known X-ray source since their positional accuracy of the many burst events they observed was insufficient. Since then bursts from about 30 sources have been detected, and an important distinction between two types of bursts was recognised by Hoffman, Marshall and Lewin (1978a). They suggested that two very different mechanisms were involved, and introduced the classification of Type I and Type II bursts. We summarise in Table 3 the characteristics of these types and list the most promising of the proposed mechanisms.

TABLE 3

CLASSIFICATION OF X-RAY BURSTS*

Type I ($\sim$ 30 sources)

- • - Burst intervals of hours - days (longer?)
- • - Distinct spectral "softening" during burst decay

    - o - Cooling of a "black body**
    - o - Thermonuclear flashes. (Helium is the most probable fuel)

Type II (Rapid Burster + others?)

- • - Burst intervals of sec - min
- • - No distinct spectral softening during burst decay

    - o - "Black-body" radiation (no significant cooling)
    - o - Instabilities in accretion flow
    - o - Gravitational potential energy

---

*This classification was introduced by Hoffman, Marshall and Lewin , 1978a. It was based on the two phenomenological differences as listed under the solid dots. The characteristics listed as open circles are interpretations (e.g. see Hoffman et al., 1978a; Joss, 1978; Joss, 1979a; and Marshall et al., 1979).

**Gray bodies with an emissivity of $\gtrsim$ 0.1 are allowed. This would increase the size of the emitting regions by a factor $\lesssim$ 3 (Hoffman, Lewin and Doty, 1977; Hoffman, Lewin and Doty, 1977a; Swank et al., 1977; Swank, Eardley and Serlemitsos, 1979; Van Paradijs, 1978a and Van Paradijs, 1979).

THE RAPID BURSTER (MXB1730-335)

The rapid burster became a Rosetta stone when it was recognised that it produced both Type I and Type II bursts (Hoffman et al., 1978a) (Figures 2 and 3). The object can produce Type II bursts of duration from a few seconds to several hundred seconds (Basinska et al., 1980; Inoue et al., 1980; Lewin et al., 1976; Marshall, 1979; Marshall et al., 1979). For Type II bursts with energies in excess of $\sim$ 4x10^{39} erg (assumed distance of 10 kpc) the integrated burst energy is linearly proportional to the "waiting time" to the next burst (Inoue et al.,1980;

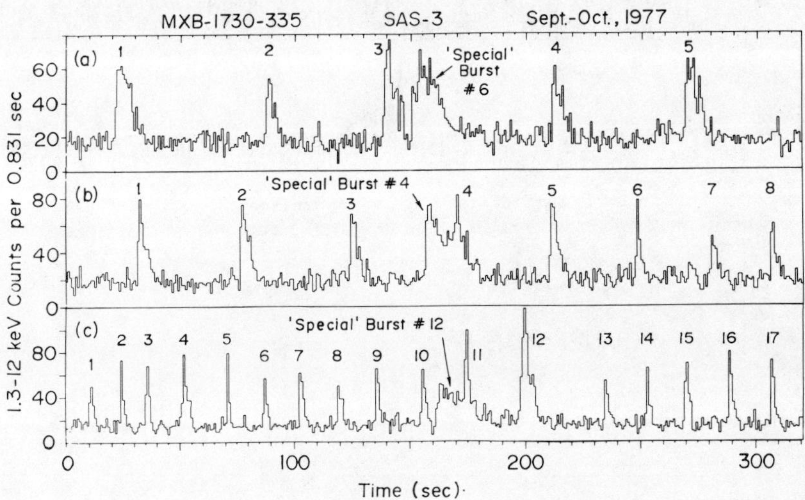

Figure 2:   Discovery of Type I and Type II bursts from the Rapid
Burster.   The Type I bursts ('special') occur independently of the
sequence of the rapidly repetitive Type II bursts (numbered indep-
endently).   This figure is from Hoffman, Marshall and Lewin (1978a).
(Reprinted by permission from Nature, Vol. 271, No. 5646, p.630,
Copyright (c) 1978 Macmillan Journals Limited).

Lewin et al., 1976; Marshall et al., 1979; White et al., 1978).
Most details of this unique object are described in two recent review
articles on burst sources (Lewin, 1979a, 1980).   We summarise here
the most important conclusions from these papers.

- The rapid burster is a recurrent transient with a period of about
6 months; the last two turn-ons came in March, 1979 (Marshall, 1979)
and near August 8, 1979 (Inoue et al., 1980; Oda, 1979b).

- The 'persistent' emission from this object is released in the form
of Type II bursts (Hoffman et al., 1978a) the energy of which is derived from the gravitational potential energy of accreting material
(Hoffman, Marshall and Lewin, 1978a).

- The rapid burster at times also produces Type I bursts at intervals of $\sim$ 3 - 4 hours (Hoffman et al., 1978a).   These are probably
the result of thermonuclear flashes of the freshly accreted material
on the surface of a neutron star (Joss, 1979a and references therein).

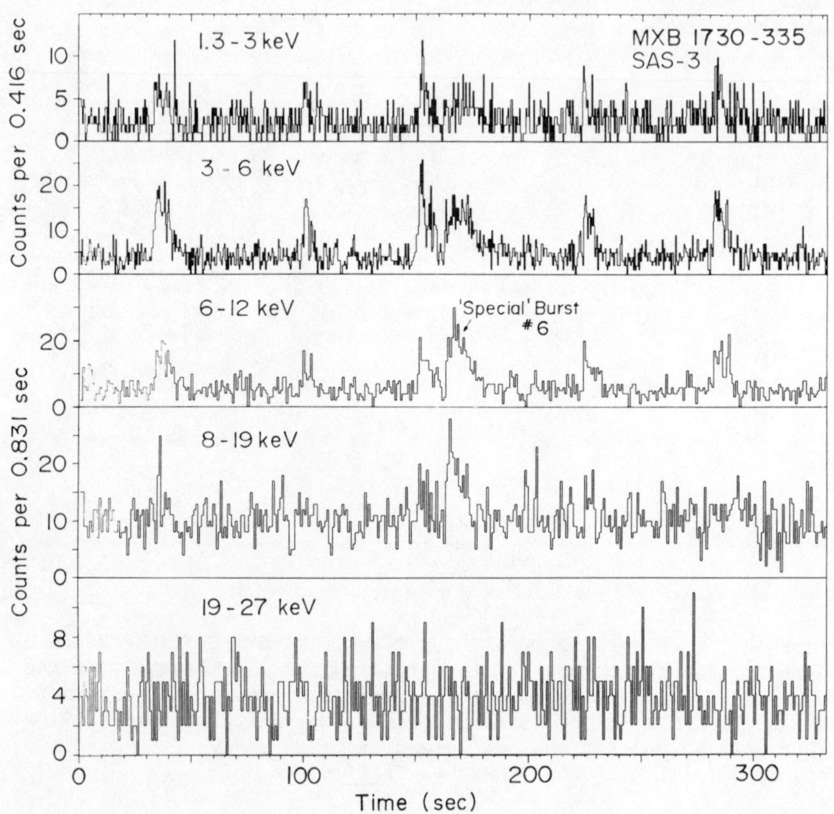

Figure 3: Spectral information for the data shown in Figure 2a. The Type I burst spectrum softens during burst decay; this softening is absent in the rapid (Type II) bursts. This figure is from Hoffman, Marshall and Lewin, (1978a). *(Reprinted by permission from Nature, Vol. 271, No. 5646, p630, Copyright (c) 1978 Macmillan Journals Limited)*

- The spectral data of both Type I and Type II bursts from the Rapid Burster can be fit reasonably well by black-body radiation (Marshall et al., 1979). For the Type I bursts a radius of ~ 8 km is found for the emitting region and this remains approximately constant throughout the bursts. During the burst decay the temperature decreases rapidly (maximum temperature $\sim 30 \times 10^6$ K). For Type II bursts a constant temperature of $\sim 18 \times 10^6$ K is found and the radius of the emitting region decreases from ~ 16 km to less than a few km. For short Type II bursts this can happen in only a few seconds (Marshall et al., 1979).

- The average power in Type II bursts is about 120 times that in
Type I bursts (when present) (Hoffman et al., 1978a).   This is consistent with the Type I bursts being due to the release of nuclear
energy (about 1 MeV per nucleon for helium-burning) and the Type II
bursts being due to the release of gravitational potential energy
(about 100 MeV per nucleon when it falls into a $\sim$1 solar mass neutron
star).   (Joss, 1979; Joss, 1978; Joss, 1979a; Lamb & Lamb, 1978; Maraschi
and Cavaliere, 1977; Woosley and Taam, 1976; and references therein).

Recently, in April 1979, infrared (Apparao et al., 1978; Kulkarni et
al., 1979) and radio (IAU Circular 3347) bursts have been reported
from the direction of the Rapid Burster. The detection of infrared
bursts were also made on September 5, 1979 by A.W. Jones (private
communication and preprint). The infrared bursts are remarkably
bright. At $\sim$ 2.2μ the averaged luminosity is $\sim$ 10^{37} erg/sec (band
width $\sim$ 0.4μ; assumed isotropic emission). The total integrated
burst energy is $\sim$ 10^{38} erg. It is still unclear whether the infrared
bursts are real and if so whether they are associated with the X-ray
bursts. The low occurrence rate of the infrared bursts perhaps
suggests that they are not associated with Type II X-ray bursts which
recur on time scales of seconds and minutes. One could perhaps think
of a mechanism whereby the infrared bursts are strongly beamed but
the X-ray bursts are not. In that case an association with Type II
X-ray bursts would seem possible. The low average occurrence rate of
the infrared bursts perhaps favours a possible association with Type
I X-ray bursts. However, of eight infrared bursts observed by the
two groups, 6 came in pairs separated by $\sim$ 1 - 2 minutes so that an
association with Type I X-ray bursts would imply the existence of a
remarkably complex mechanism to produce two infrared flashes and
only one Type I X-ray burst (the latter have never been detected in
pairs). If the infrared bursts are real and associated with Type I
bursts, bright infrared bursts would also be expected from other
Type I burst sources, and they should be much easier to detect than
optical bursts (Kulkarni et al., 1979; Apparao and Chitre, preprint).

It would be very important to perform simultaneous infrared and X-ray
observations of the Rapid Burster. This can only be done during
active periods which last only $\sim$ 2 - 6 weeks. They occur twice a
year, some time in February - April and August - October.

It is of interest to note that Type II bursts may have been observed
from several known binary X-ray sources (Hoffman et al., 1978a; Lewin,
1979a, 1980).

TYPE I BURST SOURCES

A sky map of Type I burst sources is given in Figure 4. The positions
of many more burst sources have been reported in the literature
(Belian et al., 1976; 1976a; 1976c; Evans et al., 1976; (Hoffman et
al., 1978), but we omit them because it is unclear at this time which
are Type I sources. However it is very likely that many of the
events observed by the Los Alamos group (Belian et al., 1976; 1976a;

1976b; 1976c; Hoffman et al., 1978) were of Type I and, in retrospect there seems to be little doubt that they discovered Type I X-ray bursts from XB1608-52 (Belian et al., 1976a; 1976b; 1976c).

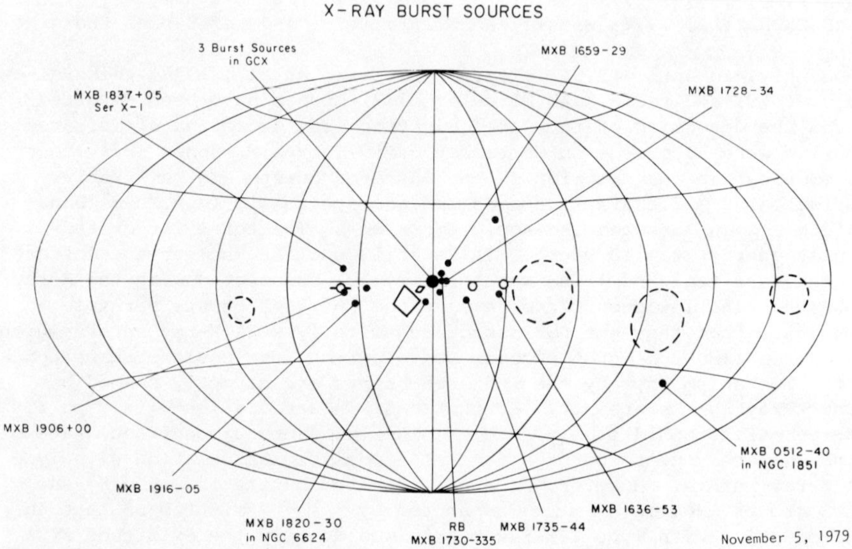

X-RAY BURST SOURCES

Figure 4: Sky map (galactic coordinates) of 26 X-ray burst sources as of November 5, 1979. Only one burst was observed from each of the 5 sources indicated with a dashed circle or a horizontal bar. There are at least 6 more burst sources probably all within 30° of the Galactic Centre. Their positions are poorly known and they are not shown here. The text refers to the many burst events observed by the Los Alamos Group; they are not shown here. At least 21, but possibly all sources shown here are of Type I.

Six of the 28 sources shown in Figure 4 are located in globular clusters (including the Rapid Burster). Figure 5 shows 5 bursts from different burst sources. The characteristic spectral softening during burst decay is clearly visible. Figures 6 - 8 show a series of bursts from various sources. They illustrate that bursts from one and the same source generally look similar although there are exceptions (e.g. see Figures 9 and the caption of Figure 10).

In Figure 10 we show a burst from MXB1850-08 in the globular cluster NGC 6712 (Li and Clark, 1980). The double-peaked burst profile is somewhat unusual. Its profile in the high-energy spectral range is reminiscent of bursts from MXB1743-29 (see Figure 8 and Lewin et al., 1976c). Double peaked bursts have also been observed from Terzan 2 and from MXB1728-34 (Grindlay et al., 1980; Hoffman et al., Ref. 230).

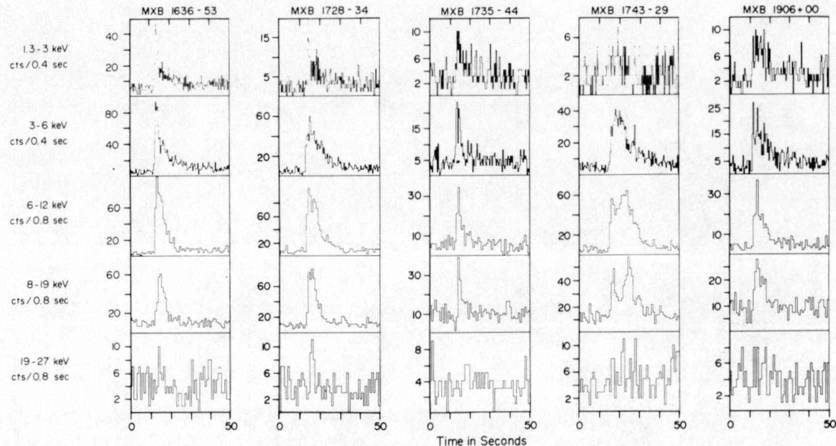

Figure 5:   Profiles (Five energy channels) of Type I bursts from five different sources (SAS-3 data). Note that in all cases the gradual decay (tail) persists longer at low energies than at higher energies. The burst profiles are reasonably distinctive for each particular source.   This figure is from Lewin and Joss (1977).   *(Reprinted by permission from Nature, Vol. 230, No. 5625, p211, Copyright (c) 1978, Macmillan Journals Limited.)*

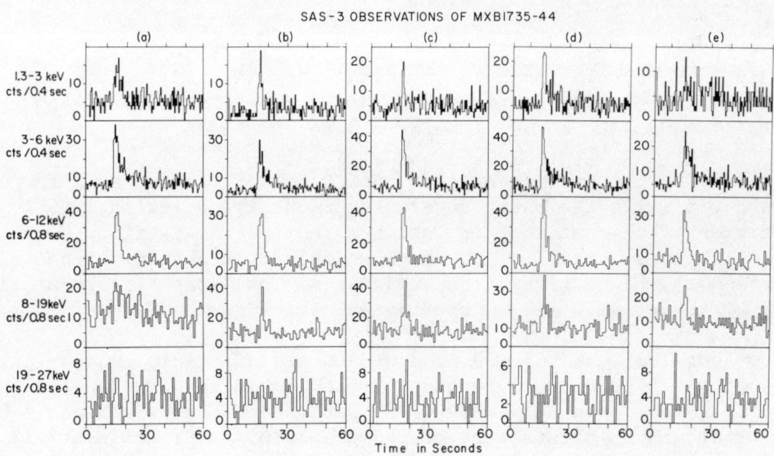

Figure 6:   Raw counting rate plots of five bursts from MXB1735-44. The burst profiles are similar.   Notice the cooling during burst decay.   Burst intervals from this source are highly irregular.   This figure is from Lewin et al., 1980. *(Reproduced by permission from the Royal Astronomical Society, Lewin et al., 177, 83P, (1976)).*

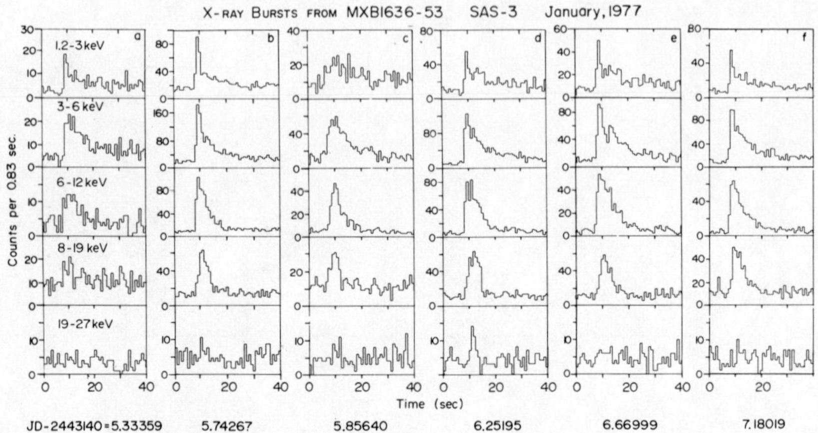

Figure 7:   Raw counting rate plots of five energy channels from the
SAS-3 horizontal tube (HT) detector system for six X-ray bursts from
MXB1636-53.   The times at the bottom, in Reduced Julian Days =
JD-2,443,140, correspond to the time bin near burst maximum.   The
bursts have similar shapes except the third one which doesn't show
the narrow feature in the 1.2 - 3 keV channel.   This burst occurred
unusually soon (2.7 h) after the preceding burst.   Intervals between
the other bursts range from 9.5 h to 12.2 h.   This figure is from
Hoffman, Lewin and Doty, 1977a).

Burst intervals can be regular or irregular on time scales of hours
to days.   A correlation between these intervals and the persistent
X-ray emission was found in two burst sources: MXB1659-29 and the
source MXB1820-30 in the globular cluster NGC 6624.

In May 1976, the persistent flux of NGC 6624 increased by a factor
of about 5 while the burst intervals gradually decreased by 50%.   The
persistent flux continued to increase and the bursts stopped completely (Clark et al., 1977).   Bursts have only been detected from this
source when the persistent flux was in a 'low' state (Clark et al.,
1976; Clark et al., 1977; Grindlay et al., 1976).

In October 1976, MXB1659-29 produced bursts at fairly regular intervals of ∿ 2.5 hours;   no persistent flux could be detected (Lewin
et al., 1976b).   In March 1978, the persistent flux was high (∿ 8% of
the Crab) and no bursts were observed (Lewin, 1978; Lewin et al.,
1978).   In several other cases (MXB1837+05, MXB1735-44 and MXB1636-53)
large fluctuations observed for burst intervals were not correlated
with the persistent X-ray flux (Hoffman et al., 1977a; Lewin et al.,
1980; Li et al., 1977).

According to a simple version of the thermonuclear flash model (Joss,
1979a and references therein) one would expect the burst intervals to

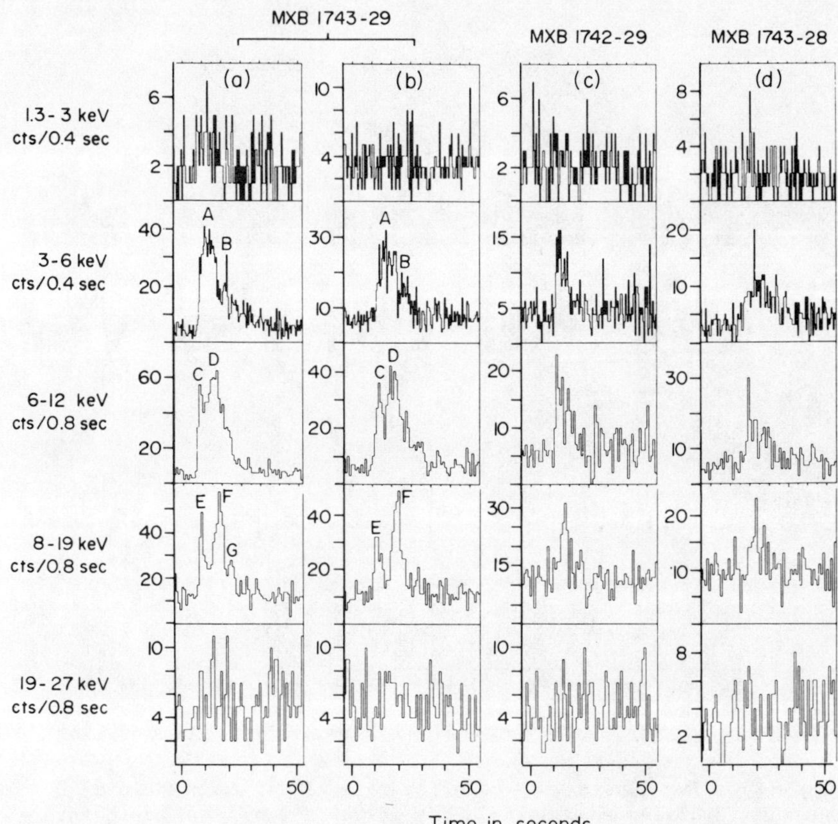

Figure 8: Bursts (raw counting rate plots) from the three burst
sources within 10° of the galactic centre. MXB1743-29 produced bursts at intervals of ∿ 35 hours. They all showed the double (sometimes triple) peaks at high energies. The burst intervals of
MXB1742-29 were ∿ 10 hours. This figure is from Lewin et al., (1976c).
*(Reproduced by permission from the Royal Astronomical Society, Lewin
et al., 177, 83P, (1976)).*

decrease if the persistent X-ray flux increases; the higher the rate
of accretion the sooner the flash should go off - except for very high
(Joss, 1978; Joss, 1979a) and very low (Lamb and Lamb, 1978) values of
the accretion rate when bursting is expected to stop. This is observed
to some extent (see above); however, more detailed calculations are
required to explain why the burst intervals can change by an order
of magnitude (and more) without an appreciable change in the luminosities of the persistent X-ray sources. If the bursts are indeed

due to thermonuclear flashes, then apparently the helium can burn
steadily for extended periods without bursts.

SAS-3 OBSERVATIONS OF MXB 1659-29

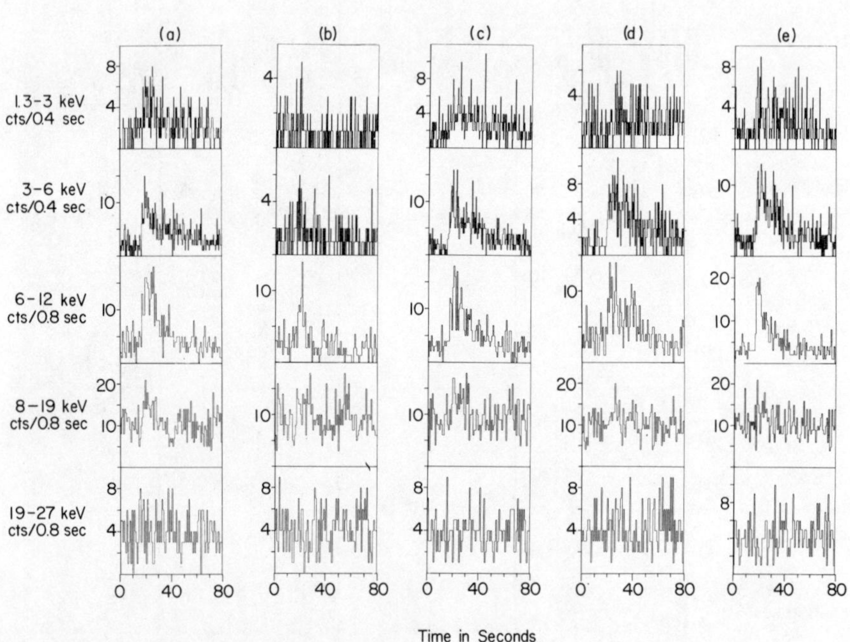

Figure 9: Five bursts (raw counting rate plots) from MXB1659-29.
The burst profiles show quite a bit of variation. The burst intervals were the most regular ($\sim$ 2.5 h) of all burst sources. This figure is from Lewin and Clark, 1980.

A possibly interesting connection between the irregular burst behaviour
of MXB1735-44 and MXB1837+05 and Joss, 1978 calculations is suggested
by the following consideration. One prediction of these calculations
is that thermonuclear flashes will not occur above a certain accretion rate, which corresponds to a few tenths of the Eddington luminosity. If the maximum burst luminosity is the Eddington limit, as is
indicated by the work of van Paradijs, (1978a), the ratio γ of the
persistent X-ray flux to the maximum burst flux measures the X-ray
luminosity in units of the Eddington limit. For 4U1735-44 and
1837+05 this ratio equals 0.2 and 0.3 respectively. These values are
the two highest among 12 burst sources for which γ was reasonably
well determined (Van Paradijs et al., 1979). This suggests that the

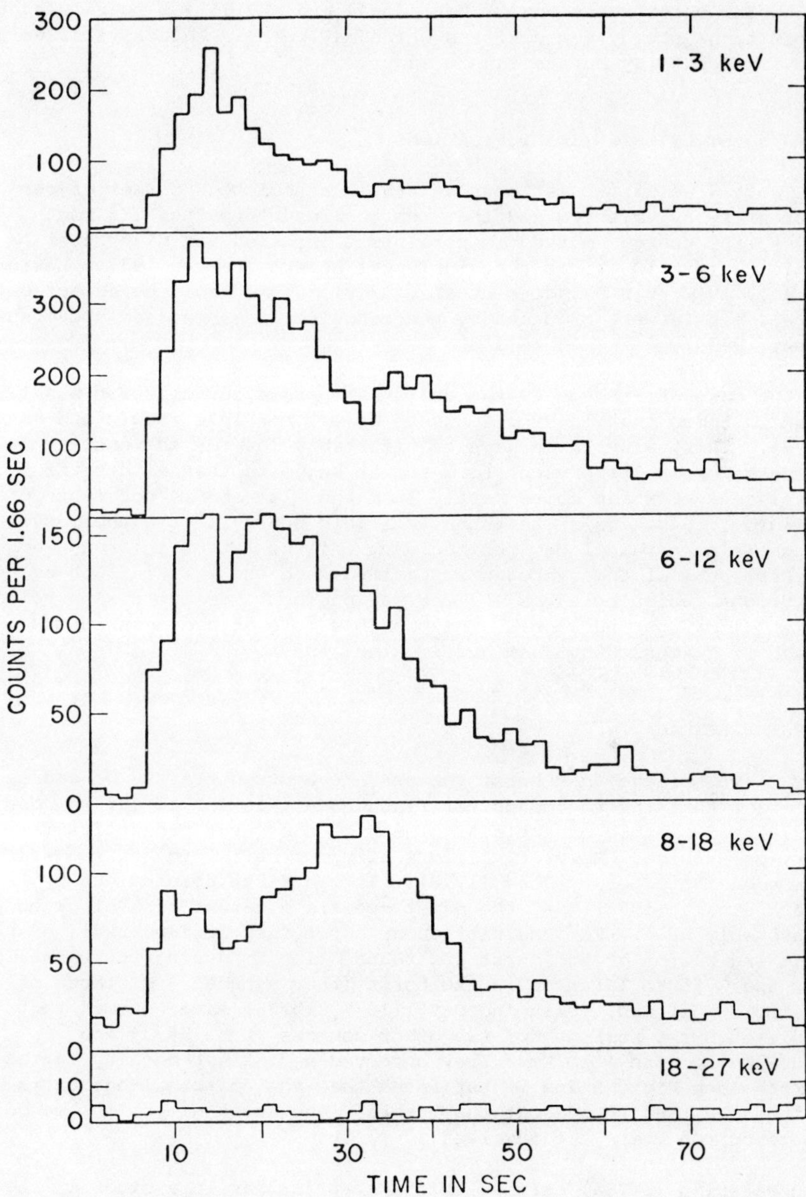

Figure 10: An X-ray burst from MXB1850-08 (in the globular cluster NGC 6712). It is quite different in shape and total luminosity from two other bursts (not shown here) also observed from this source. The double-peaked structure in the higher energy X-ray data is also observed in several other burst sources (e.g. see Figure 8). This figure is from Li and Clark, 1980; see also Ref. 230).

irregular burst behaviour of MXB1735-44 and MXB1837+05 is related to their large steady luminosities which may be near the critical values above which X-ray bursts cannot occur.

PERSISTENT EMISSION/IDENTIFICATIONS

In Table 4 we list burst sources, their associated sources of persistent X-ray emission and their optical counterparts. In most cases, the energy emitted in persistent emission (at high burst activity is $> 10^2$ times that in Type I bursts (Lewin, 1977a; Lewin et al., 1980; van Paradijs et al., 1979). This would be expected if Type I bursts were produced by thermonuclear flashes (see above and Joss, 1979a).

In the case of MXB1659-29 the energy in persistent emission was less than 25 times that in Type I bursts in October 1976 (Lewin and Joss, 1977). This value is perhaps uncomfortably low for the nuclear flash model in its present form and it suggests that either the persistent emission comes partly from the disk and is not isotropic (Milgrom, 1978) and/or that the available energy in the thermonuclear flash is more than 1 MeV per nucleon. If the latter is true, then perhaps some of the hydrogen is ignited when the helium flash occurs (hydrogen fusion releases $\sim$ 7 MeV per nucleon; helium fusion releases only $\sim$ 1 MeV per nucleon). The possible role of hydrogen in the flash is discussed by Taam and Picklum (1979).

BURST SPECTRA

Why do we believe that burst sources are neutron stars? Of course, the fact that the thermonuclear flash model, as worked out by Joss (1978), is capable of explaining many of the burst properties is very persuasive. However, historically our belief in neutron stars preceded the paper by Joss (1978). It was first pointed out by Swank et al. (1977) that the burst spectra fit that of a black body reasonably well. For one particular burst they derived radii of the emitting region of the burst and found $\sim$ 100 km during the first 15 sec and $\sim$ 15 km later on in the burst (they assumed a distance of 10 kpc). Hoffman, Lewin and Doty (1977, 1977a) later showed that radii of burst regions for two other sources (MXB1728-34 and MXB1636-53) were $\sim$ 10 km. They observed a distinct cooling in the bursts (see Figs. 5 and 7) but found that the emitting regions remained constant throughout the bursts. The above dimensions suggest that neutron stars are involved.

Van Paradijs (1978a) carried this a step further in a study of the average characteristics of bursts from each of ten sources for which good broad band spectral data are available from SAS-3. He showed that all ten objects, interpreted as black bodies emitting isotropically, have nearly the same ratio of radiating surface area to burst luminosity. If the peak luminosities are assumed to be the Eddington limit for a 1.4 $M_\odot$ star with a hydrogen envelope

$(1.8 \times 10^{38}$ ergs/sec), the radii are found to have an average value
of 7 kilometers with a scatter of only about 20%. In a later paper
van Paradijs (1979) made corrections for gravitational redshift and
adjusts the maximum burst luminosity such that the ten sources are
distributed evenly around the galactic centre (at an assumed distance
of 9 kpc). He again concluded that burst sources are neutron stars
and standard candles (at burst maximum). Swank, Eardley and
Serlemitsos (1979), assuming modified black-body spectra and
correcting for the gravitational redshift, have come to similar
conclusions.

Lewin et al. (1980) have shown that, when examined in detail, bursts
are not standard candles at burst maximum. They showed that the
standard deviation in the maximum flux of 56 bursts from MXB1735-44
(observed in 1977 and 1978) was 37% (normalized to the mean value).
The largest observed maximum flux was about seven times higher than
the lowest observed maximum flux. The observed spread (standard
deviation of $\sim$ 37%), in our opinion, is small enough to justify the
use of the average values in the analysis of global properties of
X-ray burst sources (van Paradijs 1978a). However individual bursts
are not standard candles (Lewin et al. 1980).

Grindlay and Gursky (Rf 228) proposed for the burst source in NGC 6624
a black hole with a mass greater than several hundred solar masses.
Grindlay (Rf 229) proposed as a general model for burst sources
spherical accretion on black holes with mass 10-100 $M_\odot$. The above
black-body radii seem to exclude very massive black holes independ-
dent of the thermonuclear flash interpretation.

On the basis of the spectral information discussed above, the derived
sizes of the emitting regions and the success of the thermonuclear
flash model, we conclude that it is very likely that X-ray burst
sources are neutron stars.

RADIO/OPTICAL/X-RAY BURSTS

It was our hope and expectation that the simultaneous detection of
radio, optical, infrared and X-ray bursts could make a significant
contribution to explaining the mechanism of the bursts and to
defining the geometry (e.g., their binary character) of the systems.

A world-wide coordinated burst watch was organized in 1977 by the
SAS-3 group. Forty-four observatories participated from 14
different countries. A total of 120 bursts were observed by SAS-3
in 35 days from ten different Type I burst sources. No simultaneous
positive detection was made in either the radio, optical or infrared.
Upper limits were reported for MXB1636-53, MXB1837+05 and MXB1916-05
(Abramenko et al. 1978, Bernacca et al. 1979, Takagishi et al. 1978,
Thomas et al. 1979, Ulmer et al. 1978).

On June 2, 1978 the first optical burst was observed simultaneously
with an X-ray burst from MXB1735-44 (Grindlay et al. 1978). The

TABLE 4

Burst Source				Assoc. Persistent Source				4UCatalog Intensity		Comments	References
Name	α(1950)δ (degrees)	l^{II} b^{II} (degrees)	Pos Acc (deg^2)	Names	(1950) h m s	$^\circ$ ' "	error circle radius (arcsec)	Average or Maximum Ucts	Max/Min		
MXB 0512-40 (in NGC 1851)				2S0512-400 2A0512-399 4U0513-40	05 12 28.7	-40 05 53	20	18	3		33. 35. 40. 60. 61. 102. 104. 133-135. 145. 147. 216
06?? + ??		200 -5	$\sim 8 \times 8$	4U0614 + 09? or 4U0621 + 11? star?	06 21 38 06 14 22.3	11 46 48 09 09 25	3	120 2.1	5	1 burst observed.	40. 44. 46. 60. 133. 134. 145. 196. 220: see Fig. 4
0??? – ??		263 -14	$\sim 24 \times 24$	?						Ten 4U sources in error box. 1 Type I burst observed.	45. 129. 130. 133. 134. 145. 147: see Fig. 4
14??—6?		315 -4	$\sim 24 \times 24$	?						Many 4U sources in error box. 1 Type I burst observed.	4. 135. see Fig. 4
XB 1455-31 (blue star)			~ 1	Cen x-4 star	14 55 19.5	-31 28 07	5			Transient	12. 39. 88. 117. 160. 180. 192. 205. 223. 226
1535-29?				4U1535-29?	15 35 53	-29 13 12		200	>10	Only 1 burst-like event. May not be a Type I burst source.	60
XB 1608-52 (star)				MX1608-52 4U1608-52 2S1608-523 star	16 08 51 16 08 52.2	-52 18 02 -52 17 43	20 ± 0.5	10^3	~ 25	Transient	2. 14-16. 32. 50. 53. 60. 65. 71. 73. 77. 114. 115. 116. 118. 129. 130. 133-135. 138. 151. 169. 179. 206
MXB1636-53 (blue star)				4U1636-53 2S1636-536 star	16 36 56.2	-53 39 15	3	250	2	15 optical bursts observed. 5 of which simultaneously with X-Ray bursts.	24. 60. 82. 90. 93. 96. 132. 133-135. 145. 147. 171. 173. 182. 186. 197. 207. 213. 216. 220. 221
MXB1659-29 (blue star)				4U1704-30? H1658-298 star	16 58 55.5	-29 52 26	5	3.1 80	>15	Transient. Star visible in summer 1978 (X-Ray high state) not in summer 1979 (X-Rays undetectable). Bursts not observed when persistent source in high state.	52. 64. 66. 130. 131. 133-135. 139. 145. 146. 194. 213. 216. 217
XB1702-42				4U1702-42 2S1702-424	17 02 40.4	-42 57 58	30	30	3		60. 101. 130. 133-135. 166. 198
XB1724-30 (in Terzan 2)				4U1722-30 1E 172420 -3045.6	17 24 20	-30 45 36	5	7			60. 67. 68. 70. 133-135. 199
MXB1728-34				4U1728-33 2S1728-337	17 28 39.6	-33 47 52	30	150	5	Some bursts have two peaks at high-energy X-Rays.	18. 26. 62. 63. 91-93. 98. 100. 106. 127. 129. 130. 133-135. 227
MXB1730-335 Rapid Burster (in Liller I)				H1730-333	17 30 07.4	-33 20 34		~ 50 (Type II bursts)	>10	Type I and Type II bursts. Recurrent transient with period of ~ 6 months. IR bursts reported.	3. 9. 18. 21. 26. 49. 72. 85. 86. 96. 97. 99. 103. 106. 121. 127-130. 130-138. 141. 142. 145. 147. 158. 159. 164. 165. 167. 181. 206. 211. 215-219. 221

MXB1735-44 (blue star)						4U1735-44 2S1735-444 star	17 35 19.0	-44 25 19	3	210	1.7	2 optical bursts observed. one of which simultaneously with X-Ray burst. Burst intervals very erratic.	60. 67. 78. 79. 82. 100. 129. 130. 133-135. 141. 145. 147. 150. 170. 173. 213. 216. 217. 221. 225
MXB1742-29	265.7	-29.6	359.5	-0.4	0.34	A1742-294? 2S1742-294? in GCX (4U1743-29)	17 42 53.6	-29 29 50	30	90 40	> 2 5	Association with transient ''persistent'' source uncertain in view of high source density in Gal. Center region	19. 29. 41. 42. 60. 101. 111. 119. 125. 126. 129. 130. 133-135. 140. 144. 145. 147. 190
MXB1743-28	265.9	-28.5	0.5	+0.0	0.28	in GCX (4U1743-29)				40	5	Three bursts observed in quick succession (time scale of ~10 min)	41. 42. 89. 119. 129. 130. 133. 134. 140. 144. 145. 147. 190
MXB1743-29	265.75 either one of two positions 265.58	-29.02 -29.29	359.98 359.67	-0.16 -0.17	0.09 0.09	A1742-289? in GCX (4U1743-29)	17 42 26.4	-28 59 55	25	2×10^3 40	> 200 5	Association with transient ''persistent'' source is uncertain in view of high source density in Gal. Center region	19. 29. 41. 42. 119. 126. 129. 130. 133. 134. 140. 144. 145. 190. 213
MXB1746-37 (in NGC 6441)					~ 1	4U1746-37 2S1746-370	17 46 48.8	-37 02 25	30	40	1.5	Association with persistent source probable: not certain (see Table 2)	35. 60. 80. 102. 133-135. 152
18?? - 2?			4	-4	4	?						No catalogued persistent source in error box. Globular cluster NGC 6553 is in error box	130. 133-135. 145. 147. 198. see Fig. 4
MXB1820-30 (in NGC 6624)						4U1820-30 2S1820-303	18 20 27.7	-30 23 11	20	320	3 >10	Bursts only observed when persistent source in low state	23. 25. 28. 30. 31. 34. 60. 74-76. 102. 105. 106. 129. 130. 133-135. 138. 145. 147. 161. 213. 216. 217. 221
18?? - 2?			11	-8	~8 × 8	4U1831-23	18 31 47	-23 12 18		6		4U1831-23 is the only known persistent source in the large error box of burst source. Association uncertain. 4 Globular clusters in error box.	11. 129. 130. 133-135. 145. 147. see Fig. 4
MXB1837 + 05 (blue star)						4U1837 + 04 Ser X-1 2S1837 + 049 star	18 37 29.6	04 59 21	~ 3	280	2	Blue star ($m_B \sim$ 19.2) within 2" of ~18.5 magn. star. Burst intervals very erratic.	1. 17. 43. 47. 48. 60. 84. 106. 129. 130. 133-135. 138. 145. 147. 155. 156. 163. 200. 204. 208. 210. 212. 213
MXB1850-08 (in NGC 6712)					< 0.3	A1850-08 4U1850-08 2S1850-087	18 50 21.9	-08 45 54	~ 30	9		Association almost certain (see Table 2)	36. 40. 48. 60. 102. 129. 130. 133-135. 147. 153. 188. 198
MXB1906 + 00						4U1857 + 01 A1905 + 00 2S1905 + 000	19 05 54.9	00 05 37	35	4.1 ± 1			48. 60. 106. 129. 130. 133-135. 138. 145. 147. 148. 155. 204. 213. 216
Aql MXB	287.6 either one of two positions 286.7	+3.5 +1.1			1.4 1.4	?						No known persistent source in error box of burst source. Aql X-1 is excluded.	129. 130. 133-135. 138. 145. 147. 148
MXB1916-05						4U1915-05 2S1916-053	19 16 08.5	-05 19 51	20	20	2		1. 10. 48. 60. 129. 130. 133-135. 143. 145. 147. 198. 204. 213. 216
2??? + 3?			79	-9	~ 50	?						4U 2058 + 32. 2104 + 31. and 2120 + 32 (all very weak) in large error region of burst source	129. 130. 133-135. 145. 147. see Fig. 4

340

fast rise and the small ($\sim$ 2.8 sec) delay between the X-ray and
optical bursts (McClintock et al., 1979) indicate that the optical
region must be within $\sim$ 3 lightseconds of the X-ray emitting region
(Figure 11). Recent optical studies of burst sources by Canizares,
McClintock and Grindlay (1979) indicate strongly that the persistent
optical emission comes from an X-ray heated accretion disk rather
than from a stellar companion (this does not exclude the presence of
a companion).

In the case of the simultaneously observed X-ray and optical burst
from MXB1735-44 a comparison of the steady fluxes with the fluxes at
burst maximum (Grindlay et al., 1978; McClintock et al., 1979), showed
that the X-ray signal increased by a factor 6.3 ± 1 and the optical
signal by a factor of 1.5 ± 0.3. Calculations of the effects of
X-ray heating predict that the factor by which the optical flux
increases (at burst maximum) should equal that of the X-ray increase
raised to a power β whose theoretical value for X-ray heating of a hot
black body is $\sim$ 0.25-1 (Joss and Rappaport, 1979; Milgrom, 1976;
Milgrom and Salpeter, 1975). This is consistent with the observed
value of 0.22 ± 0.1. Thus the observed value for β is consistent
with the idea that most (perhaps all) of the optical emission (both
burst and steady) comes from an accretion disk rather than a companion star.

The fact that no periodic variability is observed in the optical
emission is perhaps not surprising. If the X-ray source is a binary
system, the companion star could be so small (Joss and Rappaport,
1979; Milgrom, 1978) that its contribution to the visible light is
negligible. (The light flux due to its own nuclear burning would be
small and the light due to X-ray heating could be negligible if the
star is located in the X-ray shadow of the disk (Milgrom, 1978).)
Canizares, McClintock and Grindlay (1979) concluded from their
optical studies that a possible companion, if present, could only be
a main sequence star later than FO.

A possible optical burst was detected simultaneously with an X-ray
burst from MXB1837+05 (Hackwell et al., 1979). The optical burst was
delayed by 1.4 ± 0.5 sec relative to the X-ray burst; this delay
was $\sim$ 2.8 sec for MXB1735-44 (McClintock et al., 1979).

Recently, in June/July 1979, Pedersen et al., (1979) observed 15
optical bursts from MXB1636-53. Five of these occurred simultaneously
with X-ray bursts detected by the Japanese satellite X-ray observatory
(Hakucho). The largest increase in the stellar brightness was a
factor of $\sim$ 4, and the duration of the tail of the longest optical
burst was $\sim$ 2 minutes. In the cases that have so far been analyzed
in some detail, the optical bursts are delayed by $\sim$ 3.2 seconds
(Pedersen et al., 1981; ref. 231).

Table 5 lists the ratios of optical (B range; $\Delta\lambda$ = 1000 $\text{\AA}$) and X-ray
energy fluxes ($\sim$ 2-15 keV) observed for the persistent emission and
for the bursts. Notice that for the persistent source the value for
MXB1837+05 is 5.9 ± 0.6 times lower than for MXB1735-44; for the

bursts, the value for MXB1837+05 is 7 ± 3 times lower. These data
are therefore consistent with the idea that the two objects are very
similar but that the optical extintion for MXB1837+05 is a factor of
~ 6 times higher than for MXB1735-44.

Another world-wide burst watch is scheduled for April-August 1980.
The X-ray observatories Hakucho (Japanese for Swan) will again
participate.

TABLE 5

OBSERVED RATIOS OF ENERGY DENSITIES*

(OPTICAL/X-RAY FOR PERSISTENT RADIATION AND BURSTS)

	$\left[\dfrac{E_{opt}}{E_x}\right]_{\text{persistent}}$	$\left[\dfrac{E_{opt}}{E_x}\right]^{**}_{\text{burst}}$	References
MXB1735-44	$(1.3 \pm 0.1) \times 10^{-4}$	$(2.0 \pm 0.6) \times 10^{-5}$	Grindlay et al. 1978, McClintock et al. 1977, McClintock et al. 1979.
MXB1837+05	$(2.2 \pm 0.2) \times 10^{-5}$	$(2.8 \pm 0.9) \times 10^{-6}$	Hackwell et al. 1979.

* No correction for interstellar extinction was made.

** Integrated for ~ 10 sec.

ARE ALL GALACTIC BULGE SOURCES NEUTRON STARS?

Now that we have concluded that ~ 30 burst sources (of which at
least five lie in globular clusters) are probably neutron stars, can
we extend this conclusion to the non-bursting Galactic Bulge Sources?
There are important differences within this large class:

- A large fraction of the bulge sources do not burst.

- The persistent emission of the globular cluster X-ray
  sources seems to vary more than that of most bulge sources
  (Li, Clark and Markert, 1978).

- The most "reliable" burst sources are not located in globular
  clusters (i.e., if one wants to optimize the chance to
  detect bursts in a short observing period of days to weeks,
  one would not select any of the burst sources in globular
  clusters). The "reliable" sources (though with variable

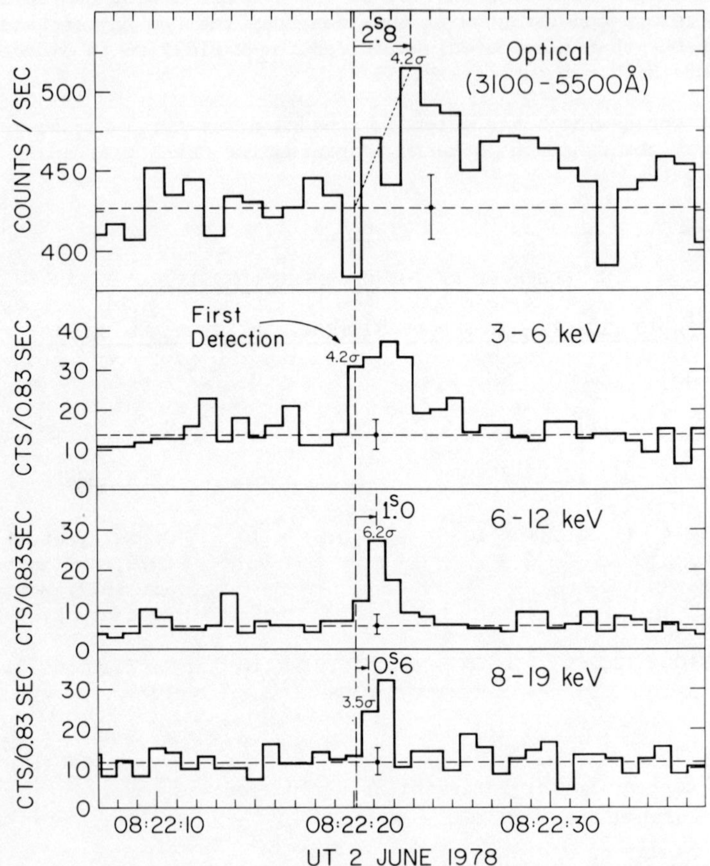

Figure 11: X-ray/optical burst from the X-ray burst source
MXB1735-44 as detected on June 2, 1978. The event was
detected first in 3-6 keV X-rays, as marked by the
vertical dashed line, and ∿ 2.8 sec later in blue light.
The background counting rates and their approximate
($\sqrt{N}$) uncertainties are indicated by horizontal dashed
lines and error bars respectively. The exact Poisson
statistical significance of the first bin to exceed a 3σ
detection threshold is also shown. The dotted line
shows that the optical data are equally consistent with
a 3 sec linear rise to maximum from the time the burst
was first detected or with a 3 sec delay followed by an
abrupt rise to maximum. A tail is apparent in the
optical data. This figure is from Mclintock et al.
(1979). *(Reprinted by permission from Nature, Vol. 270,
No. 5708, p47, Copyright (c) 1979, Macmillan Journals
Limited)*

burst intervals) are MXB1728-34, MCB1735-44, MXB1636-53 and
MXB1837+05. They are not located in globular clusters.
Incidentally, the latter three have been identified with
faint blue stars (see Table 4).

There is probably a physical connection between the higher variability of the persistent emission of the globular cluster X-ray sources
and their "unreliability" as bursters. This conclusion is supported
by the fact that the highly variable (transient) X-ray sources
MXB1659-29, XB1608-52 and Cen X-4 which are not located in globular
clusters, are also unreliable bursters. They only burst when the
persistent X-ray luminosity is within certain limits (Fabbiano and
Branduardi, 1979; Grindlay and Gursky, 1976; Lewin et al., 1978;
Oda, 1979 and 1979a; Tanaka and the Hakucho Team, 1979; Tenanbaum
et al., 1976).

Joss' thermonuclear flash calculations (Joss 1978) predict that
sources with very high luminosities do not burst (see also van
Paradijs, 1979). Thus, one would not expect to detect bursts from
the well known very bright sources such as GX5-1 (4U1758-25), GX3+1
(4U1744-26), GX9+1 (4U1758-20), GX13+1 (4U1811-17), and GX17+2
(4U1813-14) unless their X-ray luminosity would decrease.

It is likely that the formation and evolution of the bulge and
cluster systems are very different. There may well be systematic
differences in the magnetic fields would could influence the burst
behaviour (Joss and Li, 1980).

In spite of the above differences we believe that most of the non-
bursting bulge sources are neutron star systems similar to burst
sources.

FAST TRANSIENT (BURSTS?) WITH PRECURSORS

SAS-3 recorded several exceptional transient phenomena which may have
been very long X-ray bursts. Two were discussed by Hoffman et al.,
1978. We show their profiles on different time scales in figures
12-14. They resemble X-ray bursts in several respects. The
February 1977 event, which lasted $\sim$ 1500 sec, has a spectrum that
can be reasonably well represented by that of a cooling black body.
The maximum temperature is $\sim$ 29 x 10^6 K (as in the case of Type I
bursts) (Fig. 15). The calculated radius for a distance of 10 kpc
is $\sim$ 9 km (assuming that the source was near the centre of the SAS-3
Centre slat collimator in which the event was observed).

Both events are different from Type I bursts in that they have a
precursor and their duration is long (1500 sec in one case). It
will be interesting to see whether these phenomena can also be
interpreted in terms of a thermonuclear flash model. Extended SAS-3
observations (lasting for weeks), covering the large regions in the
sky from which these events were see, produced only these two.
Thus it seems that if these "bursts" are repetitive, the intervals
are quite long. This may be a clue to why they are so different.

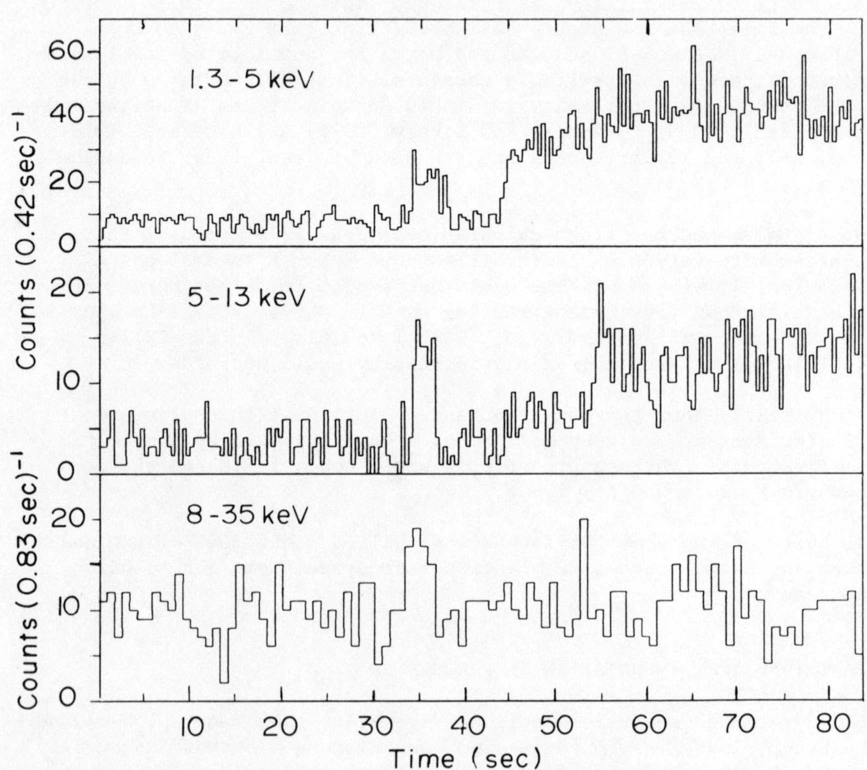

Figure 12: Light curves in three energy channels of the 1977
February 7 event as observed by the SAS-3 Centre Slat
detectors. The precursor is clearly visible and shows
no obvious spectral changes in its evolution. After
the precursor the signal drops to the pre-precursor
level before the "main" event begins. The delay between
the precursor and the rise of the "main" event increases
with energy and is $\sim$ 80 sec for the 8-35 keV channel.
Thus in this energy channel it rises near t $\sim$ 115 sec
(off scale): the rise is shown in Figure 13. The
"main" event shows significant spectral softening
during its decay (see figs. 13 and 15). This figure is
a modified version of one shown by Hoffman et al. 1978.

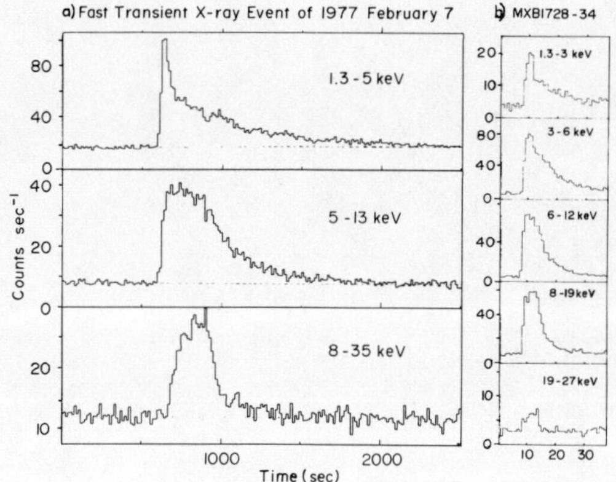

Figure 13: a) Light curves of the complete February 1977 event. Notice the distinct spectral softening during the decay of the "main" event. The precursor is not visible here because of the lumping of the data in coarse time bins.

b) A composite X-ray burst from MXB1728-34 as observed in five energy channels with SAS-3. The resemblance between the profiles of the events in a) and b) is striking. This figure is from Hoffman et al., 1978.

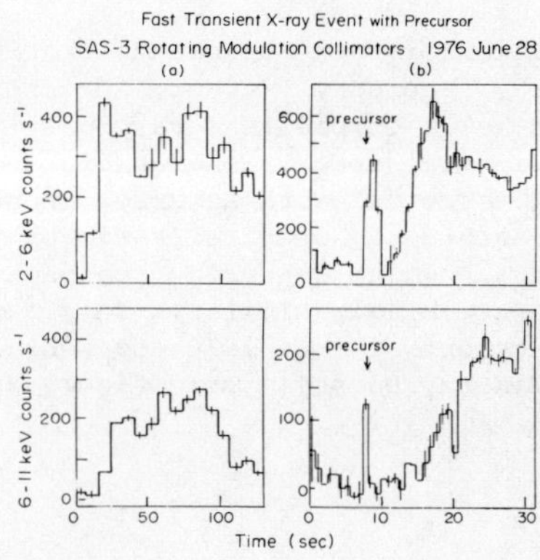

Figure 14: a) Light curves (2 energy channels) of the June 28, 1976 event observed by the SAS-3 Modulation Collimators; the data have been deconvolved for the collimator transmission.

b) The onset shown at an expanded time scale. A precursor is clearly visible (no spectral evolution). The "main" event rises later at high energies just as in the February 1977 event (see figs. 12 and 13). The 1σ error bars shown include counting statistics and the uncertainty in the deconvolution of the collimator transmission. This figure is from Hoffman et al. 1978.

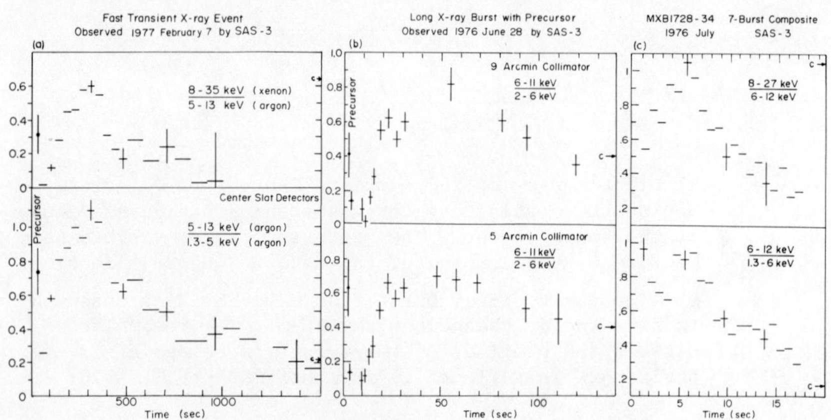

Figure 15: The evolution of spectral channel ratios with time for
(a) the 1977 February 7 event, (b) the 1976 June 28
event and (c) a composite X-ray burst from MXB1728-34.
All show initial spectral hardening and subsequent
softening. The "c" with an arrow (bottom right) in each
case represents the spectral ratio of the Crab Nebula.
The spectral evolution during the decay of these three
events is strikingly similar. This suggests that the
peculiar events of June 1976 and February 1977 may also
be Type I X-ray bursts. This figure is from Hoffman et
al., 1978.

WHY BINARIES?

We have shown above why we believe that most of the bulge sources,
and all the Type I burst sources, are neutron stars. In the absence
of eclipses, are there any reasons to believe that these neutron
stars are in binary systems?

Neutron stars could be adequately powered by accretion from the interstellar medium to produce the persistent luminosity from the burst
sources (Ostriker et al., 1970) only if they were in dense ($n > 10^6$
cm^{-3}) clouds. This seems very unlikely, however, since no intrinsic
low-energy absorption is observed in the X-ray spectra of the persistent emission from these sources. Thus the gas supply must be

nearby. The objects might be single neutron stars with a fossil
accretion disk (Paczynski and Jaroszynski, 1978). If this were so it
is difficult to see why they occur preferentially in the cores of
centrally condensed globular clusters. We therefore believe it is
most likely that the neutron stars are in binary systems (of low-mass).

ACKNOWLEDGEMENTS

The authors are grateful to L. Cominsky, P.C. Joss, D.Q. Lamb and
J. van Paradijs for their valuable comments on this manuscript and
they thank E. Basinska and S. Black for their assistance in preparing
this manuscript.

REFERENCES

1. Abramenko, A.N., Gershberg, R.E., Pavlenko, E.P., Prokof'eva,
    V.V., Lewin, W.H.G., Van Paradijs, J., Hoffman, J.A., and Li,
    F.K., 1978. MNRAS, 184, 27p.
2. Apparao, K.M.V., Bradt, H.V., Dower, R.G., Doxsey, R.E., Jernigan,
    J.G., and Li, F.K., 1978. Nature, 271, 225.
3. Apparao, K.M.V., Chitre, S.M., Ashok, N.M., Kulkarni, P.V., 1979.
    IAU Circ. 3344.
4. Apparao, K., and Narranan, 1978. Private communication, SAS-3
    data.
5. Bahcall, J.N., and Ostriker, J.P., 1975. Nature, 256, 23.
6. Bahcall, J.N., and Wolf, R.A., 1976. Ap. J., 109, 214.
7. Bahcall, N.E., and Hausman, M.A., 1977. Ap. J., 213, 93.
8. Bahcall, N.E., Lasker, B.M., and Wamsteker, W., 1977. Ap. J.
    (Letters), 213, L105.
9. Basinska, E., Lewin, W.H.G., Cominsky, L., Marshall, F., and
    Van Paradijs, J., 1980. Ap. J., 241, 787.

10. Becker, R.G., Smith, B.W., Swank, J.H., 1977. Ap. J. (Letters),
    216, L101.
11. Becker, R.G., Pravdo, S.H., Serlemitsos, P.J., and Swank, J.H.,
    1976. IAU Circ. 2953.
12. Belian, R.D., Conner, J.P., and Evans, W.D., 1972. Ap. J.
    (Letters), 171, L87.
13. Belian, R.D., Conner, J.P., and Evans, W.D., 1976. IAU Circ.
    2969.
14. Belian, R.D., Conner, J.P., and Evans, W.D., 1976a. Bull. Am.
    Astr. Soc., 8, No. 2, p.396.
15. Belian, R.D., Conner, J.P., and Evans, W.D., 1976b. Astrophys.
    J. Lett., 206, L135.
16. Belian, R.D., Conner, J.P., and Evans, W.D., 1976c. Astrophys.
    J. Lett., 207, L33.
17. Bernacca, P.L., Bianchini, A., Walker, A., Backman, D., Canizares,
    C., Van Paradijs, J., Hoffman, J.A., Doty, J., Marshall, H.,
    Wheaton, W., Jernigan, J.G., and Lewin, W.H.G., 1979. MNRAS,
    186, 287.
18. Birmingham Group (Ariel V)., 1976. IAU Circ. 2929.

19. Birmingham Group (Ariel V), 1976. IAU Circ. 2934.
20. Bradt, H., Doxsey, R.E., and Jernigan, J.G., 1979. Advances in Space Exploration, volume 3, IAU/COSPAR Symposium on X-ray Astronomy, Innsbruck, Austria, May 1978.
21. Calla, O.P.N., Bahandari, S.M., Deshpande, M.R., and Vats Hari, D.M., 1979. IAU Circ. 3347.
22. Canizares, C.R., 1975. Ap. J.(Letters), 201, 589.
23. Canizares, C.R., Grindlay, J.E., Hiltner, W.A., Liller, W., and McClintock, J.E., 1978. Ap. J., 224, 39.
24. Canizares, C., McClintock, J.E., and Grindlay, J., 1979. Astrophysics J., 234, 556.
25. Canizares, C.R., and Neighbours, J.E., 1975. Ap. J. (Letters), 199, L97.
26. Carpenter, G.F., Skinner, G.K., Wilson, A.M., and Willmore, A.P., 1976. Nature, 262, 473.
27. Clark, G.W., 1975. Ap. J. (Letters), 199, L143.
28 Clark, G.W., 1976. IAU Circ. 2907.
29. Clark, G.W., 1976. IAU Circ. 2922.
30. Clark, G.W., 1976. IAU Circ. 2932,
31. Clark, G.W., Jernigan, G., Bradt, H., Canizares, C., Lewin, W.H.G., Li, F.K., Mayer, W., McClintock, J.E., 1976. Ap. J. (Letters), 207, L105.
32. Clark, G.W., and Li, F.K., 1977. IAU Circ. 3090.
33. Clark, G.W., and Li, F.K., 1977. IAU Circ. 3092.
34. Clark, G.W., Li, F.K., Canizares, C.R., Hayakawa, S., Jernigan, G., and Lewin, W.H.G., 1977. MNRAS, 179, 651.
35. Clark, G.W., Markert, T.H., and Li, F.K., 1975. Ap. J. (Letters), 199, L93.
36. Cominsky, L., Forman, W., Jones, C., and Tananbaum, H., 1977. Ap. J., (Letters), 211, L9.
37. Cominsky, L., Lewin, W.H.G., Ossman, W., Marshall, H., and van Paradijs, J., 1980. Manuscript in preparation.
38. Cominsky, L., Jernigan, J.G., Ossman, W., Van Paradijs, J., Lewin, W.H.G., 1980a. Ap. J., Dec. 15.

39. Conner, J.P., Evans, W.D., and Belian, R.D., 1969. Ap. J. (Letters), 157, L157.
40. Cooke, B.A., Ricketts, M.J., Maccacaro, T., Pye, J.P., Elvis, M., Watson, M.G., Griffiths, R.E., Pounds, K.A., McHardy, J., Seward, F.D., Page, C.G., and Turner, M.J.L., 1978. MNRAS, 182, 489. (2A catalogue).
41. Cordova, F.A., Garmire, G.P., and Lewin, W.H.G., 1979. Nature, 278, 529.
42. Cruddace, R.G., Fritz, G., Shulman, S., Friedman, H., Margon, B., Mason, K., Hawkins, F., and Sanford, P., 1978. Ap. J. (Letters), 222, L95.
43. Davidson, A., 1975. IAU Circ. 2824.
44. Davidson, A., Malina, R., Smith, H., Spinrad, H., Margon, B., Mason, K.O., Hawkins, F., and Sanford, P., 1974. Ap. J. (Letters), 193, L25.
45. Doty, J., 1976. IAU Circ. 2922.
46. Dower, R.G., Apparao, K.M.V., Bradt, H.V., Doxsey, R.E., et al., 1978. Nature, 273, 364.
47. Doxsey, R.E., 1975. IAU Circ. 2820.

48. Doxsey, R.E., Apparao, K.M.V., Bradt, H., Dower, R., et al., 1977. Nature, 269, 112.

49. Doxsey, R.E., Bradt, H., Gursky, H., Johnston, M., Schwartz, D. A., Schwarz, J., 1978. Ap. J. (Letters), 221, L53.

50. Doxsey, R., Clark, G.W., and Li, F., 1977. IAU Circ. 3094.

51. Doxsey, R.E., Grindlay, J.E., Griffiths, R.E., Bradt, H., Johnston, M., Leach, R., Schwartz, D., and Schwarz, J., 1979. Ap. J. (Letters), 228, L67.

52. Doxsey, R.E., Grindlay, J., Griffiths, R., Bradt, H., Johnston, M., Leach, R., Schwartz, D., and Schwarz, J., 1978a. Ap. J. (Letters), 228, L67.

53. Duldig, M., Greenhill, J., Thomas, R., Haynes, R., and Simons, L., 1977. IAU Circ. 3108.

54. Evans, W.D., Belian, R.D., Conner, J.P., 1976. Ap. J. (Letters), 207, L91.

55. Fabbiano, G., and Branduardi, G., 1979. Ap. J., 227, 294.

56. Fabian, A.C., Pringle, J.E., and Rees, M.J., 1975. MNRAS, 172, 15P.

57. Fall, S.M., and Malkan, M.A., 1978. MNRAS, 185, 899.

58. Flowers, E., and Ruderman, M., 1977. Ap. J., 215, 302.

59. Forman, W., and Jones, C., 1976. Ap. J. (Letters), 207, L177.

60. Forman, W., Jones, C., Cominsky, L., Julien, P., Murray, S., Peters, G., Tananbaum, H., and Giacconi, R., 1978. Ap. J. (Suppl.), 38, No. 4, (4U Catalogue).

61. Forman, W., Jones, C., and Tananbaum, H., 1976. Ap. J., 208, 849.

62. Friedman, H., 1978. Talk presented at HEAD/AAS Meeting, San Diego, California.

63. Glass, I.S., 1978. Nature, 273, 35.

64. Griffiths, R., Johnston, M., Bradt, H., Doxsey, R., Gursky, H., Schwartz, D., and Schwarz, J., 1978. IAU Circ. 3190.

65. Grindlay, J., 1977. IAU Circ. 3101.

66. Grindlay, J., 1978. IAU Circ. 3229.

67. Grindlay, J.E., 1978a. Ap. J. (Letters), 224, L107.

68. Grindlay, J.E., 1979. X-ray Symposium in Cape Sounion, Results from the Einstein Observatory, (these proceedings).

69. Grindlay, J., Hertz, P., Branduardi, G., Lightman, A., and Murray, S., 1979. To be submitted to Ap. J.

70. Grindlay, J., Marshall, H., Hertz, P., Soltan, A., Weisskopf, M., Elsner, R., Ghosh, P., Darbro, W., and Sutherland, P., 1980. Ap. J., 240, L121.

71. Grindlay, J.E., and Gursky, H., 1976. Ap. J. (Letters), 209, L61.

72. Grindlay, J.E., and Gursky, H., 1977. Ap. J. (Letters), 218, L117.

73. Grindlay, J., and Gursky, H., 1976a. IAU Circ. 2932.

74. Grindlay, J., Gursky, H., Schnopper, H., Parisgnault, D.R., Heise, J., Brinkman, A.C., and Schrijver, J., 1976. Ap. J. (Letters), 205, L127.

75. Grindlay, J., and Heise, J., 1976. IAU Circ. 2879.

76. Grindlay, J.E., and Liller, W., 1977. Ap. J. (Letters), 216, L105.

77. Grindlay, J.E., and Liller, W., 1978. Ap. J. (Letters), 220, L127.

78. Grindlay, J.E., McClintock, J.E., Canizares, C.R., van Paradijs, J., Cominsky, L., Li, F.K., and Lewin, W.H.G., 1978. Nature, 274, 567.

79. Grindlay, J., McClintock, J., Canizares, C., and van Paradijs, J., 1978. IAU Circ. 3230.

80. Grindlay, J.E., Gursky, H., Schnopper, H., Parsignault, D.R., Heise, J., Brinkman, A.C., and Schrijver, J., 1976. Ap. J. (Letters), 205, L127.

81. Gunn, J.E., and Ostriker, J.P., 1970. Ap. J., 160, 979.

82. Gursky, H., Bradt, H., Schwartz, D.A., Doxsey, R., Schwarz, J., Dower, R., Fabbiano, G., Griffiths, R.E., Johnston, M., Leach, R., Ramsey, A., and Spada, G., 1978. Ap. J., 223, 973.

83. Hackwell, J.A., Gehrz, R.D., Grasdalen, G.L., van Paradijs, J., Cominsky, L., and Lewin, W.H.G., 1979. Ap. J., 323, L115.

84. Hackwell, J.A., Gahrz, R.D., Grasdalen, G.L., Cominsky, L., van Paradijs, J., and Lewin, W.H.G., 1979. IAU Circ. 3331.

85. Hearn, D., 1976. IAU Circ. 2925.

86. Heise, J., and Grindlay, J., 1976. IAU Circ. 2929.

87. Hills, J.G., 1975. Ap. J., 80, 1075.

88. Hjellming, R.M., 1979. IAU Circ. 3369.

89. Hoffman, J., 1976. IAU Circ. 2946.

90. Hoffman, J., Doty, J., and Lewin, W.H.G., 1977. IAU Circ. 3025.

91. Hoffman, J., Lewin, W.H.G., Doty, J., Hearn, D.R., Clark, G.W., Jernigan, G., and Li, F.K., 1976. Ap. J. (Letters), 210, L13.

92. Hoffman, J.A., Lewin, W.H.G., and Doty, J., 1977. MNRAS, 179, 57P.

93. Hoffman, J.A., Lewin, W.H.G., and Doty, J., 1977a. Ap. J. (Letters), 217, L23.

94. Hoffman, J.A., Lewin, W.H.G., Doty, J., Jernigan, J.G., Haney, M., and Richardson, J.A., 1978. Ap. J. (Letters), 221, L57.

95. Hoffman, J.A., Lewin, W.H.G., Doty, J., Hearn, D.R., and Clark, G.W., 1976. Ap. J. (Letters), 210, L13.

96. Hoffman, J.A., Marshall, H.L., and Lewin, W.H.G., 1978a. Nature, 271, 630.

97. Hoffman, J.A., Marshall, H., and Lewin, W.H.G., 1977b. IAU Circ. 3117.

98. Hoffman, J., Lewin, W.H.G., Marshall, H., Primini, F., Wheaton, W., and Cominsky, L., 1978a. IAU Circ. 3190.

99. Inoue, H., Kayama, K., Makishima, K., Matsuoka, M., Murakami, T., Oda, M., Ogawara, Y., Ohashi, T., Shibazaki, M., Tanaka, Y., Tawara, Y., Kondo, I., Hayakawa, S., Kunieda, H., Makino, F., Masai, K., Miyamoto, S., Tsunemi, H., and Yamashita, K., 1980. Nature, 283, 358.

100. Jernigan, J.G., Apparao, K.M.V., Bradt, H.V., and Doxsey, R.E., 1977. Nature, 270, 321.

101. Jernigan, J.G., Apparao, K.M.V., Bradt, H.V., and Doxsey, R.E., 1977a. Nature, 272, 701.

102. Jernigan, J.G., and Clark, G.W., 1979. Ap. J. (Letters), 231, L125.

103. Jernigan, J.G., McClintock, J., Marshall, H., Chartres, M., Hoffman, J., and Lewin, W.H.G., 1978. IAU Circ. 3204.

104. Jones, C., and Forman, W., 1976. IAU Circ. 2913.

105. Johnson, H.M., 1976. Ap. J., 208, 706.
106. Johnson, H.M., Çatura, R.C., Lamb, P.A., White, N.E., Sanford, P.W., Hoffman, J.A., Lewin, W.H.G., and Jernigan, J.G., 1978. Ap. J., 222, 664.
107. Joss, P.C., private communication.
108. Joss, P.C., 1979. Nature, 270, 310.
109. Joss, P.C., 1978. Ap. J. (Letters), 225, L123.
110. Joss, P.C., 1979a. Ann. N.Y. Academy of Sciences. Proceedings of Ninth Texas Symposium, Munich, Germany.
111. Joss. P.C., and Li, F.K., 1980. Ap. J., 238, 287.

112. Joss, P.C., and Rappaport, S.A., 1979. Astron. and Astrophys., 71, 217.
113. Joss, P.C., Ricker, G.R., Mayer, W., and Hoffman, J., 1977. IAU Circ. 3108.
114. Kaluzienski, L., and Holt, S., 1977. IAU Circ. 3108.
115. Kaluzienski, L.J., and Holt, S., Boldt, E.A., and Serlemitsos, P.J., 1975. IAU Circ. 2859.
116. Kaluzienski, L., and Holt, S., 1977. IAU Circ. 3099, 3108, 3129.
117. Kaluzienski, L., and Holt, S., 1979. IAU Circ. 3360, 3362.
118. Kaluzienski, L., and Holt, S., 1979a. IAU Circ. 3349.
119. Kellogg, E., Gursky, H., Murray, S., Tananbaum, H., and Giacconi, R., 1971. Ap. J. (Letters), 169, L99.
120. King, I.R., 1966. Astron. J., 71, 64.
121. Kleinmann, D.E., Kleinmann, S.G., Wright, E.L., 1976. Ap. J. (Letters), 210, L83.
122. Kulkarni, P.V., Ashok, N.M., Apparao, K.M.V., and Chitre, S.M., 1979. Nature, 280, 819.
123. Kylafis, N.D., and Lamb, D.Q., 1979. Ap. J. (Letters), 218, L105.
124. Lamb, D.Q., and Lamb, F.K., 1978. Ap. J., 220, 291.
125. Lewin, W.H.G., 1976. IAU Circ. 2911.
126. Lewin, W.H.G., 1976a. IAU Circ. 2918.
127. Lewin, W.H.G., 1976b. IAU Circ. 2922.
128. Lewin, W.H.G., 1977. American Scientist, 65, No. 5, p. 605.
129. Lewin, W.H.G., 1977a. MNRAS, 179, 43.
130. Lewin, W.H.G., 1977b. Annals. N.Y. Acad. Sci., 302, 210.
131. Lewin, W.H.G., 1978. IAU Circ. 3190, 3193.
132. Lewin, W.H.G., 1979. Talk at Cosmic Ray Conference, Kyoto, Japan.
133. Lewin, W.H.G., 1979a. Advances in Space Exploration, 3, IAU/COSPAR Symposium on X-ray Astronomy in Innsbruck, Austria, May 1978, eds. W.A. Baity and L.E. Petersen, Pergammon Press, Oxford: Vol. 3, P.133.
134. Lewin, W.H.G., 1980. In "Globular Clusters", eds. D. Hanes and B. Madore, Cambridge University Press. Proceedings of Globular Cluster Meeting held in Cambridge, England, August 1978.
135. Lewin, W.H.G., and Clark, G.W., 1980. Proceedings of the Ninth Texas Symposium, Munich, Germany. Annals N.Y. Acad. Sci., 336, 451.
136. Lewin, W.H.H., Cominsky, L., and Van Paradijs, J., 1978. IAU Circ. 3308.
137. Lewin, W.H.G., Doty, J., Clark, G.W., Rappaport, S.A., Bradt, H., Doxsey, R., Hearn, D.R., Hoffman, J., Jernigan, J.G., Li, F.K., Mayer, W., McClintock, J., Primini, F., and Richardson, J., 1976. Ap. J. (Letters), 207, L95.
138. Lewin, W.H.G., Doty, J., Hoffman, J.A., and Li, F.K., 1976a. IAU Circ. 2984.

352

139. Lewin, W.H.G., Hoffman, J.A., and Doty, J., 1976b. IAU Circ.
     2994.
140. Lewin, W.H.G., Hoffman, J.A., and Doty, J., 1977. IAU Circ.
     3039.
141. Lewin, W.H.G., Hoffman, J.A., Doty, J., Li, F.K., and McClintock,
     J.E., 1977a. IAU Circ. 3075.
142. Lewin, W.H.G., and Hoffman, J.A., 1977. IAU Circ. 3079.
143. Lewin, W.H.G., Hoffman, J.A., and Doty, J., 1977b. IAU Circ.
     3087.
144. Lewin, W.H.G., Hoffman, J.A., Doty, J., Hearn, D.R., Clark,
     G.W., Jernigan, J.G., Li, F.K., McClintock, J.E., and Richard-
     son, J., 1976c. MNRAS, 177, 83P.
145. Lewin, W.H.G., Hoffman, J.A., Doty, J., Clark, G.W., Swank, J.H.,
     Becker, R.H., Pravdo, S.H., and Serlemitsos, P.J., 1977c. Nature,
     267, 28.
146. Lewin, W.H.G., Hoffman, J.A., Marshall, H., Primini, F., Wheaton,
     W., Cominsky, L., Jernigan, J.G., and Ossman, W., 1978. IAU
     Circ. 3190.
147. Lewin, W.H.G., and Joss, P.C., 1977. Nature, 270, 211.
148. Lewin, W.H.G., Li, F.K., Hoffman, J.A., Doty, J., Buff, J.,
     Clark, G.W., and Rappaport, S.A., 1976d. MNRAS, 177, 93P.
149. Lewin, W.H.G., Marshall, H., and Cominsky, L., 1978a. IAU Circ.
     3211.
150. Lewin, W.H.G., Van Paradijs, J., Cominsky, L., and Holzner, S.,
     1980. MNRAS, 193, 15.
151. Li, F., 1976. IAU Circ. 2936.
152. Li, F.K., and Clark, G.W., 1977. IAU Circ. 3095.
153. Li, F.K., and Clark, G.W., 1980. Manuscript in preparation.
154. Li, F.K., Clark, G.W., and Markert, T., 1978. Nature, 275, 723.
155. Li, F., and Lewin, W.H.G., 1976. IAU Circ. 2983.
156. Li, F.K., Lewin, W.H.G., Clark, G.W., Doty, J., Hoffman, J.A.,
     and Rappaport, S.A., 1977. MNRAS, 179, 21P.
157. Lightman, A.P., Press, W.H., and Odenwald, S.F., 1978. Ap. J.,
     219, 629.
158. Liller, W., 1976. IAU Circ. 2929.
159. Liller, W., 1977. Ap. J. (Letters), 213, L21.
160. Liller, W., 1979. IAU Circ. 3366.
161. Liller, M.H., and Carney, B.W., 1978. Ap. J., 224, 383.
162. Maraschi, L., and Cavaliere, A., 1977. Highlights in Astronomy,
     4, Part I.
163. Margon, B., and Whitter, K., 1978. IAU Circ. 3246.
164. Marshall, F., 1979. IAU Circ. 3336.
165. Marshall, H., and Lewin, W.H.G., 1978. IAU Circ. 3208.
166. Marshall, H., Li, F., Rappaport, S., 1977. IAU Circ. 3134.
167. Marshall, H.L., Ulmer, M., Hoffman, J.A., Doty, J., and Lewin,
     W.H.G., 1979. Ap. J., 227, 555.
168. Markert, T., Backman, D.E., Canizares, C., Clark, G.W., and
     Levine, A.M., 1975. Nature, 257, 32.
169. Markert, T., Canizares, C., and Clark, G.W., Hearn, D.R., Li,
     F.K., Sprott, G.F., and Winkler, P.F., 1977. Ap. J., 218, 801.
170. McClintock, J.E., 1977. IAU Circ. 3084.
171. McClintock, J.E., 1977. IAU Circ. 3088.
172. McClintock, J.E., Canizares, C., and Backman, D.E., 1978.
     Ap. J. (Letters), 223, L75.

173. McClintock, J.E., Canizares, C.R., Bradt, H.V., Doxsey, R.E., 1977. Nature, 270, 320.

174. McClintock, J.E., Grindlay, J.E., Canizares, C.R., Van Paradijs, J., Cominsky, L., Li, F.K., and Lewin, W.H.G., 1979. Nature, 279, 47.

175. Milgrom, M., 1976. Astron. and Astrophys., 50, 273.

176. Milgrom, M., 1978. Astron. and Astrophys., 67, L25.

177. Milgrom, M., and Salpeter, E.E., 1975. Ap. J., 196, 583.

178. Murdin, P., Penston, M.J., Penston, M.V., Glass, I.S., Sanford, P.W., Hawkins, F.J., Mason, K.O., and Willmore, A.P., 1974. MNRAS, 169, 25.

179. Oda, M., 1979. IAU Circ. 3349.

180. Oda, M., 1979a. IAU Circ. 3366.

181. Oda, M., 1979b. IAU Circ. 3392.

182. Oda, M., 1979c. Talk presented at X-ray Astronomy Meeting, Tokyo, Japan.

183. Ostriker, J.P., Rees, M.J., and Silk, J., 1970. Ap. J. (Letters), 6, 179.

184. Ostriker, J.P., Spitzer, L., and Chevalier, R.A., 1972. Ap. J., (Letter), 176, L51.

185. Paczynski, B., and Jaroszynski, M., 1978. Acta Astron. 28, 111.

186. Pedersen, H., Oda, M., Hakucho Team, Cominsky, L., Doty, J., Jernigan, J.G., Van Paradijs, J., and Lewin, W.H.G., 1979. IAU Circ. 3399.

187. Peterson, C.J., 1976. Astron. J., 81, 617.

188. Peterson, C.J., and King, I.R., 1975. Ap. J., 80, 427.

189. Press, W.H., and Teukolsky, S.A., 1977. Ap. J., 213, 183.

190. Proctor, R.J., Skinner, G.K., Willmore, A.P., 1978. MNRAS, 185, 745.

191. Salpeter, E.E., 1973. IAU Symposium No. 55. In "X- and Gamma-Ray Astronomy". Eds. H.V. Bradt and R. Giacconi, Reidel Publishing Company, Dordrecht, Holland.

192. Seitzer, P., Smith, G., and Ross, B., 1979. IAU Circ. 3372.

193. Shapiro, S.L., 1977. Ap. J., 217, 281.

194. Share, G., Wood, K., Yentis, D., Johnson, N., Shulman, S., Meekins, J., Evans, W., Byram, E., Chubb, T., and Friedman, H., 1978. IAU Circ. 3190.

195. Silk, J., and Aarons, J., 1975. Ap. J. (Letters), 200, L131.

196. Swank, J.G., Becker, R.H., Boldt, E.A., Holt, S.S., and Serlemitsos, P.J., 1978. MNRAS, 182, 349.

197. Swank, J., Becker, R., Pravdo, S., Saba, J., and Serlemitsos, P.J., 1976. IAU Circ. 3000.

198. Swank, J.H., Becker, R.H., Pravdo, S.H., Saba, J.R., and Serlemitsos, P.J., 1976. IAU Circ. 3010.

199. Swank, J.H., Becker, R.H., Boldt, E.A., Holt, S.S., Pravdo, S.H., and Serlemitsos, P.J., 1977. Ap. J. (Letters), 212, L73.

200. Swank, J.H., Becker, R., Pravdo, S., and Serlemitsos, P.J., 1977. IAU Circ. 2963.

201. Swank, J.H., Eardley, D.M., and Serlemitsos, P.J., 1979. Preprint.

202. Taam, R.E., and Picklum, R.E., 1978. Ap. J., 224, 210.

203. Taam, R.E., and Picklum, R.E., 1979. Ap. J., 233, 327.

204. Takagishi, K., Nagareda, K., Matsuoka, M., Fujii, M., Sato, S., Van Paradijs, J., Hoffman, J.A., Jernigan, J.G., Wheaton, W., Primini, F., and Lewin, W.H.G., 1978. IASA Res. Note, ISAS RN 60, Preprint CSR-P-78-31.

354

205. Tanaka, Y., and the Hakucho Team., 1979. X-ray Symposium in
      Tokyo, Japan.
206. Tananbaum, H., Chaisson, L.J., Forman, W., Jones, C., and Mat-
      ilsky, T.A., 1976. Ap. J. (Letters), 209, L125.
207. Thomas, R.M., Duldig, M.L., Heynes, R.F., Simons, L.W., Murdin,
      P., Hoffman, J.A., Lewin, W.H.G., and Wheaton, W.A., 1979.
      MNRAS, 187, 299.
208. Thorstensen, J., Charles, P., and Bowyer, S., 1978. IAU Circ.
      3253.
209. Thorstensen, J.R., Charles, P., and Bowyer, S., and the HEAO A3
      Group. 1978. Talk presented at the HEAD/AAS Meeting. San
      Diego.
210. Ulmer, M.P., Hjellming, R.M., Lewin, W.H.G., Hoffman, J.A.,
      Jernigan, J.G., Wheaton, W., Primini, F., and Marshall, H., 1978.
      Nature, 276, 799.
211. Ulmer, M.P., Lewin, W.H.G., Hoffman, J.A., Doty, J., and Marshall,
      H., 1977. Ap. J. (Letters), 214, L11.
212. Van Paradijs, J., 1978. IAU Circ. 3197.
213. Van Paradijs, J., 1978a. Nature 274, 650.
214. Van Paradijs, J., 1979. Ap. J., 234, 609.

215. Van Paradijs, J., Cominsky, L., and Lewin, W.H.G., 1978. IAU
      Circ. 3294.
216. Van Paradijs, J., Joss, P.C., Cominsky, L., Lewin, W.H.G., 1979.
      Nature, 280, 375.
217. Van Paradijs, J., and Lewin, W.H.G., 1978. Nature, 276, 249.
218. White, N.E., and Burnell, S., 1977. IAU Circ. 3067.
219. White, N.E., Mason, K.O., Carpenter, G.F., and Skinner, G.K.,
      1978. MNRAS, 184, 1P.
220. Willmore, A.P., Mason, K.O., Sanford, P.W., et al., 1979.
      MNRAS, 169, 7.
221. Wilson, A.M., Carpenter, G.F., Eyles, C.J., et al., 1977. Ap.
      J. (Letters), 215, L111.
222. Woosley, S.E., and Taam, R.E., 1976. Nature, 263, 101.
223. Wyckoff, S., 1979. IAU Circ. 3386.
224. Van Paradijs, J., Cominsky, L., and Lewin, W.H.G., 1979a.
      MNRAS, 189, 387.
225. Bond, H.E., 1977. IAU Circ. 3085.
226. Canizares, C.R., McClintock, J.E., and Grindlay, J., 1979a. IAU
      Circ. 3362.
227. Hoffman, J.A., Lewin, W.H.G., Primini, F.A., Wheaton, W.A.,
      Swank, J.H., Boldt, E.A., Holt, S.S., Serlemitsos, P.J., Share,
      G.H., Wood, K., Yentis, D., Evans, W.D., Matteson, J.L., Gruber,
      D.E., and Peterson, L.E., 1979. Ap. J. (Letters), 233, L51.
228. Grindlay, J., and Gursky, H., 1976b. Ap. J. (Letters), 205, L131.
229. Grindlay, J., 1978b. Ap. J., 221, 234.
230. Hoffman, J., Cominsky, L., and Lewin, W.H.G., 1980. Ap. J.,
      240, L27.
231. Pedersen, et al., 1981. To be submitted to Ap. J.

Ultraviolet observations of X-ray globular clusters

L. Hartmann

Center for Astrophysics
Cambridge,
Mass. O2138.

1.   Introduction

2.   Observations

3.   Results and Discussion

1.   INTRODUCTION

The discovery of X-ray sources in the central regions of some globular clusters (Giacconi et al. 1974; Clark, Markert, and Li 1975;
Canizares and Neighbors 1975) has prompted a search for optical
counterparts. Of these optical efforts, the most successful has
been the work of Grindlay and Liller (1977), who found a blue object
that appeared to flare at the core of NGC 6624, and identified this
object with the bursting X-ray source. In general, of course, the
crowding of stars in the central regions of globular clusters
prevents the easy identification of the optical counterparts of X-ray
stars. The fact that X-ray sources tend to occur in compact clusters
has led to the suggestion of a central, massive black hole, which
accounts for both the peak in surface brightness and the X-ray
emission (Bahcall and Ostriker 1975; Silk and Arons 1975). However,
the discovery of isolated galactic X-ray bursters and estimates of
the radii of the compact objects (Lewin, this volume), and the
presence of X-ray sources in non-compact clusters (Grindlay, this
volume), make the black hole hypotheses less likely. Illingworth
and King (1977) have also shown that the sharp peak in optical surface brightness of M15 can be reproduced without invoking a central
black hole.

Here I wish to present a preliminary discussion of a search for UV
emission from X-ray sources in the central regions of six globular
clusters using the International Ultraviolet Explorer Satellite.
These observations were made as part of the CFA programme of
examining the ultraviolet spectra of strong X-ray sources. The data
are of much higher spatial and spectral resolution than previous
ultraviolet studies of globular clusters (Welch and Code 1972;
deBoer and van Albada 1976). An initial report of some of the data

356

was given by Dupree et al. (1979). The data presented here have not
been corrected for the calibration error recently found. However,
some spectra have been corrected and no major effect has been found
so far.

2.   OBSERVATIONS

The spectra reported here were taken in 1978-79 with the IUE satellite. All exposures were in low dispersion mode, which has a resolution $\sim$ 6 $\AA$ at 1500 $\AA$. For an extended source the spectral resolution
is degraded due to the size of the large aperture, which is roughly
ellipsoidal in shape with axes $\sim$ 20" and 10". Spatial resolution is
achieved in the 20 arc sec direction; the spectrum is dispersed
perpendicular to this direction. A point source has a cross-section
of FWHM $\sim$ 5"5. The short wavelength (SWP) and long wavelength (LWR)
camera exposures cover the wavelength range $\lambda 1150-\lambda 3400$ A. Details
of the instrumental performance can be found in Boggess et al. (1978).

Most of the IUE spectra were taken at the cluster centres in optical
light, as found by the satellite star tracker. In the case of NGC
6624, an offset to "star B" (cf. Grindlay and Liller 1977) was made
from a nearby star using coordinates kindly provided by W. Liller.
Experiment showed that this position agreed with the position of the
peak light to $\sim$ 1". In order to obtain more information about the
spatial distribution of the ultraviolet emission, exposures at
different aperture positions in M15, M92, and NGC 6752 were taken.
To obtain fluxes at distances greater than 10 arc sec from the centre
of light, multiple exposures were combined from the offset pointing
of the satellite. The fields of view had some overlap, and the
observed fluxes agreed well at the junction, giving a smooth and
continuous flux distribution. In addition, since the position angle
of the aperture changes with time, certain exposures, that were $\sim$ 2
months apart from the M15 and NGC 6752 data, provided nearly orthogonal aperture positions. Exposure times varied from $\sim$ 1/2 hour to
$\sim$ 4 hours.

3.   RESULTS AND DISCUSSION

In Figure 1 we exhibit the spectrum of the central region of NGC
6624. The long-wavelength spectrum ($\lambda 2400-\lambda 3400$ A) arises from a
broad distribution of stars; the energy distribution and line
blanketing features are very similar to those of the metal-rich
cluster 47 Tuc, as would be expected on the basis of the metal abundances and colour-magnitude diagrams of the two clusters. However,
47 Tuc has extremely little short-wavelength ($\lambda 1200-1900$ A) flux, and
what is present occurs in a broad distribution, which suggests that
white dwarfs account for this emission. NGC 6624 in contrast
exhibits a point source in the short-wavelength spectrum, at the
position of Grindlay and Liller (1977) "star B" within the positional
accuracy available of $\sim$ 2". The weakness of this spectrum, due in
part to the considerable extinction to the cluster ($E_{B-V} \sim$ 0.25,

Carney and Liller 1978), makes it nearly impossible to detect spectral features moderate (a few A) in strength. If we attempt to match the continuous energy distribution with a stellar spectrum after correcting for extinction using the above E_{B-V} and the standard reddening curve, we estimate a star of $T_{eff} \sim 20,000$ K and a luminosity comparable to that of a horizontal branch (HB) star. However, the unusual nature of the presence of a blue horizontal branch star in such a metal-rich cluster suggests that this UV emission is in fact the signature of the X-ray source, by analogy with our observations of other galactic sources. As Figure 1 shows, the spectrum, while uncertain for reasons of weakness and extinction uncertainties, is consistent with an optically-thin free-free ($E_\nu \sim$ const) spectral energy distribution. In fact the absolute level of the theoretical line in Figure 1 is given by a thermal bremsstrahlung fit to the X-ray flux in the high state from Canizares et al. (1978). This is reminiscent of the result of Willis et al. (1980), who found the spectrum of Sco X-1 is consistent with a bremsstrahlung model, although we would expect to have detected the C IV and N V emission at the strength relative to the continuum observed in Sco X-1. A sum of two long exposures of NGC 6624 suggests that some spectral lines are present, but more data must be accumulated before anything definite can be said.

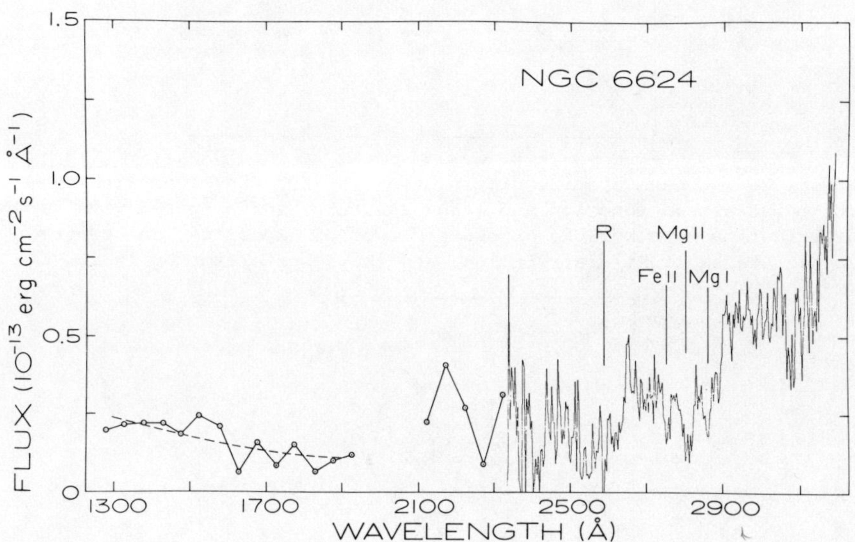

Figure 1: Unreddened spectrum of the central region of NGC 6624 in the large ($\sim 20" \times 10"$) aperture.

The other globular clusters observed have much lower metallicity than NGC 6624 and 47 Tuc, so that the blue end of their horizontal branches are much better developed, and the UV emission of an X-ray

358

source is likely to be swamped by normal stars in the central
regions. However, a surprise arose from these observations, in that
the short-wavelength emission is more spatially concentrated than the
long-wavelength flux. Figure 2a shows the situation for M15. The
adjacent cross-sections are exposures taken first centered on the
cluster, and then offset by 15". The dotted line is a calculation of
the expected cross-sectional shape based on optical data. It was
determined by using the U-magnitude surface brightness distribution
obtained by Newell and O'Neil (1979), passing it through a model
large aperture of the correct size and shape, and convolving the
result with a Gaussian approximation to the point-spread function
(FWHM $\sim$ 5."4). The calculations show that the 2900 A results are
completely consistent with the U-band data, as expected. The λ1450 A
data are inconsistent with the optical data, showing a sharper central
peak and a much more rapid fall-off. (Fig. 2b).

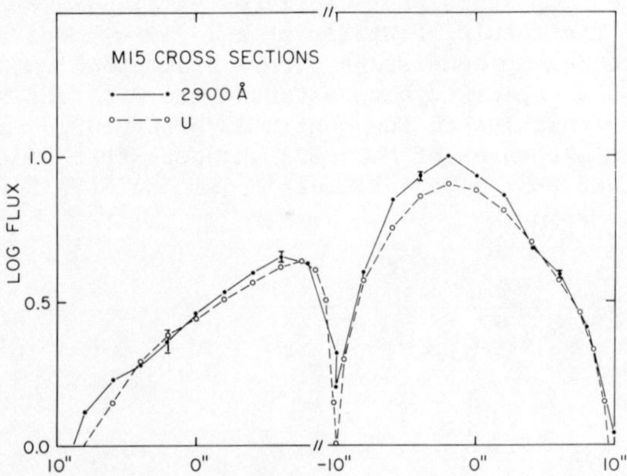

Figure 2a: U band spatial distribution in M15, convolved with the
IUE point spread function and large aperture, and compared with 2900 A
distribution seen by IUE. Shown are two IUE exposures, one centred
on the visual light distribution, and the other offset by 15".

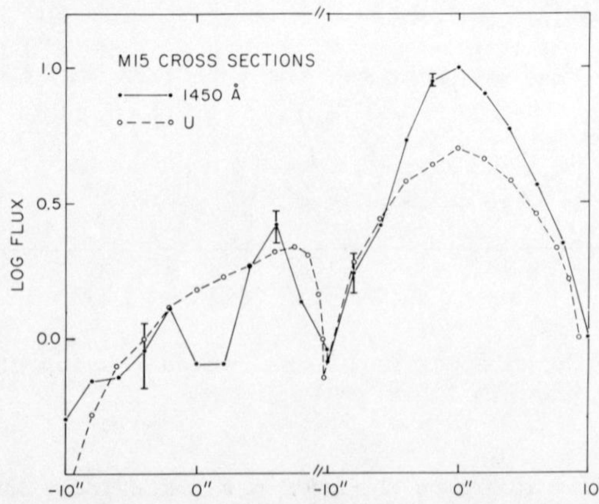

Figure 2b: U band spatial distribution in M15 as compared with 1450 A
distribution as seen by IUE.

Table 1 summarises the results of our observations, from which it is seen that the short-wavelength central concentration of light is not confined to M15. If we restrict our attention only to those metal-poor clusters expected to have blue HB stars, we find that three of the four such clusters observed have narrower short-wavelength emission than the spatial distribution of the long-wavelength flux. The only exception is NGC 6752, where the core radius of the cluster is so large as to prevent us from determining the light distribution with the small number of available exposures.

Table 1.

CLUSTER	(Fe/H)	$E_{(B-V)}$	X-RAY	F_{1500}	$\dfrac{F_{3000}}{F_{1500}}$	r_{1400}	r_{3000}
M15	-2.0	0.12	yes	1.2-13	2.4	4.2	7
M92	-2.0	0.02	no	8.3-14	1.5	8.5	14
NGC6752	-1.5	0.00	no	2.0-13	0.7	>20	>20
NGC1851	-1.0	0.14	yes	8.3-14	2.0	3.3	5.2
NGC6624	-0.7	0.25	yes	2.2-14	3.1	<4	4.5
47 Tuc	-0.5	0.02	no	2.4-14	>7	....	>10

F_{λ} = Fluxes at 1500 Å and 3000 Å, listed in units of erg cm^{-2} s^{-1} Å^{-1} and corrected for interstellar extinction using the given values of $E_{(B-V)}$.

r_{λ} = Observed half-width at half-maximum of the light distribution at λ perpendicular to the dispersion, given in units of arc seconds. The point source widths are $r^{*}_{1400} \sim 2\overset{\prime\prime}{.}7$ and $r^{*}_{3000} \sim 3\overset{\prime\prime}{.}0$.

Some mechanism clearly results in producing bluer stars near the centre of clusters. The properties of the UV excess stars appear to be consistent with the horizontal branch. The overall colours are consistent with the fluxes predicted from the number of blue HB stars expected from the luminosity function. Furthermore, in the centre of M92, where we can begin to spatially resolve the light, (as well as $\sim$ 15" from the centre of M15) there are individual "peaks" of brightness that are of the order expected from individual HB stars.

Why are these blue stars segregated towards the centre? The appearance of this effect in M92 makes it seem unlikely that a dynamical effect on stellar evolution from effects in a tightly bound core is the explanation. The simplest explanation appears to be mass segregation. Since this segregation must be accomplished relative to the red giants and HB stars which provide most of the long-wavelength light, the only way to increase the mass would be to invoke binary systems. This speculation is contrary to many previous observational efforts to detect binary systems, such as Gunn and Griffin (1979), who were unable to find a single spectroscopic binary in a sample of $\sim$ 100 red giants in M3. However, if the X-ray sources in globular clusters are not central massive black holes, then they must be binaries, so it is important to examine the implications of such speculation.

Let us work primarily in the context of M15, because the most work
has been done on this cluster. Given a distribution of one mass
class of star, it is simple to estimate the distribution of any
other mass class, assuming complete relaxation. The density distribution, in the Michie approximation, is given by

$$\rho_i(r) \propto \exp\left[\frac{2}{\sigma_i^2}(V(r)-V(o))\right] \cdot I_i \, , \qquad (1)$$

where σ_i is the velocity dispersion of mass class i, V is the
potential energy, and I_i is an integral which varies slowly near the
cluster centre; r = o denotes the cluster centre.

For different mass classes,

$$m_i \sigma_i^2 = m_j \sigma_j^2 \qquad (2)$$

Then an approximate result can be found for the density distribution.

$$\frac{\rho_j(r)}{\rho_j(r_o)} \simeq \left[\frac{\rho_i(r)}{\rho_i(r_o)}\right]^\gamma , \text{ where } \gamma = \frac{m_j}{m_i} \qquad (3)$$

Illingworth and King (1977) list a density distribution of stars for
M15 which is consistent with the optical surface brightness observations. They give the distributions of various mass classes which can
be used to demonstrate that 3) is a good approximation. Then using
equation 3) we may compute the distribution of any mass class. In
particular, if we choose $\gamma = 2$, i.e. stars twice as massive as red
giants, then such stars will have a very small core radius in M15,
$< \sim 2"$ (see Table 2). Three to ten stars, with UV luminosity
comparable to HB stars, are cleary capable of accounting for the
excess flux in M15, observed through the aperture ($\sim 3"$ HWHM) of the
IUE telescope. If the excess _is_ due to HB stars, and if all stars
have an equal probability of occurring in a binary system, then we
might expect $\sim$ a few hundred M5 binaries from the ratio of evolu-

Table 2

r (")	N_d (10^4 pc^{-3})	N_g (10^4 pc^{-3})	N_B
o	1.17	12.5	1.00
0.6	0.91	7.64	0.37
1.0	0.71	4.67	0.14
2.2	0.47	2.06	0.03
4.4	0.31	0.91	0.005

TABLE 2: N_d and N_g are the number densities x 10^4 per cubic parsec of
dwarfs and giants, respectively, ($M_d \sim 0.41\ M_\odot$, $M_g \sim 0.83\ M_\odot$); N_B is
the density of stars with $M_B = 2\ M_g$, arbitrarily scaled to unity at
the centre.

tionary lifetimes. This is an interesting number in view of the
dynamics of the central core. Newell, Da Costa, and Norris (1976)
showed that the peak in surface brightness could be explained by an
800 $M_\odot$ central black hole. Illingworth and King (1977) found that
the same result could be achieved by a comparable mass of 2 $M_\odot$
neutron stars segregating to the centre. The UV result discussed here
suggests that a complement of binaries $\sim 10^{-2} - 10^{-3}$ of the total
number of stars might also account for the central concentration as
well.

If the upcoming X-ray results from Einstein shows that the sources
are not exactly at cluster centres, then the hypothesis of binaries
must be taken seriously. Work on non-compact clusters, such as M92,
which show the UV segregation, but can be easily spatially resolved,
will be important in making observational tests of the binary
speculation, and provides new insight on the mechanisms responsible
for globular cluster X-ray emission.

REFERENCES

Bahcall, J.N., and Ostriker, J.P., 1975. Nature, 256, 23.
Boggess, A., et al., 1978. Nature, 275, 361.
Canizares, C.R., Grindlay, J.E., Hiltner, W.A., Liller, W., and
     McClintock, J.E., 1978, Ap. J., 224, 39.
Canizares, C.R., and Neighbours, J.E., 1975. Ap.J. (Letters), 199,
     L97.
Clark, G.W., Markert, T.H., and Li, F.K., 1975. Ap.J. (Letters), 199,
     L93.
deBoer, K.S. and van Albada, T.S., 1976. Astr. Ap., 52, 59.
Dupree, A.K., et al., 1979. Ap.J. (Letters), 230, L89.
Giacconi, R., Murrey, S., Gursky, H., Kellog, E., Schreier, E.,
     Matilsky, T., Koch, D., and Tananbaum, H., 1974. Ap.J. (Suppl.),
     27, 37.
Grindlay, J.E., and Liller, W., 1977. Ap.J. (Letters), 216, L105.
Gunn, J.E., and Griffin, R.F., 1979. A. J., 84, 752.
Illingworth, G., and King, I., 1977. Ap.J. (Letters), 218, L109
Newell, B., DaCosta, G.S., and Norris, J., 1976. Ap.J. (Letters),
     208, L55.
Newell, B., and O'Neil, E.J., Jr., 1978. Ap.J. (Suppl.), 37, 27.
Silk, J., and Arons, J., 1975. Ap.J. (Letters), 200, L131.
Welch, G.A., and Code, A.D., 1972. In Scientific Results from the
     Orbiting Astronomical Observatory, NASA-SP-310, p. 541.
Willis, A.J., et al., 1980. Ap.J., in press.

X-ray observations of globular clusters

Jonathan E. Grindlay[*]

Harvard-Smithsonian Center for Astrophysics,
Cambridge, USA.

1. Introduction

2. Previous Observations of X-ray Globulars

3. Preliminary Results Obtained with the Einstein Observatory

4. Discussion and Conclusions

1. INTRODUCTION

That there are luminous X-ray sources in globular clusters was a
major surprise when three examples were found in the first X-ray sky
survey conducted with the Uhuru satellite. It was surprising to
find these sources since their relatively high luminosities
($\sim 10^{36}$ - 10^{37} erg/s) and variability (factor of ~ 2 variations in
~ 10 minutes are typical) indicated from the outset, that they were
sources powered by accretion of gas onto a collapsed object -
either a neutron star or black hole. The existence of such objects
in globular clusters was unexpected: being the oldest systems in
the Galaxy, and devoid of massive stars, globular clusters would
seem to be unlikely sites for either supernovae, or direct collapse
to compact stars heavier than white dwarfs. In fact, the positions
of the first globular cluster sources (2U1820-30 = NGC6624, 2U1746-37
= NGC6641 and 2U2134+11 = NGC7078; Giacconi et al. 1972) were so
uncertain that their identification with the globular clusters was
tentative. However, additional Uhuru data, as well as results from
the OSO-7, ANS, and SAS-3 X-ray detectors quickly established the
reality of this class of galactic X-ray source, as we shall review
below. Interest has grown, therefore, in understanding the nature
and origin of the globular cluster X-ray sources and their relationship to other galactic X-ray sources, particularly those in the
galactic bulge. Interesting relationships have been found between
the occurrence of X-ray sources in globular clusters and the nature
of their X-ray temporal variability (e.g. most are bursters) as well
as between source existence and the morphology of the clusters (e.g.
most are highly centrally condensed). These relationships have been
instrumental in constructing models for the globular cluster sources

*Alfred P. Sloan Foundation Fellow

- either neutron stars in compact binaries (Clark 1975, Joss and
Rappaport 1979) or isolated black holes of mass $> 10^2 - 10^3$ $M_\odot$
(Bahcall and Ostriker 1975, Silk and Arons 1975) or of mass ~ 10-100
$M_\odot$ (Grindlay, 1978a). We shall summarize preliminary new results
from the Einstein X-ray Observatory which promise to enhance our
understanding of these objects greatly and to provide significant
tests of the neutron star vs. black hole models.

2.   PREVIOUS OBSERVATIONS OF X-RAY GLOBULARS

Only a brief summary of early observations is given here since early
results were reviewed by Grindlay (1977). After the discovery with
Uhuru of the first three globulars mentioned above, two more
(NGC1851 and NGC6440), were identified by the OSO-7, satellite
(Clark et al. 1975, Markert et al. 1975). The identification of
these first 5 globular cluster sources was based on their being
included in the X-ray error boxes, ranging in area from several
square arcminutes to a fraction of square degree, with < 1% random
probability (see Markert et al. 1975). Both NGC1851 and, apparently,
NGC 6440 were identified in the final analysis of Uhuru data culminating in the 4U Catalogue (Forman et al. 1978). Furthermore, the
source identified with NGC6440 was found to be a transient source
detected by both Uhuru and OSO-7 for only $\sim$ 1 month (Forman, Jones
and Tananbaum 1976).

Whereas all of these globulars have very high central densities (see
Grindlay 1977 for a review of their optical properties), the next
cluster to be identified was NGC6712 (Seward et al. 1976, Cominsky
et al. 1977, Grindlay et al. 1977) with a very diffuse core. This
object is anomalous in this respect, as we shall see below, and
presents both interesting challenges for source models and globular
cluster evolution.

The discovery of X-ray burst sources (Grinlay et al. 1976) with
bursts from the source in NGC6624, added a new dimension to the
globular cluster problem. Although many bursters are now known
(Lewin and Joss 1977, Lewin and Clark 1979) to be outside clusters,
we shall see in our summary below that most (at least 6 out of 8)
X-ray sources in globular clusters produce bursts. In fact, the
first source (discovered solely by virtue of its burst emission),
was the spectacular rapid burster MXB1730-335 (Lewin et al. 1976).
This was soon identified by Liller (1977) to be (very probably)
located in a previously unknown and highly reddened globular cluster.
Finally, a second such highly reddened globular cluster (Terzan 2)
was suggested by Grindlay (1978b) as the likely counterpart of
another burst source (XB1722-30) discovered by Swank et al. (1977).

Except for Terzan 2, all these globular cluster identifications had
been suggested by 1976 and, since no error boxes were entirely
contained within their respective clusters, it could only be argued
statistically (Clark et al. 1975, Grindlay 1977), globular clusters
as a class were detected. This situation changed dramatically with

the accurate X-ray positions (with 20-30" error box radii) obtained
from the RMC experiment on the SAS-3 satellite. Five clusters
(NGC1851, 6441, 6624, 7078, and 6712) were detected with this X-ray
observatory and the positions obtained were not only entirely contained within the cluster but included (in each case) the cluster
core and its centre (Jernigan and Clark 1979). The X-ray positions
were sufficiently accurate that their offsets (from the cluster
centres) could be examined as a statistical distribution for comparison with that expected for sources of a given mass. The relationship between the expected radial offset of an object (e.g., an X-ray
source) in a cluster core and the objects' mass will be discussed
below in detail; suffice it to say here that the more massive the
object, the closer to the centre it should be (in a relaxed system).
The SAS-3 positions were sufficient to show that the globular cluster
X-ray sources are more concentrated toward the cluster centres than
the visible stars in the cluster and are, therefore, more massive
than the visible stars (Jernigan and Clark 1979).

3.   PRELIMINARY RESULTS OBTAINED WITH THE EINSTEIN OBSERVATORY

The X-ray imaging capability of the Einstein (HEAO-2) Observatory
means that the study of globular clusters can now proceed with greatly
increased sensitivity and angular resolution. The Observatory and its
detectors are described by Giacconi et al. (1979). We shall describe
preliminary results obtained with the High Resolution Imaging (HRI)
detector at the telescope focus and the Monitor Proportional (MPC)
detector co-aligned with the telescope. These results are described
in full by Grindlay et al. (1979a, 1979b).

3A.   HRI SURVEY OF GLOBULAR CLUSTERS

Approximately 20 globular clusters have been surveyed with at least
one HRI exposure ($\sim 10^3$ sec, typically) in addition to the 8 previously
known clusters mentioned above. Only one new globular cluster source
has been discovered thus far (NGC104=47 Tuc) and one of the previously
suspected clusters (NGC6440) was not detected, although $\lesssim$ 2% of the
previously reported flux could have been seen. Although NGC6440 was
already identified as a transient source (Forman et al. 1976), it has
not been detected with angular uncertainty $\lesssim 0.5^\circ$ (i.e., either by
SAS-3 or yet by Einstein) so that given its location in the galactic
bulge where the density of transient sources is high, its identification as an X-ray cluster must be regarded as uncertain. The "new"
cluster, 47 Tuc, may actually have been detected by Uhuru (its flux as
detected by Einstein would have been detectable by Uhuru at $\sim$ 1 ct
s^{-1}) since 1 of the 2 "lines of position" assigned in the Uhuru processing to the source 4U0026-73 passed through 47 Tuc. The other
4U0026-73 line of position could then have been due to the relatively
bright sources SMC X-2 or SMC X-3 several degrees away but still in
the collimator. Thus, although we had previously suggested (Grindlay
1978b) that 4U0026-73 could be the globular cluster, Kron 3, since the
catalogued error box contained this diffuse SMC cluster with a chance

366

coincidence of only a few percent, the Einstein detection of 47 Tuc
(and not of Kron 3, although variability is always possible) and the
above re-interpretation of the Uhuru error box are self-consistent.

The second "new" cluster suggested in our pre-Einstein discussion
(Grindlay 1978b) has, however, proven correct. This is Terzan 2,
the highly reddened globular associated with a burster as mentioned
above. We shall discuss this object further in the next section.

In Table 1 we list all the globulars observed thus far with the HRI
in our globular cluster survey. The apparent count rate or upper
limit is given, together with the estimated cluster distance and the
corresponding luminosity in the band 0.2-4 keV for the spectral
parameters given. The spectral parameters for the detected clusters
are derived from the MPC and will be discussed in the next section.
As can be seen from the available cluster paramters given in Table 1
(the central density is from Peterson and King (1975) and the
metallicity is from Arp (1965)), a wide range of cluster types is
included. The 8 detected clusters are all, except for NGC 6712,
among the most centrally condensed known, yet others, of comparable
central density (e.g., NGC5824 or NGC6388), are not yet detected. As
Lewin and Clark (1979) point out with a Monte Carlo sampling of the
7 pre-Einstein clusters, NGC6712 is clearly anomalous as the only
X-ray cluster with a diffuse core. Most of the detected clusters
are metal-rich (relative to other globulars but not the Sun), though
NGC1851 and NGC7078 are metal-poor, and most (5 of the 8) are in the
galactic bulge. The latter two characteristics are known to be
correlated, since most of the high metallicity globulars are in the
galactice bulge region (Arp 1965).

The luminosity distribution of the X-ray globulars is plotted in
Figure 1. The soft (0.2-4 keV) X-ray luminosities detected cover
nearly 3 decades - from $\sim 5 \times 10^{34}$ erg s^{-1} (NGC104) to $\sim 3 \times 10^{37}$
(NGC6624). The mean and median luminosity is given by

$$L_x \simeq 2 \times 10^{36}$$

where the distribution has been approximated to a gaussian in log L_x
with a full-width ($\delta/\simeq 1.5$ at the $\pm 1 \sigma$ level). The fact that 10 of
the 13 upper limits are for log $L_x \leq 34.5$, or more than twice the
value, given above, and thus are below the median detected L_x, might
be significant and suggests a lower luminosity cutoff of $L_x \sim 10^{35}$
erg s^{-1} for the globular cluster sources. Obviously a larger sample
from a survey at higher sensitivity is needed to confirm this; such
a survey is now being conducted with the IPC detector on the EINSTEIN
Observatory.

3B   SPECTRA AND VARIABILITY OF GLOBULAR CLUSTER SOURCES

The spectral parameters listed in Table 1 for 7 of the 8 detected
globular clusters were obtained with the Monitor Proportional Counter
(MPC)(see Grindlay et al. 1979b). The mean spectral shapes are
similar for all the sources and are best fitted by ~ 6 keV thermal

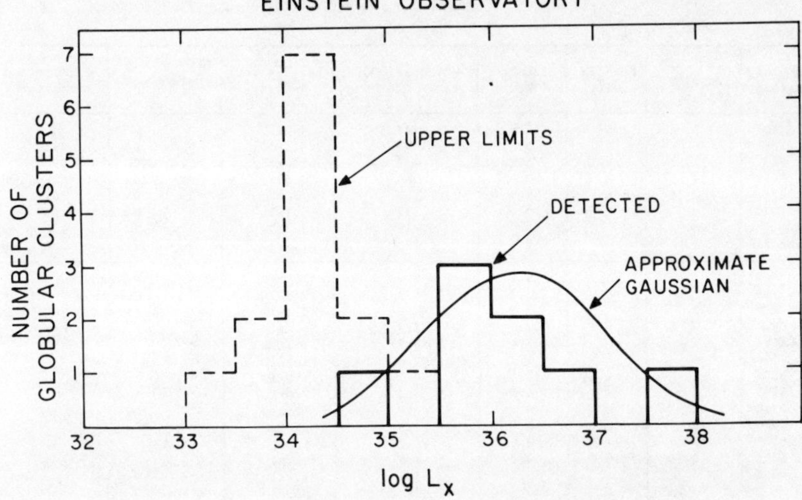

Figure 1: Distribution of detected values and upper limits for X-ray luminosities (0.3-3 keV) of sources in galactic globular clusters observed with the Einstein Observatory. A low luminosity cutoff of $\sim 10^{35}$ erg s^{-1} may be suggested for globular cluster sources (see Grindlay et al. 1979a).

bremsstrahlung spectra with moderate low energy cutoffs. It should be noted, though, that the MPC is not sensitive at energies < 1 keV or > 15 keV so that both low and high energy components could also be present. Some information on the low energy spectral shape (where interstellar absorption is important) can be obtained by comparing the observed HRI count rate with that predicted from the spectral parameters derived from the MPC and the HRI calibration, since the HRI is relatively much more sensitive at lower energies. Such a comparison for both NGC 1851 and NGC 6441 gives a predicted HRI count rate significantly lower than actually observed, indicating either that the low energy cutoff was overestimated from the MPC measurement or (more likely) that the spectra contain an additional soft component.

The latter possibility is especially interesting since the shape and temporal behaviour of a two-component spectrum may reveal much about the underlying source. For example, a relatively constant, hard, component but variable, steep, soft spectrum would resemble the black hole candidate Cyg X-1. On the other hand, black body low energy spectral components are expected for white dwarfs (Kylafis and Lamb 1979). Such objects would usually show positive correlations between apparent temperature and intensity (within the medium energy band) and although this has not been seen in earlier observations

TABLE 1

Preliminary Einstein Globular Cluster Survey Results

Globular Clusters	Central Density ($10^4 M_\odot pc^{-3}$)	Metall- icity	Distance (kpc)	HRI cts/1000s (12" x 12" box)	L_X(.2–4 keV) (erg/s)	MPC Spectral parameters kT†	E_a†	HRI Expected cts/1000s
NGC 104	5	H	5	44	4.1×10^{34}	*		
1851	8	L	9.5	1079	3.5×10^{36}	5.	.64	411
6441	10	H	5.9	837	2.0×10^{36}	6.	.96	593
6624	12	H	8	13960	3.5×10^{37}	9.	.64	13526
6712	0.15	M	7	305	4.8×10^{35}	3.	.64	381
7078	4	L	10	138	5.1×10^{35}	8.	.64	120
Liller 1	20		10	97–168	$1.7–2.9 \times 10^{36}$	10.	2.2	156
Terzan 2	5		7	160	2.9×10^{35}	8.	1.2	120
NGC 362	4		9.1	<5.9	$<1.9 \times 10^{34}$	*	*Assumed	
2298	0.06		12.4	<4.9	$<2.9 \times 10^{34}$	*	Spectrum	
2808	4		7.1	<4.3	$<8.1 \times 10^{33}$	*	kT = 6.0 keV	
5824	7	H	23.8	<4.3	$<9.1 \times 10^{34}$	*	E_a = 0.5 keV	
6093	7		7.4	<5.7	$<1.2 \times 10^{34}$	*		
6266	4		7.9	<4.8	$<1.1 \times 10^{34}$	*		
6273	2	M	3.0	<4.7	$<1.6 \times 10^{33}$	*		
6293	6		6.0	<5.2	$<7.0 \times 10^{33}$	*		
6388	12	H	8	<4.8	$<1.1 \times 10^{34}$	*		
6440	10	H	10	≤7.3	$≤2.7 \times 10^{34}$	*		
6517	18		8.8	<5.2	$<1.5 \times 10^{34}$	*		
6715	5	M	13.7	<4.4	$<3.1 \times 10^{34}$	*		
6864	3		25.1	<4.7	$<1.1 \times 10^{35}$	*		
Kron 3	<0.1	L	55	<4.7	$<5.3 \times 10^{35}$	*		

† keV

Notes to Table 1

The first 8 clusters are all the currently known globular cluster sources in the galaxy. Six of these are also burst sources; only NGC 104 (47 Tuc) and NGC 7078 have not been observed to burst. For those clusters not detected, central densities and distances are from Peterson and King (1975) and metallicities (H=High, M=medium, L=Low) are from Arp (1965). The optical parameters of the X-ray clusters are taken from references given in Grindlay (1977 or 1978b). The upper limits for the HRI count rates and luminosities given for the 14 globular clusters not detected are 3σ upper limits.

(e.g. Parsignault and Grindlay 1978), the Einstein observations are now much more sensitive. Spectral vs. intensity variability will be, therefore, studied in detail in the analysis of Einstein data on globular clusters.

The most spectacular example of both intensity and spectral varia- bility yet recorded by Einstein is the X-ray burst from the globular cluster Terzan 2. This event, which is described in detail by Grindlay et al. (1979b), was recorded by chance during a single ∿1500 sec pointing on the cluster which was planned to test its suggested identification (mentioned above) as both a persistent source and a burster. Both were confirmed. A persistent source was detected by the HRI and MPC with a 2–6 keV flux of ∿ 10 μJy (∿ 10 UFU) and spectrum best fit by bremsstrahlung with temperature kT ≃ 7 keV and a low energy cutoff implying an absorption column density $N_H \simeq 7 \times 10^{21}$ cm^{-2}. These values are similar to those found by Swank et al. (1977) for this source which they could only locate to $\sim 1^0$. The location of the Einstein source, discussed with the other globulars in the next section, is within a few arcseconds of the centre of the cluster Terzan 2. The X-ray burst observed (in the HRI) to be from the same persistent source is shown in Figure 2. The total counts recorded by the MPC in all 8 spectral channels (1.1-22.0 keV) are shown in the top plot; the other plots in Figure 1 show the energy dependence of the burst profile, by showing the differing burst profiles recorded in each group of 2 spectral channels. The integration time, or finest time resolution, for all these spectral data is 2.56 sec. Higher time resolution (to within ~ 1.6 μsec) data, but without energy information, is available from the Time Interval Processor (TIP) of the MPC. Analysis of these data show that the burst had a very rapid rise ($\lesssim 0.2$ sec) but does not have significant fast-time variability (over the entire ~ 1.1-22 keV range sampled) at the burst peak or during its decay. The peak emission lasts for ~ 20 sec and corresponds to a peak luminosity (for the black body spectral fits discussed below) of $L_x \simeq 5 \times 10^{38} d_{10}^2$ erg s^{-1}, where d_{10} is the distance to Terzan 2 in 10 kpc units and is probably ~ 0.7 (Grindlay 1978b). The burst decay is approximately exponential with time constants of 11 sec (first 20 sec) and 140 sec (next 300 sec). This is a faster decay than observed by OSO-7 (Swank et al. 1977).

Analysis of the burst detected from Terzan 2 by the OSO-7 experiment was the first to show that the decay of the burst could be fitted with a cooling black body. We also find that the Terzan 2 energy spectrum of the burst decay is best fit by a black body with temperature cooling from a peak value of kT $\simeq 3$ keV and an apparent radius of $\sim 7d_{10}$km. Modified black body spectral shapes incorporating an emissivity due to electron scattering (Swank, Eardley and Serlimitsos 1979) fit equally well but give larger apparent radii of $\sim 15d_{10}$km. During the burst itself the spectrum is much more complicated (black body or modified black body models do not fit well) and a double peaked profile of apparent temperature and radius is evident. This has interesting implications for the possible presence of adiabatically expanding shells during the burst (Grindlay et al. 1979b).

Finally, we should mention that new results have also been obtained with Einstein on the rapid burster MXB1730-335 in the globular cluster Liller 1. A single observation in March 1979 recorded during a burst active phase revealed that the burst source is within a few arcseconds of the centre of the globular cluster and that at least $\sim 95\%$ of the total emission is in the form of Type II (i.e., rapid recurrence time) bursts. A marginal detection of an underlying persistent (non-burst) emission component of the source in the globular cluster at about 5% of the total flux is being investigated further in the Einstein data by Marshall et al. (1979).

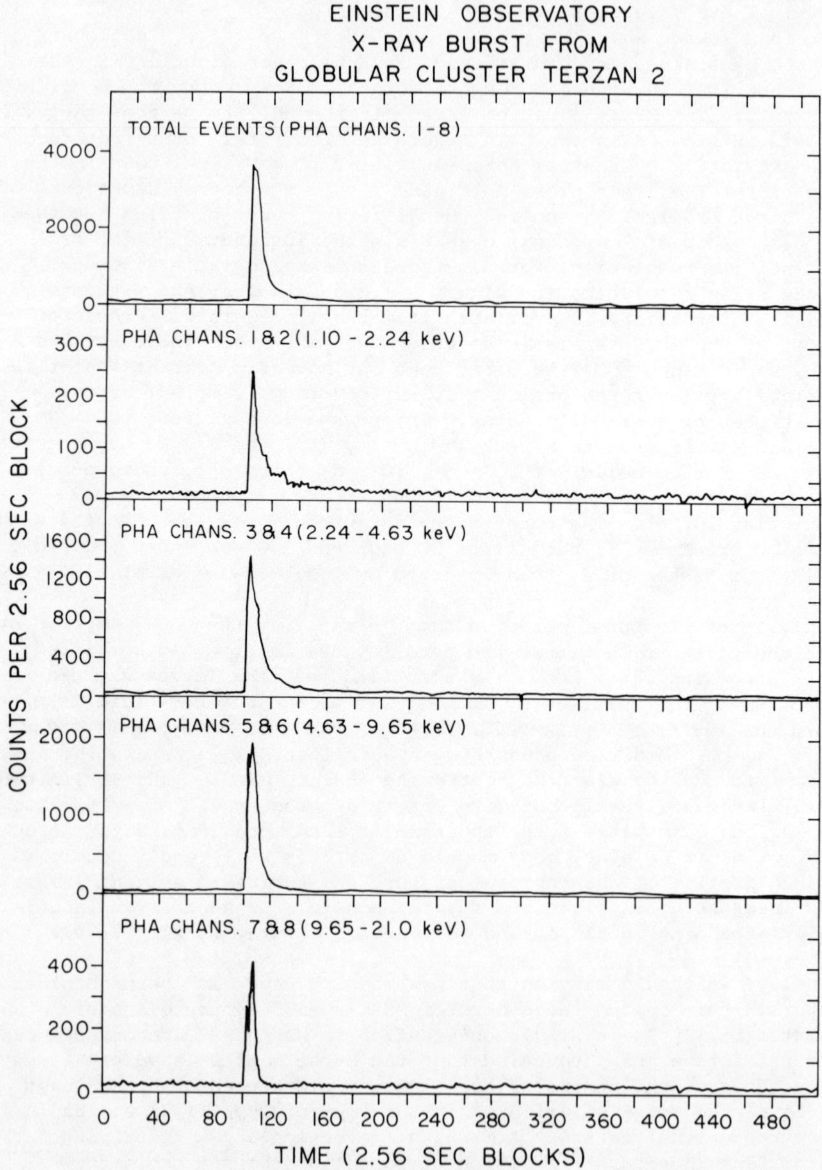

Figure 2: X-ray burst from the globular cluster Terzan 2 recorded by the MPC detector on the Einstein Observatory. The energy dependent burst profile shows two peaks of emission for which black body fits suggest an adiabatically expanding shell of approximately constant luminosity (see Grindlay et al. 1979b).

3C PRELIMINARY SOURCE LOCATIONS FROM THE EINSTEIN OBSERVATORY

The primary goal of our programme of globular cluster observations
with the Einstein Observatory has been to determine the X-ray source
positions within the clusters to arcsecond accuracy. The motivation
for this is clear, since by measuring the source displacements from
the cluster centres a statistical measure of the source masses can
be obtained (Lightman and Bahcall 1976). That is, since in a highly
relaxed stellar system like a globular cluster core (with central
relaxation time much shorter than the cluster age), a constant
"temperature" or kinetic energy per particle (star) is established.
In such an isothermal core, heavier objects would then (with their
slower mean velocities) settle towards the centre and "mass segregation" would result. The derivation of the expected radial offset
of an object (e.g., an X-ray source) of mass M from the centre of
an isothermal cluster core of stars with (uniform) mass m_* and core
radius r_c was given by Bahcall and Wolf (1976) and yields

$$r \simeq 0.9 \; r_c (M/m_*)^{-\frac{1}{2}}$$

A similar relation between r and M was obtained by Jernigan and Clark
(1979), and we present yet another derivation of Bahcall and Wolf's
result motivated by discussions with S. Tremaine. That is, for an
isothermal core the probability of an object of mass M is between r
and r + dr from the centre is

$$P(r)dr \propto r^2 \exp(-M\Phi/m_* <\Delta v^2>) dr$$

where $\Phi = 3/2 <\Delta v^2>(r/r_c)^2$ is the gravitational potential energy and
$<\Delta v^2>$ is the mean square velocity dispersion. If we now project from
the 3-D radius r to the 2-D radius R and substitute for ϕ, we find

$$P(R)dR \propto R^2 \exp(-(3/2)(M/m_*)(R/r_c)^2) dR$$

Converting to a distribution in R^2 this becomes

$$P(R^2)dR^2 \propto (dR^2 (R_x^2) \exp(-R^2/R_x^2)$$

where $R_x^2 = (2/3)(m_*/M) r_c^2$ is then the above relation between $R_x = r(M)$ and r_c. Note also that the probability distribution for R^2 can
be evaluated for 0.10, for example, to give the 90% confidence
maximum value for R_x for mass M

$$R_{max} = R_x (\ln(10))^{\frac{1}{2}} \simeq 1.5 R_x$$

This is the value exterior to which < 10% of all objects of mass M
should be found if the source offsets had no intrinsic measurement
errors.

In fact, even for the Einstein data the measurement errors are
significant and actually dominate the observed distribution of source
offsets and inferred masses. That measurement errors could be so
important is easily seen when the scales of the problem are considered.

The cluster core radius r_C sets the scale (since $R_X \leqslant r_C$ for $M \gtrsim m\star$), and this is typically $\sim$ 6-8" for the X-ray globulars (see Table 2 below). Thus if the combined error in establishing the cluster centre and the error in the X-ray position is $\sim$ 4", a $\sim 2\sigma$ excursion from any measured offset within the core can drive the inferred value of M from $m\star$ to ∞. However, even in the case when the individual measurement errors greatly exceed the core radius scale, as for the SAS-3 globular cluster results reported by Jernigan and Clark (1979), the combined results on a number of cluster sources gives significant information on the most likely range of values for M. This requires that all the objects are similar and the cluster cores are all isothermal.

TABLE 2

Preliminary Einstein Positional Offsets for X-ray Sources
in Globular Clusters

Cluster	Optical Centre	R_X	R_C	R_X/R_C
NGC 104	00 21 53.06 -72 21 29.3 ($\pm$ 2")	8.6 $\pm$ 4.5	24"	0.36 $\pm$ 0.19
NGC 1851	05 12 28.06 -40 6 11.4 ($\pm$ 1")	10.7 $\pm$ 4	7"	1.53 $\pm$ 0.57
Terzan 2	17 24 19.8 -30 45 40.0 ($\pm$ 2")	2.6 $\pm$ 4.5	7"	0.37 $\pm$ 0.64
Liller 1	17 30 07.0 -33 21 17.0 ($\pm$ 2")	5.4 $\pm$ 4.5	7"	0.77 $\pm$ 0.64
NGC 6441	17 46 48.53 -37 02 14.2 ($\pm$ 1")	7.1 $\pm$ 4	9"	0.79 $\pm$ 0.44
NGC 6624	18 20 27.62 -30 23 14.4 ($\pm$ 1")	2.6 $\pm$ 4	6"	0.43 $\pm$ 0.63
NGC 6712	18 50 20.55 - 8 46 6.1 ($\pm$ 3")	10.2 $\pm$ 5	49"	0.21 $\pm$ 0.10
NGC 7078	21 27 33.40 -11 56 48.7 ($\pm$ 1")	0.9 $\pm$ 1	10"	0.09 $\pm$ 0.10

Notes: Both the optical and X-ray positions are preliminary and errors will be reduced. The X-ray positions used to construct the radial offset R_X from the optical centre given have errors $\pm$ 1$\sigma \simeq \pm$4 arcsec. The core radii R_C are from Peterson and King (1975) or references given in Grindlay et al. (1979a).

In Table 2 we list the preliminary Einstein values for the offsets of the X-ray sources from the optical centres for the 8 globular detected. These results are preliminary since the quoted errors include systematic uncertainties that are now being reduced through in-flight calibrations and since the globular cluster measurements themselves are now being repeated for additional independent posi- tional determinations, which will then be combined. The X-ray source offsets, which all have $\pm$ 1σ uncertainties of $\sim$ $\pm$ 4", have been calculated with respect to the optical centres of the clusters as given in Table 2. These optical centres, in turn, have been determined from a centre-of-symmetry analysis of the cluster images on direct plates we obtained on the CTIO 4 m telescope. A complete discussion of these optical data and the analysis procedure is given by Grindlay and Hertz (1979). Briefly, the images CTIO U- band plates (to minimize red giants) were first digitized in 1" x 1" bins over the central 100-200" square of the cluster, then "folded" about the x and y axes (i.e., the distribution of sums $x_j = \sum_i (x_{j+i} - x_{j-i})$ for all j = 1, n bins in x gives a smooth curve with a minimum about the centre of symmetry of light in the x direction of the image; and similarly for the y direction), and a folded (in the same way) King model for the cluster surface brightness was fit to the folded data curve. The fit then defined that portion of the folded data curve minimum, or centre of light, which gave the smallest χ^2 deviation between the entire folded data vs. folded King curves. Other results of this work, such as investigating the core relaxation by determining the concentricity of giants and dwarfs in cluster cores and the colour dependence of r_c, are explored by Grindlay and Hertz (1979).

The results on all 8 clusters can be combined in a relative likelihood analysis (cf. Jernigan and Clark 1979) for the relative distribution of source masses M (or actually mass ratio $q = M/m_*$) which could give rise to the observations. Applying the relative likelihood technique of Jernigan and Clark to the data in Table 2 then leads to the relative likelihood distribution for q shown in Figure 3. Using a Kolmogoroff-Smirnoff test, we have calculated a similar distribution, but with absolute probability (that 8 objects with given q would actually yield the observed data) instead of relative likelihood in our complete discussion of this work (Grindlay et al. 1979a). The relative likelihood distribution in q shown in Figure 3 is asymmetric towards large q, and thus not very restrictive for maximum q values, because of the currently large measurement errors (relative to r_c). Nevertheless it does set interesting lower limits for q since the relative likelihood is only $\sim$ 10% that $q \lesssim 4$ or (given $m_* \simeq 0.5 M_\odot$ for typical globular cluster stars) that $M \gtrsim 2\ M_\odot$ with $\gtrsim$ 90% relative likelihood. Again the reader is referred to our more com- plete discussion and results (Grindlay et al. 1979a) for more restrictive results on the allowed range of q.

4.   DISCUSSION AND CONCLUSIONS

It is especially interesting that so far only one X-ray source

(point source) has been found in any one globular cluster. The
Einstein data have provided the first opportunity to search for
either multiple point sources or extended sources in globular clusters since the resolution of imaging is required on such small
($\leq$ 10") angular scales. The possible multiplicity of sources in a
globular cluster has a direct bearing on source models; only a
single black hole is expected near the centre of a cluster core
(smaller black holes would have probably already coalesced into a
single object), whereas if the sources are low mass binaries several
might be expected within a single cluster.

If the globular cluster sources are indeed low mass binaries containing a neutron star and a dwarf cluster star as is indicated by
several arguments discussed below (and see Lewin and Clark 1979), then
they are almost certainly capture binaries. Primordial binaries
would not now be seen in an X-ray emitting phase since that is
expected to last only $\leq 10^6$ years (van den Heuval 1977) whereas the
clusters are typically $\sim 10^{10}$ years old. The capture of isolated
neutron stars was thus suggested by Clark (1975) to explain the
globular cluster sources. Hills (1975) proposed the capture binaries
could form with much larger cross section if the neutron stars interacted instead with already-existing binaries (perhaps primordial),
thus displacing one of the members in a "charge exchange" type
collision. In either case, high central densities in the cluster core
are required and it is thus reasonable that all but one (NGC 6712) of
the X-ray globulars have very high central densities. If central
density is indeed the primary reason for the existence of 7 X-ray
sources in $\sim$ 15 clusters with comparably high central densities
($\geq 5 \times 10^4$ $M_\odot$ pc^{-3}; cf. Peterson and King 1975), then two sources
should be found in a single cluster with probability $(7/15)^2 \simeq 0.22$.
This was also pointed out by Lewin and Clark (1979). However we can
now state that none of the detected sources is double although in 7
"trials" (clusters) no doubles would be expected with probability of
only $\sim (1-0.22)^7 \simeq 0.18$. Although this is not improbable and more
clusters are needed, several interesting possibilities are suggested:
(1) high central density is not the dominant independent variable for
the formation of sources, (2) two or more binaries are not independent (i.e. they interact, possibly disrupting one pair) over the X-ray
lifetime, or (3) globular cluster sources are not low mass binaries.

The question of whether the globular cluster sources are low mass
binaries containing neutron stars or involve more massive (≥ 10 $M_\odot$)
black holes can now be reconsidered. In our earlier review (Grindlay
1977) we argued that both of these models were still equally viable.
However, in considering more recent (but pre-Einstein) results,
primarily from burster observations and interpretations, Lewin and
Clark (1979) now argue that the compact neutron star binary models
are now preferred. The principal arguments supporting this view are
that the apparent black body cooling in the spectra of X-ray bursts
suggest the burst sources all have apparent radii $\sim$ 10 km (see van
Paradijs 1979) and that the model whereby X-ray bursts are due to
thermonuclear flashes on the surface of neutron stars (Joss 1978)

accounts for many of the burster observations. The Einstein results on Terzan 2 strengthen this argument (which is otherwise based primarily on bursters not in globular clusters) by confirming the source is in the cluster and that the apparent black body radius is again $\sim$ 10-15 km. The preliminary source position results (Figure 2) are also just consistent with their all being $\sim$ 2 $M_\odot$ objects (e.g. 1.4 $M_\odot$ neutron stars plus $\sim$ 0.5 $M_\odot$ cluster stars) and the tentative evidence (Figure 1) for a low luminosity cutoff of $L_X \sim 10^{35}$ erg s^{-1} would suggest there is not a continuum of lower gas densities and accretion rates as might be expected for isolated black holes. Is there then any reason for continuing the discussion of black holes in globular cluster cores?

The answer is probably "no" (at least in this context) <u>if</u> the final Einstein positional data show that the globular cluster sources are well-described by (rather than just consistent with) objects of $\sim 2M_\odot$. This is still the crucial test, although it is of course non-binding. For example, if the final positions yield a q-distribution (see Figure 3) strongly peaked at q $\simeq$ 20 due to source positions peaked at $r_X/r_C \simeq$ 0.2, it might still be argued that the cluster cores are not isothermal and that a sub-core of heavier objects (neutron stars and binaries) has formed. This was suggested, in fact, by Illingworth and King (1977) to account for the central light spike in M15 as a $\sim$ 1% mix of neutron stars and white dwarfs in a sub-core. However, the central relaxation time is so short in such a sub-core (with parameters as given by Illingworth and King) that it would seem to be very unstable against collapse and therefore probably so short-lived that we should not see it in at least 2 X-ray clusters (M15 <u>and</u> NGC6624) (cf. Grindlay and Hertz, 1979). If this question of cluster core isothermality arises because the source positions appear to be significantly more centrally concentrated than expected for q $\simeq$ 4, the statistical analysis used (see Grindlay et al. 1979a) can provide a quantitative measure of the likelihood that the cluster cores are all isothermal and that all the source masses are equal. Thus it is really the results of this test that are needed before black hole models can be finally ruled out.

If the black hole models are <u>not</u> ruled out, the burster data might still be accommodated although not the nuclear flash model. The $\gtrsim$ 10 km radii (the inequality is due to the emissivity <1 in realistic gray body models as we have mentioned before - e.g. Grindlay 1978a, b and as discussed by Swank et al., 1979) of spherical black bodies can just as well be $\gtrsim$ 20 km radii of disks, which might form and radiate as black bodies near the Schwarzschild radii of $\sim$ 10 $M_\odot$ black holes. The bursts themselves could arise from spherical accretion insta-bilities due to X-ray pre-heating of a spherical accretion flow onto the disk if the objects are in a rather narrow mass range of $\sim$ 10-100 $M_\odot$ and the sources have underlying hard ($\gtrsim$ 100 keV) components of their spectra (Grindlay 1978a). Similar conditions were found by Cowie, Ostriker and Stark (1978) to yield burst behaviour. Although a $\sim$ 100 $M_\odot$ black hole in a cluster core could probably accrete sufficient diffuse gas distributed uniformly in the core (which would not exceed the gas limits set in many searches) to yield the observed

L_x distribution (Figure 1), this is much less likely if the sources were only $\sim$ 10 $M_\odot$. In this case accretion from the stellar winds of red giants in the core, or from a captured binary companion, would seem more likely. The latter possibility might be preferred since the recent identification of the transient X-ray source Cen X-4 (Canizares, McClintock and Grindlay 1979) with a burster may link bursters to binaries given the optical evidence that other soft X-ray transients are in binary systems. The quasi-spherical accretion flow (though perhaps down onto a disk) required for bursts from a black hole source in a globular would suggest that any binary companion be a red giant at large separation and supplying mass via a stellar wind. In any case, a $\sim$ 10 $M_\odot$ black hole would not have the formation "problem" of a massive ($\sim$ 10^3 $M_\odot$) black hole since it does not require the entire cluster core to collapse (in fact a $\sim$ 50 $M_\odot$ "seed" black hole will effectively halt collapse by evaporation - Marchant, private communication) but may result from a single massive star. Many of these questions can be dealt with, if necessary, when the final Einstein positions are available for the X-ray sources in globular clusters and the final form of the mass distribution (e.g. Figure 3) is derived.

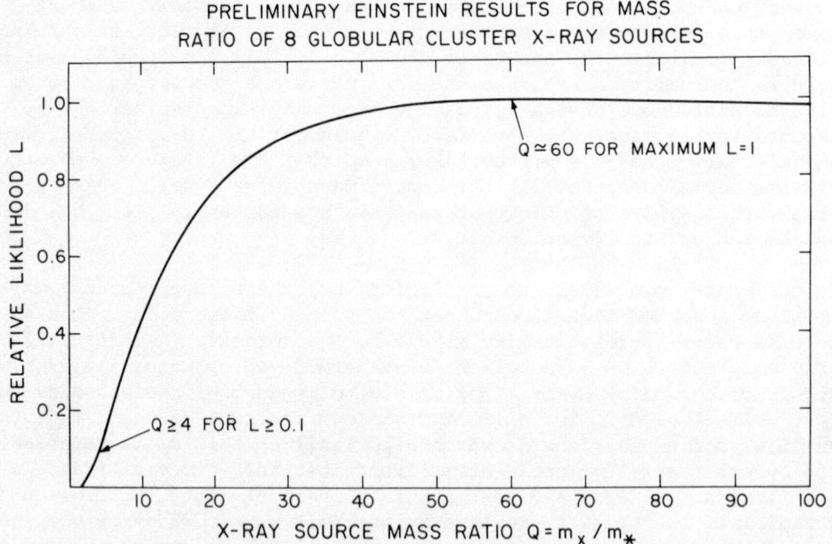

Figure 3: Relative likelihood distribution for X-ray source mass ratio q = M/m$_*$ in globular clusters derived from preliminary Einstein Observatory X-ray positions ($\pm$ 4") of sources in 8 globular clusters. The curve is normalized to unity at its maximum value (q $\simeq$ 60) and is not a probability distribution (see text).

ACKNOWLEDGEMENTS

The Einstein results described here are preliminary and are given
here as a partial progress report; they are described in detail in
the references given. I thank my co-authors on these papers –
particularly P. Hertz for the globular cluster optical position
analysis and H. Marshall for the Terzan 2 burst analysis. I also
thank S. Tremaine, A. Lightman and J. Bahcall for discussions as
well as all other members of the Einstein team at CFA, without whom
these preliminary results would not have been possible.

REFERENCES

Arp, H.C., 1965. In "Galactic Structure", Ed. A. Blaauw, and M.
     Schmidt, (Chicago: University of Chicago Press), p. 401.
Bahcall, J., and Ostriker, J., 1975. Nature, 256, 23.
Bahcall, J., and Wolf, R.A., 1976. Ap. J., 209, 214.
Canizares, C., McClintock, J., and Grindlay, J., 1979. Ap. J. (Let-
     ters), submitted.
Clark, G.W., 1975. Ap. J., (Letters), 199, L143.
Clark, G., Markert, T., and Li, R., 1975. Ap. J. (Letters), 199, L93.
Cominsky, L., Forman, W., Jones, C., and Tananbaum, H., 1977. Ap.
     J. (Letters), 211, L9.
Cowie, L., Ostriker, J., and Stark, A., 1978. Ap. J., 226, 1041.
Forman, W., Jones, C., and Tananbaum, H., 1976. Ap. J. (Letters),
     207, L25.
Forman, W., Jones, C., Cominsky, L., Julien, P., Murray, S., Peters,
     G., and Tananbaum, H., 1978. Ap. J. (Suppl.), 38, 4.
Giacconi, R., Murray, S., Gursky, H., Kellogg, E., Schreier, E., and
     Tananbaum, H., 1972. Ap. J., 178, 281.
Giacconi, R., et al., 1979. Ap. J., 230, 540.
Grindlay, J.E., 1977. Highlights of Astronomy, Vol. 4, 111.
Grindlay, J.E., 1978a. Ap. J., 221, 234.
Grindlay, J.E., 1978b. Ap. J. (Letters), 224, L107.
Grindlay, J., and Hertz, P., 1979. In preparation.
Grindlay, J., Gursky, H., Schnopper, H., Parsignault, D., Heise, J.,
     Brinkman, A., and Schrijver, J., 1976. Ap. J. (Letters), 201,
     L127.
Grindlay, J., Gursky, H., Parsignault, D., Cohn, H., Heise, J., and
     Brinkman, A., 1977. Ap. J. (Letters), 212, L67.
Grindlay, J., Hertz, P., Branduardi, G., Lightman, A., and Murray,
     S., 1979a. In preparation (Einstein survey and positions for
     Globular Clusters).
Grindlay, J., Marshall, H., Hertz, P., Soltan, A., Weisskopf, M.,
     Elsner, R., Ghosh, P., Darbro, W., and Sutherland, P., 1979b.
     In preparation (Terzan 2 burst).
Hills, J.G., 1975. A.J., 80, 1075.
Illingworth, G., and King, I., 1977. Ap. J. (Letters), 218, L109.
Jernigan, G., and Clark, G., 1979. Ap. J. (Letters), 231, L125.
Joss, P.C., 1978. Ap. J. (Letters), 225, L123.
Joss, P., and Rappaport, S., 1979. Astron. and Astrophys., 71, 217.

Kylafis, N., and Lamb, D., 1979. Ap. J., (Letters), 218, L105.
Lewin, W.H.G., and Joss, P.C., 1977. Nature, 270, 211.
Lewin, W.H.G., and Clark, G.W., 1979. These proceedings.
Lewin, W.H.G., et al., 1976. Ap. J. (Letters), 207, L95.
Lightman, A.P., and Bahcall, J., 1976. Private communication in
     Bahcall and Wolf, 1976.
Liller, W., 1977. Ap. J. (letters), 213, L21.
Markert, T., Backman, D., Canizares, C., Clark, G., and Levine, A.,
     1975. Nature, 257, 32.
Marshall, H., Grindlay, J., Weisskopf, M., and Elsner, R., 1979. In
     preparation.
Parsignault, D.R., and Grindlay, J.E., 1978. Ap. J., 225, 970.
Peterson, C.J., and King, I.R., 1975. A.J., 80, 427.
Seward, F., Page, C., Turner, M., and Pounds, K., 1976. MNRAS, 175,
     39P.
Silk, J., and Arons, J., 1975. Ap. J. (Letters), 200, L131.
Swank, J., Becker, R., Boldt, E., Holt, S., Pravdo, S., and Serlem-
     itsos, P., 1977. Ap. J. (Letters), 212, L73.
Van den Heuval, E., 1977. Ann. N.Y. Acad. Sci., 302, 14.
Van Paradijs, J., 1979. Ap. J. (Letters), In press.

PART 5
Recent results

Galactic X-ray sources as observed with the Einstein Observatory

Ethan J. Schreier

Harvard/Smithsonian Center for Astrophysics,
Cambridge, Massachusetts 02138,
USA

1.   INTRODUCTION

The advent of imaging in X-ray astronomy as manifested in the Einstein Observatory, has allowed the field to acquire a new maturity. Although many new results were expected in extragalactic areas of study - the X-ray background, quasars and other active galaxies - new results both expected and serendipitous also have much bearing on galactic X-ray astronomy. The several arcsecond resolution and three order of magnitude increase in sensitivity provides new ways of studying galactic X-ray sources.

As a result, in many respects galactic X-ray astronomy has come full circle. The study of stellar X-ray sources, a category limited to only the sun for 30 years, is now becoming a significant branch of X-ray astronomy. The canonical strong galactic X-ray sources, studied in detail throughout the 1970's, can now be observed in other galaxies. Supernova remnants can also be detected in other galaxies and can be mapped in great detail in our own galaxy.

I would like to present some Einstein observations relevant to galactic X-ray astronomy. This will in no sense be either a complete survey or a detailed study of any given area, but rather a demonstration of the observing capabilities now available to the astronomy community. Further results may be discussed informally during other presentations; Einstein observations of globular cluster sources will be discussed in detail at a later session.

382

2.   STELLAR X-RAY SOURCES

The first detected galactic X-ray source was the sun, a rather
normal G star emitting some $10^{26} - 10^{28}$ ergs s^{-1}. Many years of
detailed X-ray study followed its detection in 1948 by Burnight,
culminating in the SKYLAB X-ray telescope studies (Vaiana and Tucker
1974). We did not deal with normal stars as galactic X-ray sources
during the 1970's since we were studying the exciting and now
"classical" X-ray sources in the $10^{36} - 10^{38}$ erg s^{-1} range. However,
as detection sensitivites increased, especially at longer wavelengths, some types of stars began to be detected at significantly
lower luminosities. The dwarf nova - low mass binaries such as SS
Cygni, with hard and soft X-ray components - were not too much of an
extrapolation from the more typical higher mass binaries. However,
softer and weaker sources began to be detected - flare stars, hot
white dwarfs, AM Her types, RS CVn stars - showing a trend not
immediately recognised. Put simply, X-ray astronomy started seeing
stars in the late 1970's, even if they were the special cases. Vega
was the first extra solar solitary main sequence star seen in X-rays,
at a level of about 3×10^{28} ergs s^{-1} (Fabricant et al., 1979). The
fact that it could be seen with an imaging experiment on a rocket
flight - 7 photons in about 5 seconds - demonstrated the potential
and need for imaging in studying X-ray activity in normal stars.

The first Einstein result in this area was the observation of X-ray
emission from early type stars. The use of Cygnus X-3 as an observatory boresight permitted the viewing of the Cygnus O-B association
and the serendipitous discovery of X-ray emission from O and B stars.
Figure 1 is the Imaging Proportional Counter (IPC) image of the
region; five new X-ray sources are clearly seen, in addition to the
known intended target, Cygnus X-3, near the bottom of the field. The
X-ray positions are shown overlaid on a Palomar Survey red print in
Figure 2. Source B is in fact resolved by the High Resolution
Imager (HRI) into two discrete sources.

The X-ray sources correspond to the brightest members of the
association, with luminosities of 10^{33} to 10^{34} ergs s^{-1} and L_{Xray}/L_{opt} ratios of order 10^{-4}. The data are in fact consistent with all
the members emitting at this level. The X-ray spectra indicate far
softer emission ($kT \stackrel{<}{\sim} 1$ keV) than is typical of the stronger galactic sources.

Several possible models can be considered (Harnden et al. 1979). One
likely model invoking a warm stellar wind is discussed by Cassinelli
elsewhere in these proceedings. Whatever the model, it is apparent
that these data indicate a new class of X-ray sources associated
with early type stars has been discovered.

When η-Carina, also in a region of star formation, was observed,
again many O stars were observed. The region had been a puzzle for
some years, with inconsistencies in the results of various observers
at different X-ray energies. The complicated structure, as observed

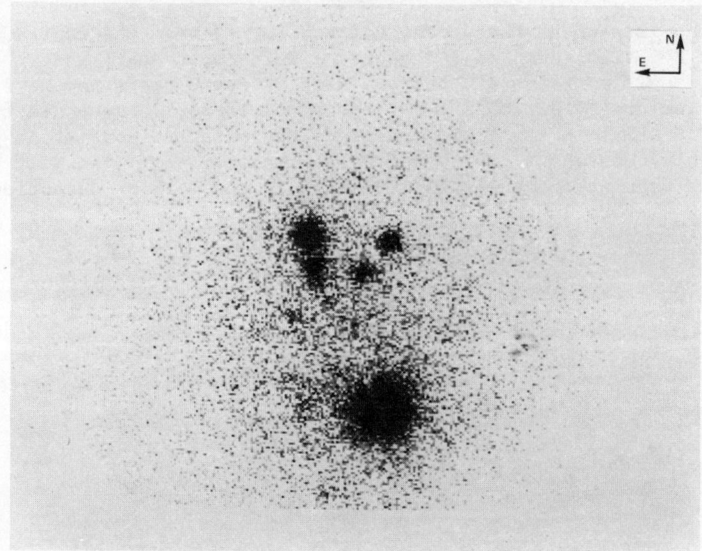

Figure 1. Einstein Observatory 10000 second IPC image of the VI Cygni OB association. The grey scale is logarithmic in intensity; the bright source in the lower right is Cygnus X-3.

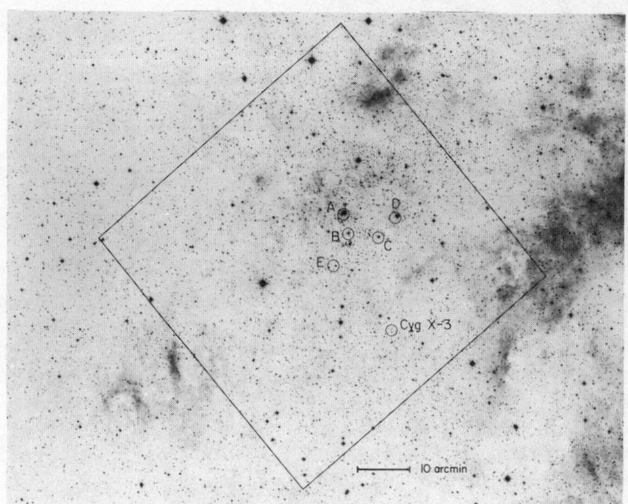

Figure 2. Palomar Sky Survey red print with 1 arcmin circles overlayed at the positions of the X-ray sources seen in Figure 1.

by the Einstein Observatory, can now help resolve the problem.

η-Carina is seen as the strong compact source near the centre of Figure 3. Also seen are 3 O stars, a Wolf-Rayet star, and Tr 14 which itself contains 3 O stars. Each of the O stars in the field has a luminosity of about 10^{33} ergs s^{-1}. In fact, since all the bright O stars in the field are seen, the data are consistent with a roughly constant X-ray to optical emission ratio. The Wolf-Rayet star is seen at about 3×10^{33} ergs s^{-1}; the lack of detection of 2 other W-R stars in the field of view indicates no such X-ray optical correlation exists.

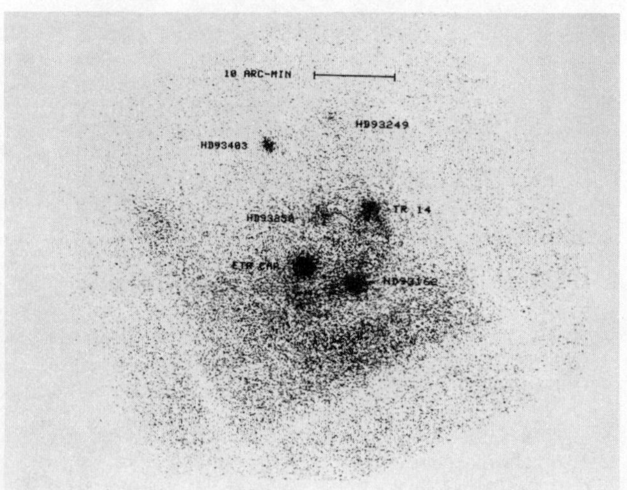

Figure 3.   X-ray image of the Eta Carinae region from an 8000 second IPC exposure. Source identifications are indicated. The shadow of the counter window support structure is seen near the edges of the field of view.

η-Carina itself, at about 3×10^{33} ergs s^{-1}, represents only some 5% of the total soft X-ray emission in the field. The diffuse emission is also apparent in Figure 4, where contours of constant X-ray surface brightness can be compared with the UV photograph. The data are consistent with X-ray emission from hot gas in front of the colder gas responsible for the bright nebula, but behind the western dust lane, seen in both optical and X-ray images. The origin of the hot gas may possibly be in the stellar winds or in supernova which are very likely in such a region.

The HRI was also used to image the region, and show η-Carina itself to have a complicated morphology. There is a compact region of X-ray emission coincident with the bright optical centre, perhaps the corona of a massive star. There is also a bright horseshoe-shaped

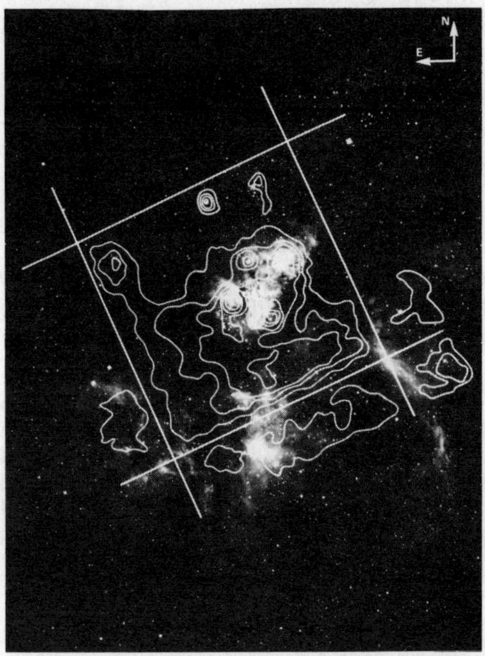

Figure 4. Contours of constant X-ray surface brightness, as seen in
          Figure 3, overlaid on a UV photograph (courtesy W.
          Liller, CTIO Curtis Schmidt).

region and a faint extension to the southeast. It might originate
from an outgoing shock (there was a famous outburst of η-Carina in
1843), but is not strong enough to suggest a supernova.

Finally, many other stars are being seen in X-rays throughout the
main sequence. These observations take many forms. There are
observations of specific stars such as Sirius, where preliminary
results show emission from both the white dwarf and the A star (with
a luminosity ratio of ∿ 100 to 1). There are specific observing
programmes to study emission from classes of stars, being performed
by our group at CFA as well as by many guest observers. Finally,
there are serendipitous sources, most long observations on a given
target detecting new sources which often turn out to be either QSOs
or F- and G-type stars.

Although the results are still preliminary, it is apparent that stars
throughout the main sequence are X-ray emitters. Emission was expected at the general level seen for solar type stars, with slow rotation rates and convective zones. However, the observed luminosities
of early and especially late type stars are far higher than would be
predicted by the existing models. It is expected that X-ray
observations of stars will lead to significant improvements in our
understanding of stellar evolution.

3.   SUPERNOVA REMNANTS

The study of supernova remnatns is relevant in many areas, ranging
from stellar evolution to the physics of neutron stars. In fact,
there are enough different phenomena involved to make each SNR
studied somewhat unique. We can expect to see synchrotron emission,
shock waves, cooling glouds of gas, and hot neutron stars. It is
also interesting to note that black body emission from hot neutron
stars was one of the earliest explanations for X-ray sources.
Although this has long been recognized as an inadequate model for
the bulk of the emission, the current observations may now allow
experimental evaluation of neutron star models.

The Crab is certainly one of the best studied SNR's, containing an
extended synchrotron source powered by a pulsar. The Einstein
observation of the Crab is shown in Figure 5. The pulsar is clearly
seen, as is extended emission of order 5 x 7 arcmin generally
correlated with the optical features. A detailed timing study of
the data is underway. At this time, it is possible to limit the
non-pulsed emission of the pulsar to less than 5% of its total
emission. This corresponds to an upper limit to its black body
temperature of about 3×10^6 K.

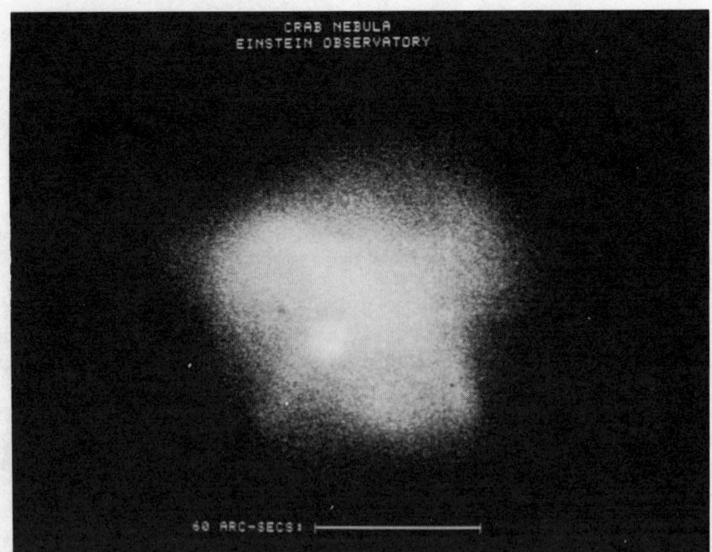

Figure 5:   An Einstein HRI exposure of the Crab, showing the pulsar
            and the extended synchrotron emission. *(Reprinted by
            permission from Science, "The Einstein Observatory: New
            Perspectives in Astronomy", Giacconi, R. & Tananbaum, H.,
            Vol. 209, pp865-876, 22 August 1980. Copyright 1980 by
            the American Association for the Advancement of Science.)*

The only other SNR currently known to contain a pulsar is Vela.
This is a very old and very close remnant; thus it is very large,
about 5° in diameter, and predominately thermal. Several 1^O fields
have been looked at with Einstein; the general emission is found to
be very irregular and splotchy. However, the region around the
pulsar is found to have a harder component. Figure 6 shows a small
extended source centered on the pulsar; this may represent a small
Crab-like synchrotron emission region. The emission is not found to
be pulsed, with a limit of about 1% of the discrete source, or about
2×10^{-12} ergs s^{-1}. The total emission associated with the pulsar
has not yet been deconvolved, but would predict a black body
temperature of order 1.5×10^6 ^{O}K. It is also interesting to note
that the Vela pulsar is observed in optical and γ-rays; the upper
limit to pulsed emission is not above the interpolation between
these. It is not obvious why pulsations should not be seen in the
X-rays.

Figure 6.   An Einstein HRI exposure of the Vela pulsar. Both a
point source and a small extended source are seen.

Many other SNR's have been observed by Einstein, including Cassiopea, Tycho, SN 1006, Puppis, and several SNR's in the LMC; no point sources have yet been found. In several cases, the upper limits on black body temperatures for any neutron stars present are already below what would be predicted from standard neutron star models. It should be noted that in the younger SNR's, the neutron star should be found not too far from the centre. This makes the existing problem of neutron star production even worse. The estimated supernova rate is too low to produce the number of pulsars observed in the radio. If we now find that most supernova do not produce neutron stars, another production mechanism might be necessary.

Just as the Crab is the canonical synchrotron source, Cas A shows many thermal X-ray emission features we might expect. Figure 7 is an HRI image of this SNR. The X-ray data show many features in common with both the optical and radio. Qualitatively there is a fuzzy outer shell, large bright areas within the shell, and weaker emission near the centre (Murray et al. 1979). The outer shell corresponds to the shock wave as seen in the radio; it is hard and thus does not appear very bright in the soft X-rays detected by Einstein. The bright spots are the soft ($\sim$ 1 keV) emission produced by gas clouds which have been heated by the shock wave and are

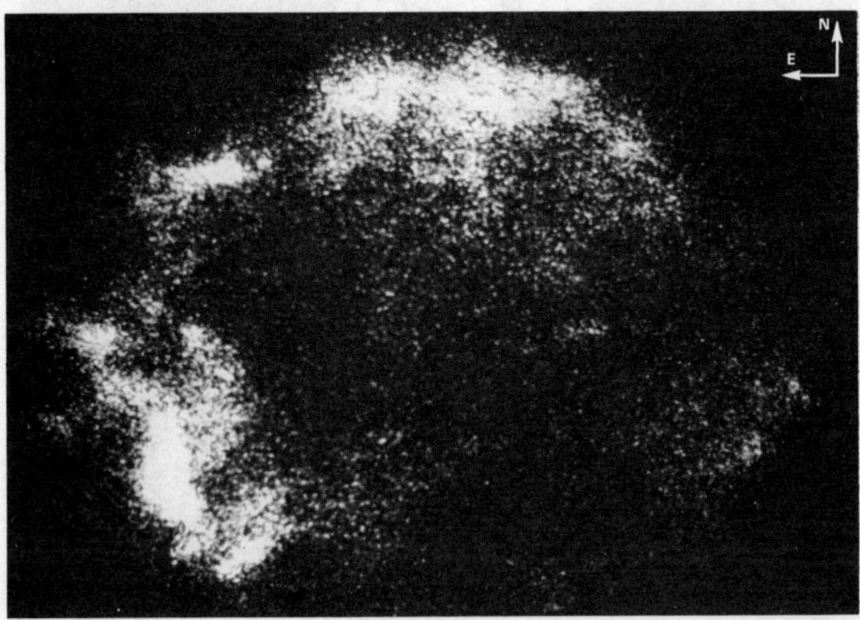

Figure 7. An Einstein HRI image of the Cas A supernova remnant representing a 32500 second exposure. The SNR is about 5 arcmin across. The outer ring of harder X-rays is very faint in the HRI band.

cooling. Nearer the centre, the gas has already cooled and thus is less luminous. It should be possible to relate the required cooling times to elemental abundances; these would complement the results obtained from the Solid State Spectrometer (elsewhere in this volume, and Becker et al. 1979).

4.   THE GALACTIC X-RAY SOURCES

As a final topic, I would like to discuss some work which is more relevant to the "classical" subject of galactic X-ray sources - the strong, discrete sources which are now known to be mass transfer binaries containing compact objects. The existence of X-ray binaries and pulsars, their time variability, their spectra, and the measured changes in the periods taught us about the nature of the compact objects and the nature of the binary systems. These studies led to constraints on the evolution of this type of star system in our galaxy. Furthermore, it became obvious that the integrated emission from these systems dominated the X-ray emission of our galaxy.

The Einstein Observatory is not the instrument of choice to perform more detailed timing studies of these objects, although better positions could facilitate some identifications. However, the natural extension of this work is in the study of "galactic" sources in other galaxies. The detection of approximately half dozen sources in the Magellanic Clouds proceeded apace with studies of our own galactic sources during the early 1970's. The LMC is currently being mapped with Einstein. Preliminary results show in excess of 50 sources down to $\sim 10^{34}$ ergs s^{-1}, including many SNRs (Long and Helfand 1979). There appear to be more bright sources per unit mass, and larger SNRs for a given luminosity. This is consistent with a lower abundance of heavy elements leading to longer cooling times.

Andromeda, or M31, is the nearest galaxy of similar type to our own. It was one of the weakest sources in the 4U catalogue, and as such is one of the more spectacular examples of the power of imaging optics in X-ray astronomy. One of the scientific factors of importance in studying the "galactic" sources in M31 is that they form a constant distance sample. Thus, the ambiguity of varying distances and low energy cutoffs present in the study of sources in our own galaxy is removed. Furthermore, the distribution of sources in the galaxy, and the categorization into bulge and spiral arm sources will be greatly simplified. Although the analysis is in its infancy, enough sources have been detected to draw some general conclusions based on group characteristics.

The galaxy is largely covered with 3 IPC fields, each over 30000 seconds effective exposure. As of this time, the northeast and central IPC fields have been reduced (Van Speybroeck et al. 1979). In addition, a 27000 second HRI image of the central region has

been studied. Figures 8 and 9 show respectively the IPC and HRI
images of the central region. The large extended region seen in the
IPC is seen to be resolved into discrete sources by the HRI.

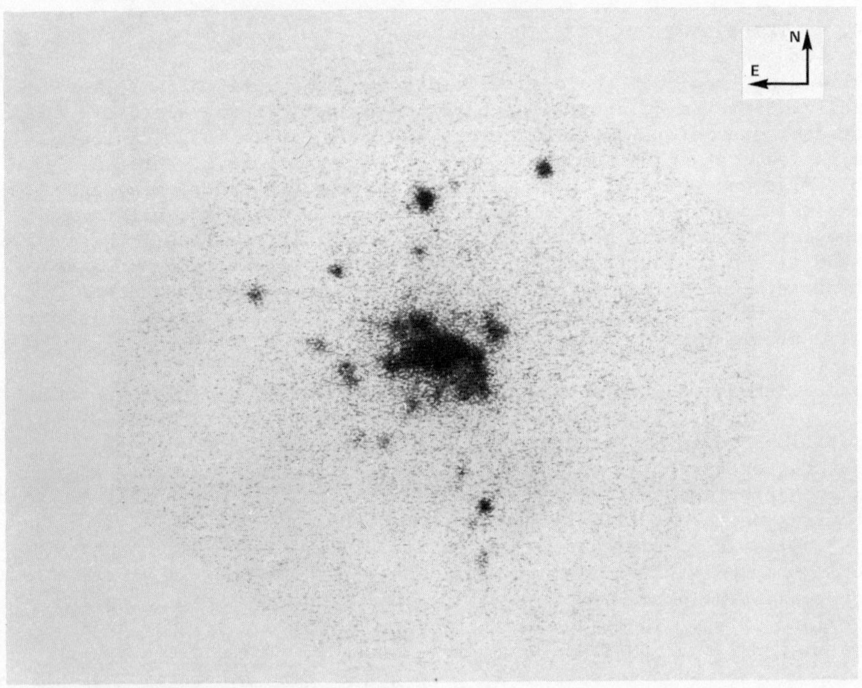

Figure 8.   Einstein IPC 36000 second exposure of the central 1^O
            field in M31. The nuclear region is unresolved with the
            arcminute resolution.

Approximately 70 sources have been found, with luminosities ranging
from about 5 x 10^{36} ergs s^{-1} to 2.5 x 10^{38} ergs s^{-1} (0.5 - 4.5 keV).
No significant intrinsic absorbtion is seen. In the central region
the sum of the luminosities of the HRI sources agrees well with the
IPC integral luminosity. Furthermore, the sum of all discrete
sources seen in the entire galaxy (assuming similar contributions
from the SW and NE regions) is consistent with the integral M31
luminosity observed by Uhuru. Thus, it appears that most of the M31
emission comes from the discrete sources we are seeing and not from
diffuse emission or large numbers of sources below the threshhold
of sensitivity. Several sources are seen to be variable on time
scales of hundreds of seconds to days. Further analysis is being
performed.

Figure 9.   Einstein HRI 27000 second exposure of the central region
of M31.

A reasonable division of the sources can be made into inner bulge,
globular clusters, and Population I categories. There is a source
within 2 arcsec of the nucleus of M31, with a luminosity of 1 x
10^{38} ergs s^{-1}. If this is indeed the nucleus, it is quite
different from any possible point source at the nucleus of our own
galaxy, which would be 10^3 or more times fainter than that of M31
in X-rays (A. Epstein, private communication), but far stronger in
radio. Thus, a thermal model might be suggested.

There are on order of 10 X-ray sources which could be associated
with globular clusters, out of about 240 in the fields of view. The
average luminosity is about 2.5 x 10^{37} ergs s^{-1}. There are some 15
to 20 sources which can be defined as inner bulge sources, with a
mean luminosity of 4.5 x 10^{37} ergs s^{-1}. They tend to be highly
concentrated within about 400 pc of the nucleus, much more so than
in our own galaxy.

The remaining 40 sources are categorized as Population I. Their
average luminosity of 2 to 3 x 10^{37} ergs s^{-1} is less than that of
the inner bulge sources. They are defined by their association with
bright visible objects, HI regions, and dust features. This population tends to extend further in towards the nucleus than does the
comparable population in our own galaxy. This may indicate a real

difference in the relative populations between the two galaxies, or may indicate that we have to reevaluate our interpretation of our own galaxy.

Whatever the explanation of the differences between M31 and our own galaxy, it is apparent that the detailed Einstein observations of sources in other galaxies is opening a new branch of galactic X-ray astronomy. Qualitatively new insights into the study of X-ray source populations, stellar evolution and galactic mass distributions should be available. The study of galactic X-ray sources will be pushed even further by the determination of galactic luminosity functions, as we accumulate data through surveys of nearby galaxies, galaxies in clusters, and even the serendipitous discovery of galaxies via the identification of X-ray sources in deep surveys of "empty" fields.

ACKNOWLEDGEMENTS

Much of the data presented above was unpublished at the time of the meeting. Where possible, references to later published papers have been added. I would like to thank the many scientists who allowed their work to be discussed prior to publication. The support of the entire Einstein Observatory staff - programmers, engineers, secretaries, data aides, and scientists - at the Consortium institutions and at NASA is also sincerely appreciated. Research has been supported under NASA contract NAS8-30751.

REFERENCES

Becker, R.H., Holt, S.S., Smith, B.W., White, N.E., Boldt, E.A., Mushotzky, R.F., and Serlemitsos, P.J., 1979. Ap.J. (Letters), 234, L73.

Fabricant, D., Topka, K., Harnden, F.R.Jr., Gorenstein, P., and Rosner, R., 1979. Ap.J., 229, 661.

Harnden, F.R.Jr., Branduardi, G., Elvis, M., Gorenstein, P., Grindlay, J., Pye, J.P., Rosner, R., Topka, K., and Vaiana, G.S., 1979. Ap.J.(Letters), 234, L51.

Long, K.S., and Helfand, D.J., 1979. Ap.J.(Letters), 234, L77.

Murray, S.S., Fabbiano, G., Fabian, A.C., Epstein, A., and Giacconi, R., 1979. Ap.J.(Letters), 234, L69.

Vaiana, G.S., and Tucker, W.H., 1974. in X-ray Astronomy, eds. R. Giacconi and H. Gursky, D. Reidel, Holland, p. 169.

Van Speybroeck, L., Epstein, A., Forman, W., Giacconi, R., Jones, C., Liller, W., and Smarr, L., 1979. Ap.J.(Letters), 234, L45.

High resolution soft X-ray spectra with the objective grating spectrometer aboard Einstein Observatory

J. Heise and A. C. Brinkman

Laboratorium voor Ruimte-Onderzoek,
Utrecht, The Netherlands.

1.   Introduction

2.   Transmission Gratings

3.   The spectrum of an optically thin source

4.   Herculis X-1

5.   AM Herculis

1.   INTRODUCTION

For the purpose of measuring high resolution soft X-ray spectra the
Einstein X-ray Observatory (Giacconi et al., 1979) has been equipped
with an Objective Grating Spectrometer (OGS). The OGS consists of
two transmission gratings, either of which may be placed in the X-ray
optical path behind the High Resolution Mirror. The gratings consist
of opaque and open strips with a density of 500 and 1000 lines/mm.
The dispersion is 0.128 and 0.063 $\overset{o}{A}$/arc sec, respectively. When the
High Resolution Imaging Detector (HRI) is used to detect the diffrac-
ted image the spectral resolution, $\Delta\lambda$, obtained between 0.3 and 2 $\overset{o}{A}$,
is limited by the mirror resolution (image quality) at short wave-
lengths and by aberrations at longer wavelengths (see Table 1). For
wavelengths greater than 44 $\overset{o}{A}$ the spectral resolution is more than an
order of magnitude better than that attainable with other means.

	500 lines/mm	1000 lines/mm
Dispersion ($\overset{o}{A}$/arcsec on HRI)	0.128	0.063
Range, HRI center on-axis 1st order	5 - 100$\overset{o}{A}$	3 - 50 $\overset{o}{A}$
Range, HRI edge-on axis 1st order	5 - 200 $\overset{o}{A}$	3 - 100 $\overset{o}{A}$
Resolution (wide mask)	2 $\overset{o}{A}$	1 $\overset{o}{A}$
Resolution (narrow mask)	0.5 $\overset{o}{A}$	0.25 $\overset{o}{A}$

TABLE 1 - GRATING CHARACTERISTICS

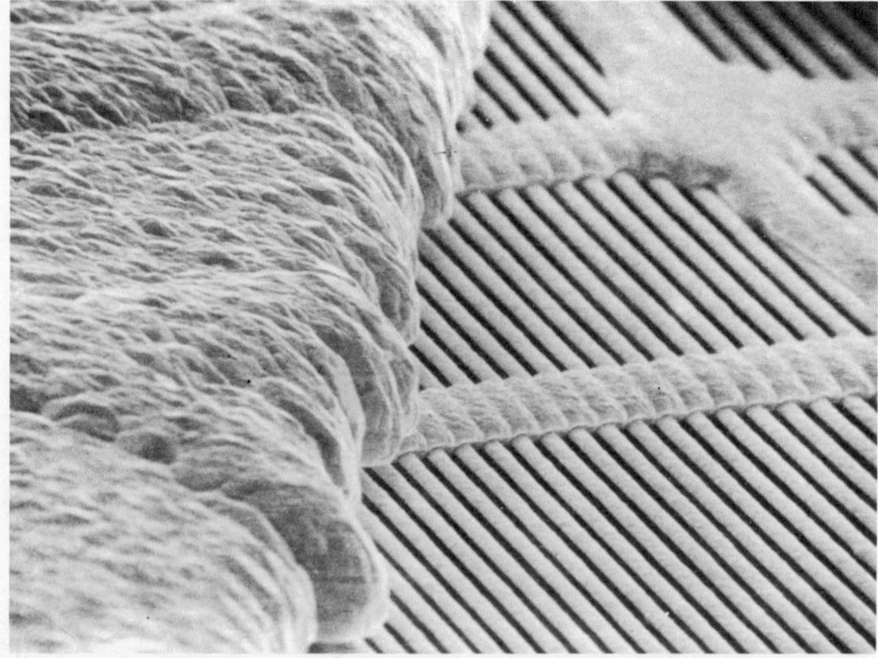

Figure 1:   An electron micrograph of a small portion of a 1000 lines/
mm grating.   Coarse and fine support structures are seen
in addition to the grating wires.

Two observations periods have taken place so far, one in March and one in May, 1979.

In this talk we will present some preliminary data obtained from the binary compact X-ray source Hercules X-1 and from the close binary white dwarf, AM Hercules.

2.   TRANSMISSION GRATINGS

The gratings, prepared at the Space Research Laboratory at Utrecht, consist of gold lines nominally 0.1 µm thick and a bar width to slit ratio approximately equal to one. A fine and coarse structure supports the thin wires (Figure 1). The gratings are produced in elements of approximately 5 x 5 cm. 160 of these elements for each grating are mounted and aligned on a support frame. The 1000 lines/mm grating has been strengthened by a coating of 2.5 µm parylene N. Two types of calibration have been carried out. First a few representative elements were measured at long wavelengths, using the SSRP 4° beam line at SPEAR (see Schnopper et al., 1977). Later, when completely assembled, the grating was measured in combination with the Einstein mirrors and High Resolution Imager, by F. Seward, in a long beam test facility.

Testing of individual grating elements over a wide range of wavelengths is necessary because the grating bars are partly transparent depending on the thickness and the incident X-ray energy. Interference between the radiation coming through the opening slit and the attenuated, phase shifted radiation coming through the grating bars, leads to a distribution over the various spectral orders which is different from the one expected for completely opaque grating bars. At some wavelengths the power in the zero order is diminished (by destructive interference) and reappears in an enhanced, higher order, diffracted, beam through constructive interference. The wavelength where this occurs is a function of the optical constant for the grating material and the thickness of the grating bars and can be used to advantage by optimising the choice of material and the thickness of the grating (Bräuniger, 1979). This has been implemented for the gratings prepared at the Space Research Laboratory in Utrecht, for future use in the EXOSAT Observatory (Brinkman et al., 1980). For the Einstein observatory the fabrication of the gratings, was limited to a maximum thickness for the gold lines to 0.1 µm. This places the region of enhanced sensitivity in the first order for the soft X-ray range between 40 and 80 Å, and the sensitivity at wavelengths lower than approximately 15 Å is degraded.

The integrated experiment (mirror, integrated grating ring and High Resolution Imaging Detector) has been calibrated with line radiation characteristic of boron (66.9 Å), carbon (44.7 Å), iron (17.6 Å), copper (13.3 Å), aluminium (8.33 Å), zirconium (6.06 Å), silver (4.15 Å) and tin (3.60 Å) (Seward, 1979). An example of a spectrum obtained in this way is shown in Figure 2, which shows the calibration spectrum of iron.

396

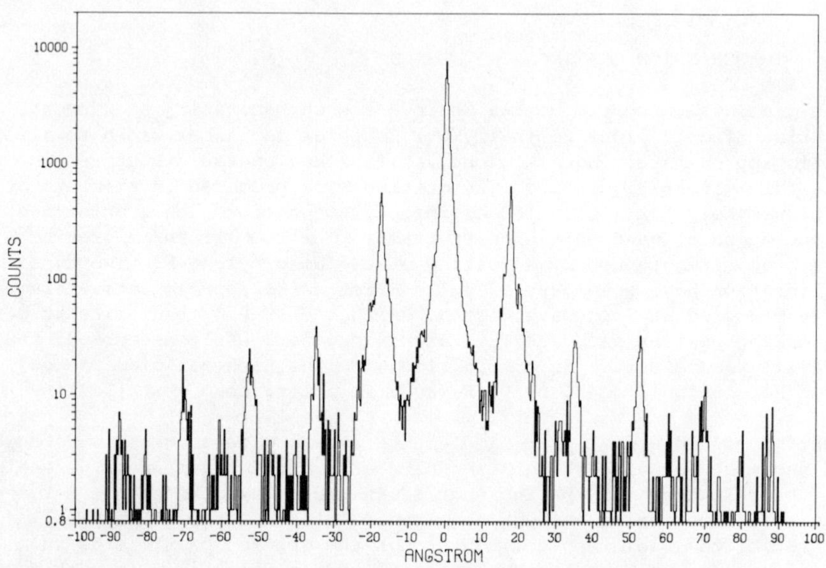

Figure 2:   Calibration spectrum of iron shows up to the fifth
            spectral order.

A calibration with beryllium continuum radiation shows that the only
instrumental feature in the spectrum is a discontinuity at 15.7 Å
caused by the nickel-L edge of the nickel coating on the mirror.

The effective area as a function of wavelength, derived from these
calibrations is given in Giacconi et al., (1979). The shape is
roughly described by a fall off at shorter wavelengths because of the
effect of increasing grating transparancy,the decreasing sensitive
area of the mirror and an overall shape determined by the X-ray
absorption caused by the UV-ion filter which covers the HRI. This
latter effect is most clearly seen as a strong absorption edge shortward of the carbon K edge at 43.7 Å and a decrease in sensitivity
towards longer wavelengths. These sensitivities are used to derive
the photon spectra given later. They are preliminary, because of
other effects which have yet to be taken fully into account. For
example the HRI used in flight is different from the one used in the
calibrations. These still have to be corrected. Also the width of
the spectrum, perpendicular to the dispersion actually used in the
observations is different from the width used in the calibrations,
which causes an underestimate in the low energy flux.

3.   THE SPECTRUM OF AN OPTICALLY THIN SOURCE

The expected X-ray spectrum of an optically thin source with solar
abundances and temperatures in the range 10^6 to 10^7 K is dominated
by spectral lines. Extensive calculations on continuum and line
emission (Mewe, 1972; Mewe, 1975) have recently been extended to
include newly available atomic data, in particular concerning the
iron complex around 17 $\overset{o}{A}$ (Gronenschild, 1979).

Figure 3 gives an example of such a spectrum multiplied with the
instrument response function but not yet convolved to the appropriate
resolution. The plasma temperature is 10^7 K. Intensities are given
for a source at a distance of 1 pc with emission measure of 10^{50} cm^{-3}.
The spectral lines are labelled with the ion species and the spectral
order.

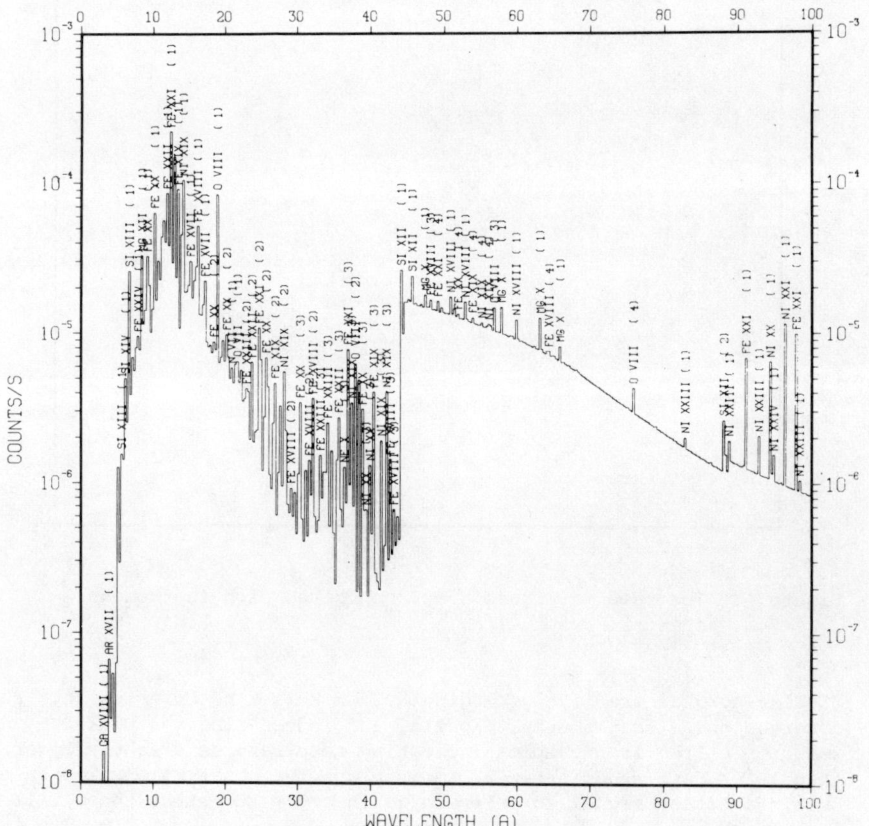

Figure 3:   Expected X-ray spectrum as a function of wavelength of an
optical thin plasma of $T = 10^7$ K multiplied by the
instrument response function of the 1000 lines/mm grating.
Ion and spectral orders are indicated.

Figure 4 shows the same spectrum now convolved with the instrument reso-
luation. Most lines are not individually resolved. The spectrum is
dominated by the iron complex, in this case Fe XVIII to Fe XXIII.
The relative contribution of the various iron lines and hence the
location of the peak in the spectrum is a sensitive function of the
temperature.

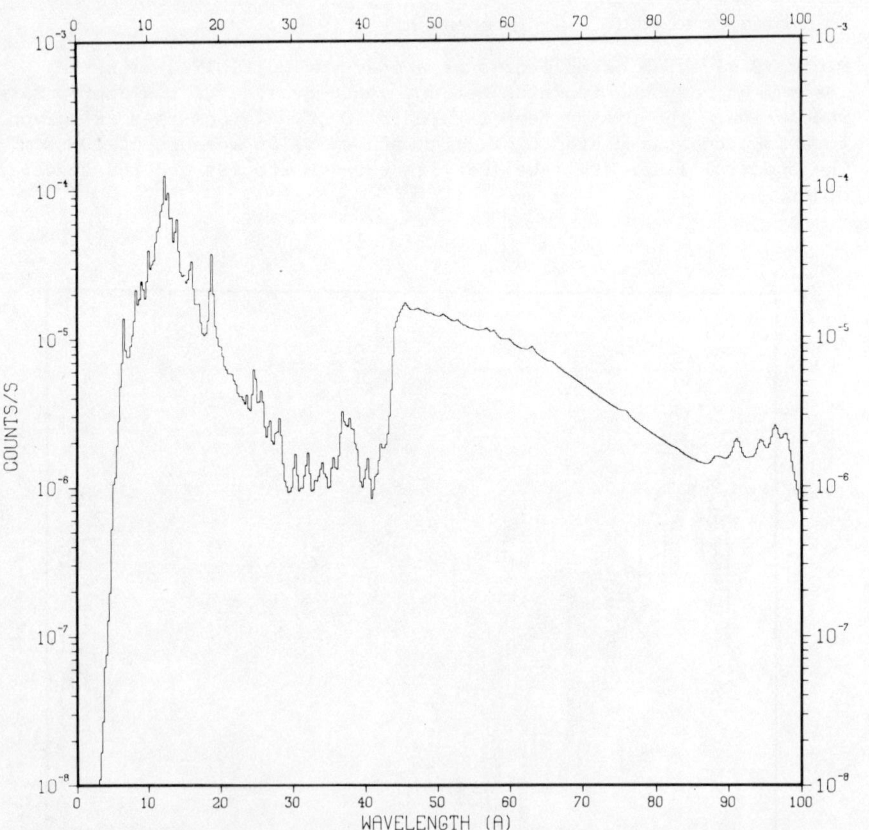

Figure 4:   The same as Figure 3 but convolved with instrument
            spectral resolution.

Stellar coronae are likely candidates for this kind of emission. We
observed the strong source, Capella, and indeed find the peaked
emission of the iron complex indicating temperatures slightly lower
than 10^7 K. We need, however, to increase the observation time to
obtain a better signal to noise ratio in order to detect the O VIII
line.

4. HERCULES X-1

Figure 5 shows an example of two 'quick-look' observations with the
500 lines/mm grating on the binary X-ray source Hercules X-1 as it
appears on the television screen at the data centre of Smithsonian
Astrophysical Observatory. A scale in Ångstroms has been added,
together with the positions for some of the calibration lines. The
bottom of the graph shows the intensity distribution for one of the
observations. A strong zero order image is seen, together with the
positive and negative first order images, each cut in two pieces
because of the jump in the absorption of the detector's UV-ion filter,
at 43.7 Å. A number of these observations of Her X-1 with the 500
lines/mm grating on have been combined to obtain Figure 6. The data,

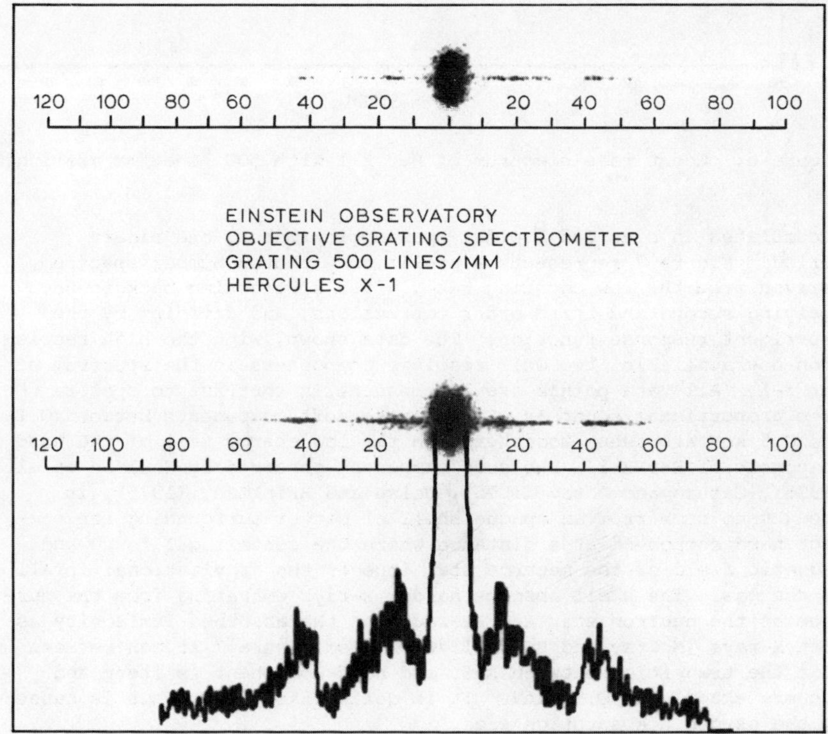

Figure 5: Two quick-look observations with the 500 lines/mm grating
on Her X-1 as it appears on the television screen at the
data centre at SAO.

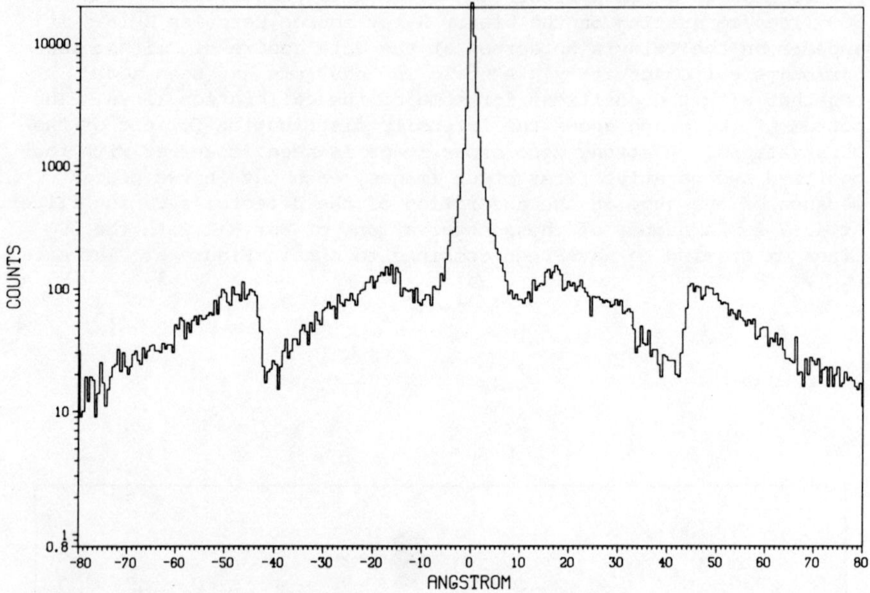

Figure 6: Count rate spectrum of Her X-1 with 500 lines/mm grating.

accumulated in bins 0.5 Å wide, span approximately one binary
period. Figure 7 represents a preliminary photon number spectrum,
derived from the raw counting rate, after subtracting background,
applying second and third order corrections, and dividing by the
experiment response function. The data shown, with the high resolution now available, two well resolved components in the spectrum of
Her X-1. All data points are independent, in contrast to spectra
from proportional counters. The excess soft components between 0.1
and 0.5 keV are seen, together with the low energy tail of the hard
component of Her X-1. The soft component observed by Shulman et al.,
(1975), Catura and Acton (1975), Heise and Brinkman, (1975), is
thought to come from an opaque shell of matter surrounding the compact hard component at a distance where the centrifugal force and
magnetic field of the neutron star impedes the gravitational infall
of the gas. The shell absorbs harder X-rays emanating from the surface of the neutron star and re-radiates the absorbed luminosity as
soft X-rays (McCray and Buff, 1976). From Figure 7 it can be seen
that the transition between soft and hard component is steep and
occurrs exactly at 0.53 keV. It is quite likely that this is caused
by the oxygen K absorption edge.

Figure 8 represents the spectrum of Her X-1 obtained a day earlier
with the 1000 lines/mm grating. Here an absorption feature is clearly
seen at 0.53 keV, possibly again the oxygen K-edge.

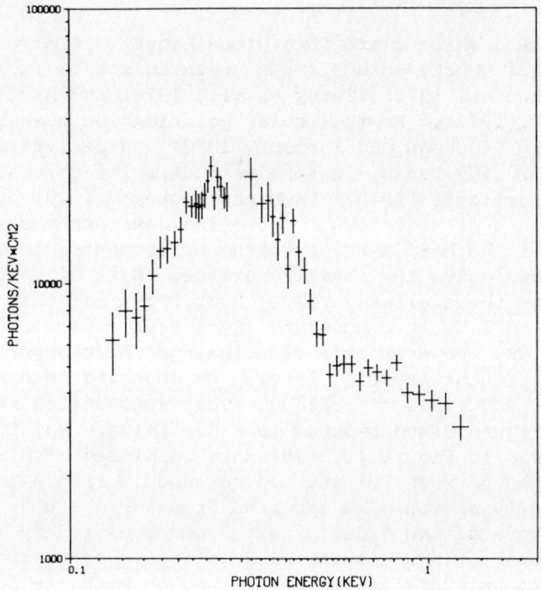

Figure 7:   Photon number spectrum of Her X-1 derived from Figure 6.

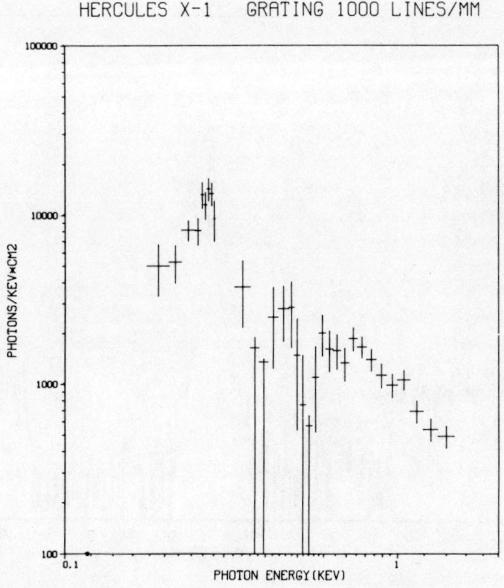

Figure 8:   Photon number spectrum of Her X-1 taken with 1000 lines/mm
            grating.

402

5. AM HERCULES

AM Hercules is a white dwarf in a close binary system with a small
period of $\sim$ 3.1 hours, which is also seen in soft X-rays (Hearn et
al., 1976; Bunner, 1978; Tuohy et al., 1978) and hard X-rays
(Swank et al., 1977). From circular polarisation measurements
(Tapia, 1977; Stockman and Sargent, 1978) one derives a large magnetic field $\sim$ 2.10^8 Gauss. Extensive models for this source (Pried-
horsky and Krzeminski, 1978; Lamb and Masters, 1979) predict three
major components of radiation. Accreting gas onto the magnetic
poles produces the hard X-ray spectrum by bremsstrahlung when a
shock is formed above the stellar surface. Half of this radiation
is absorbed at the stellar surface, heates it, and this subsequently
produces the soft X-ray component. Cyclotron emission would produce
a large UV flux. However this flux, has not been observed (Raymond
et al., 1979) at the expected level. We observed AM Her in March
1979 with the 500 lines/mm grating. The raw counting rate spectrum
is shown in Figure 9 and reduced in a preliminary way to a photon
number spectrum in Figure 10. The data is binned with a resolution
of 1 Å, plotted between 0.1 and 0.3 keV and it fits a smooth distribution very well, as would be expected from a black body with emission
at a temperature of the order of 40 eV and modified by interstellar
absorption (drawn line). When the additional corrections have been
applied the temperature can be determined to within a few eV.

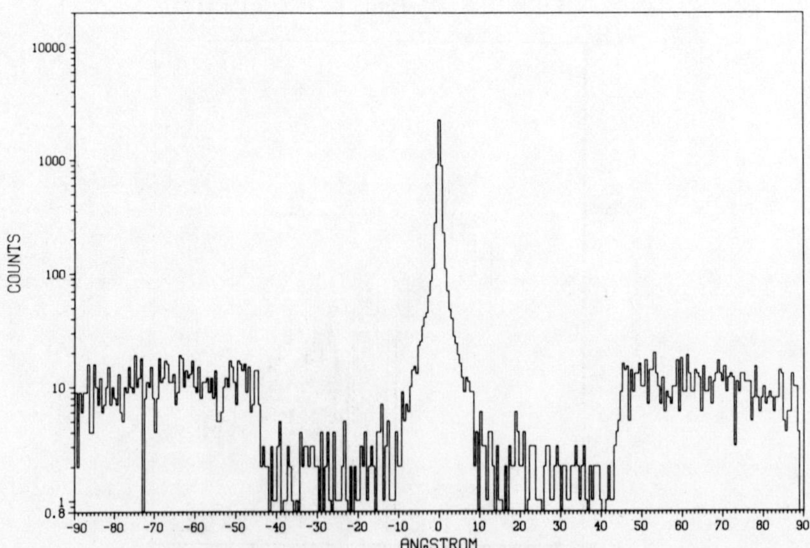

Figure 9: Count rate spectrum of AM Her taken with the 500 lines/mm
           grating.

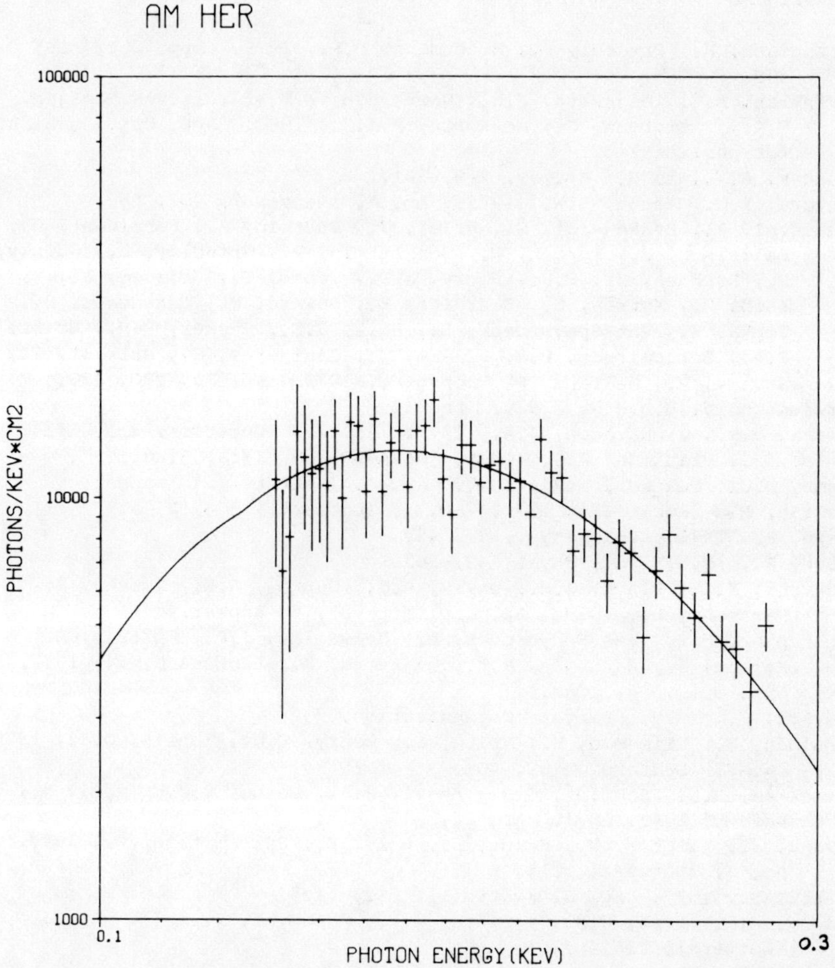

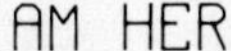

Figure 10:   Photon number spectrum of AM Her between 0.1 and 0.3 keV.

ACKNOWLEDGEMENTS

The work described here was in cooperation with Drs. J. H. Dijkstra, F. D. Seward, H. W. Schnopper, J.P. Devaille and A. Epstein. We thank Dr. R. McCray for making available his part of the 500 lines/mm grating observations of Hercules X-1

REFERENCES

Brauniger, H., Predehl, P., Beuermann, K.P., 1979. App. Opt., 18,
     368.
Brinkman, A.C., Dijkstra, J.H., Geerlings, W.F.P.A.L., van Rooijen,
     F.A., Timmerman, C., de Korte, P.A.J., 1980. App. Opt., submitted
     for publication.
Bunner, A.N., 1978. Ap. J., 220, 261.
Catura, R.C., Acton, L.W., 1975. Ap. J. (Letters), 202, L5.
Giacconi, R., Branduardi, G., Briel, U., Epstein, A., Fabricant, D.,
     Feigelson, E., Forman, W., Gorenstein, P., Grindlay, J., Gursky,
     H., Harnden, Jr. F.R., Henry, J.P., Jones, C., Kellogg, E.,
     Koch, D., Murray, S., Schreier, E., Seward, F., Tananbaum, H.,
     Topka, K., Van Speybroeck, L., Holt, S.S., Becker, R.H., Boldt,
     E.A., Serlemitsos, P.J., Clark, G., Canizares, C., Markert, T.,
     Novick, R., Helfand, D., Long, K., 1979. Ap. J., 230, 540.
Gronenschild, E.H.B.M., 1979. thesis.
Hearn, D.R., Richardson, J.A., 1977. Ap. J., (Letters), 213, L115.
Heise, J., Brinkman, A.C., 1975. NASA SP-389 "X-ray Binaries".
Lamb, D.Q., Masters, A.R., 1979. Ap. J. (Letters), to appear.
McCray, R., Lamb, J.K., 1976. Ap. J. (Letters), 204, L115.
Mewe, R., 1972. Sol. Phys., 22, 459.
Mewe, R., 1975. Sol. Phys., 44, 383.
Raymond, J.C., Black, J.H., Davis, R.J., Dupree, A.K., Gursky, H.,
     Hartmann, L., 1979. Ap. J. (Letters), to appear.
Schnopper, H.W., van Speybroeck, L., Delvaille, J.P., Epstein, A.,
     Kallne, E., Bachrach, R.Z., Dijkstra, J., Lantwaard, L., 1977.
     App. Opt., 16, 1088.
Seward, F., 1979. Private Communication.
Shulman, S., Friedman, H., Fritz, G., Henry, R.C., Yentis, D.J., 1975.
     Ap. J. (Letters), 199, 401.
Stockman, H.S., Schmidt, G.D., Angel, J.R., Liebeit, J., Tapia, S.,
     Beaver, E.A., 1977. Ap. J., 217, 815.
Swank, J., Lampton, M., Boldt, E., Holt, S., Serlemitsos, P., 1977.
     Ap. J. (Letters), 216, L71.
Tapia, S., 1977. Ap. J. (Letters), 212, L125.
Tuohy, J.R., Lamb, J.K., Garmire, G.P., Mason, K.O., 1977. Ap. J.
     (Letters), 226, L17.

Ultraviolet observations of
AM Herculis with IUE

J. C. Raymond

Harvard-Smithsonian Center for Astrophysics,
Cambridge, Mass. 02138, USA.

ABSTRACT

Spectra of AM Herculis taken with IUE show a continuum which is partially eclipsed in phase with the X-ray eclipse and can be separated
into two components: a black-body component (kT_{BB} = 25 - 30 eV) which
accounts for the $\frac{1}{4}$ keV X-rays and the eclipsed UV continuum, and a
component roughly described by $F_\nu \propto \nu^{-1}$ which is not eclipsed.

AM Herculis is a binary X-ray source believed to consist of a strongly
magnetic white dwarf and an M dwarf (cf. Tapia, 1977; Stockman et al.,
1977). On March 17, 1979 we observed AM Her with the IUE satellite
(Boggess et al., 1978) to follow up the observations reported in
Raymond et al., (1979a). Thirteen low dispersion short wavelength
spectra covering nearly two 3.1 hour orbital periods and three low
dispersion long wavelength spectra were obtained.

Continuum fluxes from the averages of these spectra are shown in
Figure 1. The line indicates $F_\nu \propto \nu^{-1}$, the spectral shape seen in
the optical (Stockman et al., 1977). It is clear that this spectral
shape accounts for the long wavelength part of the UV continuum
(2000 - 3000 $\overset{o}{A}$). The enhancement at short wavelengths is interpreted
as an additional component (with a Rayleigh-Jeans slope), which is
eclipsed in phase with the X-ray eclipse (Raymond et al., 1979a). A
similar short wavelength "turn-up" is seen in SS Cyg, however the
lack of eclipses makes the decomposition into black body and ν^{-1} components more ambiguous. On the other hand the thirteen short wavelength spectra of AM Her can be divided, almost unambiguously, into
black-body and ν^{-1} components (Raymond et al., 1979b).

Confirmation of the Rayleigh-Jeans nature of the eclipsed component
can be seen in Figure 2 which shows the difference in continuum
(averaged in intervals of 20 $\overset{o}{A}$) between two spectra. However, fluctuations of up to a factor of two in the level of the ν^{-1} component,
(Raymond et al., 1979b) and also confirmed by the long wavelength
spectra, lead to different spectral shapes for the differences between some of the pairs of spectra.

406

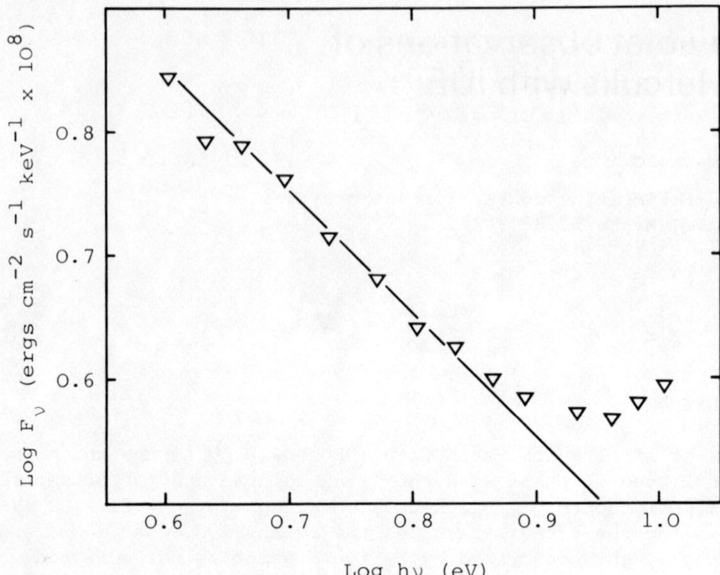

Figure 1: The average continuum flux from AM Herculis.

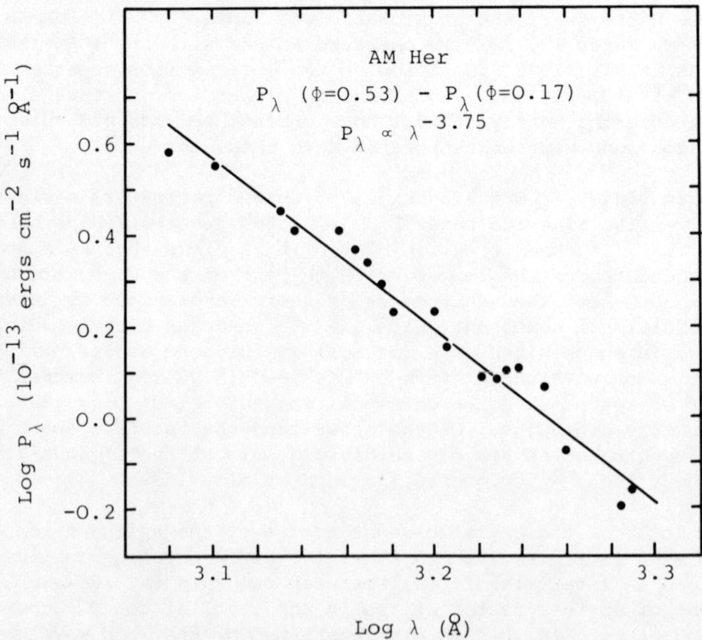

Figure 2: The component of the continuum which becomes eclipsed in phase with the X-ray eclipse of AM Herculis.

The Rayleigh-Jeans component, seen in the UV, creates difficulties
in the current models for the AM Her system. One expects from the
King and Lasota (1979) model, where bremsstrahlung rather than black
body emission accounts for the soft X-rays, a ν^0 component of about
twice the observed value for the UV continuum. The lack of optically
thick cyclotron emission can be reconciled with the Lamb and Masters
(1979) picture if the mass of the white dwarf and its magnetic
field are somewhat below earlier estimates (D. Lamb, private communication). However it is difficult to account for the black body radiation.

Lamb and Masters (1979) predict the black body emission (L_{BB}) is about
the same as the cyclotron emission (L_{cyc}) and proportional to the
bremsstrahlung emission (L_{Brems}). If a single temperature (T_{BB}) black
body emits both the soft X-rays (Tuohy et al., 1977) and the
ν^2 component in the UV

$$kT_{BB} \;=\; 25 - 30 \text{ eV},$$

$$L_{BB} \;\sim\; 20(L_{cyc} + L_{Brems})$$

(Raymond et al., 1979a).

A higher temperature black body can produce the X-rays with smaller
luminosity, but a cool black body, e.g. a halo around the base of
the accretion column with $kT \sim 15$ eV, is needed to provide the ν^2
ultraviolet emission; the discrepancy is then reduced to about a
factor of five, but not eliminated. More detailed models involving
multiple Compton scattering may resolve these problems.

We gratefully acknowledge the assistance of the IUE Observatory
staff in the acquisition and reduction of these data. This work has
been supported by the National Aeronautics and Space Administration
under Grant NSG5370 to Harvard University.

REFERENCES

Boggess, A., Carr, F.A., Evans, D.C., Fischel, D., Freeman, H.R.,
     Fuechsel, C.F., Klinglesmith, D.A., Krueger, V.L., Longanecker,
     G.W., Moore, J.V., Pyle, E.J., Rebar, F., Sizemore, K.O.,
     Sparks, W., Underhill, A.B., Vitagliano, H.D., and West, D.K.,
     1978. Nature, 274, 372.
King, A.R., and Lasota, J.P., 1979. MNRAS, 188, 653.
Lamb, D.Q., and Masters, R.A., 1979. preprint.
Raymond, J.C., Black, J.H., Davis, R.J., Dupree, A.K., Gursky, H.,
     Hartmann, L., and Matilsky, T.A., 1979a, Ap. J. (Letters),
     230, L95.
Raymond, J.C., Branduardi, G., Dupree, A.K., Fabbiano, G., and
     Hartmann, L., 1979b. to appear in IAU Symp. 88, M. Plavec, ed.
Stockman, H.S., Schmidt, G.D., Angel, J.R., Liebert, J., Tapia, S.,
     and Beaver, E.A., 1977. Ap. J., 217, 815.
Tapia, S., 1977. Ap. J. (Letters), 212, L125.
Tuohy, I.R., Lamb, F.K., Garmire, G.P., and Mason, K.O., 1977. Ap.
     J. (Letters), 226, L17.

Why DQ Her is not an X-ray source

Jacobus A. Petterson

Department of Physics,
University of Illinois at Urbana-Champaign,
IL 61801.

ABSTRACT

D Q Her is a classical nova which erupted in 1934 (Nova Herculis).
It is one of the optically brightest cataclysmic variables, and
certainly one of the best studied ones. It has been looked at in the
X-ray band by several groups, but has, contrary to a widespread
expectation, never shown any signs of emitting either soft or hard
X-rays. It is possible, of course, that D Q Her is simply different
in this respect from what we expected of a system which contains (in
all probability) a rotating, strongly-magnetic white dwarf. However,
we wish to point out that there may be another solution to the
dilemma.

D Q Her shows rapid oscillations with a period of 71 sec., probably
caused by a rotating energetic beam of radiation emitted near the
white dwarf which is reflected by the accretion disk. During
eclipses these oscillations show a striking phase-shift, while their
amplitude undergoes a smooth variation in which they essentially
disappear at eclipse-centre.

We present a numerical model which simulates the reflection process
of the beam by the accretion disk, and produces phase-shifts and
amplitude variations for the oscillations during eclipse. The
results are sensitive to the assumed inclination angle i of the
system, and reproduce the observations well only for values of i
very near 90° (i.e. almost edge on). Agreement is remarkably good
for those inclination angles at which the outer rim of the disk
(which is assumed to have the saucer-shape of the Shakura-Sunyaev
model) obscures the entire front part of the disk surface, including
the white dwarf itself. Although this requires i to lie within a
few degrees of the binary plane, it would obviously prevent us from
directly seeing any X-rays from this system. The amplitude of the
second harmonic of the oscillation during eclipse shows a rather
specific signature in the model at these edge-on inclinations, which
may serve as a test for the here proposed idea.

Supported by NSF grant No. PH7-78-04404

REFERENCES

Chanan, G.A., Nelson, J.E., and Margon, B., 1978. Ap.J., 226, 936.
Patterson, J., Robinson, E.L., and Nather, R.E., 1978. Ap.J., 224, 270.
Petterson, J.A., 1979. preprint.
Shakura, N.I., and Sunyaev, R.A., 1973. Astron. Ap., 24, 337.

Ariel V observations of a 4.8 hour periodicity in the "high-state" X-ray spectrum of Cyg X-3

R. J. Blissett*

Mullard Space Science Laboratory,
University College London,
Holmbury St. Mary, Dorking,
Surrey, U.K.

1. Introduction

2. Ariel 5 Observations

1. INTRODUCTION

The X-ray behaviour of Cyg X-3 is characterised by a smooth, quasi-sinusoidal modulation in overall flux, with a period of 4.8 hours (Parsignault et al., 1972; Sanford and Hawkins, 1972). In addition, the mean intensity is known to vary on timescales of months with associated spectral changes (Serlemitsos et al., 1975). Many observers have commented on the independence of the spectrum with the phase of the 4.8 hour cycle, however Becker et al. (1978) reported significant phase dependent spectral variability on two occasions when the source was in a low intensity state.

2. ARIEL 5 OBSERVATIONS

The Ariel 5 proportional counter spectrometer observed Cyg X-3 for 20 days in May, 1975. Preliminary analysis of the data acquired on May 25 and 26 revealed that the source was in a high state, but no significant 4.8 hour spectral variations were detected (Sanford, Mason and Ives, 1975). The data acquired between May 17 and May 24 have been subsequently analysed for spectral variability and the results are summarised here.

When measuring the spectrum of an X-ray source, the experiment integrated data while the satellite was in sunlight (see Sanford and Ives, 1976) with an energy coverage of either 1.5 to 13 keV (low energy) or 3 to 26 keV (high energy). The 4.8 hour variations in intensity of the source are clearly seen in the data. Mean data sets at phase 0.0 (X-ray minimum) and phase 0.5 were constructed by selecting only those orbits coinciding with intensity extremes. During the observations both energy ranges were used and the low energy data have been supplemented by the 13 to 26 keV high energy data giving two 48-channel pulse amplitude spectra.

*Now at SSD, ESTEC, Noordwijk, The Netherlands.

In agreement with the observations of the low state, reported by Becker et al. (1978), we also found that no simple spectral forms could represent the data adequately, in either phase. Figure 1 presents the data at both phases with the photon spectra computed by the Spectral Restoration Technique of Blissett and Cruise (1979) (see also Kahn and Blissett, 1980) which requires no a priori assumptions on the nature of the spectrum. The emission line at $\sim$ 6.5 keV is prominent in the processed spectrum. Inspection of both the pulse amplitude spectra and the subsequent restored spectra indicates additional broadening of the feature at phase 0.5. Above $\sim$ 8 keV the two spectra exhibit power-law signatures of comparable photon indexes, and on normalising the two spectra at high energies, we see that the spectrum at phase 0.0 is deficient of photons between 2 and 6 keV.

This change in the shape of the spectrum is clearly demonstrated in Figure 2 which presents a plot of ratios of the detected counts at phase 0.0, divided by those of phase 0.5 for each channel of the pulse amplitude spectrum. With the average modulation represented by the broken line, it is readily seen that the major changes in the spectrum occur in the 2 to 6 keV energy range. This demonstrates very clearly that spectral changes also occur, with phase, when the source is in a higher state of X-ray emission.

However the reduced modulation at 6.8 keV, noted by Becker et al. (1978), is not seen in our data with the higher state of the source. A multi-component model has been fitted to the data to facilitate quantitative comparisons with the low state observations. A high energy power law component with a photon index common to both phases, a low energy thermal bremsstrahlung component, and an emission line have been included in the model. Both continuum components were modified by photoelectric absorption terms using the cross-sections for a neutral, gaseous, medium given by Fireman (1974). Although this combination of components did not produce acceptable fits, substantial improvements in the quality of the fit of the data to theory were achieved, when compared to the fitting of data to a single spectral component. Table 1 summarises the numerical results of this analysis in comparison with the low state results of Becker et al. (1978).

The major difference in the spectrum between the two states is in the slope of the power law component; this is steeper in the high state. However the 4.8 hour depth of modulation for this component is similar for both states. Several common factors are noted concerning the emission line. In addition to the flux remaining constant for both states, when appropriate corrections for phase dependence are made (see Pravdo 1978), the line is also observed to shift by similar amounts ($\sim$.2 keV) to lower energies at phase 0.0 when the source is more active. Moreover, a degree of line broadening is evident in both the high and low states of the source. Becker et al. (1978) suggest that for the low state, the relative deficiency of photons observed between 2 and 6 keV in the spectra at

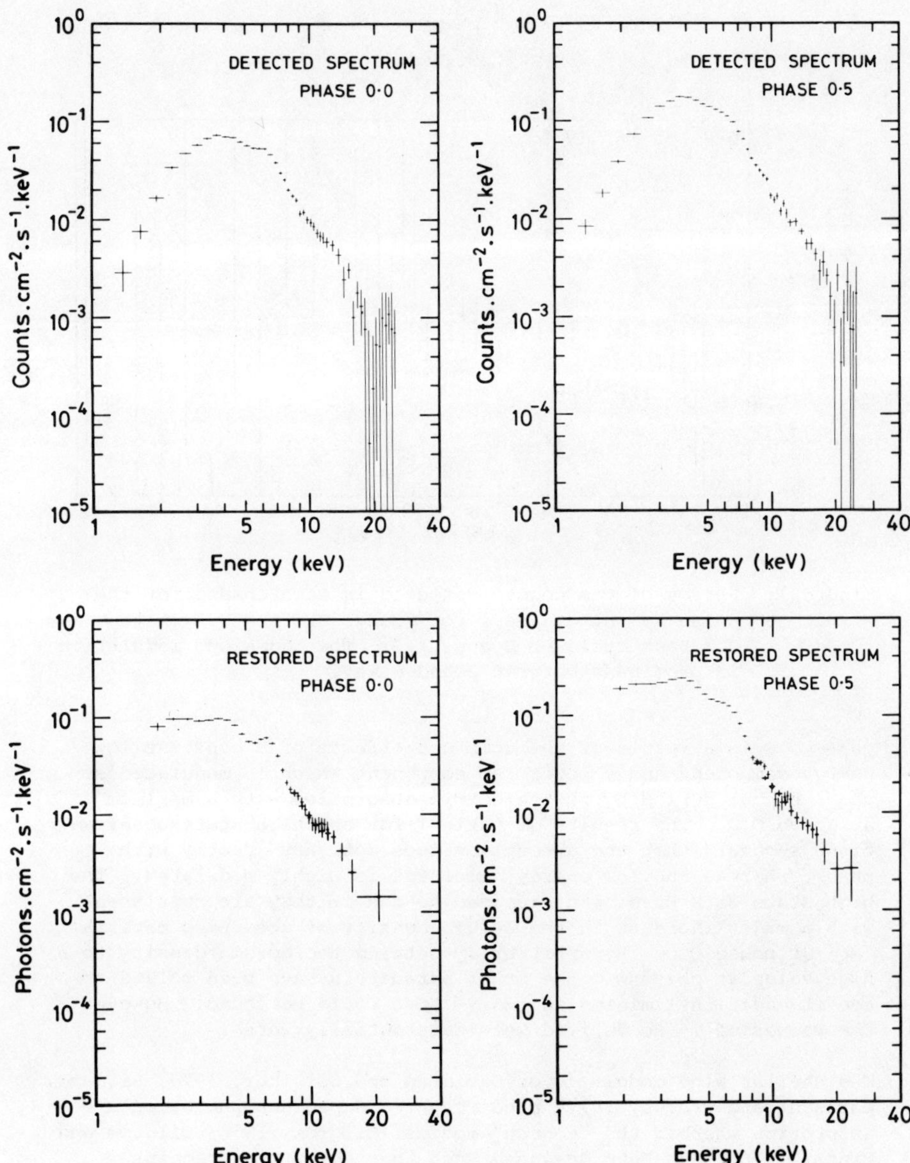

Figure 1. Pulse amplitude spectra as recorded from the detector and the spectra after restoration with the technique devised by Blissett and Cruise (1979). Phases of 0.0 (minimum intensity) and 0.5 have been used for these data.

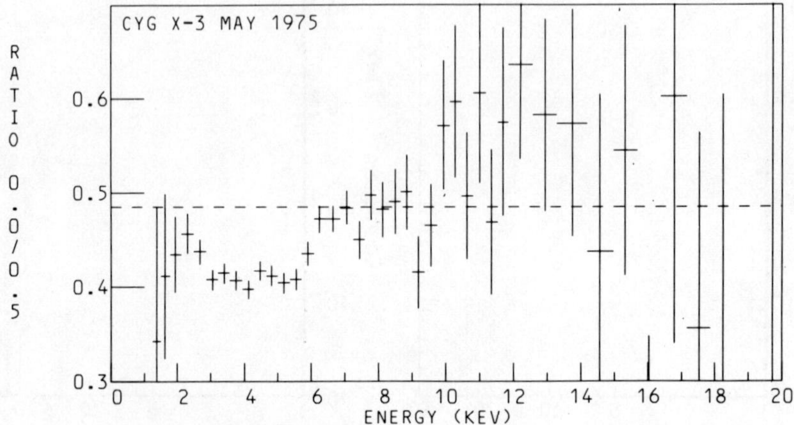

Figure 2:   Ratios of the counts detected in each channel of the
            pulse height analyser for phases 0.0 and 0.5 in the
            4.8 hour cycle of Cygnus X-3.   The anomalous modulation
            is prominent between 2 and 6 keV.

phase 0.0 is a result of the combined effects of a constant low
energy component and a power law component which is modulated at
the 4.8 hour period by photoelectric absorption with a maximum
at phase 0.0. The results of table 1 for the high state observations suggests that the absorption does not vary greatly with
phase, whereas the low energy component is highly modulated. The
high state data have been examined to see if they are consistent
with a major increase in the column density of absorbing material
(N_X) at phase 0.0. By arbitrarily setting the column density to a
high value at phase 0.0 the other parameters have been solved to
see if additional minima in the χ^2 test could be found. However
the values of χ^2 so derived were substantially worse.

The stellar wind models (e.g. Davidsen and Ostriker, 1974; Bignami,
Maraschi and Treves, 1977) predict phase dependent photoelectric
absorption whereas the "cocoon" models (Milgrom, 1976; Milgrom and
Pines, 1978) have been designed such that absorption remains
relatively constant with phase. Our observations suggest that it is
premature to dismiss either scenario. More extended observations in
the 2 to 7 keV energy range with better resolution spectrometers are
clearly needed to investigate the behaviour of this complex source
of X-rays.

This work has been developed further by Blissett, Mason, and
Culhane, (1980).

REFERENCES

Becker, R.H., Robinson-Saba, J.L., Boldt, E.A., Holt, S.S., Pravdo,
    S.H., Serlemitsos, P.J., and Swank, J.H., 1978. Ap. J. (Letters),
    224, L113.
Bignami, G.F., Maraschi, L., and Treves, A., 1977. Astr. Ap., 55, 155.
Blissett, R.J., Mason, K.O., and Culhane, J.L., 1980. MNRAS, in press.
Blissett, R.J., and Cruise, A.M., 1979. MNRAS, 186, 45.
Davidsen, A.E., and Ostriker, J.P., 1974. Ap. J., 189, 331.
Fireman, E.L., 1974. Ap. J., 187, 57.
Kahn, S.M., and Blissett, R.J., 1980. Ap. J., 238, 417.
Milgrom, M., 1976. Astr. Ap., 51, 215.
Milgrom, M., and Pines, D., 1978. Ap. J., 220, 272.
Parsignault, D.R., Kursky, H., Kellogg, E.M., Matilsky, T., Murray, S.,
    Schreier, E., Tananbaum, H., and Giacconi, R., 1972. Nature Phys.
    Sci., 239, 123.
Pravdo, S.H., 1979. In "X-ray Astronomy", Pergamon Press, Oxford.
Sanford, P.W., Mason, K.O., and Ives, J.C., 1975. MNRAS, 173, 9.
Sanford, P.W., and Hawkins, F.H., 1972. Nature Phys. Sic., 239, 135.
Sanford, P.W., and Ives, J.C., 1976. Proc. R. Soc., Lond. A., 350,
    491.
Serlemitsos, P.J., Boldt, E.A., Holt, S.S., Rothschild, R.E., and
    Saba, J.L.R., 1975. Ap. J. (Letters), 201, L9.

The peculiar object SS433

Bruce Margon

Department of Astronomy, University of California,
Los Angeles, California, U.S.A.

The strange object SS 433 is a splendid example of an astronomical
phenomenon discovered and forgotten several times. The nomenclature
refers to the appearance of the star in a catalogue of emission line
objects in the galactic plane (Stephenson and Sanduleak, 1977). Although these plates were obtained and analyzed in the early 1960's
(e.g. Krumenaker, 1975), the list was published only recently. In
view of the strange properties found subsequently in SS 433, many
people immediately ask the obvious question, "What is the nature of
SS 432 and SS 434?" or more broadly, "is the SS list a repository of
the bizarre?" A brief spectroscopic survey of other 'promising'
entries in the SS catalogue by the author and colleagues, with no
claims to completeness, has been uniformly disappointing in this
regard. Most of the objects turn out to be pre-main sequence stars
(T Tauri's, etc.) with a few OBe's, symbiotics, etc. Several show no
trace of emission at all, despite their presence in the catalogue!
Similar conclusions have been reached by Allen (1978).

The object had a second chance to emerge from obscurity quite some
time ago, but again, a twist of fate kept its importance from being
recognised. It appears in the 4C catalogue of radio sources
(Gower, Scott and Wils, 1967) as 4C04.66, but as the point source is
immersed in a bright, extended, non-uniform region of radio emission,
the position quoted in the 4C for the point source is sufficiently
inaccurate and uncertain that the correspondence with a bright optical object was not established. Later a much-improved 408-MHz map of
the region (Clark, Green and Caswell, 1975), meant to study the radio
supernova remnant W50, shows the point source very clearly and in
the correct position. However, the residual position uncertainty of
a few minutes of arc was still too large for a convenient optical
identification. Another irony is that the map of Clark et al. covered only the northern half of W50, and thus failed to reveal the
strikingly symmetric location of SS 433 with respect to the supernova remnant that has emerged in later studies. If it had done so
at the time, we might speculate that this bright, variable radio
source located in the center of a supernova remnant would have
attracted substantially more attention.

Yet another example of the source 'almost' emerging involves its
X-ray emission. The X-ray object we now know to be SS 433 was noticed

at least as early as 1975, and a report of its properties (Seward et al., 1976) presciently commented that its variability and location near a supernova remnant might imply that it was an exotic object. The X-ray source was poorly located and in a confused region of the X-ray sky, however, so again the correlation with the bright optical object was not possible. The Fourth Uhuru catalogue of X-ray sources (Forman et al., 1978) also lists the object as 4U1908+05, but again the positional accuracy is poor.

The final series of re-discoveries of the object occurred during the summer of 1978, and involved three independent groups of radio astronomers. Seaquest et al. (1979), in a survey for radio emission from pre-main sequence objects, detected intense radio emission from SS 433, and found it to be variable on timescales from fractions of a day through months. Working from entirely different motivation, Ryle et al. (1978), studying point radio sources within galactic supernova remnants, obtained a precise interferometric position for the source and noted its coincidence with a bright (14th magnitude) star. These workers were unaware that the optical object was previously catalogued as SS 433. They went on to suggest that a group of nine similar radio sources, distinguished by variability, non-thermal radio spectra, and location inside of supernova shells, might form a new class of stellar remnant. Finally, Clark and Murdin (1978), also interested in the optical identification of radio sources possibly related to supernova remnants, obtained the first reported slit spectra of SS 433, using the Anglo Australian Telescope. The spectrum they displayed shows intense Balmer and He I emission, as well as the He II $\lambda4686$ and C III/N III $\lambda\lambda4640\text{-}4650$ ubiquitous in galactic X-ray star optical counterparts. On this basis they suggested that SS 433 was in fact the optical counterpart of both the radio source and the X-ray source in the vicinity.

Some recent observations have elucidated the nature of the diffuse radio emission and its connection with SS 433. As noted previously, the radio maps of Clark et al. (1975) proved somewhat misleading concerning the overall structure of W50 and its symmetry with respect to SS 433, although the observations of Velusamy and Kundu (1974) did indicate a more spherical morphology to the remnant. Very recently an excellent new map of W50 has appeared, made by Geldzahler, Pauls, and Salter (1980) at 2.7 GHz with the Effelsberg telescope. This work strikingly indicates that the radio emission associated with W50 very symmetrically surrounds SS 433. This symmetry is the primary observational evidence connecting W50 and SS 433, so it must be kept in mind that the 'association' may actually prove to be a chance superposition on the sky. The (highly uncertain) distance estimate for the remnant, of order a few kiloparsecs, does agree with the (equally uncertain) distance estimates for SS 433, to be described below. An additional interesting feature of the radio structure of W50 is that it is one of the few 'filled shell' remnants, as opposed to the more common radio ring structure. This class of remnant, called a plerion, from the Greek πληρης (pleres, meaning 'full'), is quite rare, but also contains the Crab Nebula and 3C 58 as members (Weiler, 1979).

For completeness the more recent X-ray observations of SS 433 by Marshall et al. (1979) should also be mentioned. They find an X-ray spectrum compatible with either a power law or thermal bremsstrahlung, and evidence for an Fe emission feature at 6.8 keV. This emission feature is quite common among the compact galactic X-ray sources, so it is fair to state that the X-ray properties of SS 433 are not particularly unusual. Subsequent (as yet unpublished) imaging observations of the region made with the Einstein IPC and HRI confirm that the emission does originate from SS 433, to within an accuracy of a few seconds of arc, and that the extended remnant W50 is not a detectable source.

My colleagues and I began observations of SS 433 in the autumn of 1978, primarily expecting simply to confirm the results of Clark and Murdin (1978). In introducing these data I must stress that the very large number of observations which I will discuss are the results of contributions by almost every spectroscopist at all four University of California campuses where astronomy is conducted. A partial list of those involved in addition to myself includes Drs. Ford, Aller, Plavec, Grandi and Ulrich (UCLA), Burbidge and Smith (UCSD), Spinrad (UCB), and especially Mr. R.P.S. Stone (UCSC). These observers, plus their associated students and postdoctoral fellows, generously obtained numerous spectra of the object at the Lick Observatory, and are in a large part responsible for the conclusions to be described here.

Even our very first spectra of SS 433 showed immediately that the situation was considerably more exotic than we had expected. A sample of these data is shown in Figure 1, where spectra on three out of four consecutive nights are displayed (Margon et al., 1979a). It is an extremely lucky circumstance that at V = 14, SS 433 is (barely) bright enough for spectroscopy with the Lick 0.6-m reflector, using the Robinson-Wampler Image Tube Scanner (Robinson and Wampler, 1972). This has allowed us to obtain numerous observations of the object, with a frequency approaching 60% of all clear nights near the dark of the moon. The spectra all exhibit the enormously strong Hα emission that earned the object its place in the SS catalogue, as well as weaker emission lines of He I $\lambda\lambda$5876, 6678, 7065. There is a very red continuum, the extreme colour presumably due to interstellar extinction as the object is in the heart of the Milky Way. The strangest features, however, are readily apparent in the figure: two enormously strong, broad emission lines flanking Hα on either side. Their equivalent widths are a considerable fraction of that of Hα itself, implying that a reasonably abundant atomic or molecular species should be involved, unless the emission process is very highly selective. Yet the wavelengths of the features correspond to no such elements. Two other disquieting features of these unidentified emission lines are evident: their profiles change very drastically from night to night, and, perhaps more upsetting, their wavelengths also change by a very large amount. Regardless of the identification of the features, if these wavelength changes are interpreted as due to the Doppler effect, very large accelerations are involved: the near-infrared feature has undergone motion corresponding to +6000 km/sec in this four-day period, while the blue-flanking feature has changed by -3500 km/sec.

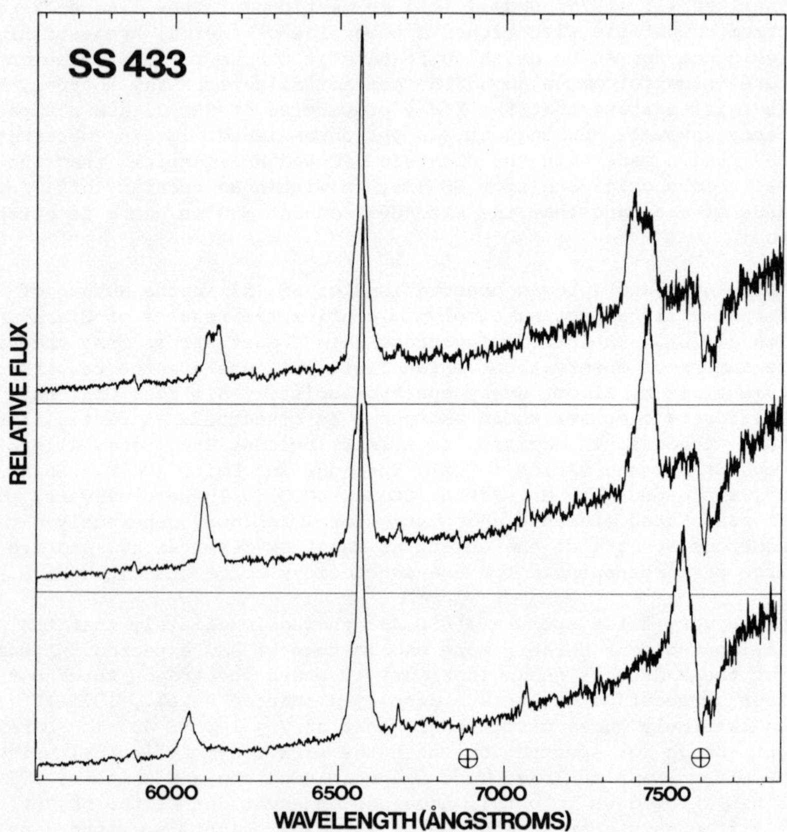

Figure 1: The red/infrared spectrum of SS 433 on three of four consecutive nights, obtained by R.P.S. Stone using the Lick Observatory
0.6-m reflector. The resolution is about 10 Å, and the dramatic
changes in both the profile and wavelength of the unidentified features flanking Hα are well-illustrated in this example. The telluric
A and B bands have also been indicated. The upper, center and lower
panels were observed on 1978 October 23, 24 and 26 respectively.

The spectra in the blue/green region are equally puzzling. At these
shorter wavelengths, the object is now much fainter due to extinction, and so one is forced to a larger instrument. In Figure 2, the
spectrum centred near λ5500 is shown, as obtained with the Lick 3-m
Shane reflector. Again, certain familiar features are seen, such as
strong Hβ emission, together with the C III/N III λλ4640/4650 multi-
plet reported previously by Clark and Murdin (1978). But again there
is at least one feature of equivalent width comparable to Hα but at
a completely miscellaneous wavelength, about λ5190. During several
days of observation, this feature was also seen to move by a considerable amount.

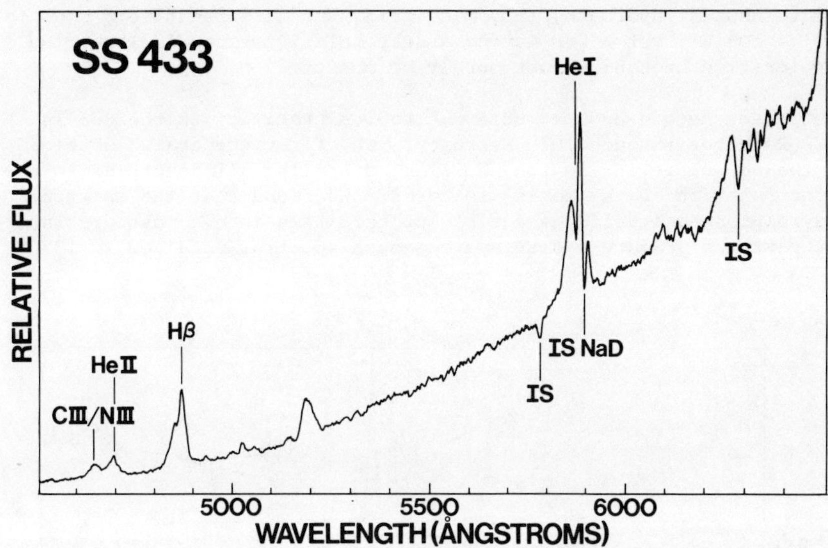

Figure 2: The blue/green spectrum of SS 433, obtained with the Lick
Observatory 3-m Shane reflector on 1978 September 30. The enormous
'stationary' Hα line has been truncated for convenience in scaling,
and is responsible for the sharp rise in flux at λ6450. Interstellar
absorption lines and bands have been marked "IS." The unidentified
intense emission line at λ5190 is easily seen. Note also the highly reddened continuum slope.

In retrospect, the interpretation of these features should have been
straightforward. The equivalent widths comparable to those of the
Balmer lines strongly argue for Doppler-shifted Balmer emission.
However, we initially rejected this interpretation for four different
reasons, all of which seemed excellent at the time, although now
known to be correct but irrelevant.

a)   - The velocities involved in such an interpretation would be
enormous, up to +50,000 km/sec of redshift and -30,000 km/sec of
blueshift, in this supposedly galactic star.

b)   - The changes in velocity would also be huge. In the first
month of observations, the near-infrared feature was found to increase
in wavelength by roughly 600 Å in 28 days, corresponding to a velocity change of 25,000 km/sec, or a constant acceleration of roughly
1 g.

c)   - The unidentified lines were seen to move in opposite directions, as is again apparent in Figure 1, where the near-infrared

feature gains wavelength while the blue-flanking features loses.
This seemed to defy simple Doppler interpretations.

d) - If the lines are identified as Balmer emission, as they are
due to neutral hydrogen, they must originate in a relatively cool
gas ($\simeq 10^4$ K), yet a gas moving a near relativistic velocity. The
acceleration mechanism must surely be obscure.

The object became unobservable due to its proximity to the Sun in
1978 December through 1979 February. The first spectra we obtained
in the new observing season, however, dispelled all doubts as to the
interpretation of the moving features. We found that the spectrum
had rearranged itself yet again, but this time into a most distinctive pattern. Figure 3 displays one such spectrum obtained in 1979

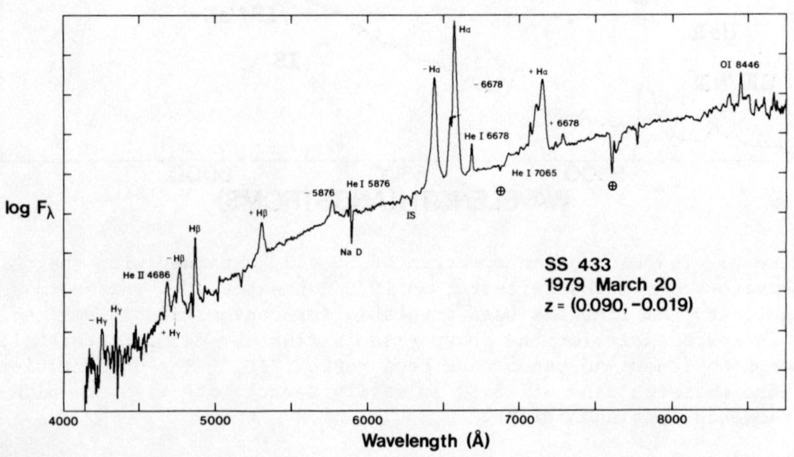

Figure 3: The spectrum of SS 433 obtained on 1979 March 20 by S.A.
Grandi, again using the Lick 3-m reflector. The principal emission
features have been identified, and the prefix "+" and "-" to these
labels denotes lines in the redshift and blueshift system, respectively. Stronger interstellar and telluric absorption features are
also labelled. Each division on the ordinate represents 0.83 mag.

March (Margon et al., 1979b). All of the principal Balmer and He I
emission lines are seen to be tripled, with a component at approximately the (rest) laboratory wavelength, one displaced to the red,
and one to the blue. Furthermore, all of the red-displaced lines
have a constant $\Delta\lambda/\lambda$, to within a few percent, and all of the blue-
displaced features also have a different but constant $\Delta\lambda/\lambda$, again to
within the measurement accuracy. As only the Doppler effect can dis-
place spectral features over a continuum of values $\Delta\lambda/\lambda$, but with constant $\Delta\lambda/\lambda$ at any one moment, and either possible sign, the identification of the shifted features now seems quite secure. This conclusion has also been reached independently by Liebert et al., (1979).

With some confidence we may now examine our entire data base of
spectra and extract a value of the redshift and a value of the blueshift on each night. We find that two such values are always apparent in the data when sufficient wavelength coverage is available,
and that in general the redshift and blueshift are quite different on
each night. As there are a half-dozen strong, broad emissions in
each of the three systems (stationary, redshift, and blueshift), confused and overlapping lines are common in the data. However, a little
ingenuity is generally sufficient to identify all of the features unambiguously. Several examples of this procedure may be instructive.
Figure 4 shows spectra obtained in the green on three consecutive
nights at the 4-m Mayall reflector of the Kitt Peak National Observatory, using the KPNO Intensified Image Dissector Scanner. The familiar Hβ and C III/N III λλ4640/4650 features can be seen at rest, as

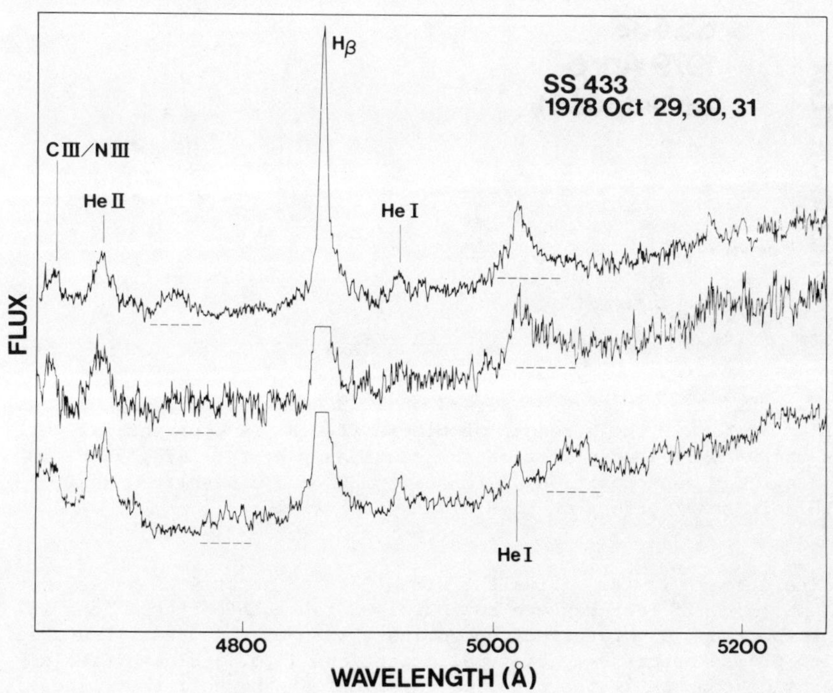

Figure 4: Observations of SS 433 obtained in 1978 October, using
the 4-m Mayall telescope of the Kitt Peak National Observatory. The
prominent moving feature just longward of λ5000 proves to be redshifted by Hγ; it finally 'uncovers' rest He I λ5015 on the third
night. Recognition of the appropriate redshift then allows redshifted Hδ to be readily located. It is found just blueward of λ4800.
The broken lines indicate the extent of both of the redshifted Balmer
features.

424

well as He I λ4921. At λ5015 is a strong feature which we initially
presumed must be He I λ5015, although its profile and width seemed
quite different from the nearby λ4921. As one watches the spectra
on three consecutive nights, however, the λ5015 feature is seen to
broaden further, become asymmetric, and finally 'calve' as it moves
further to the red, revealing what really *is* He I λ5015 left behind.
One then realizes that one is seeing redshifted Hγ gaining wavelength
and, armed with this redshift value, can proceed to locate redshifted
Hδ on the same spectra, in the λ4750 region; it yields the same redshift on each night.

Another amusing example of the problems involved in the analysis is
shown in Figure 5. This spectrum is quite typical in several respects. The enormously strong Hα line has been truncated for convenience in scaling. One can see the blueshifted Hα feature at λ6160

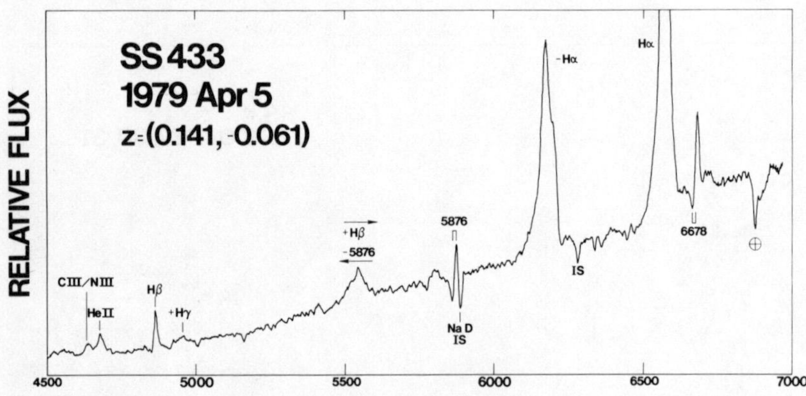

Figure 5: The spectrum of SS 433 on 1979 April 5, obtained by L.H.
Aller at Lick. The strength of blueshifted Hα is apparent, as well
as the P-Cygni absorptions on the stationary He I λλ6678, 5876, 5015
lines. The superposition of redshifted Hβ and blueshifted He I
λ5876 is near perfect on this night, near λ5515.

quite strongly; its equivalent width here of about 90 Å shows that
these 'moving' features are not barely-detectable perturbations in
the spectrum, but rather often dominate some of the 'rest' features.
Also prominent are P-Cygni type absorptions (i.e. a blueshifted absorption component superposed on emission) at the He I rest lines,
λλ6678, 5876, 5015. These are in fact the only easily-detected absorptions in the entire spectrum, although we do often also see a
line at λ5169, visible in both Figures 3 and 5, that has been identified by Crampton et al. (1980) as Fe II. This latter feature must
be highly variable as it is most usually absent. One strong feature
that puzzled us originally on this spectrum is the rather Gaussian-
shaped strong emission line at λ5530. We ultimately realised that
this emission line is in fact due to approximately equal fractions
of redshifted Hβ moving towards longer wavelengths, and blueshifted

He I λ5876 moving towards shorter wavelengths, coinciding on that
particular night to within a wavelength accuracy of approximately 1%.

As the red- and blueshifts are obviously changing with time, an obvious question is: what is the time dependence? At this writing
(Autumn 1979), we now have available data on approximately 145 separate nights during the period 1978 June through 1979 November. Most
of these data come from the Lick Observatory, but we have also supplemented the data base where appropriate with observations from the
Asiago group (Mammano et al., 1980), the Steward Observatory (Liebert
et al., 1979 and private communication), the DAO (Crampton et al.,
1980) and the original AAT data (Martin, Murdin and Clark, 1979). A
graph of these Doppler-shifts *versus* time is shown in Figure 6.

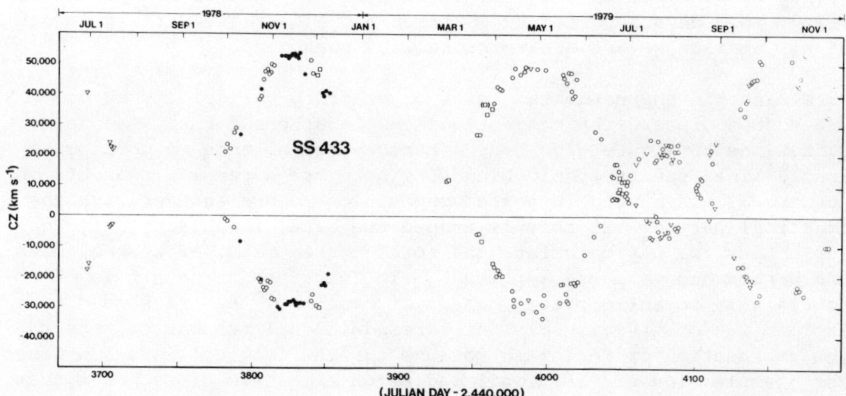

Figure 6: Values of the redshift and blueshift of SS 433 versus
time. The different plotting symbols denote contributions from different observatories, as described in the text. Note that although
the ordinate displays the Doppler shifts in velocity units cz, in
conformance with tradiational radial velocity spectroscopy, and actual kinematic velocity values are somewhat different due to the significant Lorentz factor.

Several interesting features are apparent. The values of the red-
and blueshifts obviously do change smoothly with time, as was inferred from the fragmentary individual spectra. The velocities do indeed
reach tremendous values. A maximum of more than +50,000 km/sec was
achieved in the redshift system in 1978 November, and 1979 May and
October; simultaneously, the blueshifts reached maxima of order
-30,000 km/sec. This is of course the largest blueshift ever seen in
the spectrum of any celestial object, galactic or extragalactic, by
some two orders of magnitude! The two systems of moving lines obviously know about each other. Not only do they reach extremes of
velocity more or less simultaneously, but there are also fascinating

episodes of deviation from the smooth motion which appear simultaneously (to within our one-day time resolution) in both systems.
These deviations sometimes have mirror image symmetry; an excellent
example occurs near JD 2,443,812. It is also evident that the entire
system of red- and blueshifts is asymmetric about zero velocity.
There is an obvious symmetry axis instead at $\simeq$ 11,000 km/sec, despite
the fact that this is a galactic object with one set of emission lines approximately at rest velocity! Finally, and perhaps most inter-
estingly, it is clear from even a casual inspection of the figure
that the Doppler shift variations are periodic. If the data are folded with a period chosen for best superposition, one finds this period to be 164.0 $\pm$ 0.1 days. A Fourier power spectrum of the red-
and blueshift data also shows this same period very strongly, with
absolutely no indication of higher harmonics; the variation is
apparently a remarkably pure sinusoid. It should be noted that our
period is discrepant at the 7σ level with the photometric period of
161.7 $\pm$ 0.3 days reported by Gottlieb and Liller (1979); it is not
at all obvious we are observing related phenomena.

What can this unprecedented level of velocity variability be due to?
One should discard the conventional explanations first. The velocity
variations of Figure 6 do bear a cursory resemblance to those of a
double-lined spectroscopic binary. There are after all two sets of
spectral lines moving in phased opposition to one another with the
identical period. If this is indeed the case, then using only Kep-
ler's Laws, we may calculate the total system mass, as we know both
the period and velocity amplitude. The distressing result is of
course easy to anticipate; masses of order 10^{10} $M_\odot$ are needed to e
explain the observations. There are additional reasons to reject
the explanation of Keplerian motions for the 'moving' emission lines.
For example, the gravitational radiation timescale of such a system
is very short - a few thousand years - which would be difficult,
although perhaps not impossible, to accept. However, there are even
more severe problems with a Keplerian interpretation, chiefly posed
by the 'stationary' line system. These emission lines are seen to
fluctuate in profile and equivalent width on a timescale of order
one day (Margon et al., 1979a). By the usual causality arguments,
they must therefore originate from a volume within about 100 AU of
the putative 10^{10} $M_\odot$ object. From the observed velocity limit on
the stationary lines of order 200 km/sec, one can in fact derive an
upper limit to the system mass of order 1000 $M_\odot$, suggesting strongly
that we are dealing with a stellar-type object. Finally, the diameter of a Keplerian orbit with the inferred parameters is about two
light weeks, yet we see correlated events at opposite ends of the
red- and blueshift systems with a timescale of one day. For all of
these reasons it seems quite unlikely that Keplerian motions are involved.

Most workers have turned instead to a model proposed originally by
both Fabian and Rees (1979) and Milgrom (1979a). A co-linear beam
ejected from a central object could explain both redshift and blueshift. If the beam axis then rotates with a period of 164 days, the
modulation can be simply achieved. Our observations have been analysed in detail within the framework of this simple kinematic model

by Abell and Margon (1979), and their results are shown in Figure 7.
The model has remarkably few free parameters: simply the period and
phase of the beam rotation, the beam velocity, and two inclination
angles, one expressing the inclination of the rotation axis to the
plane of the sky, and the other the inclination of the beam to the
rotation axis. The solid line in Figure 7 shows the best-fit theoretical model to the 55 nights of data available at the time that work
was completed. That fit invokes a beam velocity of $\beta = v/c = 0.27$,
and for the two angles finds values of 17° and 78°. Unfortunately,

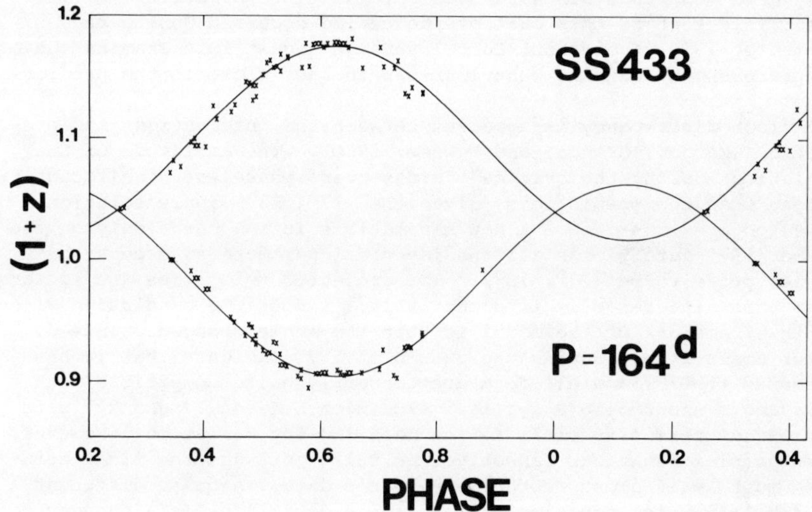

Figure 7: A simple kinematic model for SS 433, derived by Abell and
Margon (1979) using formalism similar to that of Milgrom (1979a).
The solid line is the predicted time behaviour of redshifts and blueshifts if the emitting regions are located at the ends of a co-linear
beam rotating with a 164-day period. This particular model has beam
velocity 0.27c, and inclination angles $(17^{\circ}, 78^{\circ})$. The plotted
points are the 55 nights of observation available at the time of the
initial models. *(Reproduced by permission from Nature, Vol. 279,
No. 5715, p701, Copyright (c) 1979, Macmillan Journals Limited)*

the angles are indistinguishable in this analysis; it is not clear
which is which. One remarkable aspect of the model is evident in
the figure - the systematic offset of the Dopper-shifted systems from
zero, i.e. the 11,000 km/sec symmetry axis, is reproduced. Examination of the kinematics shows that this effect is simply special relativistic time dilation, sometimes called the 'transverse Doppler
effect'. The velocity of the beams is so large that even when their
rotation axis is passing through the plane of the sky, so that

neither beam has a kinematic Doppler shift towards or away from the
observer, the 'clock ticks' of the hydrogen atom are sufficiently
slowed to give both beams a redshift. The 164-day sinusoidal variation is then superposed upon this constant offset.

As is evident in Figure 7, this kinematic model has at least one
virtue: it is easily testable. At the time the figure was drawn,
only about 65% of the 164-day orbital phase had ever been observed.
The predictions for the remaining 35% of the period are quite specific. The moving lines should join, cross through each other, and
execute a smaller amplitude variation, before rejoining to repeat
the large-amplitude sinusoid seen several times before. The first
observable test of this part of the period occurred during the
summer of 1979, from about July 1 through August 10. Examination of
Figure 6 shows that the behaviour was in fact close to the prediction.

One minor discrepancy is apparent between the observations and prediction (Margon, Grandi, and Downes, 1980). The amplitude of the
variations during the critical 'cross-over' phase was significantly
larger than the predictions, given the $(17^{\circ}, 78^{\circ})$ angle solution
based on previous data. A new kinematic solution based only on the
summer 1979 data yields all the identical parameters, except the
angles prove to be $(22^{\circ}, 78^{\circ})$. The projected velocities are so large
that this difference is irreconcilable; a possible conclusion within the framework of the model is that the angle changed. Indeed,
later observations during the autumn of 1979 indicate that it has
returned to 17°, and all data are currently quite compatible with
this angle undergoing a cyclical excursion between 17 and 22° with
period and phase identical to the main 164-day variation. However,
more complex behaviour cannot yet be ruled out, and the final solution must await about another one year's data. A quite different
possibility which can also explain these deviations is a change,
either stochastic or periodic, in the beam velocity; less than a
10% change would be sufficient to explain the observed discrepancy.
If the variation does prove to be due to a periodic change in one of
the angles, a simple explanation is that it is due to a strictly
geometric effect. If the radiation from the beam is azimuthally anisotropic, as might be expected due to optical depth effects (Rees
1979) then the angle between the bright part of the beam and the rotation axis will change slightly with orbital phase. One could predict independently the expected amplitude of this effect, as it is
simply of order of the beam width. We know this latter quantity to
be of order a few degrees, based simply on the ratio of the velocity
width of the moving lines to the beam forward velocity. This is in
fact exactly the amplitude that we observe the angle excursion to
traverse.

Regardless of the cause of this angle change, if confirmed it would
have the virtue of breaking the degeneracy between the two previous
indistinguishable inclination angles. We would know that the 78°
angle must be the inclination of the system to the plane of the sky,
as of course that quantity is unlikely to be changing on a timescale
of a few weeks! An additional interesting consequence of this change
in angle involves choices amongst specific physical models compatible

with the kinematic model, and these implications will be discussed below.

Given that the simple kinematic description (Abell and Margon, 1979) appears to fit the data well, what may be conclude about the nature of SS 433? For example, in the twin-beam description, what causes the periodic rotation of the beam axis? About the only safe speculation at this time is that the 164-day period is unlikely to be due simply to the locking of the beam axis to a slowly rotating star with 164-day period, i.e. an extremely slow pulsar. This can be argued as follows: The central object in SS 433 seems likely to be a compact object of some sort, because the observed ejection velocity of 0.27c is suspiciously close to the escape velocity from the surface of a 1 $M_\odot$ compact star. If the object were instead a much larger star, with much smaller escape velocity, why should it bother to achieve these near-relativistic expulsion velocities? If the central star is compact, then it is easy to calculate that the store of rotational energy in the object is completely trivial compared to the kinetic energy in the ejected beams. For canonical neutron star parameters, for example, the rotational energy with a 164-day period would be only about 4×10^{32} ergs. The kinetic energy in the beams depends on a number of uncertain parameters, among which is of course the distance. This latter quantity has been estimated on the grounds of interstellar absorption lines and bands (Margon et al., 1979a) as probably > 3.5 kpc. There is reasonable independent confirmation of this estimate from 21-cm absorption measurements against the radio continuum source (van Gorkom, Goss and Shaver, 1980), where the range 3.7 - 4.7 kpc is derived. It seems certain that there is > 10^{38} erg/ sec in beam kinetic energy, vastly exceeding the total store of rotational energy in the above model *every second*. Even if one limits oneself to the directly observed radiated energy in the beams, about 10^{34} erg/sec, a terrible discrepancy still exists. It seems hard to understand how the system could remain periodic, yet four separate cycles have now been observed with no obvious period change. Considerations such as this have driven most theorists to interpret the rotation of the jet axis as due to precession rather than rotation; the analogy to the 35-day period of Hercules X-1 is obvious. While most workers agree that a binary companion seems likely to be at least indirectly responsible for this precession, all of the various alternatives for the exact mechanism have vexing problems.

It is extremely important to keep in mind that the simple kinematic description of Abell and Margon (1979) is just that - kinematic and not physical. It is sadly compatible with a large variety of different physical models, and only second-order effects may be used to choose between them. Thus, it is far from certain at this time that the picture of ejected beams is a correct one. For example, a ring of matter orbiting a compact object (e.g. Terlevich and Pringle, 1979), illuminated by a 164-day rotating beam of radiation, would exhibit precisely the same kinematic properties, and thus match the observations of red- and blueshift variations quite well. Models such as these have been explored from several points of view recently, but all have serious problems. For example, if the compact star is of order 1 $M_\odot$ (e.g. Kundt, 1979), then the ring must of course be small in physical dimensions so that it may remain bound at the enormous velocities observed. There is then a terrible brightness temperature problem in generating the copious observed luminosity from this small volume. Selective processes may of course be invoked, but as the moving line spectrum looks so normal (like an H II region, except for the velocity!), there seems little evidence for such effects. The problem is circumvented by postulating a massive black hole as the object binding the ring (Amitai-Milchgrub, Piran and Shaham, 1979a, 1979b), but then the 10^6 $M_\odot$ required presents other problems. For example, there is conflict with the 10^3 $M_\odot$ upper limit inferred from the daily variability and low velocity of the 'rest' line system, as mentioned previously. In addition, this deep potential well has observable general relativistic distortions on the moving line ephemeris (Matese, Whitman, and Whitmire, 1980), which restrict the kinematic angle solutions to certain specific, non-interchangeable values if they are to be compatible with our observations. However, our discovery that one of these two angles may be changing with time then becomes awkward for this model, as the angle required to change is physically constrained to be constant in most reasonable scenarios.

The most important new breakthrough has probably come with the discovery by Crampton, Cowley and Hutchings, (1980) that the 'rest' emission line system is in fact periodically varying with amplitude 70 km/sec, and period 13.1 days. This period is suspiciously like that of the low-mass X-ray binaries like Cygnus X-1 (Cowley, Crampton and Hutchings, 1979). If one assumes the secondary is a neutron star of 1.5 $M_\odot$, and adopts an inclination of i = 78° from our kinematic data, then the observed mass function of about 0.5 $M_\odot$ yields a mass for the unseen secondary of also about 1.5 $M_\odot$. If one then insists that this object fill its Roche lobe with the observed 13.1-day period (as seems likely simply to achieve the very large mass transfer rate needed to power the beams), then one finds the star must be larger than main sequence radius. The necessary properties can be met by an F giant. Such a star would be unobservable against the intense featureless continuum, which in this interpretation would be attributed to the accretion disk. Other interences regarding the nature of the secondary have been reach from these orbital data; for example, van den Heuvel, Ostriker and Peterson (1980) prefer an Of star, attributing the continuum and 'rest' emission lines to the early-type star rather than the disk. The remarkable resemblance of the stationary spectrum to that of systems known to be dominated by a disk, for example DQ Her (Chanan, Nelson and Margon, 1978), would seem to argue against this, as well as the fact that the stationary lines are sometimes seen to be bifurcated in our data, again suspiciously like the spectra of some dwarf novae. However, a possible difficulty with the F giant interpretation is the lack of a deep photometric eclipse, which would seem difficult to avoid with i = 78°. Although the beams may be sufficiently extended that they may avoid occultation, one might expect the disk, which is the source of the majority of the observed luminosity in this model, to be eclipsed.

The 13.1-day period is also evidenced in other data. Baliunas et al. (1979) have reported that the intensity of the rest Hα line is variable,   and consistent with a 13.1-day period. Examination of about 100 spectra from our collection shows that this variability is dominated by a low-frequency trend, with timescale definitely longer than 164 days. In fact, the average intensity has been steadily increasing for our entire 18 months of observation. However, if one subtracts this low-frequency term with a polynomial (or even a straight line), the Fourier power spectrum of the Hα intensities *and* Hα equivalent widths shows a strong 13.1-day modulation. There is also a prominent first harmonic in the power spectrum, and folds of the data confirm that the phase diagram of Hα intensity or equivalent width shows two peaks. However, the 6.55-day variation is of lower amplitude than the 13.1-day modulation. The phasing is also interesting; the primary intensity maximum occurs coincident with the phase of maximum positive radial velocity quoted by Crampton et al., (1980), while the secondary intensity maximum is at phase of radial velocity minimum. This is the behaviour one might expect from observations of a hot spot in the disk, similar to that observed in dwarf novae.

If one accepts that the 13.1-day period is genuine, as now seems unavoidable, and that it quite probably does represent the orbital period of the secondary, then regardless of many remaining ambiguities, certain classes of models begin to fall into disfavour. In particular, one popular class of explanations of the 164-day period invokes general relativisitc induced precession, due to the very close proximity of another compact object. Such explanations have been discussed in some detail by Begelman et al., (1980), Martin and Rees (1979), and Faulker and Hatchett (1979), and generally require extremely short orbital periods, of order 4 - 8 minutes, to obtain the desired precession rate. The discovery that the orbital period is actually 4 orders of magnitude longer forces these models to invoke added free parameters, such as the presence of a third star, to remain viable.

On the other hand, we should not become overconfident in our current understanding of SS 433. Numerous problems remain to be explained. Among these are the following:

a)   What is the nature of the 'central' object? We have already discussed the fact that a compact star seems likely to be involved, and cited models invoking both neutron stars and black holes. A variety of other hypotheses are available, however. DeYoung and Burbidge (1979) make this object a white dwarf, for example. Milgrom (1979b) has suggested an ingenious explanation of the observed beam velocity of 0.27c that could remove the evidence for a compact star altogether. He points out that this is precisely the velocity needed to redshift the Lyman limit to Lyman α, suggesting that line-locking, i.e. radiative acceleration via sub-Lyman continuum absorptions, is the acceleration process, rather than self-regulating expulsion from the surface of the surface of a compact object.

b)   What is the nature of the energy source and acceleration mechanism? Accretion is certainly a likely candidate for the former, but one must then explain why this system looks so different from the other galactic X-ray binaries. Katz (1980) has considered some

432

fundamental constraints regarding the acceleration processes, but
far more work needs to be done.

c)   What are the details of the excitation of the beams? It is
amusing that we see only H I and He I in this near-relativistic gas.
If portions of the radiating beam had temperatures of only 30-40,000 K
we should see copious He II λ4686 in the red- and blueshift systems,
and yet we do not. Clearly detailed radiative excitation calculations are needed.

d)   What is the mechanism that collimates the beams? We know that
the ejected matter in the beam model is very well-collimated, as the
ratio of the velocity width of the moving lines to the overall system
velocity is small, about 5%. One obvious collimator is a strong magnetic field, but there is no evidence for non-interstellar linear or
circular polarization in the optical light from SS 433, either broad-
band or in the emission lines (Liebert et al., 1979). On the other
hand, this might have been expected as the strong field region can
be only very close to the star, and the beams originate from a more
extended region. Presumably due to the low sound speed at this
excitation temperature, once the collimation occurs, the beam remains
narrow throughout its extent.

e)   What is the cause of the marked daily line profile changes (see,
for example Figure 1)? The lines often have mirror-image profile
symmetry, but not always. A related question involves the exact process by which the moving lines move. Begelman et al., (1980) speculate from a few nights' spectra that the actual process involves
not a change in wavelength of an emission line, but rather the appearance and disappearance of discrete lines at monotonically changing
wavelengths. While this may be a semantic distinction, most of our
data does not show this effect. It is clear from Figure 1, for
example, that one line has not faded to be replaced by another in
that series of data.

f)   What are the correlated glitches that cause gross diversions
from the smooth ephemeris for periods of a few days, sometimes with
mirror-image symmetry in the redshift and blushift systems? It is
apparent from Figure 6 that some of these events may even recur at
a given 164-day phase, the most vivid example being the series at
JD 2,443,845, 2,444,015, and 2,444,174, all in phase to an accuracy
of about 0.01. Regardless of the cause of these events, they provide
an interesting fiducial from which one can search for time delay
effects due to the considerable difference in light travel times from
the redshifted and blueshifted beams required in most models. There
indeed do seem to be indications of anomalous events "leading" by a
day or so in the blueshifted beams (e.g. JD 2,443,810, 2,443,845,
2,444,011 and 2,444,110), but the general scatter in the ephemeris
is so chaotic that it is hard to ensure that these are significant
events.

g)   Where are the others? From the point of view of the observer,
this is perhaps the most intriguing question. Recall that Ryle et
al. (1978) suggested that nine sources were a member of this class.

One of the sources in this list, G127.11+0.54, has subsequently been proven to be a radio galaxy with redshift $z = 0.018$ (Kirshner and Chevalier, 1978; Spinrad, Stauffer and Harlan, 1979), projected accidentally through a galactic foreground supernova remnant. A second member of the group, CL4, was originally thought to be galactic not only due to its central location inside of the Cygnus Loop, but also because of lack of observable 21-cm absorption due to the foreground spiral arm (Webster and Ryle, 1976). However, this absorption has recently been detected by two independent groups (Goss, van Gorkom and Shaver, 1979; Payne and Bania, 1979), so the motivation for including this object in the Ryle et al. list is now considerably less. CL4 almost surely does have an optical counterpart (Vrba and Tapia, 1979), but this object is a 21st magnitude star located just 2" from a 12th magnitude star, so spectroscopy will be difficult. Most of the remaining radio sources in the Ryle et al. list, all of which have subarcsecond accuracy positions, are termed 'blank fields' on the Palomar Sky Survey by these authors. This fact alone tells us that they are probably unrelated to SS 433, because of the very large luminosity of the latter object. If we adopt for SS 433 $A_V > 4$, as seems unavoidable, and a Sky Survey limiting magnitude of $m = 21$, then we must conclude that for objects at similar distances to SS 433, a few kiloparsecs, *every one* of the remaining sources has $A_V > 11$ mag to remain unseen. This seems highly unlikely given that numerous different directions in the sky are involved, and we have no reason to believe that these objects have a preference for dark clouds. It seems reasonable to conclude at this time that the only object in the Ryle et al. (1980) list that is related to SS 433 is SS 433 itself.

On the other hand, again because of the great luminosity of this curious object, it could be easily detected elsewhere in the Local Group, if one knew where to look. For example, a similar object is probably brighter than $m = 15$ in the Magellanic Clouds, and brighter than $m = 20$ in M31. Sadly, a lifetime calculation is also needed, however. For particularly pessimistic calculations of the mass loss rate in the beams, 10^{-4} $M_\odot$/yr may be involved, yielding a lifetime of 10^4 yr. There may be only one SS 433 active in the Local Group at any one time.

Finally, it seems appropriate to conclude by asking what we may be saying about SS 433 five years from now. What will the significance of the system be in addition to its spectacular spectroscopic behaviour? The answer may well lie in the eventual resolution of questions b) and d) above, involving the energetics, collimation, and acceleration of the beams. We are all well aware that double-lobed structure is common in violent extragalactic events such as QSOs and radio galaxies. If the mechanisms involved in these objects and SS 433 turn out to be similar, it will be a fantastically fortuitous circumstance that gives us a bright, miniature version of this phenomenon, available quite nearby for detail study.

This work has been supported by grants from the National Science Foundation and the Alfred P. Sloan Foundation.

REFERENCES

Abell, G.O., and Margon, B., 1979. Nature, Lond., 279, 701.
Allen, D.A., 1978. MNRAS, 184, 601.
Amitai-Milchgrub, A., Prian, T., and Shaham, J., 1979a. Nature, Lond., 279, 505.
Baliunas, S., Noyes, R., Liller, W., and Tokarz, S., 1979. IAU Circ. 3410.
Begelman, M.C., Sarazin, C.L., Hatchett, S.P., McKee, C.R., and Arons, J., 1980. Astrophys. J., 238, in press.
Chanan, G., Nelson, J., and Margon, B., 1978. Astrophys. J., 226, 963.
Clark, D.H., Green, A.J., and Caswell, J.L., 1975. Austr. J. Phys. Astrophys. Suppl., 37, 75.
Clark, D.H., and Murdin, P., 1978. Nature, Lond. 276, 44.
Cowley, A.P., Crampton, D., and Hutchings, J.B., 1979. Astrophys. J., 231, 539.
Crampton, D., Cowley, A.P., and Hutchings, J.B., 1980. Astrophys. J. (letters), 235, L131.
DeYoung, D.S., and Burbidge, G., 1979. Nature, Lond. 281, 183.
Fabian, A.C., and Rees, M.J., 1979. MNRAS, 187, 13P.
Faulkner, J., and Hatchett, S.P., 1979 preprint.
Forman, W., Jones, C., Cominsky, L., Julien, P., Murray, S., Peters, G., Tananbaum, H., and Giacconi, R., 1978. Astrophys. J. (suppl.), 38, 357.
Geldzahler, B.J., Pauls, T., and Salter, C.J., 1980, Astron. Astrophys., in press.
Goss, W.M., van Gorkom, J.H., and Shaver, P.A., 1979. Astron. Astrophys., 73, L17.
Gottlieb, E.W., and Liller, W., 1979. IAU Circ. 3354.
Gower, J.F.R., Scott, P.F., and Wills, D., 1967. Mem. R. astron. Soc., 71, 49.
Katz, J.I., 1980, Astrophys. J. (Letters), in press. (236).
Kirshner, R.P., and Chevalier, R.A., 1978. Nature, Lond., 276, 480.
Krumenaker, L.E., 1975. Pub. astron. Soc. Pacific, 87, 185.
Kundt, W., 1979. Nature, Lond., 282, 52.
Liebert, J., Angel, J.R.P., Hege, E.K., Martin, P.G., and Blair, W.P., 1979. Nature, Lond., 279, 384.
Mammano, A., Ciatti, F., and Vittone, A., 1979. Astron. Astrophys., in press.
Margon, B., Ford, H.C., Katz, J.I., Kwitter, K.B., Ulrich, R.K., Stone, R.P.S., and Klemola, A., 1979a. Astrophys. J. (Letters), 230, L41.
Margon, B., Grandi, S.A., and Downes, R., 1980. Astrophys. J. (Letters), in press.
Marshall, F.E., Swank, J.H., Boldt, E.A., Holt, S.S., and Serlemitsos, P.J., 1979. Astrophys. J. (Letters), 230, L145.
Martin, P.G., Murdin, P.G., and Clark, D.H., 1979. IAU Circ. 3358.
Martin, P.G., and Rees, M.J., 1979. MNRAS, 189, 19P.
Matese, J.J., Whitman, P.G., and Whitmire, D.P., 1980. Astrophys. J., in press.
Milgrom, M., 1979a. Astron. Astrophys., 76, L3.
Milgrom, M., 1979b. Astron. Astrophys., 78, L9.

Payne, H.E., and Bania, T.M., 1979. Astron. J., 84, 611.

Rees, M.J., 1979. private communication.

Robinson, L.B., and Wampler, E.J., 1972. Pub. astron. Soc. Pacific, 84, 161.

Ryle, M., Caswell, J.L., Hine, G., and Shakeshaft, J., 1978. Nature, Lond., 276, 571.

Seaquist, E.R., Garrison, R.F., Gregory, P.C., Taylor, A.R., and Crane, P.C., 1979. Astron. J., 84, 1037.

Seward, F.D., Page, C.G., Turner, M.J.L., and Pounds, K.A., 1976. MNRAS, 175, 39P.

Spinrad, H., Stauffer, J., and Harlan, E., 1979. Pub. astron. Soc. Pacific, 91, in press.

Stephenson, C.F., and Sanduleak, N., 1977. Astrophys. J., (Suppl.), 33, 459.

Terlevich, R.J., and Pringle, J.E., 1979. Nature, Lond., 278, 719.

van den Heuvel, E.P.J., Ostriker, J.P., and Petterson, J.A., 1979. Astron. Astrophys., in press.

van Gorkom, J.H., Goss, W.M., and Shaver, P.A., 1980. Astron. Astrophys., in press.

Velusamy, T., and Kundu, M.R., 1974. Astron. Astrophys., 32, 390.

Vrba, F.J., and Tapia, S., 1979. Astron. J., 84, 470.

Webster, A.S., and Ryle, M., 1976. MNRAS, 175, 95.

Weiler, K.W., 1979. Sky and Telescope, 58, 414.

Subject Index

Source Index